MICRORHEOLOGY

Microrheology

Eric M. Furst

Department of Chemical and Biomolecular Engineering
University of Delaware

Todd M. Squires

Department of Chemical Engineering
University of California, Santa Barbara

OXFORD
UNIVERSITY PRESS

Great Clarendon Street, Oxford, OX2 6DP,
United Kingdom

Oxford University Press is a department of the University of Oxford.
It furthers the University's objective of excellence in research, scholarship,
and education by publishing worldwide. Oxford is a registered trade mark of
Oxford University Press in the UK and in certain other countries

First published 2017
First published in paperback 2020

Impression: 3

Published in the United States of America by Oxford University Press
198 Madison Avenue, New York, NY 10016, United States of America

British Library Cataloguing in Publication Data

Data available

Library of Congress Cataloging in Publication Data

Data available

ISBN 978–0–19–965520–5 (Hbk.)
ISBN 978–0–19–886709–8 (Pbk.)

Printed and bound by
CPI Group (UK) Ltd, Croydon, CR0 4YY

For my family: Teresa, Matthew, and Chloe—E.M.F.
Mine too: Nancy, Marc, Betsy, and Harry—T.M.S.

Preface

Complex fluids and soft materials surround us; they are engineered to create important consumer products like toothpastes, detergents, paints, cosmetics, inks, and fabric softeners; they are the foam, emulsion or gel of *gastronomie moléculaire*, as well as ordinary, stiffly whipped egg whites or a shaken vinaigrette; and they are the whimsical substance Silly Putty, or the high-tech electronic ink in electronic readers. We ourselves are composed of complex fluids and soft matter, albeit in a form that has learned how to self-replicate.

In studies of soft matter, we want to understand how the material structure is organized on nanometer and micrometer dimensions and how it will respond to external stimuli—in the case of paint, whether it will flow under the brush or hold still on a wall against gravity. The technical challenge is formidable and exciting, and *rheology* plays an important role in the endeavor. Sometimes rheological measurements help us to characterize a complex fluid's structure better, but just as often, rheology is the target of engineering design, specific to a material's use (*e.g.*, toothpaste) or as a means of accomplishing a technical objective—such as suspending active ingredients in a sprayable carrier fluid. Scientists and engineers throughout academia and industry have ready access to mature instruments and a tremendous knowledge base to draw from when performing and interpreting rheological measurements.

The past two decades have seen the advancement of new rheological characterization methods in the form of *microrheology*. *Passive* microrheology uses the Brownian motion of microscopic tracer or probe particles and relates this movement to the linear viscoelastic properties of the surrounding material. At the same time, there has been a renewed interest in *active* microrheology using magnetic or optical forces to drive probe particle motion. Microrheological measurements have several unique benefits that have advanced their development: They typically require mere microliters of sample, provide access to an extended range of measurement frequencies, and enable spatially resolved rheological measurements. For these and other reasons, microrheology has emerged as a powerful complement to mechanical rheometry.

We present a comprehensive overview of microrheology, emphasizing the underlying theory, practical aspects of its implementation, and current applications to rheological studies in academic and industrial laboratories. The field of microrheology continues to evolve, and its applications are expanding. We introduce the key methods and techniques, including important considerations to be made with respect to the materials most amenable to microrheological characterization and pitfalls to avoid in measurements and analysis. Microrheological measurements can be as straightforward as video microscopy recordings of colloidal particle Brownian motion; these simple experiments can yield rich rheological information. This text covers topics ranging from active microrheology using laser or magnetic tweezers to passive microrheology, such as multiple particle tracking and tracer particle microrheology with light scattering.

Our overall aim is to provide an introduction to microrheology for the industrial researcher, academic investigator, or student who wishes to become informed in this relatively new area of rheology, seeking to incorporate these methods into their own research, or who would simply like to survey and understand the growing body of microrheology literature. We consolidated many sources throughout the archival literature into an accessible framework for the rheologist and non-specialist, alike. The material covered in this text should be suitable to the biologist, chemist, materials scientist, physicist, or chemical engineer with an interest in microrheology. Indeed, the small sample sizes of many microrheology experiments have made them important methods for studying emerging and scarce materials, like cytoskeletal proteins, pharmaceutical biologics, and novel hydrogelators.

Overview of the book

This book is organized into four main sections: In the first section (Chapters 1 and 2), we cover fundamental principles of all microrheology experiments, including the nature of microscopic colloidal probes and their movement in fluids, soft solids, and viscoelastic materials. Microrheology is divided into two general areas, depending on whether the probe is driven into motion by thermal forces (passive), or by an external force (active). Following our treatment of the fundamentals, Chapters 3–6 present the theory and practice of passive microrheology. Chapters 7, 8, and 9 discuss active microrheology. The final section, Chapter 10, highlights several important applications and additional ideas about the practice of microrheology.

We begin with a presentation of general principles of rheology in Chapter 1, covering the rheological functions and principles of conventional rheometric measurements, as well as several common rheological properties that will be encountered throughout the text. Because colloidal particles are central to all microrheology measurements, in the second half of Chapter 1 we present basic concepts of colloid science, including typical probe chemistries, colloidal stability, characterization, and preparation. Likewise, the movement of colloidal particles in simple and complex fluids and viscoelastic solids is central to the microrheology endeavor. Chapter 2 lays a foundation of the fundamental mechanics of micrometer-dimension particles in fluids and soft solids. Of particular importance is the role of the Correspondence Principle, but other key concepts include mobility and resistance, hydrodynamic interactions, and both fluid and particle inertia.

In Chapter 3, we introduce the underlying theory of passive microrheology, as an in-depth examination of the Generalized Stokes–Einstein Relation. We carefully treat the assumptions that must be made for the technique to work, and what happens when these assumptions are violated. Chapters 4 and 5 discuss the general principles of two of the most important microrheology methods, multiple particle tracking microrheology using video microscopy and light scattering microrheology. Chapter 6 covers laser tracking and related techniques.

We discuss the theory of active microrheology in Chapter 7, focusing specifically on the potential and limitations of extending microrheology to measurements of non-linear properties, like yielding and shear-thinning. Chapters 8 and 9 present active microrheology techniques, with a focus on magnetic bead microrheology and optical tweezer microrheology.

Throughout the book the reader will find *application notes* highlighting areas of rheology in which microrheology can play an important role. We also discuss the operating regimes of each of the methods. These concepts will help the reader identify particular experiments of interest and plan them appropriately. We revisit the operating regimes in Chapter 10, comparing more closely the capabilities of mechanical rheology to microrheology. Several applications are covered in greater detail, including rheological screening, gelation and degradation, and viscosity measurements.

This book does not have to be read cover-to-cover, but can be used with specific interests and applications in mind. If the reader wants to learn how light scattering and diffusing wave spectroscopy instruments can be adapted to microrheology studies, then Chapter 5 will be of primarily interest, as well as discussions of colloidal probe chemistries (Chapter 1), the fundamentals of particle motion (Chapter 2), and the theoretical underpinnings of passive microrheology (Chapter 3). Likewise, readers interested in particle tracking microrheology can focus on Chapters 1–4.

We labored to be rigorous and correct, but any mistakes found in the text are certainly our own. We welcome readers to send their comments and corrections, and would especially like to hear from readers about the successes (or failures) in applying the principles and methods of microrheology to their own work.

Acknowledgements

The seed of this book started in the form of notes assembled for short courses at meetings of the Society of Rheology and the American Physical Society. Various incarnations of those lectures have been given in industry, at Delaware, and other academic institutions, including KU Leuven; IESL-FORTH, Crete; Forschungszentrum Jülich; and ETH Zürich. I am indebted to the students and colleagues who attended these lectures. Their questions and ideas helped to refine the material and its presentation while expanding its scope. I thank my short course co-instructors: Michael Solomon, Patrick Doyle, Patrick Spicer, and Roseanna Zia. Wesley Burghardt deserves special mention. He provided the initial nucleation event for this book through the generous invitation to lecture as part of a DPOLY course in 2004.

My colleagues at the University of Delaware, and in particular the world-class rheology and soft matter groups they lead, have sustained my research and inspired my interests. I benefited from discussions and advice over the years from the late Arthur Metzner, Norman Wagner, Antony Beris, Michael Mackay, Kristi Kiick, Darrin Pochan, Joel Schneider, Thomas Epps, Christopher Kloxin, and Christopher Roberts. During this time, it has also been my pleasure to work with a number of industrial collaborators on microrheology-related projects: Robert Butera, Matthew Lynch, Marco Caggioni, Seth Lindberg, William Galush, Danielle Leiske, Philip Sullivan, Yan Gao, and Alhad Pathak.

I thank William Russel, John Brady, Jay Schieber, Peter Schurtenberger, Dan Blair, Eric Dufresne, Daphne Weihs, and Randy Ewoldt for sharing their insights, and my colleagues at ESPCI—Marc Fermigier, Olivia du Roure, Anke Lindner, Julien Heuvingh, and Michel Cloitre—for their hospitality and inspiration. I am indebted to Frank Scheffold for his thorough reading of Chapter 5. This book would not have been possible without the generosity of Jan Vermant and the support of ETH Zürich, where I had the opportunity to spend a sabbatical in 2016.

To my students with whom I've had the enormous pleasure of working with on these topics—undergraduate, graduate, postdoctoral researchers, and visiting scholars: Myung Han Lee, Travis Larsen, Cecile Veerman, Becky Gable, Kelly Schultz, Indira Sriram, Alexander Meyers, Alexandra Bayles, Brian Bush, Andrew Marshall, Hyejin Han, Peter Beltramo, Lilian Josephson, Jim Swan, Matthew Shindel,

Laura Casanellas, Mahlet Woldeyes, and Kimberly Dennis. Thank you to Tamás Prileszky for help with the finishing touches and for Hojin Kim's sharp eye.

Thanks to my co-author, Todd Squires. It has been a distinct honor to work along side his bright intelligence and enthusiastic spirit.

And finally, thank you to the rheology community. It is a wonderfully global group, devoted to deep scholarship, penetrating discussions, grand camaraderie, and the pursuit of beautiful insights into the material world as it flows around us.

Eric M. Furst
January, 2017

I would start by thanking my co-author, Eric Furst for getting the ball rolling on this book, and for never even threatening to revoke my invitation to participate. It has been a fun, complementary endeavor.

I am grateful to my colleagues at UCSB for their support, for sharing their own interests and being interested enough to ask about mine. The environment at UCSB has shaped who I am today. With regards to this book, I can not count the number of conversations and ideas I've bounced around with Matt Helgeson and Gary Leal, as well as Baron Peters, Glenn Fredrickson, and long-time collaborator Joe Zasadzinski. The ideas in this book have evolved over the years, thanks to the energy and enthusiasm of many students and postdocs—SiYoung Choi, Kyuhan Kim, Zachary Zell, Ryan DePuit, Peter Chang, Ian Williams, Harishankar Manikantan, and Gwynn Elfring, and Aditya Khair. Arash Nowbahar went above and beyond, developing and perfecting demos that I now use regularly in classes, and which helped me to develop much of the feeling I have for soft matter in practice. Likewise, Alexandra Bayles (also acknowledged by Eric!) has beta-tested several chapters. It has been a joy to explore scattering and microrheology with her and Matt.

I am also exceedingly grateful to Professor Tom Mason, with whom I co-wrote a review article on microrheology several years back. Our interactions, and that effort, really kick-started my thought process more broadly and deeply about this field. I learned much from John Brady, back in the early days of our work on active and nonlinear microrheology.

I am grateful to several universities that have been particularly generous with their support: The University of Auckland allowed us to spend a heavenly three months in their realm; as did ETH Zürich (special thanks to Jan Vermant and Lucio Isa), and the University of New South Wales (thanks to Patrick Spicer). I can not overstate how liberating it has been to live in such spectacular locations, among scientifically invigorating colleagues, and the occasional quiet time to think and write.

My parents have always supported me, suffering through what must have been insufferable bouts of opinionated furor. They still do, and it's not always easy. Still, I owe much of this spirit to my mother Lin, who always taught me to think critically, to ask probing questions, and to stand for what is right. My father Clark seems happiest when he is working to help others, and it is inspiring.

Lastly, I want to reflect on something that should seem trite, and that I fear I've been taking for granted. We have been so fortunate to enjoy the open and free exchange of ideas, among a global community of scientists and thinkers. Very little that I have accomplished—and yes, it may be very little—would have been possible without the ideas, the interactions, the friendship, and the inspiration that the world-wide scientific community has shared. The resurgence of nationalism around the world gives me great concern, and I hope that we will all act to support these virtues, much like they have supported us.

Todd M. Squires
January, 2017

Contents

Introduction

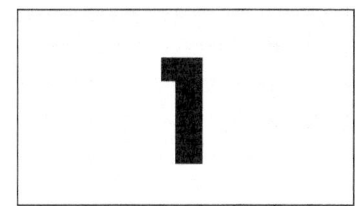

1.1 Microrheology

Imagine mixing small magnetic particles, like iron filings, into a soft material, then turning on a nearby electromagnet, and watching the particles move. If the material is a simple, viscous liquid, the particles will slowly translate through it (see Fig. 1.1). Doubling the forcing would double the migration velocity, and turning off the forcing would stop the motion. Measuring the migration velocity V in response to a range of driving forces F would reveal a strictly linear relationship, $F = \zeta V$, where the hydrodynamic resistance ζ is specific to the properties of the particles (*e.g.*, its size and shape) and of the liquid (*e.g.*, viscosity). In analogous measurements, the same particle would move more slowly in a liquid of higher viscosity, with ζ being directly proportional to the liquid viscosity η.

If the particle is instead suspended in an elastic solid, like a soft gel, a magnetic force would cause the particle to move some distance and then stop. If the field were turned off, the particle would spring back to its original (equilibrium) position. Measuring the displacement ΔX of a particle embedded in a simple-elastic solid, in response to a series of applied forces F, would reveal a linear spring constant $F = \kappa \Delta X$. The "stiffer" the solid, the higher the spring constant κ.

We have just considered a simple microrheology experiment, not that different from the first "microrheology" experiments that date to the early-twentieth century, a body of work that parallels the nascent development of colloid science and rheology.

As early as 1922–24, researchers were reporting measurements of the mechanical properties of biological samples, including cells, by tracking the motion of embedded magnetic particles. These probes were typically iron or nickel particles, tens of micrometers in diameter, that were carefully separated from powders by mechanical screening. In one early study, Heilbronn (1922) used iron filings to measure the mechanical properties of slime molds, which consist of motile, single-cell protists of the genus *Myxomycetes*. Seifriz (1924) used nickel particles to study the viscoelasticity of sand dollar eggs, *Echinarachnius parma*, having developed these methods for experiments

Microrheology. Eric M. Furst and Todd M. Squires, Oxford University Press (2017).
© Eric M. Furst and Todd M. Squires. DOI 10.1093/oso/9780199655205.001.0001

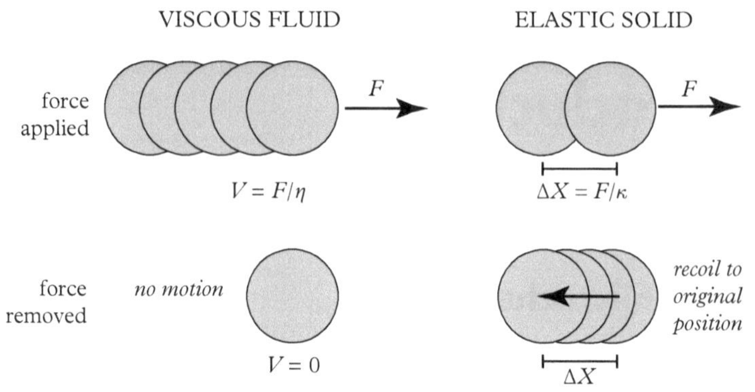

VISCOUS FLUID ELASTIC SOLID

Fig. 1.1 *Particle motion in viscous, Newtonian fluid and elastic solid when a force F is applied and removed.*

[1] Seifriz visited Herbert Freundlich, who worked in colloid chemistry and was director of the Kaiser Wilhelm Institute for Physical Chemistry and Electrochemistry from 1919 until 1933. At the time, the rheology of suspensions, especially Einstein's description of suspension viscosity, was a model for cell rheology. In this early work, the motivation to understand cell rheology didn't stem from the underlying and marvelous mechanics that arise from a microstructure of protein filaments or the action of molecular motors (see Howard (2000) and Bray (2001) to read more about the biomechanics of cells and the cytoskeleton). Early work predated our knowledge of a cell's molecular structure, genetic-heredity mechanisms, mechanics of cellular differentiation, and metabolic processes. What was clear to investigators at the time was that the rheologically *squishy* "protoplasm" of cells harbored the physical and chemical basis for life's processes. Although it appeared to be just a small, gelatinous mass, Seifriz (1928) writes, "The problem of metabolism, growth, reproduction, heredity, behavior, disease—in short, the problems of life—are the problems of the physical-chemistry of protoplasm." Presciently, Seifriz regarded the mechanics of cells as a key to understanding certain disease pathologies, including cancer.

[2] We use angle brackets ⟨.⟩ to denote an average taken over an ensemble in thermal equilibrium. Here, the ensemble consists of many realizations of a one-dimensional random walk, which tracks a particle's displacement X with time. See Fig. 4.20, for example.

involving gelatin (Freundlich and Seifriz, 1923).[1] Around the same time, Heilbrunn (1924) reported measurements of clam eggs, *Cumingia tellinoides*, using a centrifuge to force endogenous granules to move through the cytoplasm.

Active and passive microrheology

Early examples of microrheology measurements highlight their essential features—to measure probe particles embedded within soft materials as they move in response to a force, and to then deduce material-response properties from that motion. In the contemporary practice of microrheology, measurements made when the force on a probe is externally imposed—like the magnetic, gravitational, or centrifugal examples provided—fall into the class of **active microrheology** techniques. The other class, called **passive microrheology**, is a more recent development, and began with the seminal work of Mason and Weitz (1995) and Gittes *et al.* (1997).

Passive microrheology employs microrheological probe particles so small—typically a micrometer or smaller—that thermal fluctuations are strong enough to drive the probe into measurable motion. Such motion arises due to the constant bombardment by surrounding molecules, which are themselves rattling around due to thermal fluctuations. A particle thus experiences random forces, exerted over many directions and strengths and over a variety of time scales. The magnitude of the forces, and how the particle responds to those random forces, depends on the material itself. A particle randomly forced within a viscous fluid will generally wander in random directions, exhibiting diffusive trajectories with mean-squared displacement[2]

$$\langle \Delta x^2(t) \rangle = 2Dt. \tag{1.1}$$

Stokes computed the hydrodynamic resistance ζ of a sphere of radius a moving through a fluid of viscosity η (see Section 2.5.2) to be

$$\zeta = 6\pi a\eta, \tag{1.2}$$

and Einstein (1906) and Sutherland (1905) related a particle's diffusivity D to its hydrodynamic resistance, via

$$D = \frac{k_B T}{6\pi a\eta}. \tag{1.3}$$

The higher the viscosity, the more slowly the particle diffuses. This is the **Stokes–Einstein Relation**.[3]

A particle in an elastic solid, on the other hand, is effectively held in place as if by a spring with spring constant κ. In equilibrium, the equipartition theorem holds that the average energy stored within the spring in each of the three independent translational directions, $U = \frac{1}{2}\kappa \Delta X^2$, must be equal to $\frac{1}{2}k_B T$. The mean-square displacement will approach a constant value,

$$\langle \Delta x^2 \rangle = \frac{k_B T}{\kappa}, \tag{1.4}$$

unlike the linear growth in time seen in a viscous liquid (eqn 1.1). The stiffer the spring, the more tightly the particle is held in place. An elasticity calculation (Section 2.5.5) relates this spring constant to the elastic constants of the material:

$$\kappa = 6\pi aG \left[\frac{6K + 8G}{6K + 11G} \right]$$
$$\approx 6\pi aG, \text{ when } K \gg G, \tag{1.5}$$

where G is the shear modulus and K is the bulk (compressional) modulus of the material. Notably, an elastic material that is much harder to compress than shear ($K \gg G$) behaves as incompressible, with a spring constant $\kappa \approx 6\pi aG$ that looks suspiciously like Stokes drag in a liquid (eqn 1.2). This is no coincidence, as we shall see in Section 2.4.

These two limits bracket the possible responses in passive microrheology, where the forces driving the probe into motion are not imposed externally, but rather from the inherent and unquenchable thermal fluctuations within the equilibrium material. The thermal motion of small particles in a liquid or a solid, easily observed with a microscope or other means, contains a wealth of information about the properties of that material—whether viscosity, elasticity, or time scale-dependent viscoelasticity.

[3] Recently, the work of William Sutherland (1905), which paralleled Einstein's theory of Brownian motion, has been recognized. Equation 1.3 is now sometimes referred to as the Stokes–Einstein–Sutherland equation.

1.1.1   Why microrheology?

Microrheology encompasses a set of rheometric methods or techniques with unique capabilities—a part of the experimental toolbox for characterizing the rheological properties of materials to aid their understanding, or help in the design of new materials.

There are limitations to microrheology that are important to understand from the outset. Microrheology uses the movement of small particles in a material; thus, it is limited to fairly soft materials, with moduli typically no more than a few hundred pascals (not too far off from the stiffness of jello) or fluids with viscosities lower than that of honey. Many classes of materials—*e.g., polymer melts, glassy liquids, and elastomers*, for which rheological measurements played a central role in understanding—are too stiff or viscous to be amenable to microrheological methods. Despite being limited to soft materials, microrheology introduces important new capabilities for the rheologist, some of which include the following:

- **Small sample volumes**—From the studies of Heilbronn, Freundlich, and Seifriz in the early-twentieth century on, particles have been used to measure rheology in small sample volumes—down to single eukaryotic cells, with volumes $\sim$ 1 picoliter. Particle tracking (Chapters 4 and 6), magnetic bead microrheology (Chapter 8), and laser tweezer microrheology (Chapter 9) typically require sample volumes between $\sim$ 1 and 10 μl. This sample volume makes many scarce and expensive materials available to rheological characterization, and, in particular, the ability to screen material properties over a wide range of sample conditions and compositions. Formulations of protein therapeutics and emerging biomaterials are just two examples of such samples. The small sample dimensions facilitate rapid mass and heat transfer, enabling faster screening and sample preparation and manipulation using microfluidics.

- **Short acquisition times**—Microrheology data spanning several decades in time (*e.g.,* 0.01–1 s) can be acquired by multiple particle tracking in as little as a minute. This makes it possible to track the frequency-dependent response for samples that are changing with time—during gelation or degradation, for instance. The short acquisition times also aid rapid data acquisition in screening applications, enabling tens to hundreds of samples to be processed in a single day.

- **Sensitivity**—Fluids with low viscosities and solids with small-elastic moduli are within the range of microrheology. Solutions of entangled, filamentous actin (F-actin)—a principal protein of the cytoskeleton and muscle—appear almost Newtonian and are easily poured at 1 mg/ml concentrations in an aqueous buffer. Careful observation, however, reveals small bubbles to remain suspended, and to exhibit a subtle elastic recoil when the sample is twisted. Although its elastic modulus may be no more than one pascal, such weak moduli can reliably be measured with microrheology. More broadly, the "incipient rheology" of gel transitions in hydro- and organogelators, and the intrinsic viscosity of polymer and protein solutions, represent important and challenging classes of materials whose measurement is enabled by microrheology.

- **Extended range of frequencies**—Passive particle microrheology using diffusing wave spectroscopy (Chapter 5) or laser tracking (Chapter 6) measures probe motion on time scales as short as 1 μs, enabling high-frequency material response properties (kHz–MHz) to be measured directly, which is particularly useful when time-temperature superposition—commonly used for polymer melt rheology—is not applicable. The high-frequency response of polymer solutions and gels can be used to characterize the underlying nanometer-scale mechanics of the material—an application discussed in Section 5.6.

- **Local rheology**—Probe particles distributed throughout a sample can be used to map its spatial-rheological heterogeneity, clearly information that is not available to bulk rheology. We discuss this application in Section 4.10. With the use of multiple probe particles ("two-point microrheology," discussed in Section 4.11) the dependence of rheology as a function of length scale can be characterized.

- **Simple experiments**—Many microrheology experiments require little in the way of specialized equipment. Tracking particle motion with video microscopy is possible using only a microscope, video camera, and computer.

In short, microrheology opens a wide range of samples and conditions which may be difficult, if not impossible, to measure by conventional rheometry. Throughout the text, we will consider the **operating range** of microrheological methods to identify when they can be the greatest asset to a rheological study and to aid experimental design.[4] We also identify **application notes** in each of many of the chapters, highlighting areas where microrheology approaches to problems have been especially beneficial, and we discuss more applications in Chapter 10, including gelation and degradation of hydrogelators and biomaterials.

---

[4] The operating limits of passive microrheology are discussed in Section 3.11 and in the chapters on individual techniques. A comparison to the operating range of bulk rheology is made in Section 10.1.

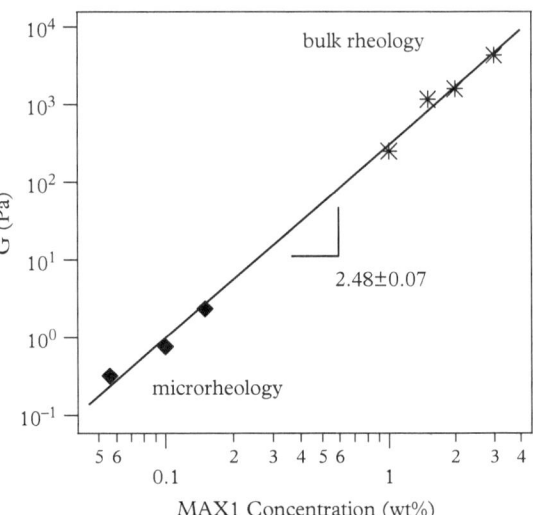

Fig. 1.2 *The modulus of a peptide hydrogelator measured using laser tweezer microrheology at low concentrations and bulk rheology at higher concentrations illustrates the complementarity of the rheological measurements. Adapted with permission from Veerman, C. et al., Macromolelcules 39, 6608–14 (2006). Copyright 2006 American Chemical Society.*

While microrheology is not a replacement for bulk rheology, one last and key benefit of microrheology is its **complementarity** to macro-rheology. The two methods can be combined to produce an understanding of a material's rheology beyond what would have been possible if only one approach were taken. This combination is illustrated by the data shown in Fig 1.2. Here, the elastic modulus of a peptide hydrogelator has been measured using oscillatory rheology and laser tweezer microrheology. The sensitivity of microrheology makes it best-suited for measurements at low peptide concentrations, when the corresponding moduli are small, whereas bulk rheology is better suited to higher concentrations and larger moduli. Together, the experiments create an interpolatable data set that spans nearly two decades in concentration and almost five decades in modulus, and is nicely consistent with the scaling with concenteration c expected for the elastic modulus of a semiflexible polymer network, $G \sim c^{5/2}$ (MacKintosh *et al.*, 1995).

The remainder of this chapter introduces background concepts that are important for microrheology, including general concepts of soft matter rheology and rheometry, rheological functions, and important aspects of colloid science.

1.2 Soft matter and rheology

The examples at the beginning of Section 1.1 described the limiting cases of particle motion in a purely viscous solvent or purely elastic solid. Many materials of interest—especially those typically studied using microrheology—fall into a more general class of **viscoelastic** fluids and solids. The way that such materials flow and deform—even on a qualitative level—depends entirely on what is done to them. Over what time scales are forces applied? How strong are the deforming forces? Are they sheared between plates, extruded through an orifice, or pulled into fibers?

We start with brief descriptions of some common rheological phenomena exhibited by everyday materials, with the goal of highlighting the rich variety that exists. We will identify the rheological property required to describe such phenomena, then follow by describing the sorts of measurements used to characterize them.

- **Honey** is a viscous, Newtonian liquid that responds as you might expect: It flows in response to applied stresses. Double the stress, and the flow rate doubles. Here, the relevant material property is the *shear viscosity η*. Viscous liquids like honey and water are usually approximated as *incompressible*.

- A **rubber** ball bounces when dropped, and bounces even more strongly when thrown. It consists of long polymer chains, each of which behaves effectively as a molecular spring, that are crosslinked chemically to form a "permanent," three-dimensional network of attached springs. As with small-molecule elastic materials (*e.g.*, steel) the energy required to deform the material is stored elastically, then recovered when the deformation is allowed to relax—in this case, with a bounce. Rubber and other elastomers can deform much more significantly (*i.e.*, to much higher strains) than steel without changing irreversibly. As with elastic solids, the shear and compressional elastic moduli G and K are relevant, as well as stress-strain curves and failure points. Unlike molecular solids, however, the elastic moduli of elastomers depend strongly on frequency, particularly at high frequencies (short time scales). This is because the polymeric springs store more energy than small-molecule crystals; rapid stress or strain pulses stretch the polymeric springs in a non-quasi-steady (non-adiabatic) fashion, exciting only some internal degrees of freedom, which dissipate energy as they relax. Therefore, frequency-dependent viscoelastic moduli $G^*(\omega)$ are required. Because of the solid-like response over long time scales, these materials are *viscoelastic solids*. Hydrogel networks similar to the one shown in Fig. 1.3 have many of the characteristics of elastic polymer networks. The rheological characterization of hydrogels is a focus of many microrheology measurements.

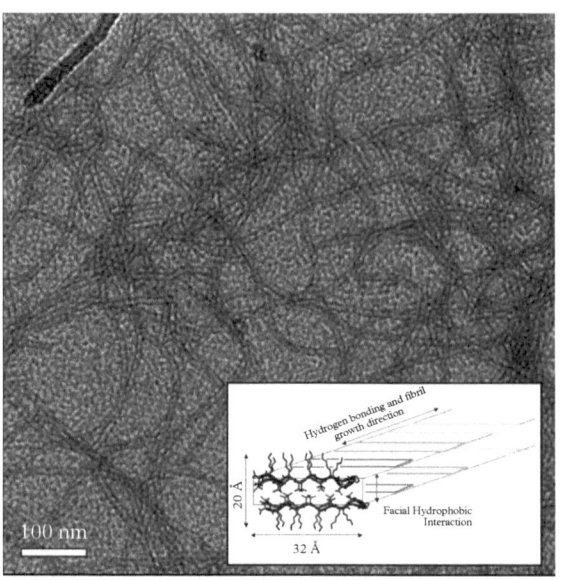

Fig. 1.3 *A cryo-transmission electron micrograph showing the highly entangled and physically cross-linked network of a peptide hydrogel. The amphiphilic peptides self-assemble into semiflexible filaments to form a viscoelastic solid. Reprinted with permission from Ozbas, B., Rajagopal, K., Schneider, J. P., & Pochan, D. J. Phys. Rev. Lett. 93, 268106 (2004). Copyright 2004 by the American Physical Society.*

- A ball of **Silly Putty** bounces when dropped, yet spreads into a pancake when left to sit for several minutes. Like a rubber ball, a ball of Silly Putty consists of long polymer chains; yet the polymers in Silly Putty are entangled without crosslinking, constantly rearranging under thermal motion. Polymers stretch and migrate when the material is deformed, but the (temporary) entanglements that exist at any given time effectively "anchor" the molecular springs in place, much like physical or chemical cross-links. If the stress is exerted over a long enough time, the entanglements eventually relax and the material flows like a liquid. Short-lived stresses, however, do not give the entanglements time to relax, and the Silly Putty springs back like an elastic solid. Whether or not Silly Putty bounces depends entirely upon the relaxation time for the entanglements—a quantity that can be measured using small-amplitude linear rheology. As with crosslinked elastomers, frequency-dependent viscoelastic moduli $G^*(\omega)$ are required to characterize Silly Putty. At medium to high frequencies, $G^*(\omega)$ may even be identical for the two materials. At low frequencies, however, the crosslinked elastomer has a finite-shear modulus $G^*(\omega \to 0) \to G_0$, whereas the elastic shear modulus of the uncrosslinked material vanishes at low frequencies. Consequently, Silly Putty is considered to be a *viscoelastic liquid*, whereas (crosslinked) rubber balls are viscoelastic solids. The crossover frequency ω_c—below which the elastic (real) component of $G^*(\omega_c)$ drops below the viscous (imaginary) component $G^*(\omega_c)$—is directly related to the longest relaxation time of the entanglements.

- **Mayonnaise** sits on a knife without flowing, despite the gravitational forces exerted on it. In this regard, mayonnaise appears to be a viscoelastic solid. Nonetheless, very little effort is required to spread mayonnaise on a piece of bread. Mayonnaise behaves as a solid under low stresses, but flows like a liquid above a critical *yield stress*. Mayonnaise consists of oil drops suspended in an aqueous solution at such a high concentration that drops can not move without rearranging (Fig. 1.4). Since rearrangement requires a finite amount of energy, a finite stress must be applied before it flows. Toothpaste, cake frosting, and yogurt also have yield stresses, but for different reasons: Each involves a weak, transient gel that takes some energy to break, but reforms—rapidly for frosting and toothpaste, and over longer time scales for yogurt (Fig. 1.5) Relevant rheological properties include $G^*(\omega)$, for insight into the equilibrium structure and relaxation processes, and the yield stress σ_y and yield strain γ_y.

- Watching **shampoo** flow in a bottle, one would assume it to be a liquid as viscous as honey. However, it is painless to spread shampoo into hair, whereas spreading honey into hair might pull it out. Shampoo also feels "slippery," indicating its *shear thinning* nature: It flows with high viscosity when sheared slowly, but at much lower viscosity when sheared rapidly. Shampoo shear thins because the structures that impart the high viscosity (*e.g.*, surfactant worm-like micelles) align with shear flows to facilitate the flow. The linear viscoelastic moduli $G^*(\omega)$ provide information about the structure and relaxation around equilibrium, but shear thinning requires the shear viscosity $\eta(\dot{\gamma})$ to be measured as a function of shear rate $\dot{\gamma}$.

- A drop of **saliva**, stretched between two fingers, develops a "beads-on-a-string" structure as it thins, like those shown in Fig. 1.6. This is characteristic of dilute polymer solutions, whose viscosity thins like shampoo under *shear* flows, but *thickens* under *extensional* flows. In extensional flows, polymers are stretched along the flow direction (Fig. 1.7), and thus act directly against the flow as they try to recoil. The stronger the flow, the further they deform, and the harder they fight the flow. This behavior is described by a rate-dependent extensional viscosity $\eta_E(\dot{\epsilon})$.

- Concentrated **cornstarch** solutions, known as oobleck to parents of young children, *shear thicken* dramatically: The apparent

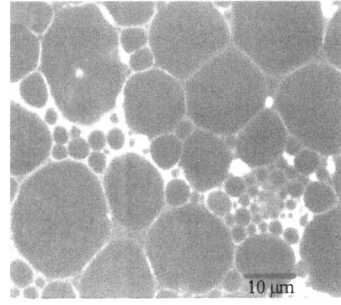

10 μm

Fig. 1.4 *Mayonnaise cools your fries without flowing off. Its yield stress is a result of jammed oil droplets, shown in a confocal micrograph. The water phase contains a fluorescent dye, while the oil droplets are dark. Some may prefer ketchup, another yield stress material. Micrograph reprinted from Food Structure, 1, Heertje, I., Structure and function of food products: A review, pp. 3–23, Copyright (2014), with permission from Elsevier.*

Fig. 1.6 *High-speed video images of "beads-on-a-string" forming on a jet of dilute polymer solution. From Clasen, C., Eggers, J., Fontelos, M. A., Li, J., & McKinley, G. H. J. Fluid Mech, 556, 283–308 2006a reproduced with permission.*

Fig. 1.5 *Electron micrograph of casein micelles forming a gel network in yogurt. Reprinted from Colloids Surfaces B Biointerfaces, 31, Aichinger, P. A. et al., Fermentation of a skim milk concentrate with Streptococcus thermophilus and chymosin: Structure, viscoelasticity and syneresis of gels, pp. 243–55, Copyright (2003), with permission from Elsevier.*

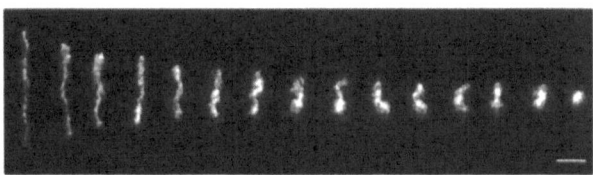

Fig. 1.7 *Flourescence microscopy images of a lambda phage DNA (48.5 kbp) molecule relaxing. The polymer chain is subjected to an extensional flow and high degrees of extension, followed by direct imaging of its relaxation into a coiled state after the cessation of flow. The scale bar is 5 μm. Image courtesy of Yuecheng (Peter) Zhou and Charles M. Schroeder.*

shear viscosity can jump a million-fold when the material is sheared above a critical rate. This reflects a competition between shear-driven cluster formation of aggregating particles (increasing the resistance to flow) and relaxation to equilibrium (reducing the resistance). As with shear thinning materials, shear thickening is described by a rate-dependent shear viscosity $\eta(\dot{\gamma})$.

- **Jello** is made by cooling an initially-heated gelatin solution. Although the solution cools relatively quickly (minutes), the material transitions from a viscous liquid to an elastic solid over a time scale of hours. A *gelation* process occurs gradually, as suspended polymers, a mix of denatured protein and peptide fragments, cross-link to form ever-larger clusters, eventually spanning the entire material. Rheological characteristics of interest include the *gel time* τ_G, and frequency-dependent viscoelastic moduli $G^*(\omega)$, especially its elastic modulus. After all, this is what gives jello its gentle jiggle—and the ability to suspend solid pieces of fruit.

- **Egg whites** are viscoelastic solutions of protein in water whose rheology enables some culinary feats (*e.g.*, merengues) while frustrating even simple tasks (*e.g.*, removing small bits of egg shell, or climbing up mixers and making a mess). This rod-climbing "Weissenberg effect" occurs due to *normal stress differences*: Shearing the solution stretches its elastic elements in the flow direction. As they recoil, they tend to raise the material's tension in the flow direction (around the rod), relative to the gradient direction (along the rod), which squeezes egg white up the spinning rod. The rheological quantity responsible for this behavior is the normal stress difference $N_1\dot{\gamma}^2$. Additionally, whip the egg white, and it forms a long-living foam. Proteins at the surface are exposed to air, denature, and aggregate to form an interfacial shell with its own **surface viscoelasticity,** which can be described by surface shear moduli $G_s^*(\omega)$.

These examples—and many more—provide some sense of the wide variety of phenomena encompassed within the rich world of rheology. Many of the materials listed elude the conventional classification of matter into "solid, liquid, or gas" phases. Instead, these soft materials usually consist of multiple components, each of which can be described individually as a liquid, solid, or perhaps as a macromolecule. These components form a mesoscopic structure within the material that is not immediately apparent to our senses of touch or sight, since they typically form on the nanometer to micrometer scale. These structures give rise to the rich set of dynamic response properties already described, that are not found in simple fluids or solids. "Macroscopic" experimental tools—*e.g.*, our fingers as they manipulate shampoo, or the rheometers described in Section 1.2.2—measure an "averaged" response of heterogeneous materials, which behave like homogeneous, continuum effective materials on those macroscopic length scales.

1.2.1 Linear and nonlinear rheology

The rheology of a material is measured by relating the stress, σ to the imposed deformation strain γ or rate of strain $\dot{\gamma} = d\gamma/dt$. A simple representation of the measurement is shown in Fig. 1.8, in which a force F_x is required to pull a plate of area A_y (the "y" denotes the direction of the outward unit normal vector). The plate, separated from a bottom plate by a distance δ_y, moves a distance Δx. The shear stress is

$$\sigma_{yx} = F_x/A_y \tag{1.6}$$

and the strain

$$\gamma = \Delta x/\delta_y. \tag{1.7}$$

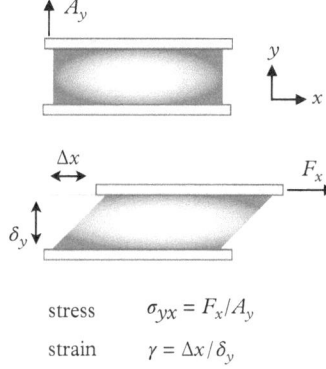

$$\sigma_{yx} = F_x/A_y$$
$$\gamma = \Delta x/\delta_y$$

Fig. 1.8 *Shear deformation of a material between two parallel plates.*

If the material between the plates were an elastic solid, the strain would reach a steady value for a given stress. If the material were a viscous fluid, the strain would represent the deformation at a finite time, and the plate would continue to move to the right at a shear rate $\dot{\gamma} = \sigma/\eta$. Both behaviors are analogous to the movement of our probe particle in Fig. 1.1.

Linear response properties, most commonly the frequency-dependent linear viscoelastic moduli $G^*(\omega)$, reflect the response of materials to negligibly small departures from equilibrium (departures which, in fact, arise spontaneously due to the material's thermal energy). These properties reflect the relaxation processes that occur within such materials in their equilibrium state. With knowledge of the microstructural elements and their organization, important static and dynamic structural features can be determined from measurements of $G^*(\omega)$. For example, the Rouse or Zimm models can be used to determine the molecular weight and concentration of polymers in solution, or relaxation times for polymer entanglements. Hydrodynamic calculations for solid particles or liquid droplets with viscosity η and interfacial tension can be used to extract the size distributions of droplets or particles from $G^*(\omega)$ measurements made on particle suspensions or concentrated emulsions.

Additionally, linear-response properties like $G^*(\omega)$ can be used to measure other material properties of evolving materials (*e.g.*, materials like yoghurt or clay that age after being sheared, or like Jello that undergoes a sol-gel transition), so long as the evolution occurs on time scales longer than is required to actually make such measurements.

Nonlinear response properties arise when the microstructure of the material is driven significantly out of equilibrium. The yield stress σ_y requires the material to be strained far enough for microstructural elements to break or rearrange. Shear thinning and shear thickening viscosities and extension thickening arise when an imposed flow alters the arrangement of microstructural elements from their equilibrium distribution, making the flow easier or harder to maintain. Normal stress differences N_1 and N_2 arise when the equilibrium microstructure is deformed enough to drive anisotropic tension within the material.

Nonlinear-rheological quantities like yield stresses, rate-dependent viscosities, and normal stress coefficients cannot be determined using linear-response measurements. In a few cases and in certain limits, correspondences may exist between linear and nonlinear properties. For example, the low-$\dot{\gamma}$ limit of the first normal stress coefficient of a viscoelastic liquid is related to the low-frequency limit of the elastic modulus $G'(\omega)$, via

$$\lim_{\dot\gamma \to 0} \frac{N_1(\dot\gamma)}{\dot\gamma^2} \Leftrightarrow \lim_{\omega \to 0} \frac{G'(\omega)}{\omega^2}. \tag{1.8}$$

The Cox–Merz rule is an empirical relation relating the frequency-dependent complex viscosity

$$\eta^*(\omega) = \frac{G^*(\omega)}{i\omega} \tag{1.9}$$

to the rate-dependent steady-shear viscosity $\eta(\dot\gamma)$, according to

$$\eta(\dot\gamma) = \left|\eta^*(\dot\gamma)\right| \quad \text{(Cox-Merz)}, \tag{1.10}$$

but does not always apply.

More generally, however, there is no way to determine nonlinear response properties from linear response measurements. On a qualitative level, toothpaste appears to behave like Jello according to linear viscoelastic measurements. Both are soft, viscoelastic solids under weak forcing. Unlike jello, given enough force, toothpaste flows—it has a yield stress—while jello will fracture and break.

1.2.2   Linear response measurements

Linear response measurements perturb a material so slightly that its equilibrium structure remains almost entirely unchanged, driving small deformations that subsequently relax. Even when a soft material is unforced, and simply sitting in equilibrium at some temperature T, it constantly experiences weak, stochastic-thermal forces that drive small-amplitude deformations of the sort used to measure linear response properties. Passive microrheology exploits these thermal fluctuations as a built-in source of small-amplitude forces to reveal (and measure) the linear viscoelastic response properties of the material.

A mechanical rotational rheometer provides a means of generating shear strains and measuring stress by the torque imposed on the tool. Mechanical rheometers employ a variety of tool geometries—*e.g.*, cone-and-plate, cylindrical, and parallel plate. These geometries are shown in Fig. 1.9. Each has its particular operating regimes of frequency, shear amplitude, shear rate, and sample properties—an operating regime—but all those shown here are designed to excite purely shear strains. The flow kinematics are determined solely by the geometry.

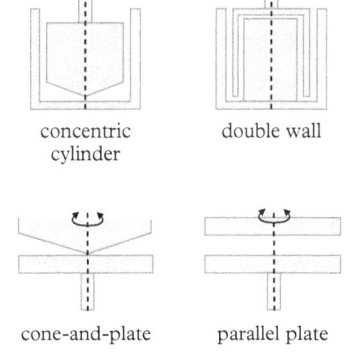

concentric cylinder   double wall

cone-and-plate   parallel plate

Fig. 1.9 *Common tool geometries of rotational mechanical rheometry.*

Complex shear modulus

A typical rheometry measurement imposes an oscillatory strain

$$\gamma(t) = \gamma_0 e^{i\omega t}, \tag{1.11}$$

with amplitude γ_0 and frequency ω, and measures the stress $\sigma(t)$ in response. In any linear response measurement, the measured stress will oscillate sinusoidally with the same frequency ω (with no harmonics),

$$\sigma(t) = \sigma_0 e^{i(\omega t + \delta)} \tag{1.12}$$

with an amplitude σ_0 and phase lag (or loss tangent) δ that encodes the rheology of the material itself. A purely elastic material is one for which $\delta = 0$: The stress is directly proportional to the strain. A purely viscous material is one for which $\delta = \pi/2$: The stress is proportional to the *rate of strain*. At the height of the strain oscillation, the shear rate is zero, and thus the stress. By contrast, the strain rate (and thus the viscous stress) is largest at zero strain.

Measuring stress as a function of strain over a range of frequencies ω gives the linear, viscoelastic moduli, encompassed in the complex shear modulus $G^*(\omega)$, defined by

$$G^*(\omega) = \frac{\sigma(t)}{\gamma(t)} = \frac{\sigma_0 e^{i\delta}}{\gamma_0}. \tag{1.13}$$

The complex shear modulus $G^*(\omega)$ is the frequency-dependent equivalent of a pure elastic modulus, defined as the shear stress divided by the shear strain. The higher G^*, the more stress is required to drive a certain strain.

We could just as easily take the same measured data, but instead compare the measured stress $\sigma(t)$ with the *strain rate*,

$$\dot{\gamma}(t) = i\omega\gamma_0 e^{i\omega t} \equiv i\omega\gamma(t) \tag{1.14}$$

as would make sense for a viscous or viscoelastic liquid. The complex viscosity is then obtained by dividing the stress by the strain rate,

$$\eta^*(\omega) = \frac{\sigma_0 e^{i\delta}}{i\omega\gamma_0}. \tag{1.15}$$

Comparison with eqn 1.13 reveals the complex shear modulus and complex viscosity to be trivially related:

$$G^*(\omega) = i\omega\eta^*(\omega). \tag{1.16}$$

This is significant, because it underscores the fact that $\eta^*(\omega)$ contains exactly the same information as $G^*(\omega)$. This should not be surprising—after all, precisely the same measurement gave rise to both quantities.

The complex modulus is often split into real and imaginary components

$$G^*(\omega) = G'(\omega) + iG''(\omega), \tag{1.17}$$

which separates out the elastic (or storage) modulus $G'(\omega)$ and the viscous (or loss) modulus $G''(\omega)$. The storage modulus $G'(\omega)$ is precisely the conventional elastic shear modulus, generalized to allow for frequency dependence, and describes the (recoverable) energy required to deform the material at a particular frequency. $G'(\omega)$ represents the portion of the shear stress that varies in-phase with the sinusoidal shear strain. By contrast, the loss modulus $G''(\omega)$ describes the (irrecoverably lost) energy that is dissipated as a material deforms at a given frequency. It is 90 degrees out-of-phase with the shear strain, or equivalently, in-phase with the shear rate. The loss modulus is intimately related to the real part of the frequency-dependent complex viscosity,

$$G''(\omega) = \omega\eta'(\omega), \tag{1.18}$$

as can be seen from eqn (1.16). Here, we have split η^* into its real and complex parts, via

$$\eta^*(\omega) = \eta'(\omega) + i\eta''(\omega), \tag{1.19}$$

by analogy with (1.17). The phase lag or phase angle δ in (1.12) is related to the storage and loss modulus via

$$\tan\delta(\omega) = \frac{G''(\omega)}{G'(\omega)}, \tag{1.20}$$

ranging from $\delta = 0$ for purely elastic materials, whose stress varies in-phase with strain, and $\delta = \pi/2$ for viscous fluids, whose stress is 90 degrees out-of-phase with the applied strain (which means, of course, that $\tan\delta$ diverges as $G' \to 0$).

More generally, any time-dependent stress $\sigma(t)$ can be decomposed into frequency-dependent components through a Fourier Transform:

$$\sigma(t) = \frac{1}{2\pi} \int_{-\infty}^{\infty} \tilde{\sigma}(\omega)e^{i\omega t} d\omega. \tag{1.21}$$

So long as the total strain is small enough for the linear response approximation to remain valid, the stress driven by each of these strain oscillations is given by eqn 1.13, and the total stress at any given time is given by the superposition of each oscillating component,

$$\sigma(t) = \frac{1}{2\pi} \int_{-\infty}^{\infty} G^*(\omega)\tilde{\gamma}(\omega)e^{i\omega t}d\omega. \tag{1.22}$$

Using the convolution theorem, eqn 1.22 can be re-expressed as

$$\sigma(t) = \int_{-\infty}^{\infty} m(t-t')\gamma(t')dt', \tag{1.23}$$

where the memory function

$$m(t) = \frac{1}{2\pi} \int_{-\infty}^{\infty} G^*(\omega)e^{i\omega t}d\omega \tag{1.24}$$

is the inverse Fourier Transform of the complex modulus $G^*(\omega)$. Physically, $m(t-t')$ expresses how much the stress at any given time t "remembers" a deformation that happened at some previous time t'. Because the stress can't "remember" a deformation that has not yet occurred, $m(t-t')$ must be zero for all $t' > t$, a property of *causal* functions. Equation 1.23 is thus often written as

$$\sigma(t) = \int_{-\infty}^{t} m(t-t')\gamma(t')dt'. \tag{1.25}$$

Alternatively, stress may be related to the previous *shear rate* history,

$$\sigma(t) = \int_{-\infty}^{\infty} G(t-t')\dot{\gamma}(t')dt', \tag{1.26}$$

where $G(t-t')$ is called the *relaxation modulus*, and expresses how well the stress at time t "remembers" the shear rate at a previous time t'. Using the convolution theorem, eqn 1.26 becomes

$$\sigma(t) = \int_{-\infty}^{\infty} \mathscr{F}\{G(t)\}i\omega\tilde{\gamma}(\omega)e^{i\omega t}d\omega. \tag{1.27}$$

Comparison with eqn 1.22 reveals the Fourier Transform of $G(t)$ to be

$$\mathscr{F}\{G(t)\} = \frac{G^*(\omega)}{i\omega} = \eta^*(\omega), \tag{1.28}$$

by definition of the complex viscosity $\eta^*(\omega)$, eqn 1.16.

This set of definitions can seem arbitrary or confusing at first. In short, two Fourier Transform pairs exist:

$$m(t) = \mathscr{F}^{-1}\{G^*(\omega)\} \tag{1.29}$$

$$G(t) = \mathscr{F}^{-1}\{\eta^*(\omega)\}. \tag{1.30}$$

One Fourier Transform pair, $G^*(\omega)$ and $m(t)$, is best suited for viscoelastic solids, but is used almost ubiquitously in rheology. The other Fourier Transform pair, $\eta^*(\omega)$ and $G(t)$, is better suited for viscoelastic liquids. Unfortunately, this convention can be quite confusing, since one might naturally expect $G^*(\omega)$ to represent the Fourier Transform of $G(t)$, from a purely notational standpoint. This is not true, so take note.

Kramers–Kronig relations

The storage and loss moduli are not independent functions, since the dynamic response they encode is *causal*—a material can not respond to a stimulus that has not yet occurred. Consequently, the memory function $m(t)$ *must* be zero for all $t < 0$. From complex analysis, this implies that $G^*(\omega)$ is analytic in the lower-half plane. Moreover, the real and imaginary parts of $G^*(\omega)$—namely, $G'(\omega)$ and $G''(\omega)$—are related exactly by the Kramers–Kronig relations (McQuarrie, 2000; Landau *et al.*, 1986)

$$G'(\omega_0) = -\frac{1}{\pi}\mathscr{P}\int_{-\infty}^{\infty}\frac{G''(\omega)}{\omega - \omega_0}d\omega \tag{1.31}$$

$$G''(\omega_0) = \frac{1}{\pi}\mathscr{P}\int_{-\infty}^{\infty}\frac{G'(\omega)}{\omega - \omega_0}d\omega, \tag{1.32}$$

where the $\mathscr{P}$ denotes the Cauchy Principle Value of the integral.[5] These are derived in Appendix A.3. Booij and Thoone (1982) derived various alternative forms of the Kramers–Kronig relations that are of particular benefit to rheologists, including

$$G''(\omega) = \frac{2\omega}{\pi}\int_{0}^{\infty}\frac{G'(u) - G'(\omega)}{u^2 - \omega^2}du \tag{1.33}$$

$$G'(\omega) = G'(0) - \frac{2\omega^2}{\pi}\mathscr{P}\int_{0}^{\infty}\frac{G''(u)/u - G''(\omega)/\omega}{u^2 - \omega^2}du \tag{1.34}$$

$$G'(\omega) = G'(\infty) - \frac{2}{\pi}\mathscr{P}\int_{0}^{\infty}\frac{uG''(u) - \omega G''(\omega)}{u^2 - \omega^2}du. \tag{1.35}$$

[5] The Cauchy principal value integral accounts for the singularity at $\omega' = \omega$,

$$\mathscr{P}\int_{0}^{\infty} f(\omega')d\omega'$$

$$= \lim_{\varepsilon \to 0^+}\left[\int_{0}^{\omega-\varepsilon} f(\omega')d\omega' + \int_{\omega+\varepsilon}^{\infty} f(\omega')d\omega'\right].$$

The Kramers–Kronig relations provide an important validation of measurements or calculations of the viscoelastic moduli and are the basis for calculating the moduli in techniques such as laser tracking microrheology, which is discussed in Chapter 6.

As we have seen, the memory function $m(t)$ contains precisely the same information as $G^*(\omega)$, which should be obvious given that they form a Fourier Transform pair. The complex viscosity $\eta^*(\omega)$ and re-laxation modulus $G(t)$ likewise contain the same information. Since $G(t) = dm/dt$, however, $m(t)$ can only be determined from $G(t)$ up to an additive constant, as seen directly in eqns 1.34 and 1.35. Either the zero-frequency $G'(0)$ or infinite-frequency $G'(\infty)$ must be supplied to fully determine the elastic modulus $G'(\omega)$ from the loss modulus $G''(\omega)$.

Creep Compliance

By now it should be clear that there are a great many equivalent representations of a material's linear-viscoelastic response, each of which contains the same fundamental information, yet some arise more naturally in particular contexts, and are therefore more natural to interpret or manipulate than others.

The creep compliance $\mathcal{J}(t)$ is one functional representation of a material's linear viscoelastic properties that will be particularly useful for passive microrheology. The creep compliance $\mathcal{J}(t)$ is the strain that results following a suddenly-imposed stress of unit magnitude,

$$\sigma(t) = H(t), \tag{1.36}$$

where $H(t)$ is the Heaviside step function, then measuring the strain response

$$H(t) = \int_{-\infty}^{t} m(t-t')\mathcal{J}(t')dt'. \tag{1.37}$$

Fourier Transforming gives

$$\frac{1}{i\omega} = G^*(\omega)\tilde{\mathcal{J}}(\omega), \tag{1.38}$$

revealing the transformed creep compliance to be related to the complex modulus via

$$\tilde{\mathcal{J}}(\omega) = \frac{1}{i\omega G^*(\omega)}, \tag{1.39}$$

or equivalently

$$\mathcal{J}(t) = \mathscr{F}^{-1}\left(\frac{1}{i\omega G^*(\omega)}\right). \tag{1.40}$$

This relation will facilitate the interpretation of microrheology measurements. In particular, the mean squared displacement $\langle \Delta r^2(t)\rangle$ of a tracer particle in an equilibrium material turns out to be directly proportional to the creep compliance $\mathcal{J}(t)$.

Two limiting cases are revealing. A Newtonian fluid with viscosity η, and complex modulus $G^* = i\omega\eta$, has creep compliance

$$\mathcal{J}(t) = \frac{t}{\eta} \quad \text{(Newtonian fluid)} \tag{1.41}$$

that grows linearly (and unbounded) in time. An elastic solid with shear modulus G has a creep compliance

$$\mathcal{J}(t) = \mathcal{J}_e = \frac{1}{G} \quad \text{(Elastic solid)} \tag{1.42}$$

that is constant in time. Examples for viscoelastic liquids and solids are shown in Fig. 1.10: At long times, each $\mathcal{J}(t)$ asymptotes to the appropriate limit of a viscous fluid or an elastic solid.

1.2.3   Nonlinear-rheology measurements

Shear thinning and thickening

Nonlinear measurements are fundamentally different and require different techniques, since the material is typically driven far out of equilibrium. Here we will review a few examples of nonlinear-rheological behavior to give the reader a sampling of the phenomena that are of interest and issues that arise in their measurement. A large body of work in the rheology literature deals with nonlinear phenomena that arise in polymer processing, but materials like polymer melts are generally far outside the operating regime of microrheology. We will consider a few examples of materials that have been investigated in microrheology experiments and are discussed later in the book: The shear thinning of suspensions and measurements of yield stresses.

Shear-dependent viscosities are measured using a continuous deformation at different shear rates $\dot{\gamma}$. In a material that exhibits **shear thinning**, the viscosity decreases with increasing shear rate. Figure 1.11 shows viscosity measurements from the classic study of Choi and Krieger (1986b), who measured the shear thinning of polymer-stabilized PMMA nanoparticles suspended in silicone oil.

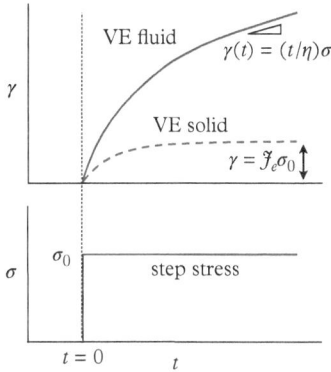

Fig. 1.10 *The material strain $\gamma(t)$ from an applied step stress σ_0 for a viscoelastic liquid and viscoelastic solid.*

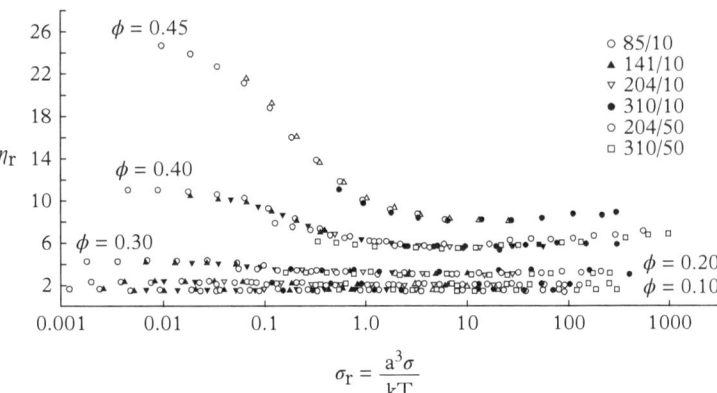

Fig. 1.11 *The relative viscosity* $\eta_r = \eta/\eta_s$ *of colloidal suspensions exhibits* shear thinning *as the shear stress increases. Reprinted from J. Colloid Interface Sci., **113**, Choi, G. N. & Krieger, I. M., Rheological studies on sterically stabilized dispersions of uniform colloidal spheres. II. Steady-Shear Viscosity, pp. 101–13, Copyright (1986), with permission from Elsevier.*

In suspensions, shear thinning occurs due to the reorganization of particles along the shear gradient. Two limits are observed: At low shear rates, the material has a viscosity η_0 that reflects the equilibrium structure of the suspension. At high shear rates, the nonequilibrium structure is fully formed, and the high-shear viscosity is η_∞. The empirical Cross model can faithfully describe data such as that shown in Fig. 1.11,

$$\frac{\eta - \eta_\infty}{\eta_0 - \eta_\infty} = \frac{1}{1 + K\dot{\gamma}^{1-n}} \tag{1.43}$$

but detailed microscopic models that accurately account for the Brownian and hydrodynamic origins of the high- and low-shear viscosities have also been developed (Brady, 1993).

At still higher shear rates, suspensions sometimes exhibit **shear thickening**, in which the viscosity again increases. Modest shear thickening is expected to occur due to lubrication hydrodynamic forces between particles, which causes them to form "hydroclusters" that disrupt the high-shear nonequilibrium microstructure (Egres and Wagner, 2005; Wagner and Brady, 2009). At high concentrations, suspensions shear thicken strongly, or "discontinuously" (D'Haene *et al.*, 1993).

Yield stress

A material with a yield stress behaves as a soft solid under weak stress, but flows like a fluid at high-enough stress (Møller *et al.*, 2006; Denn and Bonn, 2011). Foods like mayonnaise and ketchup have yield stresses, as do foams, toothpaste, and many paints (the paint flows from the brush onto the wall, but not down to the floor!) Yield stresses are frequently engineered into materials to suspend particles, like droplets of silicon oil in conditioning shampoos, solid rock cuttings in drilling muds, or crystallites of active crop protectant for agricultural treatments. Common yield stress fluids include suspensions of asso-ciative hydrocolloids such as xanthan and other biopolymer "gums," cellulose fibers, swellable microgel particles (like Carbopol), associa-tive colloids, and natural or synthetic clays—bentonite, kaolin, and the synthetic clay Laponite.

To model a fluid with a yield stress σ_y, several variants on the general constitutive relation,

$$\begin{aligned} \dot{\gamma} &= 0 & \sigma &\leq \sigma_y \\ \sigma &= \sigma_y + f(\dot{\gamma}) & \sigma &> \sigma_y. \end{aligned} \tag{1.44}$$

are commonly used. If the flowing state behaves as approximately Newtonian, then this gives the Bingham fluid,

$$f(\dot{\gamma}) = \eta_p \dot{\gamma}. \tag{1.45}$$

Yield stress fluids that behave as power-law fluids when flowing are described by the Herschel–Bulkley model,

$$f(\dot{\gamma}) = k\dot{\gamma}^n. \tag{1.46}$$

Both models satisfy the conditions that $f(0) = 0$, which defines a con-sistent yield stress, and $df/d\dot{\gamma} > 0$, which is required for mechanical stability.

Yield stress fluids present several vexing problems in bulk rheom-etry, as discussed by Møller *et al.* (2006), which may be expected to complicate microrheology experiments as well. Wall slip is a com-mon artifact, and frequently accomodated by using roughened tools or vane geometries, although the latter do not provide a direct meas-urement of the shear strain (Dzuy and Boger, 1983; Nguyen and Boger, 1992; Barnes and Nguyen, 2001). Additionally, shear banding or even fracture may occur within the material.

Different strategies have been employed to measure yield stresses. One set of techniques starts with a stationary material, then gradu-ally increasing the applied stress until the material flows. Alternatively, the strain or strain rate may be imposed, *e.g.*, starting from a steadily flowing system, and gradually reducing the strain rate. Typically, the resulting stress approaches a constant value (Fig. 1.12) as the strain rate approaches zero. Such measurements suggest an apparent viscos-ity that diverges with decreasing (Moller *et al.*, 2009). Corresponding linear, frequency-sweep measurements of $G^*(\omega)$ show the material does indeed behave as a soft solid.

Fig. 1.12 *Carbopol (0.5 wt%), a soft yield stress fluid. Reprinted from* J. Non-Newtonian Fluid Mech., *142, Oppong, F. K. & de Bruyn, J. R., Diffusion of microscopic tracer particles in a yield stress fluid, pp. 104–11, Copyright (2007), with permission from Elsevier.*

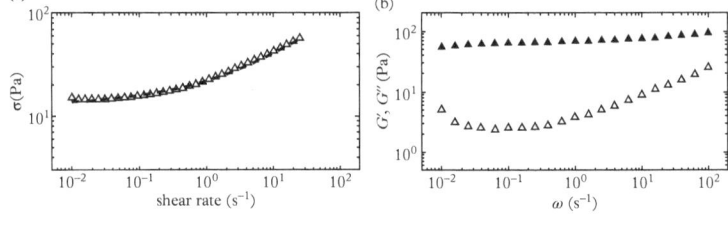

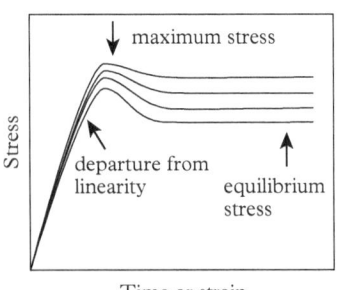

Fig. 1.13 *Different definitions of the yield stress.*

The yield stress measurements described typically report different yield stress values, depending on whether measurements start from flowing or quiescent states (Fig. 1.13). In some cases, the yield stress is defined as the value where the stress departs from linearity, or as the maximum measured stress (if a maximum occurs). Those two quantities are called the *static* yield stress, since they are measured starting from a quiescent state. They stand in contrast to the *dynamic* yield stress, which are measured by starting with a flowing state, decreasing the shear rate to $\dot{\gamma} = 0\ s^{-1}$, and extrapolating the measured stress.

Thixotropy

Closely related to yield stress is **thixotropy**, which refers to a history-dependence in the measured rheology. In fact, thixotropy and yield stresses are often (but not always) found together. Thixotropy arises due to reversible (and irreversible) microstructural changes in the material that grow an ever-stiffening mesostructural network over time. The yield stress of such materials increases gradually (often logarithmically in time) as they are left to rest. When forced to flow, these networks are broken to an extent that depends upon their strength, the strength of the flow, and time scales for aging (network rearrangement or build-up) to occur.

Experimentally, thixotropic behavior is detected by imposing an increasing set of shear stresses on the material, and measuring the resulting shear rate, then reversing the stress ramp to return back to the non-flowing state. Thixotropic materials produce a strong hysteresis during this cycle, whereas non-thixotropic materials do not (Fig. 1.14). One can thus draw a distinction between simple, or "ideal" yield stress materials like foams, emulsions, and microgel suspensions like Carbopol, which exhibit little or no hysteresis during this stress ramp cycle, and thixotropic yield stress materials like colloidal or fibrous gels, associative polymers, and clays (such as bentonite), which are strongly thixotropic (Moller *et al.*, 2009).

Extensional rheology

Unfortunately, a material's nonlinear-shear rheology cannot be used to predict its nonlinear extensional rheology. In fact, materials with appreciable elasticity may *extension thicken* significantly under extensional flows, despite having viscosities that *shear thin* just as dramatically. Strong extension thickening makes a material very difficult to pump through porous media or filters, and causes "beads-on-string" morphologies in viscoelastic fluid threads (Fig. 1.6), whereas Newtonian fluid threads would break into drops (Clasen *et al.*, 2006*a*).

Different techniques are required to probe the extensional rheology of materials, since the material must be subjected to a controlled extensional flow (Fig. 1.15). During this flow, the strain on the material grows exponentially in time in one direction, while contracting exponentially in the other direction(s). Extensional rheometry is far more challenging than shear rheometry in this regard: While a shear rheometer can impose arbitrarily large strain by simply rotating a cone or concentric cylinder indefinitely, geometric and practical constraints limit the spatial extent over which extensional rheometers can stretch a material with exponentially-growing strain. Strategies employed to do so include filament stretching rheometry (FISER) and capillary breakup extensional rheometry (CABER), which impose controlled extensional strains (McKinley and Sridhar, 2002; Bhardwaj *et al.*, 2007). Microfluidic methods have also been introduced (Pipe and McKinley, 2009). Our point is not to delve into a detailed discussion of extensional rheometry, but instead to highlight the difficulties associated with precise measurements of even the second paradigmatic type of flow for complex fluids. As with nonlinear shear rheometry measurements, nonlinear extensional rheology measurements require a system that has been carefully designed to excite only a particular flow type, to measure the response, and to interpret the results in terms of an intrinsic material property.

1.3 Colloidal particles

Microrheology, whether passive or active, is based on measurements of the motion of colloidal probe or tracer particles ranging in size from roughly 0.1–10 μm (Fig. 1.16). In Chapter 2, we discuss the mechanics of probe motion that underlie all microrheological methods. In this section, we will briefly review the chemical and physical properties of probe particles.

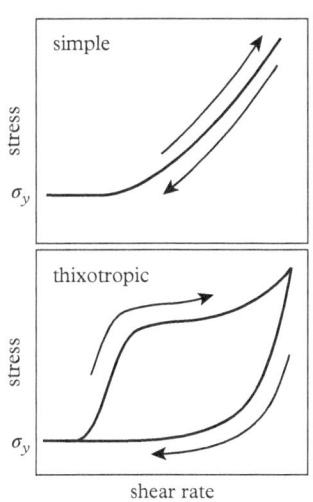

Fig. 1.14 *Stress ramp experiments, in which applied shear stress increases in steps from zero to a maximum value, then is decreased through the same series, reveal no hysteresis in simple yield stress materials, but strong hysteresis in thixotropic materials like suspensions of clay particles. Both materials have a finite yield stress σ_y.*

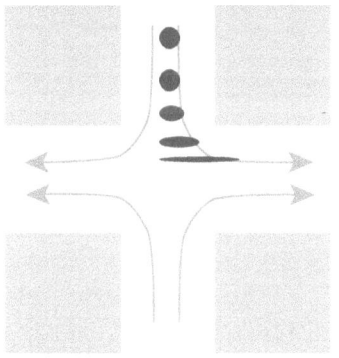

Fig. 1.15 *Purely extensional deformations, like those near the center of a cross-slot geometry of two converging flows, probe extensional rheology. A material element is shown deforming in the extensional flow.*

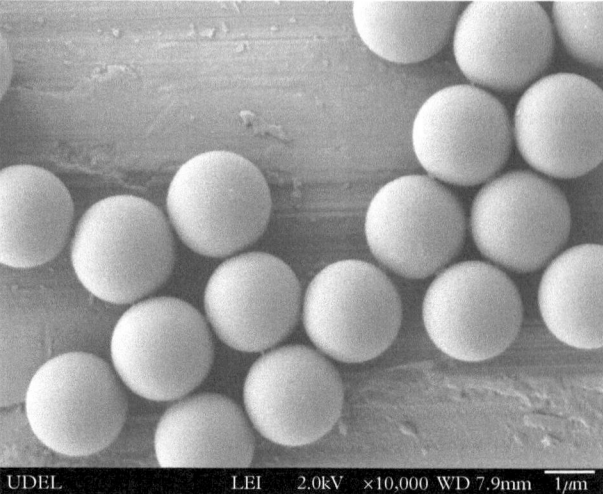

Fig. 1.16 *Scanning electron micrograph of uniform polystyrene latex particles with diameter 1.6 μm.*

Colloid refers to the Greek word for glue, κολλα, and is attributed to the Scottish chemist Thomas Graham (1805–69).[6] In his studies of dialysis, Graham observed that some of his solutions were unable to pass through a parchment membrane (Graham, 1861; Graham, 1864). These solutions, which he recognized as suspensions of microscopic particles, stayed put on one side of the membrane, bound as if held together or "glued." During the same period, Robert Brown (1773–1858) strengthened the connection between colloids and a characteristic length scale of matter (Brown, 1828). Brown did not adopt Graham's term, but instead referred to these small, seemingly animated particles as *molecules* during his own careful microscopy observations of their random motion. His work is why we now call the random thermal agitation of colloids *Brownian motion*, as seen in Figure 1.15.

A meticulous experimentalist, Brown, concluded that the random motion he observed was due to the small size of the particles. Brown ruled out that the composition or origin of the particles gave rise to their animation. It was not a characteristic of organic matter only— an important idea at the time, since others who had made similar observations had speculated that the spontaneous motion of organic particles was a manifestation of the "vital force" that distinguished animated matter (living things) from inanimate matter.[7] Brown carefully ground all sorts of materials into fine particles, some derived from organic matter and transformed organic matter (such as coal), and others from minerals, rocks, and even a small piece of the great Sphinx of Giza. He concluded that particles on the order of $1/30,000^{th}$ to

[6] Graham served as a professor in Glasgow, then later at University College, London. In 1855 he became Master of the Mint, a position that Isaac Newton held.

[7] To place Brown's work in perspective, it came not long after the 1818 publication of Mary Shelley's *Frankenstein*, in which the protagonist is a scientist who develops a technique to impart life to non-living matter.

1/20,000th of an inch (0.85–1.3 μm), which he meticulously measured using a micrometer, will exhibit random motion in solution regardless of their origin. It was not until the work of Einstein, Sutherland, Langevin, Smoluchowski, and their contemporaries that the Brownian motion of colloids would be definitively explained by the thermal motion of the surrounding fluid molecules through the kinetic theory of heat. We return to this history when we discuss passive microrheology in Chapters 3–5.

Today we recognize colloids as a special division of matter—a length scale that exhibits molecular-scale processes like Brownian motion, but are many times larger than atomic or molecular dimensions. Accordingly, the IUPAC definition of colloidal is a "state of subdivision such that the molecules or polymolecular particles dispersed in a medium have at least one dimension between approximately 1 nm and 1 μm, or that in a system discontinuities are found at distances of that order" (Slomkowski *et al.*, 2011).

1.3.1 Colloidal probe chemistries

Colloidal particles are defined by their length scale, yet the applicable dimensions span a wide range, from nanometers to several micrometers, and can encompass different shapes, chemistries (inorganic, organic), and even phases (from fluid emulsion droplets to solid polymer particles). Here we will review the key attributes of colloidal probe particles used in microrheology experiments.

Colloids must exhibit three principal characteristics to be suitable for use as microrheological probes. First, the particles should be uniform in size and shape. Second, the probes should be stable against aggregation or chemical degradation, and must disperse well into the medium of interest. Third, the probe surface chemistry should not alter the local microenvironment. The first two issues are addressed here; the effect of surface chemistry on the probe microenvironment will be discussed further in Section 3.10.

The choice of particle chemistry depends on the material to be probed. Since many materials of interest to microrheology are aqueous, polymer latex microspheres, especially polystyrene, are a good and common choice. They are available from many commercial vendors and straightforward to synthesize. In general, polystyrene is stable, and its density ($\sim$1.05 g/cm^3) is close to that of room temperature water, which reduces (but does not eliminate) gravitational sedimentation of the probes.

In organic solvents, inorganic particles such as silica grafted with an organophilic layer or more solvent resistant resins like melamine and polymer latex such as poly(methylmethacrylate) (PMMA), are a better choice. However, the density difference between inorganic particles and many organic solvents can lead to rapid probe sedimentation.

The microrheologist should assume that particles received as delivered from a manufacturer will contain impurities. These impurities will normally be surfactant stabilizers that keep particles dispersed during manufacture and storage. Surfactants and other residual impurities should be removed by a repeated centrifugation, decantation, and redispersion steps. The particles should be redispersed by gentle agitation into a solution, such as buffer, that closely matches the solution conditions of the final sample (washing methods are discussed further in Section 1.3.4). Care must be taken that the particles do not aggregate, especially if the ionic strength of the solution is high or the pH towards the extreme end. Most suspensions are stabilized, at least in part, by charges on their surface.

In Table 1.1, we summarize several physical properties of probe colloids that will be described in more detail.

Polystyrene

The synthesis of highly-uniform polystyrene latex is well established and many commercial vendors supply a variety of sizes as well as particles with modified surface chemistries. The particles are easily labeled with fluorescent dyes for particle tracking using fluorescence microscopes, and unlabeled particles readily scatter light due to the high contrast between indices of refraction ($n = 1.58$ for polystyrene, 1.33 for water). Finally, the surface chemistry of polystyrene particles can be controlled, by the adsorption of polymers and proteins, the addition of co-monomers during their synthesis, or through the reaction of chemically active sites on the particle surface. Commercial vendors supply particles ready for use as aqueous solutions at concentrations between 2.5–10 wt%. Because of the importance of polystyrene colloids, we will summarize their synthesis and chemical properties.

Table 1.1 *Common probe-particle chemistries and their physical properties.*

chemistry	density (g/cm^3)	refractive index
polystryene (PS)	1.05	1.58
polymethylmethacrylate (PMMA)	1.19	1.49
melamine	1.57	1.68
silica	2.2	1.46
titania (anatase)	3.78	2.49

Polystyryene latex is synthesized by emulsion polymerization (Piirma and Gardon, 1976; Poehlein *et al.*, 1985; Candau and Ottewill, 1990). The monomer is dispersed as an emulsion, stabilized with surfactant, in a non-solvent (typically water). Adding an initiator (water soluble, in this case) generates an initial surge of free radicals, which begins to polymerize the monomer that has partitioned into the swollen micelles. As the polymerization reaction continues, adsorbed surfactant provides stability to the growing particles and bulk monomer emulsion providing a reservoir for growth until the monomer is depleted. The rapid particle-nucleation stage followed by a longer growth phase ensures a narrow range of particle diameters. It is also common to synthesize polystyrene latex using a surfactant-free polymerization process (Goodwin *et al.*, 1973).

Surface chemistry—The polymer chains in the particle begin and end with a functional group derived from the initiator, and, thus, the initiator used in the reaction imparts properties to the final particle, such as charged groups. The use of a sodium or potassium persulfate initiator, for instance, results in a significant coverage of negatively charged surface groups in the form of sulfonates, which are weak bases (conjugates to sulfonic acids, they are largely deprotonated, with pK_a values in the range of ~ 2). The typical charge densities on the particles are $\sigma = -1$ to -5 $\mu C/cm^2$, which confers good colloidal stability in water at low ionic strengths, as we will discuss in Section 1.3.3.

Other surface chemistries can be incorporated on polystyrene particles by using co-monomers with different functional groups (Table 1.2). Carboxylate (COOH) surface chemistries are introduced by the inclusion of acrylic acid monomer (typically $< 5\%$) in the particle synthesis (Poehlein *et al.*, 1985). Care must be taken when describing the surface chemistry of such particles, since it can be composed of the monomer-derived groups in addition to surface groups from the initiator, like sulfonates. This fact is often ignored in the microrheology literature, as surface chemistries are rarely as pure as envisioned. Water soluble polymer and monomer left over from the polymerization reaction may also be present on the particles, which requires careful cleaning before they are used. Another common chemistry is polystyrene with primary amine surface groups (Cousin and Smith, 1994; Voorn *et al.*, 2005). These are used either as positively charged particles or for amine reaction coupling chemistries. Because latex spheres are used in a number of biotechnology applications, such as immunodiagnostic assays and agglutination tests (Pichot, 2004; Tadros, 1993), particles are available with a number of other reactive surface chemistries, including epoxy, chloromethyl, chlorosulfonyl, aldehyde, and mercapto groups.

Table 1.2 *Polystyrene latex surface chemistries used in microrheology. Some arise during the chemical synthesis itself, whereas others are attained through modification steps, including adsorption or grafting.*

sulfonate	$(-SO_3^-)$
carboxylate	$(-COOH)$
amine	$(-NH_2)$
PEG	
BSA	
poly-lysine	

Adsorption—The hydrophobic nature of polystyrene latex enables a number of surface modifications by physical adsorption in aqueous media, most prominently by the adsorption of polyelectrolytes and proteins (Fig. 1.17). For microrheological studies of F-actin, McGrath *et al.* (2000) adsorbed poly-L-lysine, a cationic polymer, onto polystyrene probes. The adsorbed polyelectrolyte reverses the particle charge (Blaakmeer *et al.*, 1990) and increases the interactions between probes and the negatively charged proteins of the entangled network. Conversely, the protein bovine serum albumin (BSA) has been pre-adsorbed to probes to reduce their interaction with F-actin filaments by blocking the surface (McGrath *et al.*, 2000; Valentine *et al.*, 2004; Chae and Furst, 2005).

Adsorbed or grafted polymer layers can also be used to improve the stability of colloids at high ionic strengths or in organic solvents (Napper, 1983).

Covalent coupling—Covalent coupling reactions are another method for modifying probe surface chemistry (Ikada, 1994). One common chemistry is the covalent coupling reaction of poly(ethylene glycol) (PEG) to surface chemical moieties (McGrath *et al.*, 2000; Valentine *et al.*, 2004). Typically, PEG molecules, usually with molecular weights in the range of several thousand daltons, are grafted by N-hydroxysuccinimide ester-amine reactions using PEG-succinimidyl carboxyl methyl ester. The resulting PEG-decorated probes exhibit lower protein adsorption and have been used in microrheology studies of protein filaments and filamentous viruses (Valentine *et al.*, 2004; He and Tang, 2011; Sarmiento-Gomez *et al.*, 2012). Others have attached PEG by physically trapping an adsorbed triblock co-polymer, poly(ethylene glycol)-*b*-poly(propylene glycol)-*b*-poly(ethylene glycol), by swelling the particles with toluene, which allows the hydrophobic blocks to migrate into the probes, then removing the swelling solvent (Kim *et al.*, 2005; Sato and Breedveld, 2006). These particles remain stable even at high-ionic strength, which, as we will see in Section 1.3.3, indicates that the grafted polymer provides sufficient steric forces between the particles in addition to reducing the adsorption of species like proteins.

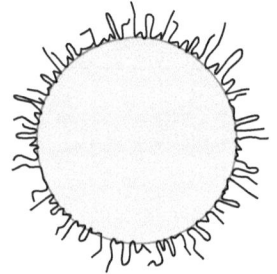

Fig. 1.17 *Adsorbed polymers and proteins or chemically grafted polymers can be used to tailor the probe surface chemistry. The polymer chains are not shown to scale, but typically extend only nanometers from the surface.*

Silica

Like polystyrene polymer latices, inorganic silica particles are also commercially available in the micrometer-diameter size range as highly uniform suspensions in water. Silica can be rendered organophilic, and is of course impervious to swelling and dissolution in organic solvents. The chief drawback of silica and other metal oxide particles (like titania or zinc oxide) for microrheology is their high density relative to water and many organic compounds, $\rho \approx 2.2$ g/cm^3, which leads to relatively rapid sedimentation in fluid samples.

Using the Stöber method, monodisperse silica is synthesized by a combined hydrolysis and condensation of a silicon alkoxide precursor in a mixture of water, ethanol, and ammonia (Stöber *et al.*, 1968; Van Helden *et al.*, 1981; Bogush *et al.*, 1988; Bogush and Zukoski, 1991). The hydrolysis of tetraethylorthosilica (TEOS) forms silanols while the condensation polymerization reaction produces siloxane bridges. In the reaction, ethanol serves a co-solvent for the mixture of alkoxide and water, which are otherwise immiscible. Ammonia acts as a catalyst to initiate the rapid nucleation of particles.

Silica particle surfaces are rich in silanol groups that are readily de-protonated in water, tyipcally giving silica a negative surface charge. The surface density of silanol groups is about 4.6/nm^2 (Bergna, 1994). Silica sols exhibit an increasingly negative electric poten-tial with increasing pH above the isoelectric point, $pH_{iep} \sim 2-3$ (Healy, 1994). Particles can be rendered fluorescent by incorporat-ing a silanized dye, such as fluorescein isothiocyanate (FITC) that has been treated with (3-amino-propyl)triethoxysilane (APS), during the particle synthesis (van Blaaderen and Vrij, 1992).

Like latex particles, the surface chemistry of silica can also be al-tered by physical adsorption of polymers and proteins, but a common modification is to render the particles organophilic by an esterifica-tion reaction with stearyl alcohol according to R. K. Iler's method (Iler, 1979; Van Helden *et al.*, 1981).

Alkoxides of other metals, including titania, can be used as precur-sors in the Stöber synthesis. The alkoxide reaction can also be used to coat silica onto these and other particles to create core-shell particles for optical trapping, for instance (Viravathana and Marr, 2000).

Other particle chemistries

Microrheology is not limited to the use of polystryrene or silica particles. Any colloidal particle that satisfies the criteria of probe microrheology— larger than the material microstructure, uniform in size and shape, and stably dispersable in the material of interest— can be used. Melamine resin (urea formaldehyde) is a thermosetting plastic that remains stable without swelling or degrading in a variety of organic solvents like decalin and mixtures of decalin and cyclo-hexylbromide (CHB) (Meyer *et al.*, 2006). Many biological samples naturally contain various particles (*e.g.*, granules or organelles) that may be used as microrheological probes, much like Heilbrunn (1924) did nearly a century ago.

1.3.2   Probe size uniformity

Quantitative measurements of rheology using embedded probe particles requires an accurate knowledge of the probe size. When using methods that measure the motion of an ensemble of particles, such as multiple particle tracking and light scattering, each particle should be roughly identical. Even microrheology methods like laser tracking and magnetic and optical tweezers that track the motion of individual particles generally require uniform particles, due to the difficulty of accurately determining particle sizes *in situ*. Fortunately, the methods of particle synthesis described in the previous section lead to narrow size distributions.

Particle size polydispersity can be measured directly by electron microscopy, or by the motion of the particles in a medium of known viscosity (*e.g.*, by dynamic light scattering). The average particle diameter will be taken as the *number average*,

$$\bar{d} = \frac{1}{N} \sum_i d_i, \qquad (1.47)$$

where d_i is the diameter of the ith measured particle in a sample of N particles. The standard deviation of the particle diameter will then be

$$\sigma = \left[\frac{1}{N} \sum_i (d_i - \bar{d})^2 \right]^{1/2} \qquad (1.48)$$

which is often reported in terms of the *coefficient of variation*

$$\text{C.V.} = (\sigma/\bar{d}) \times 100\%. \qquad (1.49)$$

Typical C.V. values for monodisperse particles are 1–2%.

1.3.3   Colloid stability

Our chief concern is the stability of the colloids used in a microrheology experiment, which depends critically on the interaction forces experienced between the particles. Colloidal particles interact with each other primarily through van der Waals attractions, electrostatic interactions, and steric forces due to polymers, proteins, or surfactants adsorbed or grafted to their surfaces.

Because of the inherent and ubiquitous van der Waals attractive forces between pieces of condensed matter, the lowest energy state of a colloidal dispersion is an aggregated mass that forms as particles fall into their energy potential minima. Once aggregated, particles are difficult, if not impossible, to redisperse. Additional forces are thus required to render colloidal suspensions *kinetically* stable, by introducing potential energy barriers, typically an electrostatic or a steric repulsion, to keep the colloids apart despite the van der Waals attraction.

Unfortunately, the balance of forces that imparts kinetic stability to suspensions may be upset, leading to probe aggregation, as conditions vary. Examples of interest to microrheology include the ionic strength and pH of the samples or sample precursors, or bridging interactions by macromolecules and proteins. The microrheologist must therefore check for probe aggregation to ensure meaningful results. Aggregation is almost always immediately obvious in microrheology experiments that use microscopy, like particle tracking (Chapter 4). In Fig. 1.18, images taken using a fluorescence microscope of polystyrene show particles in a dispersed state, and two samples in which they have formed aggregates and large clumps. Such aggregation may be more difficult to discern when using light scattering techniques (Chapter 5). Most active microrheology methods use particle concentrations that are dilute enough that probe stability is less of an issue.

We will now briefly describe several typical colloidal interactions in more detail to understand conditions that might result in probe aggregation. Our treatment represents a sliver of the extensive knowledge concerning the interactions and stability of colloidal dispersions. We will introduce equations without the nuances of their assumptions or details of their derivations. For more in-depth discussions, the reader is referred to the many excellent colloid and surface chemistry texts available on the subject, including those by Russel, Saville, and Schowalter (1989), Hunter (2001), Hiemenz and Rajagopalan (1997), Adamson and Gast (1997), and Israelachvili (2011).

van der Waals forces

The van der Waals interaction is a nearly ubiquitous attractive force that arises, from a classical standpoint, due to fluctuations of electrons in a material, but it is ultimately quantum mechanical in nature. Only under special conditions is this inherent attraction minimized, such as when colloids are dispersed in a solvent having an identical index of refraction.

The van der Waals interaction potential between two spherical particles of micron-scale radius a, depends on the distance h separating the particle surfaces,

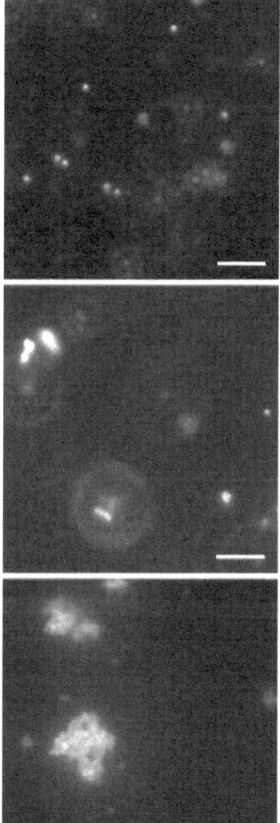

Fig. 1.18 *Fluorescence microscopy images of dispersed probe particles (top image) and samples with mild and strong aggregation. The scale bars are 10 µm.*

$$\Phi_{\mathrm{vdw}}(h) = -\frac{A_H}{12}\left\{\left(\frac{4a^2}{4ah+h^2}\right)+\left(\frac{2a}{2a+h}\right)^2\right.$$

$$\left.+2\ln\left[1-\left(\frac{2a}{2a+h}\right)\right]\right\} \tag{1.50}$$

from which the force is calculated by $F = -d\Phi/dh$. The Hamaker constant A_H depends on the materials involved, and is typically of order 10^{-20} J for polymer particles such as polystyrene and poly(methyl methacrylate) dispersed in water, and an order of magnitude higher for metals like gold, for which $A_H \approx 3 \times 10^{-19}$ J (Hough and White, 1980).

Electrostatic interactions and the electric double layer

Surface charges on colloids arise from a number of sources depending on the particle chemistry, and may include dissolution of ionic species, the dissociation of acidic sites, and the adsorption of charged species like polyelectrolytes and surfactants. These mechanisms lead to typical surface charge densities on the order of $\sigma_q = 0.005 - 0.1$ C/m^2 (often expressed in the convenient units 0.5–10 μC/cm^2) for colloidal particles of interest to microrheology. The larger of these values reflects, on average, about one charge in 1.6 nm^2.

Rather than a direct Coulombic force, charges on neighboring particle surfaces interact through a solvent, which very often contains dissolved ionic species. Ions in solution re-arrange in response to charged surfaces, forming an oppositely-charged cloud (called the *electric double layer*) that screens the surface charge. Figure 1.19 depicts an electric double-layer around a negatively-charged surface, which attracts positively-charged counter-ions, and repelling negatively charged co-ions. The distribution of charge is captured by the Gouy–Chapman model of the electric double layer, which is based on the Poisson–Boltzmann equation for the electrostatic potential $\psi(x)$ (Israelachvili, 2011). The Poisson equation

$$-\epsilon\epsilon_0\nabla^2\psi = \rho, \tag{1.51}$$

describes how a charge density ρ impacts the electrostatic potential $\psi(x)$. Here, ρ is established by the imbalance between positively- and negatively-charged ions, via

$$\rho(x) = ez\left(n_+(x) - n_-(x)\right). \tag{1.52}$$

Assuming each ion species responds to the local electrostatic potential $\psi(x)$ via the Boltzmann relation,

$$\rho = n_0 ez\left(e^{-ez\psi/k_B T} - e^{ez\psi/k_B T}\right) \tag{1.53}$$

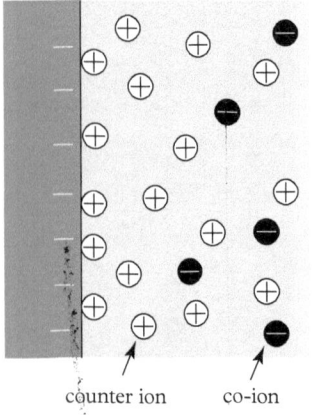

Negatively charged surface

counter ion   co-ion

Fig. 1.19 *The diffuse double-layer near a charged interface.*

gives the Poisson–Boltzmann equation,

$$\nabla^2 \psi = \left(\frac{2zen_0}{\epsilon\epsilon_0} \right) \sinh \left(\frac{ez\psi}{k_B T} \right). \tag{1.54}$$

Further insight follows by scaling electrostatic potentials by the thermal potential

$$\psi_T = \frac{k_B T}{ze}, \tag{1.55}$$

which is roughly 26 mV at room temperature for monovalent ions, resulting in

$$\nabla^2 \tilde{\psi} = \left(\frac{2z^2 e^2 n_0}{\epsilon\epsilon_0 k_B T} \right) \sinh \tilde{\psi} \equiv \kappa^2 \sinh \tilde{\psi}, \tag{1.56}$$

where κ^{-1} is the Debye length. A more general expression, allowing for multiple ion species of various valences, is

$$\kappa^{-1} = \left(\frac{\epsilon\epsilon_0 k_B T}{\sum_i n_i e^2 z_i^2} \right)^{1/2}. \tag{1.57}$$

Detailed solutions to the Poisson–Boltzmann equation for interacting particles is beyond our current interest, but it is useful to consider at least one description of the *interparticle interaction* between charged colloids in solution. For spheres with constant potential surfaces, the double-layer interaction is (Russel *et al.*, 1989)

$$\Phi_{el}(h) = 2\pi\epsilon\epsilon_0 (k_B T/ze)^2 a\tilde{\psi}_s^2 \ln \left[1 + e^{-\kappa h} \right] \tag{1.58}$$

where $\tilde{\psi}_s = \psi_s/\psi_T$ is the scaled surface potential of the particle.

Equation 1.58 applies to conditions in which the double layer is thin relative to the particle size, $\kappa a \gg 1$, and is approximate but accurate enough for our purposes. Thin double layers form at sufficiently high-ionic strengths. Table 1.3 lists Debye lengths for the monovalent salt sodium chloride. A useful rule of thumb is $\kappa^{-1} = 0.307 \, \text{nm}/\sqrt{c_s}$ for monovalent salts, where c_s is the molar salt concentration.

The surface charge density and the type and concentration of ions in solution have the greatest effects on the double layer interactions. As we can see in Table 1.3, high-ionic strength solutions lead to a compact double layer very close to the particle surface, and comparable to the range of the van der Waals attraction. At modest ionic strengths, the Debye length is on the order of 10 nm or less. As we will

Table 1.3 *Debye lengths calculated for an aqueous solution of a monovalent salt.*

NaCl (mM)	κ^{-1} (nm)
0.1	31
1	9.7
10	3.1
100	0.97

soon show, aggregation between particles occurs as the ionic strength increases. Because the Debye length depends on the square of the ion charge, divalent ions like Ca^{2+} and Mg^{2+} are potent inducers of aggregation. In microrheology samples, the presence of buffer salts will also contribute to double layer screening, and should be accounted for when calculating the Debye length using eqn 1.57.

It is difficult to definitively relate the surface charge density of a colloid to its electrostatic potential, or to measure the surface potential. Electrophoretic measurements of the so-called ζ-potential are often used, which reflect the potential drop across the diffuse cloud of counter-ions predicted by the Poisson–Boltzmann equation. The ζ-potential may differ from the true electrostatic potential at the surface of the colloid, which is where the colloid's surface charge density resides. Various factors are hypothesized to play key roles, including a "Stern" layer of immobilized counter-ions, possibly physically adsorbed to the surface, hydrophobic effects and water structuring around solvated ions and surfaces, and others. The difference between the electrostatic potential at the colloidal surface and the electrokinetic potential ζ depends strongly on the identity of the counter-ion, for example (Brown *et al.*, 2016).

A common approximation relating an effective colloidal surface charge density (*i.e.*, outside the Stern layer) to the electrokinetic potential ζ is given by the Graham equation,

$$\sigma_q = 2(2\epsilon\epsilon_0 k_B T n_b)^{1/2} \sinh \tilde{\zeta}, \tag{1.59}$$

which follows from solutions to the nonlinear Poisson–Boltzmann equation for monovalent electrolytes. Here, n_b is the number of ions in the bulk, summed over all ion species $n_b = \sum_i n_i$.

DLVO theory

The classic theory of colloid stability is attributed to two teams of co-workers who independently derived it, Derjaguin and Landau (1941) and Verwey and Overbeek (1948). The DLVO theory combines the van der Waals and electrostatic double layer interactions described previously to calculate the total interaction potential between colloids

$$\Phi(h) = \Phi_{vdw}(h) + \Phi_{el}(h) \tag{1.60}$$

as a function of the separation between the particle surfaces h.

An example of the DLVO potential energy is shown in Fig. 1.20. The combined van der Waals attraction and double layer repulsion

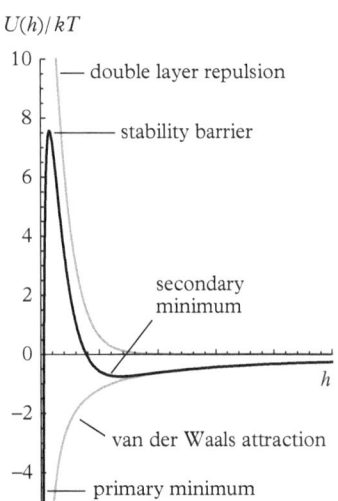

Fig. 1.20 *A schematic of the DLVO interaction potential.*

leads to three general features in the energy profile: A primary mini-mum, a stability barrier, and sometimes a secondary minimum. The primary minimum represents the deep, attractive energy well between two colloidal particles, set by the strong van der Waals attraction and physical repulsion at contact. The repulsive barrier at longer separations confers the kinetic stability to colloidal suspensions. Its height Φ_{max} and range sets the kinetics of aggregation between the probe particles. Two particles approach close enough to cross over the energy barrier only rarely, requiring a characteristic time scale

$$t \sim (3\pi a^3 \eta / k_B T) \exp(\Phi_{max}/k_B T). \qquad (1.61)$$

For a large stability barrier, the rate of aggregation is infinitesimal, as expected.

Depending on the range and magnitude of the double layer repul-sion, a secondary minimum, beyond the range of the stability barrier, may be present. If the secondary minimum is sufficiently deep, on the order of the thermal energy $k_B T$, then particles may aggregate in this energy well. In contrast to particles that fall into the primary minimum, particles that aggregate in the secondary minimum may redisperse with a modest input of energy by weak shaking, stirring, or with the use of a mixer. Otherwise, an aggregated suspension must be subjected to a significant energy using, for instance, a probe sonicator.

Let's examine the interactions between micrometer diameter poly-styrene particles at several ionic strengths of a monovalent salt (NaCl) more quantitatively. In Fig 1.21, we plot the interaction potential be-tween particles with a surface potential of $\psi_s = -50$ mV in 1, 10, and 100 mM NaCl solutions. As illustrated in the inset, the repulsive bar-rier is large, on the order of several hundred $k_B T$. At low salt (1mM), the repulsive barrier extends to tens of nanometers separation. Nev-ertheless, a modest secondary minimum on the order of ~ 1 $k_B T$ is evident at $h = 100$ nm. At 10 mM, the secondary minimum is more pronounced, and could result in aggregation into a secondary mini-mum. As the NaCl concentration increases to 100 mM, the secondary minimum is now deep, and although the calculation shows a stability barrier, the separation of the barrier is on the order of nanometers. Under these conditions, the stability of the particles is likely to be compromised.

Grafted polymers and steric stabilization

Chemically grafted polymer brushes are effective stabilizers for col-loids and can mitigate adsorption that leads to bridging interactions between probes, especially in protein solutions (Napper, 1983). The

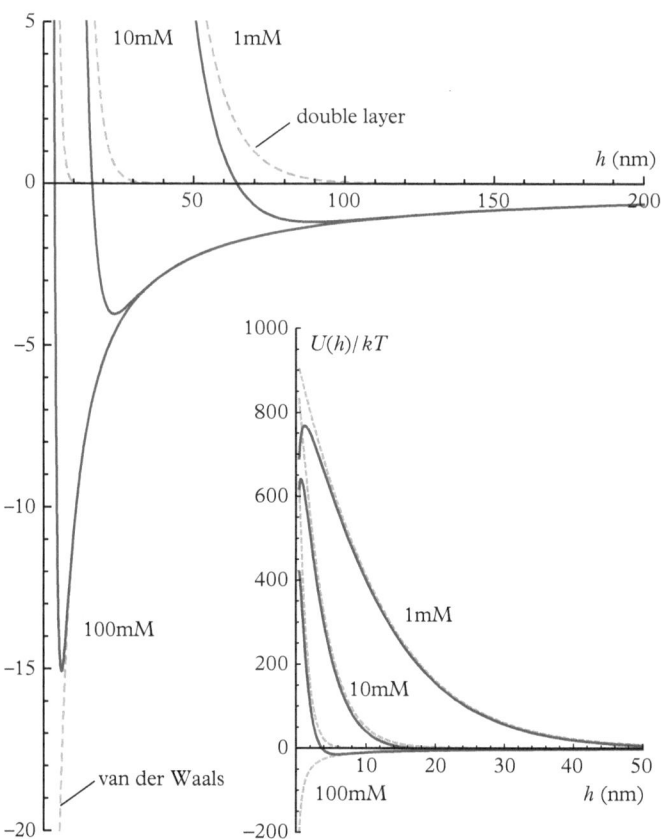

Fig. 1.21 *DLVO interaction potential as a function of surface separation $h = r - 2\,a$ calculated for 1 μm diameter polystyrene particles with a surface potential $\psi_s = -50$ mV in NaCl solutions from 1–100 mM. Each solid line is the total potential calculated from the sum of electrostatic double layer and van der Waals interactions. The individual contributions to the potential are indicated by dashed lines.*

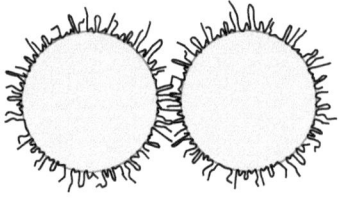

Fig. 1.22 *Grafted or adsorbed polymers produce steric interactions that can stabilize colloidal particles.*

steric repulsion produced by two adsorbed polymer layers are illustrated in Fig. 1.22. The polymer contribution to the interaction potential is

$$
\Phi_p = \begin{cases} \Phi_0\left[-\ln y - \frac{9}{5}(1-y) + \frac{1}{3}(1-y^3) - \frac{1}{30}(1-y^6)\right], & 0 < y < 1 \\ 0 & y \geq 1 \end{cases} \tag{1.62}
$$

where $y = h/2L = (r - 2a)/2L$ and

$$
\Phi_0 = \left(\frac{\pi L \sigma_p k_B T}{12 N_p l^2}\right) a L^2. \tag{1.63}
$$

Here, L is the contour length, N_p is the degree of polymerization, l is the segment length, and ϕ_p is the surface graft density.

As we noted earlier, another important role for grafted polymers is to control the surface chemistry of the probes—either to tailor the interaction with the surrounding medium or to block these interactions.

Bridging interactions

When the surface concentration of adsorbed polymers or proteins is low, it is possible for molecules on one particle to stick simultaneously to a bare patch on a neighboring particle, as illustrated in Fig. 1.23. Such phenomena are called bridging interactions (Evans and Napper, 1973; de Gennes, 1987; Dickinson and Eriksson, 1991) and they represent another potential destabilization mechanism in microrheology experiments. The adsorbed material acts as a bridge that causes coagulation of the probes (Healy and LaMer, 1962; de Gennes, 1982).

Bridging interactions are normally mitigated through the use of surface chemistries that prevent or minimize adsorption, like the PEG chemistries discussed in Section 1.3.1 for polystryene probes. Competing adsorption, by pre-treating probes with protein solutions of bovine serum albumin for instance, can block the surface and provide adequate colloidal stability, but in turn may affect the probe-material interaction, as was reported for BSA-coated polystyrene probes in entangled dispersions of F-actin (McGrath *et al.*, 2000).

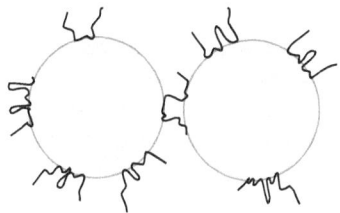

Fig. 1.23 *Bridging interactions caused by adsorbed polymer or protein can destablize colloidal particles.*

Depletion interactions

Related to steric stabilization and bridging interactions are depletion interactions (Lekkerkerker and Tuinier, 2011). In this case, the depletion attraction occurs when larger particles like probes are dispersed in a solution of a smaller non-adsorbing species, such as polymer coils, surfactant micelles, or small particles. Depletion interactions occur as larger particles come together and provide more free volume for the smaller particles (Asakura and Oosawa, 1954). An equivalent view, illustrated in Fig. 1.24, is that an osmotic pressure imbalance occurs on the large particles when they are at a separation that excludes the smaller species. The osmotic pressure of the small particles, polymer coils, or proteins, pushes the larger particles together.

The depletion interaction is calculated to a first order by considering the osmotic pressure for ideal particles with radius R_g

$$\Pi = nk_B T \tag{1.64}$$

where n is the number density $n = \left(\frac{4}{3}\pi R_g^3\right)^{-1}$. Integrating this osmotic pressure over the available surface area of the larger particles,

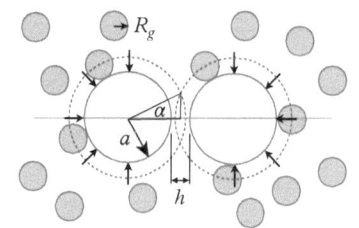

Fig. 1.24 *An osmotic pressure imbalance of the depletion interaction occurs when the excluded volume of the larger particles, indicated by the dashed lines, overlaps. The overlapping volume is highlighted and defines the angle α.*

$$F = -2\pi a^2 nk_B T \int_0^\alpha \cos\phi \sin\phi d\phi, \tag{1.65}$$

with a range of angles given geometrically,

$$\cos\alpha = \frac{a + h/2}{a + R_g}, \tag{1.66}$$

leads to

$$\frac{F}{\pi a^2 nk_B T} = \left[\left(\frac{1 + h'\Delta/2}{1 + \Delta}\right)^2 - 1\right] \tag{1.67}$$

with

$$\Delta = R_g/a \tag{1.68}$$

$$h' = h/R_g. \tag{1.69}$$

From the force, the dimensionless depletion potential is

$$\Phi' = \frac{-\Delta(h' - 2)^2[6 + \Delta(h' + 4)]}{12(\Delta + 1)^2} \tag{1.70}$$

where $\Phi' = \Phi/\pi a^2 R_g nk_B T$. The range of the attraction is the diameter of the depletant $2R_g$ and the attraction at particle contact can reach several to tens of $k_B T$.

1.3.4   Probe sedimentation, washing, and concentration

Having reviewed the chemistry and stability of colloidal particles, here we will make a few comments on practical issues of their use, including a short discussion of probe sedimentation, the preparation of probes by washing, and aspects related to probe concentration.

Probe sedimentation

The buoyant force exerted on a colloid is

$$F_b = \frac{4}{3}\pi a^3 \Delta\rho g \tag{1.71}$$

where $\Delta\rho = \rho_m - \rho_p$ is the density difference between the medium and the particle, and g is the acceleration due to gravity. The sedimentation velocity of a colloid in a viscous Newtonian fluid is

$$V_b = \frac{2a^2 \Delta\rho g}{9\eta} \tag{1.72}$$

where the hydrodynamic drag exerted on the particle is $F_d = 6\pi a\eta V_p$. For a complex, viscoelastic fluid, η is the zero shear viscosity.

The sedimentation Peclet number Pe_s is a dimensionless quantity that characterizes the magnitude of sedimentation. It is the ratio of the characteristic time scale of a particle to diffuse its radius, $a^2/D_s = 6\pi a^3\eta/k_B T$, with respect to the characteristic time scale to sediment the same distance, a/V_s,

$$Pe_s = \frac{2\pi a^4 \Delta\rho g}{3k_B T}. \tag{1.73}$$

Sedimentation becomes significantly stronger as the particle size increases. The sedimentation Peclet number for a 1 μm diameter polystyrene probe particle in water is about $Pe_s = 0.03$. Probe particles that are just twice this size exhibit values of $Pe_s \sim O(1)$.

Probe washing

Surfactants are often added by manufacturers to colloidal suspensions to stabilize them and improve their shelf life. Common non-ionic surfactants include Tween-20 (a polysorbate surfactant) and Tergitol (a secondary alcohol ethoxylate). Anionic surfactants like sodium dodecyl sulfate are also common stabilizers. Surfactants can potentially alter the sample through complexation or change the interactions of the probe particles with the material, and should be removed by washing the probe particles before use.

The preferred method of washing is by multiple centrifugation and redispersion steps. The probe suspension is centrifuged to form a loose pellet of particles and the supernatant pipetted off. The probes are redispersed and the process repeated three to five times. Because the surfactants confer stability, the centrifugation must be performed lightly to prevent probes from aggregating. Other methods, such as dialysis and mixing particles with an ion exchange resin, can also be used.

Probe concentration

The concentration of probe particles depends on the method of microrheology being used. In passive microrheology, the particle concentration, given by the volume fraction ϕ, will vary between 10^{-2} for experiments that employ diffusing wave spectroscopy (light scattering in the highly multiple scattering regime, which is discussed in Chapter 5) and 10^{-4} for particle tracking microrheology (see Chapter 4). Obviously, it is important that the probe particles do not influence the rheology being measured. For particles dispersed in a Newtonian medium with viscosity η_0, Einstein showed that the viscosity of the suspension changes as the particle volume fraction increases by

$$\eta/\eta_0 = 1 + \frac{5}{2}\phi + O(\phi^2),\tag{1.74}$$

which is valid below about $\phi < 0.05$.

Einstein's formula (and suspension viscosity formulae more general) reflect the viscosity of the suspension, as would be measured with a macroscopic rheometer. The (tracer) diffusivity of a spherical particle in a dilute suspension of identical particles, on the other hand, was computed by Batchelor (1976) and Rallison and Hinch (1976) to be

$$D_s^0 \sim D_0(1 - 1.81\phi)\tag{1.75}$$

for short times, and

$$D_s^\infty \sim D_0(1 - 2.06\phi)\tag{1.76}$$

for long times. As will be seen shortly, the (tracer) self-diffusivity is what is measured in many microrheology measurements to extract material rheology. In Newtonian liquids, finite probe concentrations, if not properly accounted for, would appear to give an apparent viscosity between $(1 - 1.8\phi)^{-1}$ and $(1 - 1.2\phi)^{-1}$ too high, and with a weakly non-Newtonian character.

Finite probe concentrations thus change several aspects: They directly affect the actual, macroscopically-measurable rheology of the material (by Einstein's correction in the dilute limit), as well as the self-diffusivity of each probe. Probe concentrations should therefore be as low as is feasible.

- -

EXERCISES

(1.1) **Sedimentation.** A tracer particle microrheology experiment uses 1 μm polystyrene probe particles dispersed in a fluid with viscosity $\eta = 1.1 \times 10^{-3}$ Pa $\cdot$ s.

   (a) What is the sedimentation Peclet nuber?

   (b) If the sample chamber is 200 μm thick, calculate the time required for the probe particles to sediment to the bottom wall.

(c) The image plane for the experiment is positioned half-way between the sample walls. Assuming that the probe volume fraction is initially $\phi = 10^{-5}$ and that the particles are initially evenly distributed through the chamber, calculate the probe concentration in the image plane with time.

(1.2) **Probe stability.** A tracer particle microrheology experiment uses 1 μm polystyrene probe particles dispersed in a fluid with viscosity $\eta = 1.1 \times 10^{-3}$ Pa $\cdot$ s. Calculate the DLVO interaction potential between these particles for a surface charge density $\sigma_q = -3$ μC/cm^2 in 10 μM, 1 mM, and 100 mM aqueous NaCl solutions. Are the particles stable under these conditions?

2

Particle motion

2.1 Introduction

All microrheology experiments measure the resistance of a probe particle forced to move within a material, whether that probe is forced externally or simply allowed to fluctuate thermally. For example, the viscosity of a Newtonian liquid could be measured microrheologically, using a spherical colloid as a probe. In an active microrheology experiment, a colloid of radius a is driven externally with a specified force $\mathbf{F}$ (*e.g., magnetic, optical, or gravitational*), and moves with a velocity $\mathbf{V}$ that is measured. The rheology of the liquid (*i.e.*, the viscosity η) may be extracted from Stokes' classic formula for the drag on a sphere moving through a viscous fluid,

$$\zeta = \frac{F}{V} = 6\pi\eta a, \tag{2.1}$$

which will be computed in Section 2.5.2.

In passive microrheology experiments, on the other hand, the position of a thermally-fluctuating probe is tracked and analyzed to determine its diffusivity, which Einstein (1906) and Sutherland (1905) related to the hydrodynamic resistance ζ according to

$$D = \frac{k_B T}{\zeta} = \frac{k_B T}{6\pi\eta a}. \tag{2.2}$$

The interpretation of such experiments in purely viscous liquids is deceptively straightforward, as they rely upon hydrodynamic calculations by Stokes, Einstein, and Sutherland that are now taken for granted. To determine rheological properties (*e.g.*, G^*) from the probe resistance ζ in more complex materials, however, solutions to the analogous continuum-mechanics problem are required. Herein lies the difficulty: One must know the material's rheological properties in order to even pose the continuum-mechanical problem that must be solved, yet the solution of that problem is required to determine the material rheology! Fortunately, the Correspondence Principle (Section 2.4) cuts this Gordian Knot for linear response measurements. No such simplification occurs for nonlinear microrheology experiments, however, complicating their interpretation significantly.

Microrheology. Eric M. Furst and Todd M. Squires, Oxford University Press (2017).
© Eric M. Furst and Todd M. Squires. DOI 10.1093/oso/9780199655205.001.0001

In this chapter, we briefly derive and discuss the fundamental equations governing continuum materials as they deform, and will specifically focus on the mechanics of probe particles moving within these materials.

2.2 The mechanics of deformable continua

Many readers are no doubt familiar with the Navier–Stokes equations, which govern the flow of viscous liquids. Some will also be familiar with the equations of motion for elastic solids. Both require the **continuum hypothesis**, which relies upon fictitious "material elements" that must satisfy two competing demands. Material elements must be large enough, and contain enough micro-structural elements (here atoms or molecules) to behave as the macro-scale material does. At the same time, material elements must be significantly smaller than any length scale associated with a flow or deformation field, so that gradients can be well-resolved.

The continuum approximation is easily satisfied with simple materials like water, glass, and steel on all but molecular length scales,

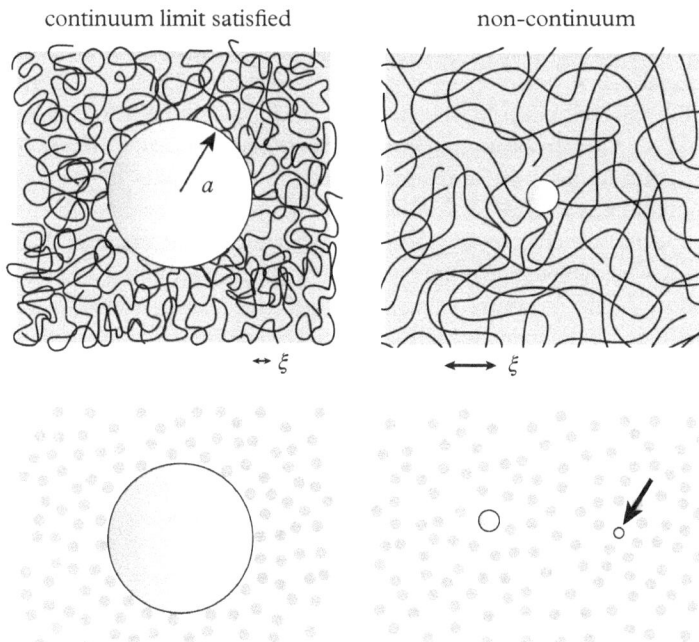

Fig. 2.1 *The continuum approximation is satisfied when probe particles are larger than the characteristic length scale ξ of the material. In materials like suspensions, the probe particle must be much larger than the dispersed particles. The arrow points to a probe that is smaller than the surrounding bath particles.*

depletion

accumulation

patchy

Fig. 2.2 *A second class of non-continuum effects occur in probe microrheology when the material structure is affected by the probe, in which case depleted layers, accumulation, or patchy interactions can arise.*

yet can be violated in microrheology of soft materials. For example, a grape embedded in Jello sees its material environment as a continuum: When forced, the grape deforms the surrounding Jello as a continuum. A sugar molecule or salt ion, however, is far smaller than the pores within the gel network comprising the Jello, and so diffuses through the Jello as though it were water. If this were a microrheology experiment, grape probes would (correctly) determine Jello to be a viscoelastic solid with the same $G^*(\omega)$ measured in a rheometer, whereas experiments using dye molecules as probes would reveal Jello to be a viscous liquid. Each experiment is meaningful in its own way—the grape correctly identifies the macroscopic rheology of Jello, whereas the dye reveals information about the mesostructure that would be inaccessible to macrorheometry. This example highlights both opportunities and challenges for the microrheologist.

Naturally, if a soft material has microstructural elements on the order of length scale ξ, then probe particles must exceed this dimension for the continuum approximation to hold. If the material contains dispersed polymers or particles, then those elements must be smaller than the probe. Both situations are represented in Fig. 2.1. An entangled biopolymer network, for instance, should have a mesh size $\xi \ll a$. Dispersed protein solutions, for which the individual molecules are on the order of tens of nanometers in size, naturally satisfy the continuum limit. But the microstructures in some gelators and rheology thickeners, including peptides, microgel particles, or clays, can often exceed normal probe dimensions of a few micrometers (Lu and Solomon, 2002; Oppong and de Bruyn, 2007; Savin and Doyle, 2007*a*; Rich *et al.*, 2011*b*). In Section 3.10, we discuss methods for verifying that the continuum approximation of probe microrheology is being met in an experiment. A second non-continuum effect occurs when the probe particle changes the local microstructure of the material. Material can be depleted near the particle, bunched up around it, or take on a more patchy structure, as we depict in Fig. 2.2. Later, we will discuss methods for detecting local heterogeneity, including manipulating the probe surface chemistry (Section 3.10), two-point microrheology experiments and the probe mechanics in locally heterogeneous materials (Section 4.11).

2.2.1   The Cauchy Stress Equation: F = *M*a for continuum materials

When a material is treated as a continuum, rather than as some discrete object, Newton's equations must be "smeared out", with masses and forces distributed on a per-volume basis. The Cauchy stress equation,[1]

$$\rho \frac{\partial^2 \mathbf{u}}{\partial t^2} = \nabla \cdot \boldsymbol{\sigma} + \mathbf{f}_b, \tag{2.3}$$

represents a continuum version of $\mathbf{F} = M\mathbf{a}$, which must be obeyed at each point $\mathbf{r}$ within the material. Note that ρ is the material density and $\mathbf{u}(\mathbf{r})$ represents the *displacement* of the material element at $\mathbf{r}$.

Three forces (per unit volume) appear in Eqn 2.3. The first term $\rho\ddot{\mathbf{u}}$ describes the inertial force density that arises at each point $\mathbf{r}$ within the material due to the unsteady acceleration of the material element. The third term, $\mathbf{f}_b$, represents a **body force** that is exerted throughout the volume of each material element. Common examples of body forces include gravitational, electrical, magnetic, and van der Waals forces. Unless otherwise noted, however, body forces will play little or no role in our discussion, and so will be omitted.

In Section 2.3, each material element accelerates due to **stresses $\boldsymbol{\sigma}$** exerted on its surfaces by neighboring elements. The stress $\boldsymbol{\sigma}$ within a continuum material has units [force/area], and is a tensor quantity: The force $\mathbf{t}$ that is exerted per unit area of any particular surface depends on the orientation of that surface. The stress vector[2] exerted on a surface with outward-directed normal vector $\hat{\mathbf{n}}$ (Fig. 2.3) is given by

$$\mathbf{t} = \boldsymbol{\sigma} \cdot \hat{\mathbf{n}}, \tag{2.4}$$

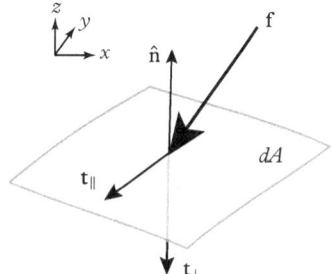

Fig. 2.3 *A force f exerted on an area segment dA with outward unit normal vector $\hat{n}$ defines the stress vectors parallel $t_\parallel$ and perpendicular $t_\perp$ to the surface.*

and has components both normal to the surface (like pressure) and tangential to the surface (like viscous stresses). For example, the stress on a material located beneath $z = 0$, with outer normal $\hat{\mathbf{n}} = \hat{\mathbf{z}}$, is given by

$$\mathbf{t} = \sigma_{xz}\hat{\mathbf{x}} + \sigma_{yz}\hat{\mathbf{y}} + \sigma_{zz}\hat{\mathbf{z}}, \tag{2.5}$$

where, *e.g.*, $\sigma_{xz} = \hat{\mathbf{x}} \cdot \boldsymbol{\sigma} \cdot \hat{\mathbf{z}}$, and so on. σ_{xz} and σ_{yz} are shear stresses, and σ_{zz} is a normal stress.

The convective (nonlinear) derivative $\rho(\dot{\mathbf{u}} \cdot \nabla)\dot{\mathbf{u}}$ has been omitted from the left-hand side of eqn (2.3). As is familiar from fluid mechanics, two phenomena give rise to inertial forces. The first is the *unsteady* inertia $\rho\ddot{\mathbf{u}}$ that appears in (2.3). The other source arises even under steady flows (*i.e.*, when $\partial \mathbf{v}/\partial t = \ddot{\mathbf{u}} = 0$), when fluid elements are accelerated as they move along streamlines. The second (nonlinear) inertia gives rise to turbulence, whereas the first (linear) inertia gives rise to transverse waves. In microrheology, the *unsteady*

---

[1] Throughout this book, we will use lower-case variables u and v = u̇ to represent displacement and velocity fields, respectively, within a continuum material. We will use upper-case variables **U** and **V** to denote the displacement and velocity of a particle within the material.

[2] Also called the traction.

inertia can be significant at high frequencies, but the *convective* inertia is generally not.

2.2.2   Linear-constitutive relations

The Cauchy Stress Equation (2.3) is exceedingly general, and governs the dynamics of all continuum materials—whether liquids, solids, gels, emulsions, solutions, powders, or foams. While powerful, it simply accounts for momentum conservation at every point in a material. To derive equations of motion for a particular material, we must know how the stress σ is related to the material's deformations: *e.g.*, strain (for elastic materials), strain-rate (for viscous materials), strain-rate history (for viscoelastic materials), metastable states (for powders and granular materials), and so on. These are **constitutive relations** and are specific to each material. Broad classes of constitutive relations distinguish between different classes of materials (*e.g.*, liquids versus solids), and material parameters within each constitutive relation distinguish between materials within each class (*e.g.*, viscosity for liquids, shear and bulk moduli for solids, and coefficients of restitution for granular materials).

   The constitutive relation for a viscous liquid is particularly simple, and has proven remarkably successful. Other materials are not so simple, with stress tensors that are nonlinear functions of the strain and rate-of-strain tensors ϵ and $\dot{\epsilon}$, and of the deformation history of the material. Examples of rate-dependent responses include **shear thinning** in shampoo, and **shear thickening** in concentrated suspensions of cornstarch. Examples of strain-dependent responses include **strain hardening**: A rubber band pulls back gently when stretched slightly, but stiffens when stretched more. Some materials **yield**, like mayonnaise and toothpaste: They sit like elastic solids under gravitational stresses, yet flow like liquids under sufficiently large strains or stresses. Moreover, the stress in a material can depend on the type of deformation experienced by the material. Polymer solutions generally **shear thin**, such that they feel slippery, but **extension thicken** such that threads are hard to break. The constitutive relations for even simple elastic solids are generally only linear in the small-strain limits: Deform any solid significantly, and its response will change (*e.g.*, via ductile plasticity, or brittle fracture).

2.2.3   Constitutive relations in the linear response limit

Considerable simplifications arise in the **linear response** limit, which is found when deformations are so small or slow that the stress is simply proportional to the strain (or strain rate).

Viscous, or Newtonian, Liquids

Molecular liquids and gasses almost always operate in the linear response limit, as they are inherently disordered in a way that is not strongly disrupted by flow, except under exceptional circumstances, like appreciable Mach-number flows close to the speed of sound. When the fluids can be considered incompressible, $\nabla \cdot \mathbf{v} = 0$ must be imposed to conserve mass, and the stress is then given by

$$\sigma = -p\delta + 2\eta\dot{\epsilon}, \tag{2.6}$$

where δ is the identity tensor and

$$\dot{\epsilon} = \frac{1}{2}\left(\nabla\dot{\mathbf{u}} + (\nabla\dot{\mathbf{u}})^T\right) \tag{2.7}$$

is the rate of deformation tensor. Equivalently, (2.7) can be expressed using index notation,

$$\dot{\epsilon}_{ij} = \frac{1}{2}\left(\frac{\partial\dot{u}_i}{\partial x_j} + \frac{\partial\dot{u}_j}{\partial x_i}\right), \tag{2.8}$$

and

$$\delta_{ij} = 1 \ \text{ for } i = j \tag{2.9}$$
$$\delta_{ij} = 0 \ \text{ for } i \neq j. \tag{2.10}$$

The stress in a viscous fluid depends only on the *rate* of deformation $\dot{\epsilon}$ at a given time, rather than the total deformation ϵ or any past history of deformation.

Compressible elastic solids

We will focus on elastic solids that are *isotropic*, as is appropriate for many soft materials. The conceptual differences that arise when treating anisotropic materials are relatively few, yet the mathematical complications are substantial, and would unnecessarily confuse the development of the core principles of microrheology presented here.[3] While viscous fluids are almost always incompressible, elastic materials often have a finite compressibility.

Two independent moduli are required to describe stress in isotropic elastic media—one for shear, and for compression. One way to write the stress in a compressible media is

$$\sigma = \lambda(\nabla \cdot \mathbf{u})\delta + 2G\epsilon, \tag{2.11}$$

[3] Anisotropic-elastic solids (*e.g.*, crystals) generally require a fourth-rank stiffness tensor C for their description, $\sigma = \mathbf{C} : \epsilon$, or $T_{ij} = C_{ijkl}\epsilon_{kl}$.

wherein λ is Lamé's first coefficient, which involves material compressibility, and G is the standard shear modulus. Although this form of the stress tensor is the simplest to write, it is not necessarily the clearest form conceptually, in terms of differentiating between shear and compressive properties. To see this, note that the trace of ϵ is $\nabla \cdot \mathbf{u}$, which is not necessarily zero in compressible media. Consequently, the stress arising from compressive deformations includes contributions from both the shear modulus G and λ. Specifically, the (isotropic) stress in response to a pure compressive strain is given by

$$\sigma_{ii} = (3\lambda + 2G)\nabla \cdot \mathbf{u}. \tag{2.12}$$

Thus, it is often convenient to explicitly define a bulk modulus K,

$$K = \lambda + \frac{2}{3}G, \tag{2.13}$$

so that

$$\sigma_{ii} = 3K\nabla \cdot \mathbf{u}. \tag{2.14}$$

When written in terms of the bulk and shear moduli K and G, the stress tensor explicitly separates into two components: One associated with volume-preserving deformations, and the other with compressive deformations:

$$\sigma = K(\nabla \cdot \mathbf{u})\delta + 2G\left(\epsilon - \frac{1}{3}(\nabla \cdot \mathbf{u})\delta\right). \tag{2.15}$$

Various choices are thus available to describe isotropic, compressible media. We have seen three: The shear modulus G, the bulk modulus K, and Lamé's first coefficient λ. Another common choice is the Poisson ratio ν, which gives the ratio of how much a material *expands* in a direction *transverse* to the direction in which it is compressed. An incompressible material, for example, has $\nu = 1/2$: If compressed with a strain Δ in the z-direction, it must expand with strains $\Delta/2$ in the x and y directions to preserve volume. The Poisson ratio can be derived from the shear and bulk moduli according to

$$\nu = \frac{3K - 2G}{2(3K + G)}, \tag{2.16}$$

giving a stress tensor of the form

$$\sigma = G\left[\frac{2\nu}{1 - 2\nu}(\nabla \cdot \mathbf{u})\delta + 2\epsilon\right]. \tag{2.17}$$

The final two moduli in common use are Young's modulus E, which relates uniaxial strain (stretching) to unaxial stress, and the P-wave or longitudinal modulus, which describes axial stress in response to strains that are *purely* axial (*e.g.*, as occurs in pressure (P) waves).

To summarize, two independent moduli are required to describe the stress-strain relationship in compressible, isotropic media. There are six such moduli in common use, every one of which can be expressed in terms of two others, giving 15 superficially distinct expressions for σ. Any of these expressions (*i.e.*, any pair of moduli) can be used to pose and solve a given elasticity problem. Which pair is best depends on the natural geometry of the problem and to a large extent on one's taste.

In rheology and microrheology, the shear modulus G is almost always the property of interest, and so will be retained throughout this text. Various choices are often chosen for the second modulus— usually λ, K or ν. We will generally present results in terms of G and K.

Incompressible elastic solids

Some elastic materials (*e.g.*, Jello) are much harder to compress than to shear, and can often be approximated as *incompressible*. This occurs when the bulk modulus K is much larger than the shear modulus G, so that deformations with non-zero divergence $\nabla \cdot \mathbf{u}$ would give rise to stresses $K\nabla \cdot \mathbf{u}$ that are enormous compared to shear stresses $\sim G\nabla\mathbf{u}$.

Approximating a material as incompressible is mathematically subtle, requiring the limit $K \to \infty$ to be taken while simultaneously imposing $\nabla \cdot \mathbf{u} \to 0$. It is not immediately obvious whether the compressive stress $-K\nabla \cdot \mathbf{u}$ should be infinity, or zero, or something finite. The standard approach, familiar in fluid mechanics, is to define a pressure $p = -K(\nabla \cdot \mathbf{u})$ as a separate field whose function is to enforce the incompressibility condition $\nabla \cdot \mathbf{u} = 0$, which is imposed as a separate equation. In this case, the linear response constitutive equations for an isotropic, incompressible elastic solid are given by

$$\sigma = -p\delta + 2G\epsilon. \tag{2.18}$$

$$\nabla \cdot \mathbf{u} = \mathrm{Tr}\,\epsilon = 0. \tag{2.19}$$

Notably, this constitutive relation is identical in form to that for an incompressible viscous liquid (eqn 2.6), but depends on strain ϵ rather than rate of strain $\dot{\epsilon}$.

Incompressible, isotropic viscoelastic materials

Finally, we turn to linear viscoelastic (LVE) materials, which exhibit both viscous and elastic responses to deformations. LVE materials have frequency-dependent moduli, reflecting the relaxation of different structural modes that occur at different time scales. We will focus on incompressible LVE materials, since most soft materials consist of some meso-structure suspended in a viscous liquid, and viscous liquids are essentially always treated as incompressible.

When an LVE material is subjected to a gentle oscillatory deformation

$$\epsilon(t;\omega) = \epsilon_0 e^{i\omega t}, \tag{2.20}$$

at frequency ω, the resulting stress,

$$\sigma(t;\omega) = \sigma_0 e^{i(\omega t+\delta)}, \tag{2.21}$$

need not be in-phase with the strain. Instead, the stress is given by

$$\sigma_0(\omega) = -p_0(\omega)\delta + G^*(\omega)\epsilon_0(\omega), \tag{2.22}$$

where

$$G^*(\omega) = G_0 e^{i\delta} \tag{2.23}$$

is the complex-storage modulus, and δ is the phase angle between shear stress and shear strain. The phase angle δ is zero for elastic solids, where stress and strain are in phase, and is $\delta = \pi/2$ for viscous liquids, for which the stress is in-phase with the strain rate.[4]

Under linear response conditions, the stress tensor $\sigma(t)$ in a linear viscoelastic liquid at any time t due to a general (but gentle) strain history,

$$\epsilon(t) = \frac{1}{2\pi} \int_{-\infty}^{\infty} \tilde{\epsilon}(\omega) e^{i\omega t} d\omega, \tag{2.24}$$

is then simply given by superposing the responses at each frequency, with the appropriate amplitude:

$$\sigma(t) = \frac{1}{2\pi} \int_{-\infty}^{\infty} \left[-\tilde{p}(\omega)\delta + G^*(\omega)\tilde{\epsilon}(\omega) \right] e^{i\omega t} d\omega, \tag{2.25}$$

giving

$$\sigma(t) = -p(t)\delta + \int_{-\infty}^{t} m(t-t')\epsilon(t')dt'. \tag{2.26}$$

Here $m(t)$ is the **memory function**, defined by

$$m(t) = \frac{1}{2\pi} \int_{-\infty}^{\infty} G^*(\omega) e^{i\omega t} d\omega. \tag{2.27}$$

[4] Note that some authors use $e^{-i\omega t}$ rather than $e^{i\omega t}$ in defining viscoelastic materials. Both are equivalent, although it is common for confusion and errors to arise when results from both conventions are used.

Notably, eqn 2.27 reveals that the stress at time t depends upon the strain history at previous times $t' < t$, weighted by the memory function $m(t)$.

2.3   Equations of motion for isotropic continua

It is now straightforward to derive equations of motion for various materials by simply evaluating the Cauchy Stress equation (2.3) using the relevant constitutive equation for each material.

We start with incompressible materials, which obey

$$\nabla \cdot \dot{\mathbf{u}} = \nabla \cdot \mathbf{u} = 0. \tag{2.28}$$

The momentum equations differ from material to material, and are given by

$$\rho \ddot{\mathbf{u}} = -\nabla p + \eta \nabla^2 \dot{\mathbf{u}} \tag{2.29}$$

for incompressible viscous liquids,

$$\rho \ddot{\mathbf{u}} = -\nabla p + G \nabla^2 \mathbf{u} \tag{2.30}$$

for isotropic, incompressible elastic solids, and

$$\rho \ddot{\mathbf{u}} = -\nabla p + \int_{-\infty}^{t} m(t - t') \nabla^2 \mathbf{u}(t') dt' \tag{2.31}$$

for isotropic, incompressible viscoelastic media.

Equations (2.29) and (2.30) for (incompressible) viscous liquids and elastic solids appear quite similar, differing by one mere time derivative. The momentum equation for an incompressible, isotropic-viscoelastic material (eqn 2.31) at first glance appears quite different. However, computing the Fourier time-transforms of eqns 2.29–2.31 yields

$$-\rho \omega^2 \tilde{\mathbf{u}} = -\nabla \tilde{p} + i \omega \eta \nabla^2 \tilde{\mathbf{u}} \tag{2.32}$$

$$-\rho \omega^2 \tilde{\mathbf{u}} = -\nabla \tilde{p} + G \nabla^2 \tilde{\mathbf{u}} \tag{2.33}$$

$$-\rho \omega^2 \tilde{\mathbf{u}} = -\nabla \tilde{p} + G^*(\omega) \nabla^2 \tilde{\mathbf{u}}. \tag{2.34}$$

Remarkably, the momentum equations for viscous fluids, elastic solids, and viscoelastic materials are essentially identical. These equations differ only in their scalar shear moduli, which are purely imaginary ($i\omega\eta$) for liquids (eqn 2.32), purely real (G) for solids (eqn 2.33), and generally complex $G^*(\omega)$ for viscoelastic materials (eqn 2.34). Moreover, stress tensors for all three transform similarly:

$$\tilde{\sigma} = -\tilde{p}(\omega)\delta + G^*(\omega)\tilde{\epsilon}, \tag{2.35}$$

where $G^*(\omega)$ can now be viewed as a general shear modulus: $G^*(\omega) = i\omega\eta$ for a purely viscous liquid, and $G^*(\omega) = G$ for a purely elastic solid.

These equations can equally well be expressed in terms of velocity fields $\tilde{v} = i\omega\tilde{u}$, in which case eqn 2.34 becomes

$$i\rho\omega\tilde{v} = -\nabla\tilde{p} + \eta^*(\omega)\nabla^2\tilde{v}. \tag{2.36}$$

When writing equations of motion in terms of velocity fields $\tilde{v}$, rather than displacement fields $\tilde{u}$, it is often sensible to introduce the **complex viscosity**

$$\eta^*(\omega) \equiv \frac{G^*(\omega)}{i\omega}, \tag{2.37}$$

rather than the shear modulus $G^*(\omega)$. Both approaches are valid, and are entirely equivalent, although $G^*(\omega)$ can can only be determined from $\eta^*(\omega)$ to within a single, additive constant, typically $G'(\omega \to 0)$.

Analogous results hold when the Laplace transform is employed, rather than the Fourier Transform, although subtleties exist regarding initial conditions, since Laplace Transforms single out a particular $t = 0$. For simplicity's sake, we will assume homogeneous initial conditions: $\mathbf{u}(t \le 0) = \dot{\mathbf{u}}(t \le 0) = \mathbf{0}$. In that case, the Laplace-Transformed equations of motion for viscous fluids, elastic solids, and viscoelastic media become

$$\rho s^2\hat{\mathbf{u}} = -\nabla\hat{p} + s\eta\nabla^2\hat{\mathbf{u}} \tag{2.38}$$

$$\rho s^2\hat{\mathbf{u}} = -\nabla\hat{p} + G\nabla^2\hat{\mathbf{u}} \tag{2.39}$$

$$\rho s^2\hat{\mathbf{u}} = -\nabla\hat{p} + \hat{G}(s)\nabla^2\hat{\mathbf{u}}. \tag{2.40}$$

or, using velocity rather than displacement fields,

$$\rho s\hat{v} = -\nabla\hat{p} + \hat{\eta}(s)\nabla^2\hat{v}, \tag{2.41}$$

where

$$\hat{\eta}(s) = \hat{G}(s)/s \tag{2.42}$$

is the Laplace-Transformed complex viscosity.

Equations (2.32–2.41) have profound implications for microrheology, as developed in Section 2.4.

2.4   Correspondence Principle

A remarkable feature of the time-transformed equations of motion for isotropic and incompressible materials—given by eqns 2.32–2.34—is that they are essentially identical, no matter whether the material is a viscous liquid, an elastic solid, or more generally viscoelastic. All such materials obey the same time-transformed equations of motion, with a shear modulus $G^*(\omega)$ that would be purely imaginary for a viscous liquid, purely real for an elastic solid, or complex for a viscoelastic material.

This observation forms the basis for the **Correspondence Principle** (Pipkin, 1986), which provides a simple way to solve linear viscoelastic flow or displacement problems, by simply adapting a solution to a corresponding Stokes flow or elasticity problem. Its traditional formulation holds for viscoelastic materials that

- can be treated as **continuum**;
- are spatially **homogeneous**;
- are spatially **isotropic**;
- can be approximated as **incompressible**; and
- are deformed gently enough that the **linear response** approach remains valid.

The time-transformed equations of motion for LVE materials (eqn 2.34) are identical to the time-transformed Stokes equations (eqn 2.32) for viscous flow when η is replaced by $\eta^*(\omega)$. One can therefore take a time-transformed solution $\mathbf{v}_{\text{Stokes}}$ to a Stokes flow problem with a given geometry, and replace the Stokes viscosity η with a complex viscosity $\eta^*(\omega)$, to obtain a valid time-transformed solution to the corresponding problem for an LVE material with complex viscosity $\eta^*(\omega)$,

Correpondence Principle: Time-transformed LVE flows can be obtained from analogous solutions to the Stokes flow or elasticity equations, by replacing the Newtonian viscosity η is replaced by the complex viscosity $\eta^*(\omega)$, or elastic shear modulus G by the complex shear modulus $G^*(\omega)$.

$$\tilde{\mathbf{v}}_{\text{LVE}}(\omega) = \tilde{\mathbf{v}}_{\text{Stokes}}(\omega)\big|_{\eta \to \eta^*(\omega)}, \qquad (2.43)$$

so long as the time-transformed boundary conditions of the LVE problem are also identical to those of the time-transformed Stokes flow problem. Similarly, one can take a time-transformed displacement field $\mathbf{u}_{\text{Elasticity}}$ computed for an incompressible solid, and replace the shear modulus G with the complex shear modulus $G^*(\omega)$, to obtain a valid time-transformed LVE displacement field $\tilde{\mathbf{u}}_{\text{LVE}}(\omega)$,

$$\tilde{\mathbf{u}}_{\text{LVE}}(\omega) = \tilde{\mathbf{u}}_{\text{Elasticity}}(\omega)\big|_{G \to G^*(\omega)}. \qquad (2.44)$$

Analogous results hold for Laplace-Transformed fields.

Consider the example of a plate at $z = 0$ executing transverse oscillations of amplitude U_0 and frequency ω. The plate excites elastic shear waves that propagate through an elastic medium, giving a displacement field

$$u_x(z) = \text{Re}\left(U_0 e^{i\omega(t-z/c)}\right) = U_0 \cos\left[\omega\left(\frac{z}{c} - t\right)\right], \qquad (2.45)$$

where

$$c = \sqrt{G/\rho} \qquad (2.46)$$

is the transverse wave speed in the medium. An alternative form,

$$u_x(z) = \text{Re}\left(U_0 e^{i(\omega t - q_T z)}\right) = U_0 \cos\left[q_T z - \omega t\right], \qquad (2.47)$$

highlights the transverse-wave number,

$$q_T = \frac{2\pi}{\lambda_T} = \sqrt{\frac{\rho\omega^2}{G}}. \qquad (2.48)$$

Direct substitution confirms that eqns 2.45 and 2.47 obey eqn 2.33.

The analogous problem for a viscous fluid—where a plate at $z = 0$ oscillates in the x-direction with amplitude U_0 and frequency ω—can be obtained directly from eqn 2.45 using the Correspondence Principle. Replacing the elastic shear modulus G in eqn 2.47 with the modulus for a viscous fluid ($G \to i\omega\eta$) gives

$$-iq_T z \to -iz\sqrt{\frac{\rho\omega^2}{i\omega\eta}} = \pm(i+1)\frac{z}{\lambda_V}, \qquad (2.49)$$

where we have used $\sqrt{i} = \pm(1+i)/\sqrt{2}$, and where

$$\lambda_V = \sqrt{\frac{2\eta}{\rho\omega}} = \sqrt{\frac{2\nu}{\omega}} \qquad (2.50)$$

is the oscillatory boundary layer thickness. The displacement field for a *Stokes* flow, obtained from eqn 2.45 using the Correspondence Principle, is then

$$u_x(z) \to \text{Re}\left[U_0 \exp\left(i\omega t \pm (i+1)\frac{z}{\lambda_V}\right)\right]. \qquad (2.51)$$

Choosing the negative root ensures the displacement field decays as $z \to \infty$, giving

$$u_x(z) = U_0 \cos\left(\omega t - \frac{z}{\lambda_V}\right) e^{-z/\lambda_V}. \tag{2.52}$$

The *flow* field $\mathbf{v}$ follows from the *displacement* field $\mathbf{u}$ via $\mathbf{v} = i\omega\mathbf{u}$, giving the expected oscillatory boundary layer velocity field

$$v_x(z) = V_0 \cos\left(\omega t - \frac{z}{\lambda_V}\right) e^{-z/\lambda_V} \tag{2.53}$$

in terms of the velocity V_0 of the oscillating plate.

Solutions for shear waves in LVE materials may be obtained similarly, using a complex modulus,

$$G^*(\omega) = G_0 e^{i\delta}, \tag{2.54}$$

giving

$$u_x(z) = U_0 \cos\left(\omega t - q_T^* z\right) e^{-z/\lambda_V^*}, \tag{2.55}$$

with wavenumber q_T^*,

$$q_T^* = \sqrt{\frac{\rho\omega^2}{G_0}} \cos\frac{\delta}{2} \tag{2.56}$$

and attenuation length λ_V^*,

$$\lambda_V^* = \sqrt{\frac{G_0}{\rho\omega^2}} \frac{1}{\sin\frac{\delta}{2}}. \tag{2.57}$$

Equations 2.55–2.57 recover elastic shear waves (eqn 2.45) in the $\delta \to 0$ limit appropriate for pure elastic materials, and viscous shear waves (eqns 2.50 and 2.52) in the $\delta \to \pi/2$ limit relevant for viscous fluids. For all linear viscoelastic materials (with phase angles $0 \leq \delta < \pi/2$), the thickness λ_V^* of the oscillatory-boundary layer exceeds the wavelength λ_V of the oscillatory shear waves, becoming equal only in the purely viscous limit $\delta = \pi/2$. These shear waves are depicted in Fig. 2.4.

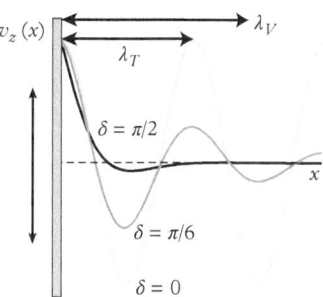

Fig. 2.4 *Shear waves near an oscillating plate for viscous, elastic, and viscoelastic materials.*

2.5 Particle motion

Earlier, we posed an apparent conundrum central to microrheology: In order to infer the material rheology from measurements of a probe's response, one must solve a continuum-mechanics problem. In order to even pose this continuum-mechanical problem in the first place, however, one must know the material rheology! As we shall shortly see the Correspondence Principle circumvents this difficulty. So long as material's constitutive equations are consistent with the Correspondence Principle, one can simply take solutions to the viscous-flow (or elastic displacement) field around the force probe, and replace the Newtonian viscosity with the corresponding complex viscosity appropriate for the material. The Correspondence Principle, then, paves the way for the widespread success of microrheology.

We therefore turn to the continuum-mechanics of particle motion.

2.5.1 Mobility and resistance

We start by discussing the hydrodynamic **resistance** ζ of a probe forced to move within a liquid, as well as its **mobility** b. In a Newtonian liquid, the mobility and resistance are *linear response* properties. The resistance ζ gives the drag force $\mathbf{F}_d$ on a probe translating with velocity $\mathbf{V}$ through the liquid, whereas the mobility b gives the probe velocity $\mathbf{V}$ in response to a driving force $\mathbf{F}$:

$$\mathbf{F}_d = -\zeta \mathbf{V}, \quad \mathbf{V} = b\mathbf{F}. \tag{2.58}$$

We have assumed a spatially isotropic probe, for which ζ and b are scalar quantities; more generally, mobility and resistance *tensors* are required for anisotropic particles, as explored in Section 2.8.

The mobility and resistance of probes take more complex forms in viscoelastic media,

$$\mathbf{F}_d(t) = -\int_{-\infty}^{t} \zeta(t - t')\mathbf{V}(t')dt' \tag{2.59}$$

$$\mathbf{V}(t) = \int_{-\infty}^{t} b(t - t')\mathbf{F}(t')dt'. \tag{2.60}$$

meaning that the velocity of a probe depends on its past force history, and vice versa. The mobility and resistance are thus not simple inverses of each other, as they are for Newtonian fluids (cf. eqn 2.58). Their time-transformed versions, however, are:

$$\tilde{\mathbf{F}}_d(\omega) = -\zeta^*(\omega)\tilde{\mathbf{V}}(\omega) \tag{2.61}$$

$$\tilde{\mathbf{V}}(\omega) = b^*(\omega)\tilde{\mathbf{F}}(\omega), \tag{2.62}$$

from which it follows that, for $\mathbf{F}_d = -\mathbf{F}$,

$$\zeta^*(\omega)b^*(\omega) = 1. \tag{2.63}$$

These equations are central to understanding the response of probe particles to applied or inherent (thermal) forces. On a practical level, it is often easier to pose the resistance problem than the mobility problem if one needs to compute these quantities (*e.g.*, for a complex probe shape, or in a complex geometry). This is because the resistance problem involves a standard boundary condition, in which the *velocity* is specified on every point of the probe surface. By contrast, the mobility problem imposes the *total* force (and torque) on the particle, without specifying how the stress is distributed over the probe surface.

While it is more natural to solve the resistance problem, and then invert the resistance to obtain the mobility, this inversion can be subtle. As discussed in Sections 2.6.2 and 2.8, anisotropic and multi-particle systems generally have tensor mobilities and resistances, for which one cannot simply invert one component (*e.g.*, ζ_{xx}) of the resistance tensor to obtain that component (b_{xx}) of the mobility tensor. Rather, the full-resistance tensor must be inverted to obtain the mobility tensor.

Lastly, in keeping with the concept of the Correspondence Principle, the mobility and resistance relations between the force on a probe and its velocity may alternately be expressed in the form of a spring constant (possibly complex) that relates the displacement to the force:

$$\mathbf{F}(t) = -\int_{-\infty}^{t} \kappa(t-t')\mathbf{U}(t')dt', \tag{2.64}$$

which when Fourier transformed becomes

$$\tilde{\mathbf{F}}(\omega) = -\tilde{\kappa}^*(\omega)\tilde{\mathbf{U}}(\omega). \tag{2.65}$$

Since $\tilde{\mathbf{F}} = i\omega\tilde{\mathbf{U}}$, the complex spring constant is related to the complex resistance via

$$\tilde{\kappa}^*(\omega) = i\omega\tilde{\zeta}^*(\omega). \tag{2.66}$$

2.5.2   The Stokes resistance and mobility of a translating sphere

We start with the simplest, most well-known, and most important example—the Stokes resistance of a solid sphere of radius a, centered at $\mathbf{r} = 0$ and translating with velocity $\mathbf{V}_0$ in a fluid of viscosity η. The fluid is set into motion, with velocity, pressure, and stress fields given by $\{\mathbf{v}, p, \boldsymbol{\sigma}\}$. The Stokes equations 2.29 are solved subject to no-slip boundary conditions on the sphere surface ($\mathbf{v} = \mathbf{V}_0$ for $r = a$) and no-disturbance far away ($\mathbf{v} \to 0$ as $r \to \infty$). The stream function,

$$\psi(\mathbf{r}, \theta) = \left(\frac{3r}{2a} - \frac{a}{2r} \right) \frac{a^2 \sin^2 \theta}{2} V_0, \qquad (2.67)$$

gives a compact form of the solution as a function of position $\mathbf{r}$ around the translating sphere, from which the velocity fields are obtained via

$$v_r = \frac{1}{r^2 \sin \theta} \frac{\partial \Psi}{\partial \theta} \qquad (2.68)$$

$$v_\theta = -\frac{1}{r \sin \theta} \frac{\partial \Psi}{\partial r}. \qquad (2.69)$$

The velocity and pressure fields around a sphere translating with velocity $\mathbf{V}_0$ are then given by[5]

$$\mathbf{v}(\mathbf{r}) = \frac{3a}{4r} \left(\mathbf{V}_0 + (\mathbf{V}_0 \cdot \hat{\mathbf{r}})\hat{\mathbf{r}} \right) + \frac{a^3}{4r^3} \left(\mathbf{V}_0 - 3(\mathbf{V}_0 \cdot \hat{\mathbf{r}})\hat{\mathbf{r}} \right) \qquad (2.70)$$

$$p(\mathbf{r}) = \frac{3\eta}{2a} \frac{a^2}{r^2} \mathbf{V}_0 \cdot \hat{\mathbf{r}} \qquad (2.71)$$

and stress tensor

$$\boldsymbol{\sigma}(\mathbf{r}) = -\frac{9\eta a}{2r^2} \mathbf{V}_0 \cdot \hat{\mathbf{r}}\hat{\mathbf{r}}\hat{\mathbf{r}} + \frac{3\eta a^3}{2r^4} \left(5\mathbf{V}_0 \cdot \hat{\mathbf{r}}\hat{\mathbf{r}}\hat{\mathbf{r}} - \hat{\mathbf{r}}\mathbf{V}_0 - \mathbf{V}_0\hat{\mathbf{r}} - \mathbf{V}_0 \cdot \hat{\mathbf{r}}\boldsymbol{\delta} \right) \quad (2.72)$$

where

$$\hat{\mathbf{r}} = \frac{\mathbf{r}}{r} \qquad (2.73)$$

is the unit vector in the radial direction. The velocity field in a fixed-reference frame and in the reference frame of the translating probe are shown in Fig. 2.5 and Fig. 2.6, respectively. The fixed-reference frame will be useful later when we consider the interactions between neighboring particles, while the sphere's reference frame provides a sense of the mix of deformation modes (*e.g.*, shear *versus* extension) and Lagrangian unsteadiness material elements experience as they move around it.

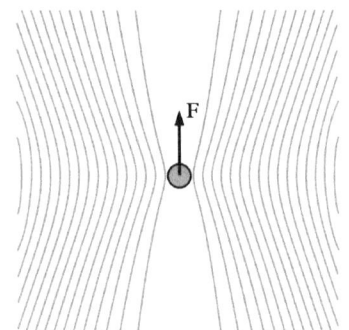

Fig. 2.5 *Flow generated by a translating sphere. In a fixed reference frame, the fluid flow around a sphere consists of a Stokeslet (Fig. 2.7) and Source Dipole (Fig. 2.8). At long distances, the Stokeslet flow field dominates, which decays with distance like r^{-1}.*

[5] In Appendix A.4 we use a vector-harmonic functions to arrive at the same solution.

Finally, we compute the force $\mathbf{F}_0$ exerted by the fluid on the particle via

$$\mathbf{F}_0 = \int_{r=a} \hat{\mathbf{r}} \cdot \boldsymbol{\sigma}\, dA = 6\pi\eta a \mathbf{V}_0, \qquad (2.74)$$

revealing the hydrodynamic resistance ζ_T^{sphere} of a translating sphere to be given by

$$\zeta_T^{\text{sphere}} = 6\pi\eta a. \qquad (2.75)$$

The mobility b_T^{sphere} of a translating sphere can now be easily determined by simply inverting ζ_T^{sphere} (eqn 2.75),

$$b_T^{\text{sphere}} = \frac{1}{6\pi\eta a}. \qquad (2.76)$$

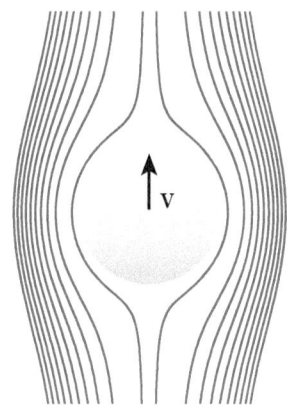

Fig. 2.6 *Streamlines in the reference frame of the translating sphere highlight the material deformation.*

The translation of viscous liquid drops and gas bubbles can be treated in a similar fashion, with the no-slip boundary condition replaced by the relevant stress-matching boundary condition. The translation of a liquid drop is described by the Hadamard–Rybczinski formula,

$$\zeta_T^{\text{drop}} = 4\pi\eta a \frac{3\lambda_\eta + 2}{2(\lambda_\eta + 1)} \qquad (2.77)$$

where $\lambda_\eta = \eta_d/\eta$ is the viscosity ratio of the drop to that of the surrounding medium. As $\lambda_\eta \to \infty$, the Stokes resistance of a rigid sphere is recovered,

$$\zeta(\lambda_\eta \to \infty) \to 6\pi\eta a. \qquad (2.78)$$

The limit $\lambda_\eta \to 0$, corresponding to an inviscid bubble, gives a resistance

$$\zeta(\lambda_\eta \to \infty) \to 4\pi\eta a \qquad (2.79)$$

that is lower than for the rigid sphere, but only by 50%.

The fluid velocity field (eqn 2.70) naturally splits into two distinct components: The first term decays slowly ($\sim r^{-1}$) and represents the flow due to a point force, or **Stokeslet**, at the origin. The Stokeslet flow gives rise to the slowest-decaying (r^{-2}) first term in eqn 2.72, and is entirely responsible for the drag on the sphere, as shown in Fig. 2.7 and Fig. 2.8. The second term in the flow field decays more quickly ($\sim r^{-3}$) and represents a potential dipole, or "point source" dipole.

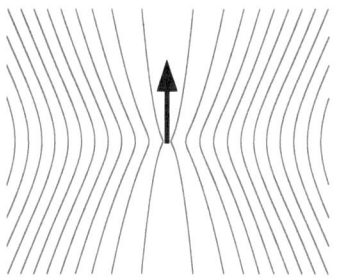

Fig. 2.7 Stokeslet *flow field. The fluid flow established by a point force, called a Stokeslet, decays like* $1/r$.

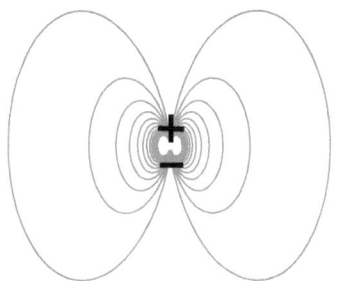

Fig. 2.8 **Potential** *dipole field. The fluid flow established by a point-source dipole decays like* $1/r^3$.

The source dipole flow is irrotational, and gives rise to the second ($\sim r^{-4}$) term in the stress tensor (2.72) and does not contribute to the drag. Equation 2.70 can be written

$$v(r) = 6\pi\eta a\left(G^{St}(r) - \frac{a^2}{3}G^{PD}(r)\right)\cdot V_0, \qquad (2.80)$$

$$= \left(G^{St}(r) - \frac{a^2}{3}G^{PD}(r)\right)\cdot F_0, \qquad (2.81)$$

where $G^{St}(r)$ and $G^{PD}(r)$ represent the Green's functions for a point force and a (potential) source dipole, respectively, located at the origin:

$$G^{St}(r) = \frac{1}{8\pi\eta}\left(\frac{\delta}{r} + \frac{\hat{r}\hat{r}}{r}\right), \qquad (2.82)$$

$$G^{PD}(r) = \frac{1}{8\pi\eta}\left(-\frac{\delta}{r^3} + 3\frac{\hat{r}\hat{r}}{r^3}\right) \equiv \frac{1}{8\pi\eta}\nabla\nabla\left(\frac{1}{r}\right). \qquad (2.83)$$

The Stokeslet tensor G^{St} is also known as the **Oseen tensor**. The flow field at large distances r from the probe is dominated by the Stokeslet flow, which depends only on the total force, rather than the size (or even shape) of the particle. The second (potential) component decays more quickly, and is shape-dependent.

The flow around a translating sphere is special in that exactly two terms (G^{St} and G^{PD}) are required for its description. An infinite number of terms (comprised of all multipoles of point forces and sources) are generally required to describe the flow around more general shapes. Several key features are preserved even for complex-shaped probes, however. The Stokeslet flow G^{St} depends only upon the force on the probe, and represents the only component of the flow that decays with distance like r^{-1}.

The higher-order multipoles depend on the detailed probe shape, and are essential in determing the self-mobility and self-resistance of the probe. Hydrodynamic interactions between well-separated particles, on the other hand, are dominated by the slowest-decaying components of the flow, as discussed in Section 2.6. The "coupling mobilities" between particles, then, are dominated by the Stokeslet flow. This motivates *two-point* microrheology techniques (see Section 4.11), which measure the cross-correlated fluctuations of two different probes. These cross correlations are proportional to the *coupling mobility* (which gives the velocity of one particle in response to a force on the other particle), which depends almost exclusively on the Stokeslet term, and is therefore essentially independent of the shape of each probe.

2.5.3 Stokes resistance of a probe undergoing oscillatory translations

Thus far, we have examined *steady* probe motion. Diffusing particles, by contrast, execute stochastic, fluctuating motions, which can be decomposed (through the Fourier Transform) to oscillatory motions at every frequency. Oscillations at sufficiently low frequencies are well-described by the steady Stokes flow solutions computed (*i.e.*, the *quasi-steady* approximation). Above a characteristic inertial frequency, however, the response of a sphere to an oscillatory force changes qualitatively, significantly impacting the interpretation of a microrheology experiment. One thus needs to know the relevant frequency where fluid inertia (and thus transient flow behavior) becomes important, and how this frequency scales with probe size and material properties.

We now compute the frequency-dependent resistance to oscillatory translations. To demonstrate the impact that fluid inertia can have on probe motion, we consider a sphere that oscillates with velocity $\mathbf{V}_0 e^{i\omega t}$, such that the fluid velocity, pressure, and stress fields have the form

$$\{\mathbf{v}, p, \boldsymbol{\sigma}\} = \{\mathbf{v}_0, p_0, \boldsymbol{\sigma}_0\} e^{i\omega t} \tag{2.84}$$

The Stokes equations (2.29) for viscous flow would then take the form

$$i\omega\rho\mathbf{v}_0 = -\nabla p_0 + \eta\nabla^2\mathbf{v}_0 \tag{2.85}$$

$$\nabla \cdot \mathbf{v}_0 = 0, \tag{2.86}$$

subject to boundary conditions

$$\mathbf{v}_0|_{r=a} = \mathbf{V}_0 \tag{2.87}$$

$$\mathbf{v}_0(r \to \infty) \to 0. \tag{2.88}$$

The stress field $\boldsymbol{\sigma}_0$ can be computed from the pressure (p_0) and velocity ($\mathbf{v}_0$) fields,

$$\boldsymbol{\sigma}_0 = -p_0\boldsymbol{\delta} + \eta\left(\nabla\mathbf{v}_0 + (\nabla\mathbf{v}_0)^T\right), \tag{2.89}$$

so that the drag force

$$\mathbf{F}^f(t) = \mathbf{F}_0^f e^{i\omega t} \tag{2.90}$$

exerted by the fluid on the probe is given by

$$\mathbf{F}_0^f = \int_{|\mathbf{r}|=a} \hat{\mathbf{n}} \cdot \boldsymbol{\sigma}_0 dA. \tag{2.91}$$

Stokes (1850) solved this problem while studying the influence of fluid inertia on the oscillations of pendula. Schieber *et al.* (2013) detail the history and solution of this problem, and place it within the context of microrheology.

The stream function

$$\psi(r, \theta, t) = \psi_0(r, \theta, \omega)e^{i\omega t} \tag{2.92}$$

for a sphere oscillating with frequency ω is given by

$$\frac{\psi_0(r, \theta, \omega)}{V_0 \sin^2 \theta} = \frac{3a}{2\Gamma^2 r}\left((1 + \Gamma r)e^{-\Gamma(r-a)} - (1 + \Gamma a)\right) - \frac{a^3}{2r}, \tag{2.93}$$

where

$$\Gamma(\omega) = (1 + i)\sqrt{\frac{\rho\omega}{2\eta}} = \frac{1 + i}{\lambda_V}, \tag{2.94}$$

and where

$$\lambda_V = \sqrt{\frac{2\eta}{\rho\omega}} \tag{2.95}$$

is the oscillatory boundary layer thickness as in eqn 2.50.

Velocity and pressure fields can be derived from eqn 2.93 using (2.68–2.69), giving

$$\frac{v_r}{V_0 \cos\theta} = 3\frac{a}{r}\left(\frac{1 + a\Gamma - (1 + \Gamma r)e^{-\Gamma(r-a)}}{\Gamma^2 r^2}\right) + \frac{a^3}{r^3} \tag{2.96}$$

$$\frac{v_\theta}{V_0 \sin\theta} = 3\frac{a}{r}\left(\frac{1 + \Gamma a - (1 + \Gamma r + \Gamma^2 r^2)e^{-\Gamma(r-a)}}{2\Gamma^2 r^2}\right) + \frac{a^3}{2r^3} \tag{2.97}$$

$$\frac{p}{V_0 \cos\theta} = i\omega\rho a\frac{3 + 3\Gamma a + \Gamma^2 a^2}{2\Gamma^2 r^2}, \tag{2.98}$$

from which the force exerted by the fluid on the sphere can be computed as

$$\mathbf{F}_0^f = -6\pi\eta a\left(1 + \Gamma a + \frac{\Gamma^2 a^2}{9}\right)\mathbf{V}_0, \tag{2.99}$$

or

$$\mathbf{F}_0^f = -\zeta_0\left(1 + \frac{a}{\lambda_V} + i\frac{a}{\lambda_V}\right) - iM_f\omega\mathbf{V}_0. \tag{2.100}$$

Here

$$M_f = \frac{1}{2} \frac{4\pi a^3 \rho_f}{3} \tag{2.101}$$

is the so-called added mass of the fluid, which represents the equivalent mass of fluid that must be accelerated to make way for the oscillating sphere.

The inertia of the fluid thus changes the hydrodynamic resistance $\zeta^*(\omega)$,

$$\zeta^*(\omega) = 6\pi \eta a \left(1 - \Gamma(\omega)a\right) + iM_f\omega \tag{2.102}$$

$$\zeta^*(\omega) = 6\pi \eta a \left(1 + \frac{a}{\lambda_V} + i\frac{a}{\lambda_V}\right) + iM_f\omega \tag{2.103}$$

giving it both real and imaginary components.

Physically, λ_V corresponds to the distance vorticity (momentum) diffuses into the fluid during one oscillation period. The resistance (eqn 2.103) shows qualitatively distinct limits, depending on the relative size of the sphere radius a compared with the oscillatory penetration depth λ_V. Since λ_V depends on ω, a natural "inertial" frequency emerges,

$$\omega_I = \frac{2\eta}{\rho a^2}, \tag{2.104}$$

which corresponds to the oscillation frequency above which inertia dominates the resistance to oscillation, and below which viscous stresses dominate the resistance. For reference, a microrheological probe of order $a \sim 1\ \mu$m in water (for which $\nu \sim 10^{-2}\mathrm{cm}^2/\mathrm{s}$) has an inertial frequency $\omega_I \sim 2 \times 10^6/\mathrm{s}$.

At low frequencies ($\omega \ll \omega_I$), which corresponds to $a/\lambda_V \ll 1$, the sphere moves quasi-steadily, with a resistance

$$\zeta^*(\omega \ll \omega_I) \rightarrow \zeta_0 \left(1 + \frac{a}{\lambda_V}\right) + i\zeta_0 \frac{a}{\lambda_V}, \tag{2.105}$$

that predominantly reflects Stokes drag, with a minor correction due to inertia.

In the opposite limit of high frequencies ($\omega \gg \omega_I$), so that $a/\delta \gg 1$, the hydrodynamic resistance becomes predominantly imaginary,

$$\zeta^*(\omega) \sim \frac{4\pi \eta a^3 i}{3\lambda_V^2} = iM_f\omega. \tag{2.106}$$

Finally, the force on a sphere moving with an arbitrary (but "small") velocity history $\mathbf{V}(t)$ can be constructed by Fourier-transforming $\mathbf{V}(t)$,

$$\tilde{\mathbf{V}}(\omega) = \int_{-\infty}^{\infty} \mathbf{V}(t) e^{-i\omega t} dt, \tag{2.107}$$

then using eqn 2.99 to determine the force due to each frequency component $\tilde{V}(\omega)$, and computing the inverse transform. This gives

$$\mathbf{F}(t) = -6\pi \eta a \mathbf{V}(t) - M_f \frac{d\mathbf{V}}{dt} - 6a^2 \sqrt{\pi \eta \rho} \int_{-\infty}^{t} \frac{d\mathbf{V}(\tau)}{d\tau} \frac{d\tau}{\sqrt{(t-\tau)}}. \tag{2.108}$$

The first term is the standard, quasi-steady Stokes drag; the second term represents the added mass, accounting for the inertia of the fluid that must be accelerated as the velocity changes. The third term is the Basset "memory" term (Basset, 1888), and shows the resistance depending on the sphere's previous acceleration history.

Exercise 2 concerns an analogous problem—a sphere oscillating in a purely elastic medium—for which

$$\mathbf{F}_0 = -6\pi Ga(1 - a\Gamma_E)\mathbf{U}_0 + \frac{1}{2}M_f\omega^2 \mathbf{U}_0, \tag{2.109}$$

where

$$\Gamma_E = \sqrt{\frac{\rho\omega^2}{G}} = \frac{\omega}{c}, \tag{2.110}$$

or equivalently

$$\mathbf{F}_0 = -6\pi Ga\left(1 - \frac{a}{c}\omega\right)\mathbf{U}_0 + \frac{1}{2}M_f\omega^2 \mathbf{U}_0. \tag{2.111}$$

The complex spring constant becomes

$$\kappa^*(\omega) = 6\pi Ga\,(1 - a\Gamma_E) - \frac{1}{2}M_f\omega^2. \tag{2.112}$$

Exercise 2.3 asks the reader to show that the Correspondence Principle can be used to derive these results from the analogous results for viscous fluids, and vice versa. Because the spring constant and resistance are related via

$$\kappa^*(\omega) = i\omega\zeta^*(\omega), \tag{2.113}$$

eqn 2.112 can be written

$$\zeta^*(\omega) = \frac{6\pi\, Ga}{i\omega}(1 - a\Gamma_E) + \frac{i}{2}M_f\omega. \tag{2.114}$$

Finally, incompressible, viscoelastic materials have complex shear moduli

$$G^* = G_0 e^{i\delta} \tag{2.115}$$

where δ ranges between 0 for elastic media and $\pi/2$ for purely viscous fluids, and where both G_0 and δ depend on frequency ω. The Correspondence Principle immediately yields the relevant results for spheres oscillating in such materials, *e.g.*, by substituting

$$\Gamma_E^* = \omega\sqrt{\frac{\rho}{G^*}} = \Gamma_E e^{-i\delta/2} \tag{2.116}$$

for Γ_E in eqn 2.112, giving

$$\tilde{\kappa}^*(\omega) = 6\pi\, G_0 e^{i\delta} a\left(1 - a\Gamma_E e^{-i\delta/2}\right) - \frac{1}{2}M_f\omega^2, \tag{2.117}$$

or alternatively

$$\zeta^*(\omega) = \frac{6\pi\, G^*(\omega)a}{i\omega}\left(1 - a\Gamma_E e^{-i\delta/2}\right) + \frac{i}{2}M_f\omega. \tag{2.118}$$

2.5.4 Particle inertia

Thus far, we have neglected the inertia of the probe in treating its dynamics. In this case, the force $\mathbf{F}^p$ driving a probe into motion is exactly balanced by the drag force $\mathbf{F}^f$ exerted by the medium on the particle

$$\mathbf{F}^p + \mathbf{F}^f = 0. \tag{2.119}$$

Probes accelerate during unsteady motion, however, which are balanced by the inertia of the probe, giving

$$\mathbf{F}^p + \mathbf{F}^f = m_p \ddot{\mathbf{U}}_p. \tag{2.120}$$

Probes that oscillate with frequency ω obey the force balance

$$\mathbf{F}_0^p + \mathbf{F}_0^f = -M_p\omega^2\mathbf{U}_0 = i\omega M_p\mathbf{V}_0. \tag{2.121}$$

Using eqn 2.99 for the drag force from the fluid gives

$$\mathbf{F}_0^p = \zeta_0\mathbf{V}_0\left(1 + \frac{a}{\lambda_V}\right) + i\left(\frac{a}{\lambda_V}\zeta_0 + \omega(M_f + M_p)\right)\mathbf{V}_0. \tag{2.122}$$

2.5.5 Spheres forced within compressible elastic media

The elastic-displacement field around a sphere of radius a subject to a force $\mathbf{F}$ in a compressible elastic medium with shear and bulk moduli G and K is given by

$$\mathbf{u}(\mathbf{r}) = \frac{\mathbf{F}}{8\pi\,Gb} \cdot \left[(2b-1)\frac{\delta}{r} + \frac{\mathbf{rr}}{r^3} \right] - \frac{a^2}{3}\frac{\mathbf{F}}{8\pi\,Gb} \cdot \left[-\frac{\delta}{r} + 3\frac{\mathbf{rr}}{r^3} \right], \quad (2.123)$$

or, in index notation,

$$u_i(\mathbf{r}) = \frac{F_j}{8\pi\,Gb} \left[(2b-1)\frac{\delta_{ij}}{r} + \frac{r_i r_j}{r^3} \right] - \frac{a^2}{3}\frac{F_j}{8\pi\,Gb} \left[-\frac{\delta_{ij}}{r} + 3\frac{r_i r_j}{r^3} \right],$$

$$(2.124)$$

where

$$b = 2(1-\nu) = \frac{3K+4G}{3K+G}. \quad (2.125)$$

Note that $b \to 1$ in the incompressible limit $(K \gg G)$.

Just like for Stokes flow (Section 2.5.2), the displacement field around a forced sphere can be decomposed into two contributions. The first is Thomson's solution (Thomson, 1848) for the field due to a point force in a compressible elastic medium,

$$\mathbf{u}(\mathbf{r}) = \frac{\mathbf{F}}{8\pi\,Gb} \cdot \left[(2b-1)\frac{\delta}{r} + \frac{\mathbf{rr}}{r^3} \right]. \quad (2.126)$$

which is the analog of the Oseen Tensor (eqn 2.82) for compressible elastic media. For incompressible materials $(K \gg G$, for which $b = 1)$, in fact, eqn 2.126 reduces exactly to the Oseen tensor (eqn 2.82) when the Correspondence Principle is used to convert between elastic and viscous solutions, as the reader is asked to show in Exercise 2.5). The second term in eqn 2.123 is a point-source dipole—irrotational and incompressible—just as in a viscous fluid.

Evaluating the displacement field at the sphere's boundary $r = a$ gives the sphere displacement $\mathbf{U}$,

$$\mathbf{U} = \mathbf{u}(a) = \frac{\mathbf{F}}{12\pi\,Gab}(3b-1) = \frac{\mathbf{F}}{6\pi\,Ga}\left(\frac{6K+11G}{6K+8G}\right). \quad (2.127)$$

Notably, from the incompressible limit $K/G \to \infty$, we recover

$$\mathbf{U}(K \gg G) \to \frac{\mathbf{F}}{6\pi\,Ga}, \quad (2.128)$$

as required by the Correspondence Principle.

In their F-actin microrheology studies, Schnurr *et al.* (1997) discuss the small impact that finite compressibility plays in eqn 2.127. In the limit of a highly compressible material, $K \ll G$ (or in terms of the Poisson ratio, as it approaches $\nu = -1$), the sphere displacement becomes

$$\mathbf{U}(K \ll G) \rightarrow \frac{\mathbf{F}}{6\pi\, Ga} \cdot \frac{11}{8}, \tag{2.129}$$

which is only about 40% larger than it would move within an incompressible medium with the same shear modulus. This relatively small contribution explains, in part, why compressibility has generally been ignored in the microrheology literature. Moreover, most soft materials probed in microrheology tend to consist of an elastic meso-structure immersed in an incompressible fluid, which must "drain" through for the material to deform compressibly, as discussed by Schnurr *et al.* (1997), Gittes *et al.* (1997), and Levine and Lubensky (2001), and will be explored in Section 2.7.

For reference, we note that Oestricher (1951) computed the resistance of a sphere to oscillatory translations in compressible viscoelastic media (for which $G^*(\omega)$ and $K^*(\omega)$ are both frequency-dependent, which Norris (2006) generalized to allow for particle/medium slip. In fact, Oestricher recognized the Correspondence Principle in his study as well.

2.6 Hydrodynamic interactions

We have thus far examined the behavior of individual particles in infinite media. In practice, however, experimental sample cells are finite, with *e.g.*, glass slides and cover slips bounding the material. Hydrodynamic interactions between the probe and fluid boundaries changes the probe's response to applied forces (*i.e.*, the mobility and resistance of the probes). Since a common goal of microrheology is to use measured probe mobilities to extract rheological properties intrinsic to the material, it is important to quantify the impact of these interactions (*e.g., probe-wall hydrodynamic interactions*), so as to avoid misinterpreting nearby walls as material rheology.

Hydrodynamic interactions need not only be deleterious, but may be specifically exploited. For example, "two-point" microrheology (Section 4.11) uses correlations between two Brownian probes, which depend explicitly on their hydrodynamic interactions, to measure the rheology of the material located between them.

We will next discuss how to treat these hydrodynamic interactions, and will focus specifically on the two systems mentioned here:

Hydrodynamic interactions between two spherical probes, which forms the basis of *two-point microrheology*, and hydrodynamic interactions between a probe and a wall, which must be accounted for to correctly infer the rheology.

2.6.1 Method of reflections

Simple, closed-form solutions (like those presented for the motion of an isolated sphere) are no longer available once walls or multiple spheres are present. In typical microrheology experiments, however, probe particles are generally well-separated (from walls or from each other). This separation enables a powerful approximation technique, generally called the method of reflections (Kim and Karilla, 1991; Leal, 2007; Pozrikidis, 1992), which can produce a series expansion that accounts for hydrodynamic interactions in the resistance or mobility of probes.

The essence of the strategy is to recognize that well-separated particles behave approximately as isolated particles, and therefore establish velocity fields that are very nearly like those in infinite space. These "isolated probe" velocity fields violate the boundary conditions on other particles or walls, however. In order to "fix" this violated-boundary condition, a "reflected" velocity field is computed as though that particle (or wall) were alone in the world. This first reflected velocity field, however, violates the no-slip boundary condition on the original probe. A second reflection is thus computed to fix this violation, again for an isolated sphere, which once again violates the boundary condition on the wall or second probe, and so on. Ultimately, the method of reflections produces a power series expansion in powers of a/d, where d is the distance between the probe and the wall, or between the probe and a second particle.

The full method of reflections requires additional concepts and results from viscous hydrodynamics. In many cases, however, only the leading-order correction of hydrodynamic interactions to the probe mobility is required. We will therefore detail this first reflection here, and leave the advanced treatment to Section 2.6.5.

The key question concerns how a particle responds when the fluid around it is moving. Remember that inertia is usually negligible in the low-Re limit relevant to typical microrheology experiments. Rather than $\mathbf{F} = M\mathbf{a}$, then, the probe typically responds via $\mathbf{F} = \zeta \mathbf{V}$ (eqn 2.58). That is, a force $\mathbf{F}$ must be exerted on a probe in order for that probe to move *through* the local fluid. If no force is exerted on the probe, then the probe simply moves with the local-fluid velocity.

To the leading order, then, a probe immersed in a fluid with some velocity field $\mathbf{v}_\infty(\mathbf{r})$ at some position $\mathbf{r}_p$ moves with approximate velocity

$$V_p \sim \mathbf{v}_\infty(\mathbf{r}_p) \tag{2.130}$$

unless some force prevents it from doing so.

This simple notion now allows HI to be computed directly for the key situations we have described.

2.6.2 Hydrodynamic interactions between spheres in incompressible media

We first start by computing the hydrodynamic coupling between two well-separated spherical probes, each of radius a, located at the origin ($\mathbf{r}_1 = 0$) and $\mathbf{r}_2 = d\hat{\mathbf{r}}$, respectively (Fig. 2.9).[6] We will assume that the distance of separation d between the probes is much larger than the probe radii, so that $a/d \ll 1$. This is the key computation required for two-point microrheology (Section 4.11).

If a force $\mathbf{F}_1$ is exerted on particle 1, what is the velocity $\mathbf{V}_2$ of particle 2 in response? This relation is called the *coupling mobility* b_{21} between the spheres, and will be computed here.

The force on probe 1 drives it into motion with velocity given approximately by

$$\mathbf{V}_1 \approx \frac{\mathbf{F}_1}{6\pi\eta a}, \tag{2.131}$$

and establishes a velocity field $\mathbf{v}_1(\mathbf{r})$ that can be well-approximated as that around an isolated sphere in an infinite fluid eqn 2.70,

$$\mathbf{v}_1(\mathbf{r}) = \frac{1}{8\pi\eta r}\left(\mathbf{F}_1 + (\mathbf{F}_1 \cdot \hat{\mathbf{r}})\hat{\mathbf{r}}\right) + \frac{a^2}{24\pi\eta r^3}\left(\mathbf{F}_1 - 3(\mathbf{F}_1 \cdot \hat{\mathbf{r}})\hat{\mathbf{r}}\right). \tag{2.132}$$

How does the second particle respond to the forced motion of the first particle? Strictly speaking, the second particle does not "know" that the first particle even exists, nor that it is moving. Rather, the second particle is immersed in a fluid that has been set into motion by the force on the first. If no force is exerted on particle 2, then it simply moves with the local velocity of the fluid in its immediate vicinity,

$$\mathbf{V}_2 \approx \mathbf{v}_1(\mathbf{r}_2). \tag{2.133}$$

Since we are only treating the leading-order correction due to hydrodynamic interactions, we must keep only the leading-order term in

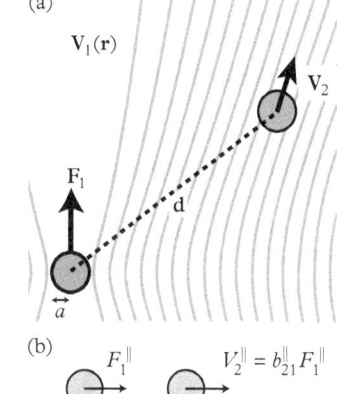

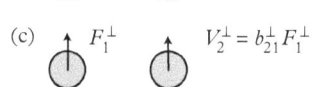

Fig. 2.9 *Hydrodynamic interactions between spheres (a) Two well-separated spheres, each of radius a, are separated by a distance d in a viscous fluid, where $d \gg a$. A force F_1 on sphere 1 drives the surrounding fluid to flow with velocity field $v_1(r)$, which causes sphere 2 to move with velocity $V_2 = b_{21} \cdot F_1$, where b_{21} is the coupling mobility tensor. Analysis is simplified by decomposing this system into force-components parallel (b) and perpendicular (c) to the separation vector d. Sphere 2 moves with velocities $V_2^\parallel$ and $V_2^\perp$ in response to parallel and perpendicular force components $F_1^\parallel$ and $F_1^\perp$ on sphere one, which defines the parallel and perpendicular coupling mobilities $b_{21}^\parallel$ and $b_{21}^\perp$, respectively. Both are functions of the relative-separation d/a, and given by eqns 2.136–2.137.*

[6] Exercise 8 treats the problem with different-sized spheres.

the far-field velocity field driven by sphere 1,[7] which is the Stokeslet term

$$\mathbf{v}_1(\mathbf{r}_2) \approx \mathbf{G}^{St}(\mathbf{r}_2) \cdot \mathbf{F}_1. \tag{2.134}$$

A freely-suspended particle 2 thus responds to a force $\mathbf{F}_1$ on particle 1 by moving with approximate velocity

$$\mathbf{V}_2 = \frac{\hat{\mathbf{x}} \cdot \mathbf{F}_1}{4\pi\eta d}\hat{\mathbf{x}} + \frac{\hat{\mathbf{y}} \cdot \mathbf{F}_1}{8\pi\eta d}\hat{\mathbf{y}} + \frac{\hat{\mathbf{z}} \cdot \mathbf{F}_1}{8\pi\eta d}\hat{\mathbf{z}}. \tag{2.135}$$

Several features in eqn 2.135 are noteworthy. First, the leading-order approximation to the coupling mobility does not depend on the size of either probe! In fact, it does not even depend on the shape of either probe, so long as the separation distance d between particles significantly exceeds the longest dimension of either particle. This reflects two key facts. The first is that the far-field flow around the forced particle is dominated by the Stokeslet—which depends only on the force that is exerted, rather than the shape or size of the parti-cle to which it is exerted. Second, particle 2 is not forced through the fluid, but rather simply moves along with whatever velocity the fluid is moving. That is—particle 2 does not move because it is forced to; it moves because it is *not* forced *not* to move, and simply moves with whatever velocity its surroundings move. Neither the forced flow, nor the advection velocity, cares about the size or shape of either probe, to the leading order.

Second, the velocity (eqn 2.135) is anisotropic: Particle 2 moves twice as fast when the force $\mathbf{F}_1$ is directed *toward* particle 2 (*i.e.*, $\mathbf{F}_1 = F\hat{\mathbf{x}}$),

$$V_2^{\parallel} = \frac{1}{4\pi\eta d}F_1^{\parallel} \equiv b_{21}^{\parallel}F_1^{\parallel} \tag{2.136}$$

than when the force $\mathbf{F}_2$ is directed *perpendicular* to the vector separat-ing the particle pair,

$$V_2^{\perp} = \frac{1}{8\pi\eta d}F_1^{\perp} \equiv b_{21}^{\perp}F_1^{\perp}. \tag{2.137}$$

The coupling mobility is therefore an *anisotropic* tensor,

$$b_{21} = \mathbf{G}^{St}(\mathbf{r}_2). \tag{2.138}$$

Likewise, a force $\mathbf{F}_2$ on particle two drives it to move with velocity

$$\mathbf{V}_2 \approx \frac{\mathbf{F}_2}{6\pi\eta a} \tag{2.139}$$

[7] The faster decay of the source dou-blet flow in 2.132 makes its contribution smaller than the leading-order (Stokeslet) contribution by an amount of order $(a/d)^2$. Correctly computing this correction requires the curvature of $\mathbf{v}_1(\mathbf{r})$ to be treated prop-erly, using Faxen's law, and is described in Section 2.6.5.

and drives particle 1 to move with velocity

$$\mathbf{V}_1 = \frac{\hat{\mathbf{x}} \cdot \mathbf{F}_2}{4\pi \eta d}\hat{\mathbf{x}} + \frac{\hat{\mathbf{y}} \cdot \mathbf{F}_2}{8\pi \eta d}\hat{\mathbf{y}} + \frac{\hat{\mathbf{z}} \cdot \mathbf{F}_2}{8\pi \eta d}\hat{\mathbf{z}}. \tag{2.140}$$

One can thus construct a multiparticle mobility tensor,

$$\begin{pmatrix} \mathbf{V}_1 \\ \mathbf{V}_2 \end{pmatrix} = \begin{pmatrix} \mathbf{b}_{11} & \mathbf{b}_{12} \\ \mathbf{b}_{21} & \mathbf{b}_{22} \end{pmatrix} \cdot \begin{pmatrix} \mathbf{F}_1 \\ \mathbf{F}_2 \end{pmatrix}, \tag{2.141}$$

where each $\mathbf{b}_{ij}$ represents a 3×3 mobility tensor, diagonal blocks representing self-mobilities,

$$\mathbf{b}_{ii} = \frac{1}{6\pi \eta a}\boldsymbol{\delta}, \tag{2.142}$$

and off-diagonal blocks represent coupling mobilities by the Oseen tensor:

$$\mathbf{b}_{i \neq j} = \frac{1}{8\pi \eta d}\left(\boldsymbol{\delta} + \hat{\mathbf{d}}\hat{\mathbf{d}}\right), \tag{2.143}$$

where

$$\hat{\mathbf{d}} = \frac{\mathbf{r}_2 - \mathbf{r}_1}{|\mathbf{r}_2 - \mathbf{r}_1|}. \tag{2.144}$$

To illustrate, consider forces $\mathbf{F}_1$ and $\mathbf{F}_2$ directed parallel to $\hat{\mathbf{d}}$ (the line between the particles), as in Fig. 2.10. In this case, the two-particle mobility tensor, valid to $\mathcal{O}(\frac{a}{d})$, is given by

$$\begin{pmatrix} V_1^\| \\ V_2^\| \end{pmatrix} = \frac{1}{6\pi \eta a}\begin{pmatrix} 1 & \frac{3}{2}\frac{a}{d} \\ \frac{3}{2}\frac{a}{d} & 1 \end{pmatrix} \cdot \begin{pmatrix} F_1^\| \\ F_2^\| \end{pmatrix}. \tag{2.145}$$

Diagonalizing shows how hydrodynamic interactions affect multiparticle dynamics, as two distinct eigenmodes appear.

$$\begin{pmatrix} V_C^\| \\ V_R^\| \end{pmatrix} = \frac{1}{6\pi \eta a}\begin{pmatrix} 1 + \frac{3}{2}\frac{a}{d} & 0 \\ 0 & 1 - \frac{3}{2}\frac{a}{d} \end{pmatrix} \cdot \begin{pmatrix} F_C^\| \\ F_R^\| \end{pmatrix} \tag{2.146}$$

One mode (denoted C) is "collective", in which forces on particles point in the same direction, so that hydrodynamic interactions contribute to each sphere's own force-driven velocity. By contrast, forces on the particles in the relative mode (denoted R) are oppositely-directed, so that hydrodynamic interactions act against the velocity

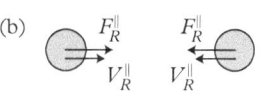

(a)

(b)

Fig. 2.10 *Collective and Relative motion of two spheres with hydrodynamic interactions. Hydrodynamic interactions give rise to two eigenmodes: (a) Collective mode, where the force on each sphere is equal in magnitude and direction, and (b) Relative mode, where forces are equal in magnitude, but oppositely directed. In the collective mode, hydrodynamic interactions impart particle velocities in the direction each was forced, giving an eigenmobility* $b_C^\| = b_0(1 + 3a/2d)$ *that is higher than the isolated-particle mobility. In the relative mode, hydrodynamic interactions contribute a velocity directed against the velocity each particle is forced to move, giving a lower eigenmobility* $b_R^\| = b_0(1 - 3a/2d)$.

with which each sphere would move if in isolation. The two eigenmo-
bilities,

$$b_C^\| = b_0(1 + 3a/2d) \qquad (2.147)$$

$$b_R^\| = b_0(1 - 3a/2d), \qquad (2.148)$$

are thus enhanced and reduced for collective and relative modes,
respectively.

The analogous calculation for particles forced *perpendicular* to $\hat{\mathbf{d}}$
gives weaker hydrodynamic interactions:

$$b_{R,C}^\perp = b_0(1 \mp 3a/4d), \qquad (2.149)$$

as shown in Exercise 2.9.

2.6.3 Hydrodynamic interactions in compressible media

Compressibility affects the hydrodynamic interactions between suspended particles. As in Fig. 2.9, we consider two probes separated by
a distance d, and compute the displacement $\mathbf{U}_2$ of probe 2 in response
to a force $\mathbf{F}_1$ on sphere 1. It is most convenient to decompose the ap-
plied force into components that are parallel $F_1^\|$ and perpendicular
$F_1^\perp$ to the line between the two particles (Fig. 2.10).

From the displacement field around a spherical probe eqn 2.123,
we identify the slowest-decaying component as that due to the point
force (Thomson's solution, eqn 2.126). As described in Section 2.6.2,
a particle immersed in a medium will simply move along with its local-
material environment (to leading order), unless some force is exerted
on it to make it act otherwise. We must therefore simply evaluate the
point-force displacement field eqn 2.126 at the center $\mathbf{r}_2$ of the second
probe, to determine the leading-order approximation for the velocity of probe 2 in response to a force on probe 1. The parallel and
perpendicular velocities are

$$U_2^\| = \frac{F^\|}{8\pi Gd}\left(\frac{3K + 7G}{3K + 4G}\right) \qquad (2.150)$$

$$U_2^\perp = \frac{F^\perp}{4\pi Gd}. \qquad (2.151)$$

Notably, the velocity *perpendicular* to the line of centers does not
depend on material compressibility. The velocity *parallel* to the line

of centers, on the other hand, depends on both moduli. In the incompressibility limit $K/G \rightarrow \infty$, these results are consistent with incompressible Stokes flows, as the Correspondence Principle would suggest. Because the parallel and perpendicular resistances depend on K and G in distinct ways, independent measurements of these two quantities would enable the two moduli to be extracted. This statement can also be read in reverse: Because this material has two distinct material parameters, two "linearly independent" measurements are required to properly characterize the material.

Lastly, note that the coupling mobility decays with separation like d^{-1}, whether or not the material has finite compressibility. As with incompressibility, the coupling mobility does not depend on the size or shape of either probe, so long as they are well-separated.

2.6.4 Particle-wall hydrodynamic interactions: Confinement effects

In practice, all experimental systems are finite in extent, and are typically bounded by either solid walls. Walls interact hydrodynamically with particles, which changes the probe mobility in a way that could be misinterpreted as rheology. It is therefore important to understand the magnitude of confinement effects inherent to practical sample cells, and their impact on the interpretation of microrheology experiments.

To do so, we follow a similar strategy as we did for interparticle-hydrodynamic interactions. A forced particle sets up a flow that is approximately that of an isolated sphere. This flow, however, violates the no-slip condition on a rigid wall. We then must compute a new fluid velocity field that "corrects" this error on the boundary conditions at the wall. The probe forced particle thus "sees" its environment moving at the velocity set up by the wall, and simply moves along with its world.

We first consider a sphere of radius a, located a distance $z = h$ from a solid wall located at $z = 0$ (Fig. 2.11). The no-slip condition is imposed on the solid wall. A force $\mathbf{F}$ exerted on an isolated probe would set up a velocity field

$$
\mathbf{v}(\mathbf{r}) = \frac{1}{8\pi \eta R_p} \left[\left(\mathbf{F} + (\mathbf{F} \cdot \hat{\mathbf{R}}_p)\hat{\mathbf{R}}_p \right) \right.
$$
$$
\left. + \frac{a^2}{3R_p^2} \left(\mathbf{F} - 3(\mathbf{F} \cdot \hat{\mathbf{R}}_p)\hat{\mathbf{R}}_p \right) \right], \quad (2.152)
$$

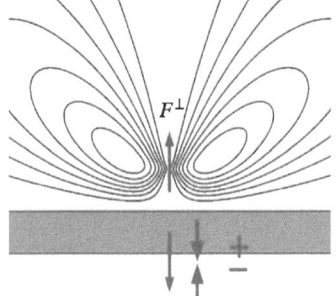

Fig. 2.11 *Probe-wall hydrodynamic interactions.* (a) The flow field due to a point force located a distance h from a no-slip surface consists of (b) the Stokeslet flow due to the point force at hẑ, and (c) a wall flow v_w established by a Stokeslet, Stokeslet doublet, and Source Dipole located at the "image" location $r_i = -h\hat{z}$. (d) The velocity of a forced sphere near a wall is given approximately by the self-mobility $b_0\mathbf{F}_1$ due to the force (b), with a correction $v_w(r_p - r_i)$ given by the velocity with which the wall flow advects the sphere, given by eqns 2.161–2.162.*

where $\mathbf{R}_p$ is the vector between the observation point (at $\mathbf{r}$) and the probe (at $h\hat{\mathbf{z}}$):

$$R_p = \sqrt{x^2 + y^2 + (z - h)^2} \tag{2.153}$$

$$\hat{\mathbf{R}}_p = \frac{\mathbf{r} - h\hat{\mathbf{z}}}{R_p}. \tag{2.154}$$

As mentioned earlier, this force would drive the probe to move with velocity

$$\mathbf{V} = (6\pi\eta a)^{-1}\mathbf{F} \tag{2.155}$$

if it were isolated.

We here assume the sphere is located far from the wall, meaning that $h \gg a$. Far from the particle ($R_p \gg a$), the fluid velocity field (2.152) is given approximately by the Stokeslet (point force) flow,

$$\mathbf{v}(\mathbf{r}) = \frac{1}{8\pi\eta R_p}\left(\mathbf{F} + (\mathbf{F} \cdot \hat{\mathbf{R}}_p)\hat{\mathbf{R}}_p\right) \equiv \mathbf{F} \cdot \mathbf{G}^{\mathrm{St}}(\mathbf{R}_p) \tag{2.156}$$

which does not vanish on the wall (at $z = 0$), in violation of the no-slip boundary condition.

Blake (1971) showed that a simple set of image singularities, located behind the wall at $z = -h$, fixes the no-slip boundary condition on the wall (Fig. 2.11). The image flow field to correct the no-slip condition for a point force $\mathbf{F}_\perp = F_\perp\hat{\mathbf{z}}$ *perpendicular* to a nearby wall is given by

$$\mathbf{v}_w^\perp(\mathbf{r}) = \left[-\mathbf{G}^{\mathrm{St}}(\mathbf{R}_i) + 2h^2\mathbf{G}^{\mathrm{PD}}(\mathbf{R}_i) - 2h\mathbf{G}^{\mathrm{StD}}(\mathbf{R}_i)\right] \cdot \mathbf{F}_\perp, \tag{2.157}$$

where

$$\mathbf{R}_i = \mathbf{r} + h\hat{\mathbf{z}} \tag{2.158}$$

is the vector from the image position ($-h\hat{\mathbf{z}}$) and the observation point $\mathbf{r}$, $\mathbf{G}^{\mathrm{PD}}(\mathbf{R}_i)$ represents the flow at $\mathbf{r}$ due to a potential dipole (eqn 2.83) located at $-h\hat{\mathbf{z}}$, and $\mathbf{G}^{\mathrm{StD}}$ is the flow due to a Stokeslet doublet, defined by

$$\mathbf{G}^{\mathrm{StD}}(\mathbf{r}) = \frac{\partial}{\partial z_0}\mathbf{G}^{\mathrm{St}}(\mathbf{r} - \mathbf{r}_0). \tag{2.159}$$

The image flow field $\mathbf{v}_w^\perp$, given by eqn 2.157, represents the flow set up by the wall in response the action of the forced probe. This image flow field thus causes the fluid environment around the probe to move, and advects the probe. Evaluating $v_w^\perp(\mathbf{r} = h\hat{z})$ at the probe location $\mathbf{r} = h\hat{z}$ gives the correction to the probe velocity due to hydrodynamic interactions with the wall,

$$v_w^\perp(\mathbf{r} = h\hat{z}) = -\frac{3F_\perp}{16\pi\eta}. \tag{2.160}$$

The probe mobility, perpendicular to a wall, is thus given by

$$V_\perp = b_0 \left(1 - \frac{9a}{8h}\right) F_\perp. \tag{2.161}$$

Using the image system for a Stokeslet force $\mathbf{F}_\parallel$ oriented parallel to the wall, the mobility of a sphere forced parallel to a wall can be shown to be

$$V_\parallel = b_0 \left(1 - \frac{9a}{16h}\right) F_\parallel. \tag{2.162}$$

In similar fashion, hydrodynamic interactions between a probe and other surfaces may be computed using the method of reflections, including liquid/gas interfaces (where a no-shear stress condition is imposed), for which the perpendicular mobility is reduced, but parallel mobility is enhanced; planar interfaces between two viscous interfaces, and partial-slip boundaries.

Key points to remember from this section are: (i) Hydrodynamic interactions with solid walls *reduce* probe mobility; (ii) the probe radius a is the "unit" distance over which hydrodynamic interactions decay, and (iii) hydrodynamic interactions are fairly long-ranged, decaying like (a/h).

2.6.5 Higher-order corrections: Faxen's law, and multiple reflections

Equations 2.143 and 2.161–2.162 give the *leading-order* approximations to the hydrodynamic interactions between two spheres and between a sphere and a well, respectively. This level of approximation will suffice for almost all results relevant to microrheology. More accurate expressions can be obtained using the *method of reflections*. We include this section primarily for those readers interested in taking the next step; more extensive discussions can be found in advanced texts in fluid mechanics and suspension mechanics—*e.g.*, Kim and Karilla (1991), Leal (2007), and Pozrikidis (1992).

We will illustrate this explicitly for the coupling mobility between spheres. Notably, eqn 2.143 only depends on the slowest-decaying (Stokeslet) component of the velocity field around the forced sphere. Including the source dipole field eqn 2.83 in the calculation is simple enough, and gives a correction that is smaller than the Stokeslet contribution by an amount of $\mathcal{O}(a^2/d^2)$.

When considering corrections this small, however, one must also account for the fact the Stokeslet velocity field (which we had evaluated at $\mathbf{r}_2$, the center of particle 2) is itself heterogeneous. To do so, we turn to Faxen's laws, which give the force $\mathbf{F}$ and torque $\mathbf{L}$ required to make a sphere of radius a, located at $\mathbf{r}_p$, translate at velocity $\mathbf{V}_p$, and rotate at angular velocity $\mathbf{\Omega}_p$ while immersed in a background flow $v_\infty(\mathbf{r})$. Specifically, Faxen's laws reveal

$$\mathbf{F} = 6\pi\eta a \left(\mathbf{V}_p - \mathbf{v}_\infty(\mathbf{r}_p) - \frac{a^2}{6}(\nabla^2 \mathbf{v}_\infty)|_{\mathbf{r}=\mathbf{r}_p} \right) \qquad (2.163)$$

$$\mathbf{L} = 8\pi\eta a^3 \left(\mathbf{\Omega}_p - \frac{1}{2}\nabla \times \mathbf{v}_\infty |_{\mathbf{r}=\mathbf{r}_p} \right). \qquad (2.164)$$

The coupling mobility of interest here relates the velocity $\mathbf{V}_2$ of sphere 2, which is force- and torque-free, in response to a force $\mathbf{F}_1$ on sphere 1. The force $\mathbf{F}_1$ on sphere 1 establishes a flow which advects sphere 2, so that the velocity field $\mathbf{v}_\infty(\mathbf{r}) = \mathbf{v}_1(\mathbf{r})$ in Faxen's laws (2.163–2.164). Because there is no force ($\mathbf{F}_2 = 0$) or torque ($\mathbf{L}_2 = 0$) on particle 2, Faxen's laws (2.163–2.164) can be re-arranged with $\mathbf{F} = \mathbf{L} = 0$ to reveal to obtain the advection velocity of particle 2,

$$\mathbf{V}_2 = \mathbf{v}_1(\mathbf{r}_2) + \frac{a^2}{6}\nabla^2 \mathbf{v}_1 |_{\mathbf{r}=\mathbf{r}_2} \qquad (2.165)$$

$$\mathbf{\Omega}_2 = \frac{1}{2}\nabla \times \mathbf{v}_1 |_{\mathbf{r}=\mathbf{r}_2}. \qquad (2.166)$$

Evaluating the full (isolated) velocity field $\mathbf{v}_1$ at $\mathbf{r}_2 = d\hat{\mathbf{x}}$ gives

$$\mathbf{v}_1(d\hat{\mathbf{x}}) = \frac{1}{6\pi\eta a}\left[\left(\frac{3a}{2d} - \frac{a^3}{2d^3} \right)\mathbf{F}_1^\| + \left(\frac{3a}{4d} + \frac{a^3}{4d^3} \right)\mathbf{F}_1^\perp \right] \qquad (2.167)$$

and

$$\frac{a^2}{6}\nabla^2 \mathbf{v}_1|_{d\hat{\mathbf{x}}} = \frac{1}{6\pi\eta a}\left[-\frac{a^3}{2d^3}\mathbf{F}_1^\| + \frac{a^3}{4d^3}\mathbf{F}_1^\perp \right] \qquad (2.168)$$

so that

$$\mathbf{V}_2 = \frac{1}{6\pi\eta a}\left[\left(\frac{3a}{2d} - \frac{a^3}{d^3} \right)\mathbf{F}_1^\| + \left(\frac{3a}{4d} + \frac{a^3}{2d^3} \right)\mathbf{F}_1^\perp \right]. \qquad (2.169)$$

The parallel and perpendicular coupling mobilities are thus

$$b_\| = \frac{1}{4\pi \eta d} \left(1 - \frac{2a^2}{3d^2} \right) \qquad (2.170)$$

$$b_\perp = \frac{1}{8\pi \eta d} \left(1 + \frac{2a^2}{3d^2} \right), \qquad (2.171)$$

valid to $\mathcal{O}(a^2/d^2)$.

Hydrodynamic interactions may also lead to relative rotations between spheres: A force $\mathbf{F}_1$ causes particle 2 to rotate via

$$\mathbf{\Omega}_2 = -\frac{1}{8\pi \eta d^2} \hat{\mathbf{F}}_1 \times \hat{\mathbf{d}}. \qquad (2.172)$$

This basic strategy holds more generally: To compute hydrodynamic interactions, one must first compute the flow field established by the forced sphere, accurate to whatever order in a is required. In the particle-wall case, for example, this would require computing the higher-order ($\sim a^2$) correction to the flow field established by the wall, which would require the image system for a potential dipole. One must then use this flow field in Faxen's law eqn 2.165 to compute the advection velocity of the sphere.

2.7 Elastic networks in viscous liquids: The two-fluid model

Soft materials often consist of a compressible elastic microstructure immersed in an incompressible viscous liquid. Polymer gels, swollen by a good solvent, provide an illustrative example: The polymer network itself is elastic and compressible, yet the surrounding solvent is not. Consequently, regions of the elastic network may only be compressed (or dilated) if the solvent flows out of (or into) those regions, respectively. Viscous forces resist this flow, and set a time scale for the fluid to drain from the elastic network.

In the microrheology context, any probe motion that excites compressional deformations to the elastic microstructure, then, can cause problems. At sufficiently long times (or low frequencies) for the viscous liquid to drain freely from the elastic microstructure, the elastic structure does indeed compress (and dilate) around the probe, whereas the liquid simply redistributes to maintain its own incompressibility (Schnurr *et al.*, 1997; Gittes *et al.*, 1997; Levine and Lubensky, 2001). The material around the probe thus *becomes* inhomogeneous—invalidating the Correspondence Principle.

Swollen polymer gels are often described using the two-fluid model (Milner, 1993), where a displacement field $\mathbf{u}$ describes the (compressible) deformation of the elastic network, with bulk and shear moduli K and G, respectively (eqn 2.15), and a velocity field $\mathbf{v}$ describes the flow of the solvent. The momentum equation for each obeys the respective Cauchy stress equation 2.3, wherein a body force is included to account for the force that each phase exerts on the other. If the velocities of the two phases are equal everywhere, there is no body force; if, however, the fluid moves relative to the elastic network, then the fluid exerts a force on the network,

$$\mathbf{f}_b = \Gamma_\xi (\mathbf{v} - \dot{\mathbf{u}}), \tag{2.173}$$

which is equal and opposite to the force that the network exerts on the fluid. The parameter Γ_ξ describes the solvent/network coupling, which for a polymer network can be approximated by

$$\Gamma_\xi \sim \frac{\eta}{\xi^2}, \tag{2.174}$$

where ξ is a characteristic mesh spacing for the network. This form for Γ_ξ follows from treating the network as a porous medium, through which the fluid must flow (akin to Darcy flow).

The two-fluid model for a homogeneous, polymer gel is then given by

$$\rho_e \frac{\partial \dot{\mathbf{u}}}{\partial t} = \left(K + \frac{1}{3}G\right) \nabla(\nabla \cdot \mathbf{u}) + G\nabla^2\mathbf{u} + \Gamma_\xi \left(\mathbf{v} - \frac{\partial \mathbf{u}}{\partial t}\right) \tag{2.175}$$

$$\rho_f \frac{\partial \mathbf{v}}{\partial t} = -\nabla p + \eta\nabla^2\mathbf{v} - \Gamma_\xi \left(\mathbf{v} - \frac{\partial \mathbf{u}}{\partial t}\right), \tag{2.176}$$

where ρ_e and ρ_f represent the mass densities of the elastic and fluid phases, respectively. As we will see, finite compressibility impacts probe dynamics at low frequencies, and so we will neglect the transient-inertial terms. We will consider oscillations at frequency ω, for which the two-fluid equations become

$$0 = \left(K + \frac{1}{3}G\right) \nabla(\nabla \cdot \mathbf{u}_0) + G\nabla^2\mathbf{u}_0 + \Gamma_\xi (\mathbf{v}_0 - i\omega\mathbf{u}_0) \tag{2.177}$$

$$0 = -\nabla p_0 + \eta\nabla^2\mathbf{v}_0 - \Gamma_\xi (\mathbf{v}_0 - i\omega\mathbf{u}_0). \tag{2.178}$$

Compressional deformations may be isolated by taking the divergence of both equations, accounting for the incompressibility ($\nabla \cdot \mathbf{v}_0 = 0$) of the liquid,

$$0 = \left(K + \frac{1}{3}G\right) \nabla^2(\nabla \cdot \mathbf{u}_0) - i\omega\Gamma_\xi \nabla \cdot \mathbf{u}_0 \tag{2.179}$$

$$0 = -\nabla^2 p_0 + i\omega\Gamma_\xi \nabla \cdot \mathbf{u}_0. \tag{2.180}$$

The second equation reveals how compression impacts the dynamic pressure p in the fluid, whereas the first governs the dyanamics of compressive deformations. Scaling gradients with a probe radius a to treat the displacement field around a spherical probe gives

$$i\omega\Gamma_\xi(\nabla \cdot \mathbf{u}_0) = \frac{\left(K + \frac{1}{3}G\right)}{a^2}\tilde{\nabla}^2(\nabla \cdot \mathbf{u}_0), \qquad (2.181)$$

revealing a natural "free-draining" frequency

$$\omega_c \sim \frac{\left(K + \frac{1}{3}G\right)}{\Gamma_\xi a^2} \sim \frac{\xi^2}{a^2}\frac{\left(K + \frac{1}{3}G\right)}{\eta} \sim \frac{\xi^2}{a^2}\frac{G}{\eta}\frac{1}{1-2v}. \qquad (2.182)$$

At low frequencies ($\omega \ll \omega_c$), the fluid drains freely from the network, effectively decoupling from the dynamics. In this limit, the probe moves quasi-statically within a compressible medium, as described in 2.5.5. At high frequencies ($\omega \gg \omega_c$), the fluid does not have time to drain through the gel, and instead forces the network to deform as an effectively incompressible medium. For frequencies sufficiently above ω_c, both fields behave as incompressible, isotropic media, and therefore the correspondence principle holds. For low frequencies, on the other hand, the two "fluids" decouple, with different compressibilities, and so the Correspondence Principle breaks down. Notably, however, the displacement of a probe in a highly-compressible medium, with the $K \gg G$ limit given by eqn 2.129, differs from the displacement in an incompressible medium by only 40%. In this case, even thought the CP fails, it still makes reasonable predictions.

2.8 Non-isotropic probes

Axisymmetric probes. The isotropic shape of a sphere gives rise to its isotropic mobility and resistance. By contrast, the response of more generally-shaped particles depends on which direction they move in, and is described using mobility and resistance *tensors*. For example, the resistance of long, slender rods with velocity *perpendicular* to the rod axis is *twice* that for velocity *parallel* to the axis,[8]

$$\zeta_{\text{rod}}^{\perp} = 2\zeta_{\text{rod}}^{\parallel}. \qquad (2.183)$$

The resistance and mobility tensors of a rod, whose axis is directed along $\hat{z}$, is given by

$$\mathbf{b} = b_\parallel \begin{pmatrix} 1/2 & 0 & 0 \\ 0 & 1/2 & 0 \\ 0 & 0 & 1 \end{pmatrix}, \quad \boldsymbol{\xi} = \zeta_\parallel \begin{pmatrix} 2 & 0 & 0 \\ 0 & 2 & 0 \\ 0 & 0 & 1 \end{pmatrix}, \qquad (2.184)$$

[8] Without this anisotropic mobility, flagella could not be used to propel microorganisms, or cilia to drive fluid flows!

where $\zeta_\| = 1/b_\|$. The anisotropy in *rotational* mobility and resistance is much stronger.

Translation-rotation coupling. Rotationally symmetric particles with fore-aft *asymmetry* (*e.g.*, egg-shaped particles, or asymmetric dumbells) will generally rotate when forced in any direction other than along its symmetry axis. Alternately, a screw-like body (rod-like, with chiral asymmetries) will rotate about the primary rod axis, in response to a force directed along the rod axis. Such translation-rotation coupling as

$$\mathbf{L} = \boldsymbol{\xi}_{\mathrm{RT}} \cdot \mathbf{V}, \tag{2.185}$$

appears as an off-diagonal block in a more general-resistance tensor:

$$\begin{pmatrix} \mathbf{F} \\ \mathbf{L} \end{pmatrix} = \begin{pmatrix} \boldsymbol{\xi}_T & \boldsymbol{\xi}_{TR} \\ \boldsymbol{\xi}_{RT} & \boldsymbol{\xi}_R \end{pmatrix} \cdot \begin{pmatrix} \mathbf{V} \\ \boldsymbol{\Omega} \end{pmatrix}, \tag{2.186}$$

so that $\boldsymbol{\xi}_{TR}$ and $\boldsymbol{\xi}_{RT}$ are 3x3 tensors that give the (drag) *force* on the particle when it rotates with angular velocity $\boldsymbol{\Omega}$, and the drag torque $\mathbf{L}$ on the particle when it translates with velocity $\mathbf{V}$. Moreover, it can be shown that the entire tensor is symmetric, implying[9]

$$\boldsymbol{\xi}_T = (\boldsymbol{\xi}_T)^T \tag{2.187}$$
$$\boldsymbol{\xi}_{TR} = (\boldsymbol{\xi}_{RT})^T \tag{2.188}$$
$$\boldsymbol{\xi}_R = (\boldsymbol{\xi}_R)^T. \tag{2.189}$$

The mobility tensor is given by the inverse of the resistance tensor:

$$\mathbf{b} = \begin{pmatrix} \mathbf{b}_T & \mathbf{b}_{TR} \\ \mathbf{b}_{RT} & \mathbf{b}_R \end{pmatrix} = \begin{pmatrix} \boldsymbol{\xi}_T & \boldsymbol{\xi}_{TR} \\ \boldsymbol{\xi}_{RT} & \boldsymbol{\xi}_R \end{pmatrix}^{-1}. \tag{2.190}$$

It is important to note that (2.190) does *not* imply that each component of the resistance tensor is given by the reciprocal of the equivalent component of the mobiity tensor. A generally-shaped particle that translates *without rotating* in the $\hat{z}$-direction experiences a drag force and torque given by $\boldsymbol{\xi} \cdot \hat{z}$, of which the $\hat{z}$-component is ζ_{zz}. By contrast, if the same particle is allowed to settle under a force $\hat{z}$, it does so with translational and rotational velocities given by $\mathbf{b} \cdot \hat{z}$, of which b_{zz} gives the $\hat{z}$ component of the velocity. Physically, the two situations are distinct.

---

[9] More detailed descriptions can be found in Kim and Karilla (1991), Leal (2007), Happel and Brenner (1983), and Guazzelli and Morris (2012).

. . . .

EXERCISES

(2.1) **Rotational mobility of a sphere**. Show that the flow field around a sphere of radius a, rotating in a viscous fluid with angular velocity Ω about the $\theta = 0$ axis is

$$v_\phi = \Omega a \sin\theta \left(\frac{a^3}{r^3}\right). \qquad (2.191)$$

Now, relate the torque $\mathbf{L}$ on the same sphere to its rotational velocity Ω

$$\mathbf{L} = \zeta_R \Omega \text{ and } \Omega = b_R \mathbf{L}, \qquad (2.192)$$

to derive the rotational resistance (or mobility) of the sphere,

$$\zeta_R^{\text{sphere}} = 8\pi\eta a^3 = (b_R^{\text{sphere}})^{-1}. \qquad (2.193)$$

Compare the decay of the flow field around a rotating sphere to that of a translating sphere. Compare how resistance (or mobility) depend on probe size a for translation vs. rotation.

(2.2) **Displacement field around an oscillating sphere.** Consider a sphere of radius a oscillating with displacement $U_0 e^{i\omega t}$ in an isotropic, incompressible elastic medium with shear modulus G. Using the elastic analog of the stream function, show that the elastic displacement field obeys $E^4\psi + \Gamma_E^2 E^2\psi = 0$, where $\Gamma_E = \omega\sqrt{\rho_m/G} = \omega/c$ is the frequency divided by the shear wave speed in the medium. Show the solution to be

$$\frac{\psi_0(r,\theta,\omega)}{U_0\sin^2\theta} = -\frac{a^3}{2r} +$$

$$\frac{3a}{2\Gamma_E^2 r}\left((1 + \Gamma_E r)e^{-\Gamma_E(r-a)} - (1 + \Gamma_E a)\right), \qquad (2.194)$$

keeing only outgoing waves $(e^{i(\omega t - \Gamma_E r)})$. Derive the displacement and pressure fields, the stress tensor, and ultimately show the force on the sphere to be

$$\mathbf{F}_0 = -6\pi G a \mathbf{U}_0 \left(1 - a\Gamma_E - \frac{a^2\Gamma_E^2}{9}\right). \qquad (2.195)$$

(2.3) **Correspondence Principle for oscillating spheres.** Show
that the Correspondence Principle can be used to derive the
stream functions for a sphere oscillating in an incompressible
elastic solid (2.194) from the solutoin in an incompressible
viscous liquid (2.93), and vice versa. Hint: You will need to
compute $\sqrt{1/i}$, for which there are two choices, only one of
which behaves well far from the sphere. Similarly, show that
the force on a sphere oscillating in an elastic medium (2.195)
can be obtained from the force on a sphere oscillating in a
viscous liquid (2.100), and vice-versa.

(2.4) **Energy balance for sphere oscillating in elastic medium.**
Using (2.195), show the force on a sphere of radius a in an
elastic medium, undergoing a general displacement $\mathbf{U}(t)$ to
be given by

$$\mathbf{F} = -6\pi\, Ga\mathbf{U} - \frac{\mathbf{V}}{c}6\pi\, Ga^2 - \frac{1}{2}M_a\ddot{\mathbf{U}}. \tag{2.196}$$

where $M_a = 4\pi a^3 \rho_m/3$ is the equivalent mass of the elastic
material occupied by the sphere.

Show that the power exerted by the sphere $P(t)$ =
$-\mathbf{F}\cdot\mathbf{V}$ on the material during an oscillatory displacement
$\mathbf{U}_0 \sin \omega t$ is

$$P(t) = \frac{6\pi\, Ga^2}{c}U_0^2\omega^2 \cos^2\omega t + \ldots \tag{2.197}$$

$$+ \left(6\pi\, Ga\omega - \frac{M_f\omega^3}{2}\right) U_0^2 \sin\omega t \cos\omega t. \tag{2.198}$$

Show that a sphere oscillating in an elastic medium exerts a
time-averaged power on the medium,

$$\bar{P} = 3\pi\rho a^2 cU_0^2\omega^2. \tag{2.199}$$

Even in a purely elastic medium, the elastic energy of a
displaced sphere is lost over time. Where does it go?

(2.5) **Correspondence Principle: Point forces in incompressible viscous and elastic media.** Evaluate Thomson's solution
(2.126) for the elastic displacement field $\mathbf{u}$ around a point
force $\mathbf{F}$, in the incompressible limit $K/G \to \infty$. Use the Correspondence Principle to replace G with $i\omega\eta$, and verify that
the result is consistent with the Stokeslet (Oseen Tensor, eqn
2.82) flow $\mathbf{v}$ due to a point force.

(2.6) **Rotational oscillations in an elastic medium.** Consider a sphere of radius a executing oscillatory rotations with strain amplitude $\Theta(t) = \Theta_0 e^{i\omega t}$ about the $\theta = 0$ axis in an isotropic elastic medium, with shear modulus G and Poisson ratio ν. Show that the displacement field is given by

$$\mathbf{u}(t) = \Theta_0 \times \mathbf{r} \left(\frac{a}{r}\right)^3 \frac{1 + i\Gamma_E r}{1 + i\Gamma_E a} e^{i(\omega t - \Gamma_E(r-a))}, \qquad (2.200)$$

where

$$\Gamma_E = \sqrt{\frac{\omega^2 \rho}{G}} = \frac{\omega}{c} \qquad (2.201)$$

and where only outgoing shear waves are kept. Show that the torque on the sphere is given by

$$L_0 = 8\pi Ga^3\Theta_0 \left(1 - \frac{a^2\Gamma_E^2}{3(1 + i\Gamma_E a)}\right), \qquad (2.202)$$

to give a rotational spring constant

$$\kappa_R = 8\pi Ga^3 \left(1 - \frac{a^2\Gamma_E^2}{3(1 + i\Gamma_E a)}\right), \qquad (2.203)$$

or resistance

$$\zeta_R^* = \frac{8\pi Ga^3}{i\omega} \left(1 - \frac{a^2\Gamma_E^2}{3(1 + i\Gamma_E a)}\right). \qquad (2.204)$$

(2.7) **Rotational oscillations in a viscous fluid.** Consider a sphere of radius a executing oscillatory rotations with angular velocity $\Omega(t) = \Omega_0 e^{i\omega t}$ about the $\theta = 0$ axis in a Newtonian liquid with viscosity μ. Show that the velocity field is given by

$$\mathbf{v}(t) = \Omega_0 \times \mathbf{r} \left(\frac{a}{r}\right)^3 \frac{1 + \Gamma r}{1 + \Gamma a} e^{-\Gamma(r-a) + i\omega t} \qquad (2.205)$$

where $\Gamma = (1 + i)/\lambda_V$, and where $\lambda_V = \sqrt{2\nu/\omega}$ is the oscillatory boundary-layer thickness. so that

$$\zeta_R^* = 8\pi\eta a^3 \left(1 - \frac{a^2\Gamma^2}{3(1 + i\Gamma a)}\right), \qquad (2.206)$$

(2.8) **Coupling mobility between two different-sized spheres.**
Consider the leading-order approximation to the coupling
mobility between two spheres of radii a_1 and a_2, located at
$r_1 = 0$ and $r_2 = d\hat{x}$, respectively. Start with the case where
forces are parallel to the line of centers ($F_i = F\hat{x}$). Given a
force $F_1 = F\hat{x}$ on sphere 1, what are the velocities V_1 and V_2
of spheres 1 and 2? Given a force $F_2 = F\hat{x}$ on sphere 2, what
are the velocities V_1 and V_2 of spheres 1 and 2? Construct
the mobility tensor

$$\begin{pmatrix} V_1 \\ V_2 \end{pmatrix} = \begin{pmatrix} b^{11} & b^{12} \\ b^{21} & b^{22} \end{pmatrix} \cdot \begin{pmatrix} F_1 \\ F_2 \end{pmatrix}, \tag{2.207}$$

for different-sized spheres. What are the two eigenmodes of
this system? What happens when $a_1 \gg a_2$?

(2.9) **Coupling mobility between two identical spheres, forced
perpendicular to line of centers.** Compute the leading-
order coupling mobility $b_{21}^{\perp}$ for two identical spheres of radius
a, separated by $d = d\hat{x}$. Verify (2.149).

(2.10) **Coupling resistance between two identical spheres.**
Now, consider the the leading-order approximation to the
coupling *resistance* between two spheres of radius a, located
at $r_1 = 0$ and $r_2 = d\hat{x}$, respectively. Start with the case where
velocities are parallel to the line of centers ($V_i = V\hat{x}$). Given
a force $V_1 = V\hat{x}$ on sphere 1, what are the forces F_1 and F_2
on spheres 1 and 2? Given a velocity $V_2 = F\hat{x}$ on sphere 2,
what are the forces F_1 and F_2 of spheres 1 and 2? Construct
the resistance tensor

$$\begin{pmatrix} F_1 \\ F_2 \end{pmatrix} = \begin{pmatrix} \xi^{11} & \xi^{12} \\ \xi^{21} & \xi^{22} \end{pmatrix} \cdot \begin{pmatrix} V_1 \\ V_2 \end{pmatrix}. \tag{2.208}$$

Invert this tensor to find the mobility tensor,

$$b = \xi^{-1}, \tag{2.209}$$

and show it agrees with (2.145).

(2.11) **Sphere near a free surface.** Section 2.6.4 computed the hydrodynamic mobility of a sphere of radius a located a distance
h from a planar, no-slip wall (*e.g.*, a glass slide). Now, compute the hydrodynamic mobility of a sphere in the vicinity of
a free surface (*e.g.*, a *liquid-gas interface*), where a no-stress
condition ($\tau_{xz} = 0$) holds at the wall. Show that "wall flow"

for a sphere forced towards the wall can be expressed simply by a Stokeslet,

$$u_w^\perp = -\mathbf{G}^{St}(\mathbf{r} + h\hat{z}) \cdot F^\perp \hat{z} \tag{2.210}$$

$$u_w^\| = \mathbf{G}^{St}(\mathbf{r} + h\hat{z}) \cdot F^\| \hat{x} \tag{2.211}$$

located behind the wall at the image location $\mathbf{r}_i = -h\hat{z}$, similarly directed for parallel forces $\mathbf{F}_\|$ and oppositely directed for perpendicular forces $\mathbf{F}_\perp$. That is, show that

$$u^\perp = \left[\mathbf{G}^{St}(\mathbf{r} - h\hat{z}) - \mathbf{G}^{St}(\mathbf{r} + h\hat{z})\right] \cdot F^\perp \tag{2.212}$$

$$u^\| = \left[\mathbf{G}^{St}(\mathbf{r} - h\hat{z}) + \mathbf{G}^{St}(\mathbf{r} + h\hat{z})\right] \cdot F^\| \tag{2.213}$$

obeys the no-flux and no-stress conditions

$$\left.\frac{\partial u_x}{\partial z}\right|_{z=0} = 0 \tag{2.214}$$

$$\left.\frac{\partial u_y}{\partial z}\right|_{z=0} = 0 \tag{2.215}$$

$$u_z(x, y, z = 0) = 0. \tag{2.216}$$

Given this, show that the leading-order correction to the sphere's mobility is given by

$$b^\| = \frac{1}{6\pi \eta a}\left(1 + \frac{3a}{8h}\right) \tag{2.217}$$

$$b^\perp = \frac{1}{6\pi \eta a}\left(1 - \frac{3a}{4h}\right). \tag{2.218}$$

3

Passive microrheology

Passive microrheology is distinct from other micro- and macro-rheological measurements, in that it relies on the inherent thermal motion of probe particles that are dispersed within the viscoelastic material of interest. Random thermal forces displace the particles, and the statistics of their subsequent motion encode the surrounding material rheology. We will see in later chapters that this thermal motion can be measured by a number of experimental techniques, including microscopy and light scattering. For now, we will focus on the theoretical basis of passive microrheology because this analysis leads to insight into its strengths and a few important limitations.

The Generalized Stokes–Einstein Relation (GSER) is the principal defining equation of passive microrheology. It is a physical relation between the thermal motion of probe particles and the material rheology. Specifically, it relates the observable *displacement* of the probe particles to the surrounding material's rheological response.

The derivation of the GSER consists of two important components: First is the Einstein relation, which states that the thermally fluctuating motion of probe particles is related to the resistance imposed on the probe by the surrounding material. The second component is the generalized Stokes drag (Chapter 2), which is used to calculate the stresses exerted by the material on the probe. Both the Einstein relation and the Stokes equation make assumptions regarding the material that warrant explicit discussion, since these impose limitations on the samples that can be measured using passive microrheology.

We begin this chapter by discussing the Langevin equation, the equation of motion from which the GSER is derived. After deriving the Stokes–Einstein relation and the GSER, we discuss the interpretation of passive microrheology experiments and its operating regime.

3.1 The Langevin equation

A discussion of the Langevin equation precedes our detailed development of the Stokes–Einstein relationship in Section 3.2. Our primary interest is to develop the equation of motion for probe particles and

Microrheology. Eric M. Furst and Todd M. Squires, Oxford University Press (2017).
© Eric M. Furst and Todd M. Squires. DOI 10.1093/oso/9780199655205.001.0001

understanding the contributions of the random thermal force and the dissipative forces.

Consider a tracer particle suspended in a viscoelastic medium. When a force **f** is exerted on the particle, we expect it to move, subject to the resistance of the surrounding material. The equation of motion—Newton's second law—is[1]

$$M_p\dot{\mathbf{V}}(t) = \mathbf{f} - \int_{-\infty}^{t} \zeta(t - t')\mathbf{V}(t')dt', \tag{3.1}$$

where $M_p = (4\pi/3)a^3\rho_p$ is the particle mass.

The second term in eqn 3.1 reflects the resistance exerted on the particle by the surrounding material, written as a convolution of the instantaneous velocity $\mathbf{V}(t)$ with the microscopic resistance $\zeta(t)$. We will derive this function in Section 3.5 when we discuss the Stokes component in detail. The resistance function accounts for both the viscous and elastic stresses exerted on the probe. Obviously, a particle suspended in a viscous medium will stop moving in the absence of an applied force **f**; the velocity must decay eventually to zero.

Before we proceed, note the limits of the resistance function integral in eqn 3.1. Specifying the lower limit of integration as $t = -\infty$, effectively states that the particle is in thermodynamic equilibrium at $t = 0$. Consequently, the resistance function must obey

$$\zeta(t) = 0, \ t < 0. \tag{3.2}$$

to ensure that causality is not violated. The particle cannot be subjected to resistance forces generated by future velocities!

For now, consider the resistance function for a purely viscous fluid, accelerating slowly enough that inertial forces may be neglected. The resistance is solely due to the viscous drag force, which depends only on the instantaneous velocity. The memory function in this case is

$$\zeta(t) = \zeta_0\delta(t), \tag{3.3}$$

where ζ_0 is a constant and $\delta(t)$ is the Dirac delta function, and eqn 3.1 becomes

$$M_p\dot{\mathbf{V}}(t) = \mathbf{f} - \zeta_0\mathbf{V}(t). \tag{3.4}$$

In Section 2.5.2, the Stokes drag on a sphere translating in a Newtonian liquid of viscosity η was shown to be $\zeta = 6\pi a\eta$ for no-slip (solid) spheres, and $\zeta = 4\pi a\eta$ for perfectly slipping spheres (*e.g.*, bubbles). Note, however, that the quasi-steady Stokes drag eqn 3.4 is only valid

[1] Kubo *et al.* (1991) and Zwanzig and Bixon (1970) are excellent references for their detailed treatments of the Langevin equation and Brownian motion.

on time scales greater than the viscous-relaxation time scale $\rho_f a^2/\eta$, where ρ_f is the fluid density, as shown in section 2.5.3.

Under a constant imposed force such that the particle is not accelerating, the simplified equation of motion can be solved for the velocity by a straightforward rearrangement,[2]

$$\mathbf{V}(t) = \mathbf{f}/\zeta_0. \tag{3.5}$$

In fact, under most circumstances, the particle inertia is so small that the first term in eqn 3.1 is negligible. Only on very short time scales ($t \sim M_p/\zeta_0$ in the viscous fluid) do we need to consider particle (and fluid) inertia. We can demonstrate this by considering a time-dependent force $\mathbf{f}(t)$. Equation 3.1 can be solved formally to give the velocity

$$\mathbf{V}(t) = \mathbf{V}(0)e^{-\zeta_0 t/M_p} + \frac{1}{M_p}\int_0^t e^{-\frac{\zeta_0}{M_p}(t-t')}\mathbf{f}(t')dt'. \tag{3.6}$$

An impulsive force,

$$\mathbf{f}(t) = \frac{M_p}{\zeta_0}\mathbf{f}_0\delta(t), \tag{3.7}$$

exerted on a particle initially at rest, $\mathbf{v}(0) = \mathbf{0}$, drives the particle with velocity

$$\mathbf{V}(t) = (\mathbf{f}_0/\zeta)e^{-\zeta_0 t/M_p}. \tag{3.8}$$

The particle moves initially with velocity f_0/ζ_0, which then decreases exponentially until the particle comes to rest, over a relaxation time scale $\tau = M_p/\zeta_0$. For a one micrometer diameter particle dispersed in water, the relaxation time scale is quite small, $M_p/\zeta_0 \sim 10^{-8}$ s—on the order of only a hundredth of a microsecond!

The **Langevin equation** is no more than eqn 3.4 with one peculiarity—that the force is a random, fluctuating force $\mathbf{f}_B$ that results from the thermal motion of the surrounding molecules. In a similar way, inserting a random force $\mathbf{f}_B$ in eqn 3.1, which does not assume a form for $\zeta(t)$, gives rise to the **Generalized**-Langevin equation. The random force is assumed to have random direction and magnitude (over sufficiently long time scales), so that its time average is zero,

[2] See Exercise 1 for the response in an elastic solid.

$$\langle \mathbf{f}_B(t)\rangle = \mathbf{0}. \tag{3.9}$$

For purely viscous fluids that obey eqn 3.4, the random force is also assumed to be *uncorrelated*[3] with velocity,

$$\langle \mathbf{f}_B(t) \cdot \mathbf{V}(t') \rangle = 0. \tag{3.10}$$

The form and time-dependence of the random force $\mathbf{f}_B(t)$ are determined by the details of the collisions of the particle with the surrounding fluid. For viscous fluids obeying eqn 3.4, the random force is only correlated over molecular collision time scales—generally much shorter than time scales for particle motions—and is therefore generally approximated by a delta function,

$$\langle \mathbf{f}_B(t) \cdot \mathbf{f}_B(t') \rangle = F_0 \delta(t - t'), \tag{3.11}$$

where the constant F_0 is proportional to the mean-squared magnitude of the Brownian force. Thermal forces within complex fluids that obey eqn 3.1 exhibit a more complicated time correlation, as discussed by (Kubo *et al.*, 1991) and in Section 3.3.

In this simplest case, the Fourier Transform[4] of eqn 3.11,

$$\langle \tilde{\mathbf{f}}_B(\omega) \cdot \mathbf{f}_B(t') \rangle = \int_{-\infty}^{\infty} e^{-i\omega t} \langle \mathbf{f}_B(t) \cdot \mathbf{f}_B(t') \rangle dt = F_0 e^{-i\omega t'}, \tag{3.12}$$

so that Fourier transforming over t' gives

$$\langle \tilde{\mathbf{f}}_B(\omega) \cdot \tilde{\mathbf{f}}_B(\omega') \rangle = 2\pi F_0 \delta(\omega + \omega'), \tag{3.13}$$

or,

$$\langle \tilde{\mathbf{f}}_B(\omega) \cdot \tilde{\mathbf{f}}_B(-\omega') \rangle = \langle \tilde{\mathbf{f}}_B(\omega) \cdot [\tilde{\mathbf{f}}_B(\omega')]^* \rangle = 2\pi F_0 \delta(\omega - \omega'). \tag{3.14}$$

The power spectral density expresses how much of the distribution is contained within $d\omega$ of a given frequency ω,

$$S_f(\omega) = \langle |\tilde{\mathbf{f}}_B(\omega)|^2 \rangle = \int_{\omega-d\omega/2}^{\omega+d\omega/2} \langle \tilde{\mathbf{f}}_B(\omega) \cdot [\tilde{\mathbf{f}}_B(\omega')]^* \rangle d\omega' \tag{3.15}$$

which in this case is independent of frequency

$$\langle |\tilde{\mathbf{f}}_B(\omega)|^2 \rangle = 2\pi F_0, \tag{3.16}$$

a characteristic of *white noise*. The magnitude of the Brownian force is determined by the requirements of thermal equilibrium, and will now be discussed in Section 3.2.

[3] See Chapter 5 for a discussion of time-correlation functions.

[4] The Fourier Transform is discussed in Appendix A.1.

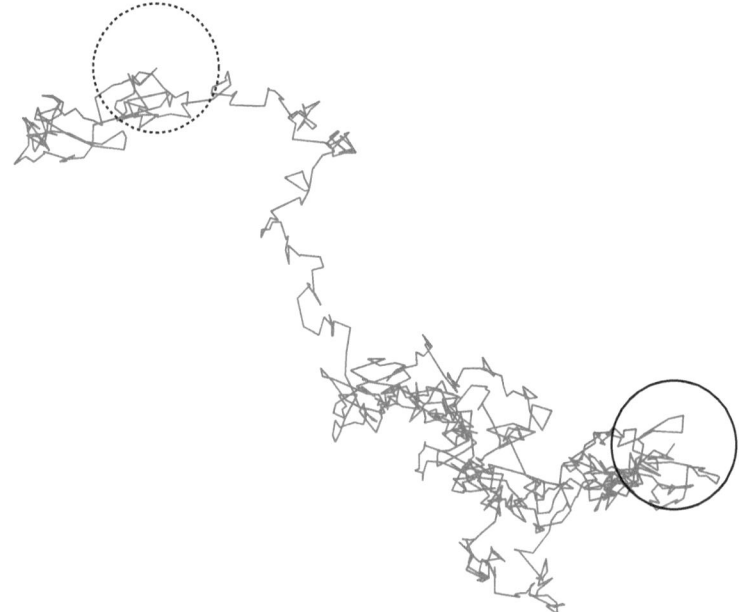

Fig. 3.1 *The trajectory of a Brownian particle is a discontinuous, random walk. The dashed circle is the particle's starting point and the solid circle is its end point.*

Because the Langevin equation is driven by a random, fluctuating force f_B—owing to the stochastic, thermal motion of the surrounding molecules—solutions to the Langevin equation are non-deterministic. Particles that obey the Langevin equation exhibit random walks, an example of which is shown in Fig. 3.1.

As a *stochastic* equation, solutions to the Langevin equation will have the form of statistical (ensemble) averages over many realizations of probe particle trajectories. Before we discuss the general solution of the (generalized) Langevin equation for a viscoelastic material, let us first consider the Brownian motion of particles in a viscous Newtonian fluid.

3.2 Brownian motion

The thermal, or Brownian, force is well known through the perpetual random motion of small particles and, historically, provided direct evidence of the molecular motion inherent in the microscopic understanding of the nature of matter (Maiocchi, 1990; Bigg, 2008).[5] As Einstein demonstrated, the thermal force is also related to the frictional drag force, which arises due to molecules impacting the moving particle (Einstein, 1905). These molecular collisions produce both tangential (shear) forces and normal forces that slow the motion of

[5] Named for the botanist and talented microscopist Robert Brown (1773–1858), who famously described the perpetual random motion of pollen organelles (amyloplasts and spherosomes) and finely ground inorganic particles suspended in water. Brown made his observations in 1827. His account was published in 1829, nearly a century before Einstein published his molecular theory (Brown, 1828).

the particle, which gives rise to the resistance ζ_0 used in the previous section. The relationship between the random thermal force and frictional drag force is a manifestation of the *fluctuation-dissipation theorem* (Kubo, 1966; Kubo *et al.*, 1991).

   Solutions to the Langevin equation are not deterministic, because the force $\mathbf{f}_B$ acting on the particle is random and fluctuating. Two approaches are commonly used to solve the Langevin equation, each with distinct advantages and disadvantages. We will explicitly discuss each for the simplest system introduced in the Langevin equation (3.4) for quasi-steady Stokes flow, tracking motion in only one direction (*e.g.*, the *x*-direction) for simplicity:

$$M_p \dot{V}(t) = f_B - \zeta_0 V. \tag{3.17}$$

We warn, however, that this approach omits the inertia of the fluid, and therefore makes incorrect predictions for any material whose density is not substantially smaller than that of the probe. Nonetheless, this simpler system is clearer pedagogically, and illustrates the conceptual and logical strategies employed to solve the Langevin equation, without many of the mathematical difficulties or subtleties that arise for more general (non-Newtonian, or inertial) materials. Once we have this basic framework in place, we will then describe the proper generalization for more general systems—both non-Newtonian and inertial.

3.2.1   Laplace Transform solutions

We begin by taking the *Laplace* Transform of the Langevin equation (3.4) for quasi-steady Stokes flow, giving

$$sM_p \hat{V}(s) - M_p V(0) = \hat{f}_B(s) - \zeta_0 \hat{V}(s). \tag{3.18}$$

Solving for $\hat{V}(s)$ gives

$$\hat{V}(s) = \frac{\hat{f}_B(s) + M_p V(0)}{\zeta_0 + sM_p}. \tag{3.19}$$

The Laplace Transform naturally introduces the initial velocity $V(0)$, which makes the equipartition theorem relatively easy to invoke. Multiplying eqn 3.19 by the initial velocity $V(0)$, then taking the ensemble average, gives

$$\langle \hat{V}(s) V(0) \rangle = \frac{\langle \hat{f}_B(s) V(0) \rangle + M_p \langle V(0) V(0) \rangle}{\zeta_0 + sM_p}. \tag{3.20}$$

Because the Brownian force has zero average (eqn 3.9) and is assumed to be uncorrelated with particle velocity (eqn 3.10), the first term vanishes, leaving

$$\langle \hat{V}(s) V(0) \rangle = \frac{M_p \langle V(0) V(0) \rangle}{\zeta_0 + s M_p}. \tag{3.21}$$

The equipartition theorem governs the kinetic energy of particles in thermal equilibrium with their surroundings, stating that each independent degree of freedom has energy $\frac{1}{2} k_B T$. Since we are tracking only one-dimensional displacement in this particular example, this yields

$$\frac{1}{2} M_p \langle V(0) V(0) \rangle = \frac{1}{2} k_B T, \tag{3.22}$$

so that

$$\langle \hat{V}(s) V(0) \rangle = \frac{k_B T}{\zeta_0 + s M_p}. \tag{3.23}$$

Taking the inverse transform gives the velocity autocorrelation function (VAC),

$$\langle V(t) V(0) \rangle = \frac{k_B T}{M_p} e^{-\zeta_0 |t| / M_p}. \tag{3.24}$$

Although the mean velocity must be zero for a particle experiencing a random fluctuating force, $\langle V \rangle = 0$, eqn 3.24 shows that the corresponding velocity fluctuations decay on the time scale M_p / ζ_0 identified previously, for the particle's deterministic response to an impulsive force.

Strictly speaking, eqn 3.24 only holds for probes whose density signifiantly exceeds that of the fluid. Much more common in microrheology is when probe and medium densities are of the same order, in which case fluid inertia decays on the same time scale as the particle inertia ($\sim a^2 / \nu$, where $\nu = \eta / \rho_f$ is the kinematic viscosity). In that case, the assumption of a constant ζ_0 is therefore flawed, and will be treated properly in Section 3.4.

3.2.2   Fourier Transform solutions

An alternative solution strategy (Kubo *et al.*, 1991; Indei *et al.*, 2012*b*) is to decompose $V(t)$ via a Fourier Transform,

$$V(t) = \frac{1}{2\pi} \int_{-\infty}^{\infty} \tilde{V}(\omega) e^{i\omega t} d\omega, \tag{3.25}$$

which gives

$$i\omega M_p \tilde{V}(\omega) = \tilde{f}_B - \zeta_0 \tilde{V}, \tag{3.26}$$

and therefore

$$\tilde{V}(\omega) = \frac{\tilde{f}_B(\omega)}{\zeta_0 + i\omega M_p}. \tag{3.27}$$

An advantage (but also disadvantage) to the Fourier Transform approach is that it does not single out any time $t = 0$, even if that time is ultimately arbitrary.[6]

To find the velocity autocorrelation function, we compute the ensemble average

$$\langle \tilde{V}(\omega)\tilde{V}^*(\omega') \rangle = \frac{\langle \tilde{f}_B(\omega)\tilde{f}_B^*(\omega') \rangle}{(\zeta_0 + i\omega M_p)(\zeta_0 - i\omega' M_p)}. \tag{3.29}$$

Using eqn 3.14 for the statistics of thermal force exerted on a probe within a quasi-steady, Newtonian liquid gives

$$\langle \tilde{V}(\omega)\tilde{V}^*(\omega') \rangle = 2\pi F_0 \frac{\delta(\omega - \omega')}{(\zeta_0 + i\omega M_p)(\zeta_0 - i\omega' M_p)}. \tag{3.30}$$

Inverse Fourier transforming over ω, using a time $t + \tau$, gives

$$\langle V(t + \tau)\tilde{V}^*(\omega') \rangle = F_0 \frac{e^{i\omega'(t+\tau)}}{\zeta_0^2 + \omega'^2 M_p^2}. \tag{3.31}$$

Taking the complex conjugate and inverse transforming over ω' gives

$$\langle V(t + \tau)V(t) \rangle = \frac{F_0}{2\pi} \int_{-\infty}^{\infty} \frac{e^{-i\omega'\tau}}{\zeta_0^2 + \omega'^2 M_p^2} d\omega', \tag{3.32}$$

which can be integrated (*e.g.*, using residue calculus) to yield

$$\langle V(t + \tau)V(t) \rangle = \frac{F_0}{M_p \zeta_0} e^{-\zeta_0 |\tau|/M_p}. \tag{3.33}$$

Notably, the velocity autocorrelation function depends only on the lag time τ, not on t; which follows from the fact that the thermal force was assumed to be stationary.

The final step is to invoke known properties of thermal equilibrium in order to determine the magnitude of F_0. Computing the VAC at

[6] There is nothing unique about the initial velocity $V(0)$ in this derivation. Because the equilibrium ensemble average is stationary, and the classical-mechanical equations of motion are time reversible, the correlations between dynamical variables like the velocity should depend only on the separation between times, and not the absolute value of time (McQuarrie, 2000; Chandler, 1987). Therefore, we may write

$$\langle V(t')V(t'') \rangle = \langle V(t' - t'')V(0) \rangle$$
$$= \langle V(t'' - t')V(0) \rangle. \tag{3.28}$$

zero time lag allows the equipartition theorem eqn 3.22 to be invoked, so that

$$\langle V(t) V(t) \rangle = \frac{k_B T}{M_p} = \frac{F_0}{M_p \zeta_0},$$ (3.34)

meaning

$$F_0 = k_B T \zeta_0,$$ (3.35)

so that

$$\langle V(t+\tau) V(t) \rangle = \frac{k_B T}{M_p} e^{-\zeta_0 |\tau|/M_p},$$ (3.36)

in agreement with eqn 3.24, found using Laplace Transforms.

Note that eqn 3.35 reveals that the random, fluctuating force exerted on a particle by the collective action of individual, thermally agitated molecules actually "knows" about the determinstic resistance on that particle being forced to move through its environment. Why and how would that information be transmitted to individual molecules, each crashing into its neighbors (and the probe)? The connection is subtle and profound—and reflects the consequences of the fluctuation-dissipation theorem. In short, the forces exerted on the particle by the surrounding molecules perform work on the particle as they do so. In order for the probe to remain in thermal equilibrium with its surroundings, that energy must be dissipated back into the medium, to maintain a net energy balance. The latter (dissipative) step involves the (deterministic) drag resistance ζ that dissipates the energy; and the former introduces $k_B T$. This topic is explored in Exercise 3.3.

It should come as no surprise that these two methods of solving the Langevin equation give the same answer. It is worth noting, however, how the two approaches differ. By its very nature, the Laplace Transform only incorporates positive times $t > 0$, and therefore the initial condition $V(0)$ enters the Laplace-Transformed Langevin equation 3.18 explicitly. Computing the dot product with $V(0)$, and ensemble averaging, naturally caused the Brownian force term to vanish from the equation, and left the (ensemble-averaged) initial kinetic energy alone in the equation. One need not compute or even consider the magnitude of the Brownian forces explicitly in this approach, since the ensemble-averaged kinetic energy appears directly, and can immediately be related to $k_B T$ via the equipartition theorem.

By contrast, the Fourier-Transform approach incorporates *all* times $-\infty < t < \infty$ in its analysis, and therefore no "initial condition" $V(0)$ is singled out, or ever appears, in its solution. Instead, one computes $\langle \tilde{V}(\omega) \tilde{V}^*(\omega') \rangle$ directly, then inverting both Fourier Transforms.

In so doing, the autocorrelation of the (Fourier-Transformed) Brownian forces $\langle \tilde{\mathbf{f}}_B(\omega)\tilde{\mathbf{f}}_B^*(\omega')\rangle$ is introduced. The statistics of $\mathbf{f}_B(t)$ must therefore be employed directly in the Fourier Transform solution.

The *unilateral*, or *one-sided*, Fourier Transform,

$$\mathscr{F}_u[V(t)] = \int_0^\infty e^{-i\omega t} V(t)\,dt, \tag{3.37}$$

is sometimes used to solve these problems. Like the Laplace Transform, the unilateral Fourier Transform singles out a particular time $t = 0$ as an "initial condition," and only incorporates times $t > 0$ in its analysis. The unilateral Fourier Transform and Laplace Transform can be related via *analytic continuation*, using techniques from Complex Analysis, so long as the transformed functions meet criteria common for microrheology conditions. In particular, the probe response must be causal, meaning that a probe can not respond to a force that has not yet occurred. In practice, analytic continuation involves replacing s in the Laplace Transform with $i\omega$ for the unilateral Fourier Transform

$$\hat{V}(s \to i\omega) = \tilde{V}(\omega) \tag{3.38}$$
$$\tilde{V}(\omega \to -is) = \hat{V}(s). \tag{3.39}$$

See Appendix A.2 for a discussion.

3.2.3 Relating VAC to MSD

We have related the *statistical* properties of probe velocities to the (deterministic) probe resistance ζ_0 for a probe moving in a quasi-steady viscous fluid. In practice, however, it is difficult to measure velocity autocorrelations. A little additional analysis, however, relates the velocity autocorrelation function to quantities that are more amenable to measurement.

For example, particles effectively move via random walks over long time scales, with a diffusivity that can be determined from the velocity autocorrelation function via

$$D_0 = \int_0^\infty \langle V(t)V(0)\rangle\,dt = \lim_{s\to 0} \mathscr{L}\{\langle V(t)V(0)\rangle\} \tag{3.40}$$

$$= \lim_{s\to 0} \frac{k_B T}{\zeta_0 + M_p s}, \tag{3.41}$$

so that

$$D_0 = \frac{k_B T}{\zeta_0} \equiv \frac{k_B T}{6\pi \eta a}, \tag{3.42}$$

as expected.

More generally, it is far easier in light scattering, particle tracking, or other experiments to measure the statistical *displacement* of the particles over time using the relation

$$\langle \hat{V}(s) V(0) \rangle = \frac{1}{2} s^2 \langle \Delta \hat{x}^2(s) \rangle. \tag{3.43}$$

If eqn 3.43 feels like a slight of hand, the reader is encouraged to derive this useful relation in Exercise 3.2. Equation 3.23 can thus be written in terms of the Laplace Transform of the mean-squared displacement,

$$\langle \Delta \hat{x}^2(s) \rangle = \frac{2k_B T}{s^2 (\zeta_0 + s M_p)}. \tag{3.44}$$

Equation 3.43 has an analog in Fourier space, when the *unilateral* Fourier Transform (eqn 3.37) is computed:

$$\langle \tilde{V}(\omega) V(0) \rangle_u = \frac{1}{2} (i\omega)^2 \langle \Delta \tilde{x}^2(\omega) \rangle_u. \tag{3.45}$$

For a particle diffusing in a quasi-steady Newtonian fluid, the Laplace Transform can be inverted explicitly, giving

$$\langle \Delta x^2(t) \rangle = 2D_0 t - 2D_0 \frac{M_p}{\zeta_0} \left(1 - e^{-\zeta_0 t / M_p} \right), \tag{3.46}$$

where D_0 is given by eqn 3.42. Indeed, the MSD grows linearly in time for times $t \gg M_p/\zeta_0$, with diffusivity D_0. At very short times ($t \ll M_p/\zeta_0$), by contrast, the MSD evolves via

$$\langle \Delta x^2(t \ll M_p/\zeta_0) \rangle \sim \frac{k_B T}{M_p} t^2, \tag{3.47}$$

reflecting ballistic probe motion with thermal velocity $V = \sqrt{k_B T / M_p}$. Notably, the fluid viscosity has no impact on the MSD over these extremely short times; and instead determines the time scale M_p/ζ_0 beyond which fluid rheology begins to dominate.

To successfully measure material rheology (here, fluid viscosity), measurements should focus on sufficiently low Laplace frequencies ($s \ll M_p/\zeta_0$). Under these conditions, eqn 3.44 may be simplified by neglecting the inertia of the probe ($ms \ll \zeta_0$), giving an approximate form

$$\langle \Delta \tilde{x}^2(s) \rangle \approx \frac{2k_B T}{s^2 \zeta_0}, \tag{3.48}$$

with inverse transform

$$\langle \Delta x^2(t) \rangle \approx \frac{2k_B T}{\zeta_0} t \qquad (3.49)$$

for times $t > 0$. This is the famous Einstein equation, which is often written in terms of the particle diffusivity

$$\langle \Delta x^2(t) \rangle = 2D_0 t, \qquad (3.50)$$

where D_0 is given by eqn 3.42. Using eqn 2.75 for the steady-translational resistance ζ_0 of a sphere in a viscous fluid gives the Stokes–Einstein formula for the particle diffusivity,

$$D_0 = \frac{k_B T}{6\pi \eta a}. \qquad (3.51)$$

This calculation may be generalized to track displacements in two or three dimensions. Velocity autocorrelation functions, and mean square displacements, can be computed for each of the three dimensions in exactly the same way. One can track (and compute) each individually; alternatively, one can pose and solve the vector equivalent of eqn 3.17, and form the scalar product $\langle \mathbf{V}(t) \cdot \mathbf{V}(0) \rangle$ for the VAC and MSD. In that case, eqn 3.44 reads

$$\langle \Delta \hat{x}^2(s) \rangle = \frac{2 \mathbb{D} k_B T}{s^2(\zeta_0 + sM_p)}, \qquad (3.52)$$

where $\mathbb{D}$ is the number of dimensions tracked that contribute to the MSD:

$$\langle \Delta \hat{\mathbf{r}}^2(s) \rangle_{\mathbb{D}=2} = \langle \Delta \hat{x}^2(s) + \Delta \hat{y}^2(s) \rangle \qquad (3.53)$$

$$\langle \Delta \hat{\mathbf{r}}^2(s) \rangle_{\mathbb{D}=3} = \langle \Delta \hat{x}^2(s) + \Delta \hat{y}^2(s) + \Delta \hat{z}^2(s) \rangle. \qquad (3.54)$$

The calculation described made several restrictive assumptions: That the fluid and particle inertia were negligible, and that the resistance of the particle in the fluid has no "memory," meaning $\zeta(t - t') = \zeta_0 \delta(t-t')$. The latter assumption does not hold for the viscoelastic materials of interest to microrheologists, and the assumptions regarding particle inertia may or may not hold, depending on experimental conditions. In what follows, we will relax all of these assumptions to derive the *Generalized Stokes–Einstein Relation*, which is central to the entire endeavor. In fact, the core strategy and reasoning used in this simpler derivation will hold, with only minor modifications, for the more general case. We will start with the Generalized Einstein Relation, which represents one component (and assumption) of the GSER.

3.3 The Generalized Einstein Relation

The Einstein Relation previously derived relates a deterministic transport coefficient (the diffusivity) to the absolute temperature, which describes the stochastic fluctuations inherent in thermodynamic equilibrium. We have thus far limited our derivation to quasi-steady motion in purely viscous fluids. Here, we derive the Generalized Einstein Relation (GER) for more general viscoelastic fluids and solids. We thus assume no specific form for the resistance or memory function $\zeta(t - t')$, other than causality.

The derivation of the GER follows the approach taken in Section 3.2 for quasi-steady viscous fluids, now employing the Generalized Langevin Equation (Mason and Weitz, 1995; Kubo, 1966; Zwanzig and Bixon, 1970)

$$M_p \dot{\mathbf{V}}(t) = \mathbf{f}_B - \int_{-\infty}^{t} \zeta(t - t')\mathbf{V}(t')dt'. \tag{3.55}$$

The viscoelastic (and inertial) response properties of the medium are contained within $\zeta(t - t')$. The lower limit of integration, $-\infty$, represents the fact that the force exerted by the medium on the particle depends on the particle's past velocity history. Any time may be identified as an initial time ("$t = 0$"), since the system is in equilibrium.

3.3.1 Fourier Transform

Fourier-Transforming eqn 3.55 and using the convolution theorem gives

$$i\omega M_p \tilde{\mathbf{V}} = \tilde{\mathbf{f}}_B - \tilde{\zeta}\tilde{\mathbf{V}}, \tag{3.56}$$

which can be solved via

$$\tilde{\mathbf{V}}(\omega) = \frac{\tilde{\mathbf{f}}_B(\omega)}{\tilde{\zeta}(\omega) + i\omega M_p}. \tag{3.57}$$

Computing the scalar product with $\tilde{\mathbf{V}}^*(\omega')$ gives

$$\langle \tilde{\mathbf{V}}(\omega) \cdot \tilde{\mathbf{V}}^*(\omega') \rangle = \frac{\langle \tilde{\mathbf{f}}_B(\omega) \cdot \tilde{\mathbf{f}}_B^*(\omega') \rangle}{(\tilde{\zeta}(\omega) + i\omega M_p)(\tilde{\zeta}^*(\omega') - i\omega' M_p)}, \tag{3.58}$$

so that if the statistics of the random forcing $\mathbf{f}_B$ are known, the statistics of $\mathbf{V}$ can be determined.

Both viscoelasticity and inertia impart "memory" to the material, and therefore to the probe response—as evident from the convolution in eqn 3.55. The Brownian force in a quasi-steady Newtonian fluid (eqn 3.11) was proportional to the instantaneous probe resistance ζ_0 and had delta-function time correlation. Likewise, the time correlation of the Brownian force in a complex or inertial medium is proportional to the probe resistance, although it is not delta-correlated in time (Kubo *et al.*, 1990),

$$\langle \mathbf{f}_B(t) \cdot \mathbf{f}_B(t') \rangle = 2\mathbb{D}k_B T \zeta(|t - t'|). \tag{3.59}$$

The absolute value in eqn 3.59 reflects the fact that the autocorrelation function must be even in time—since either force may appear "first" in the averaging product—whereas causality requires $\zeta(t - t')$ to be zero for all $t' > t$. Here, we will take $\mathbb{D} = 3$ for simplicity, meaning $\mathbf{f}_B$ and $\mathbf{V}$ are three-dimensional vectors. We will give results for general dimensions after deriving the key results.

Fourier Transforming over both t and t' gives

$$\langle \tilde{\mathbf{f}}_B(\omega) \cdot \tilde{\mathbf{f}}_B(\omega') \rangle = 6k_B T \int e^{-i\omega(t-t')} e^{-i(\omega+\omega')t'} \zeta(|t - t'|) \, dt \, dt' \tag{3.60}$$

$$= 6k_B T \mathrm{Re}\left[\tilde{\zeta}(\omega)\right] \int e^{-i(\omega+\omega')t'} \, dt', \tag{3.61}$$

corresponding to

$$\langle \tilde{\mathbf{f}}_B(\omega) \cdot \tilde{\mathbf{f}}_B^*(\omega') \rangle = 12\pi k_B T \mathrm{Re}\left[\tilde{\zeta}(\omega)\right] \delta(\omega - \omega'). \tag{3.62}$$

The Brownian noise is not "white" but depends on frequency in the same way that the probe resistance $\tilde{\zeta}(\omega)$ does.

With this result, eqn 3.58 becomes

$$\langle \tilde{\mathbf{V}}(\omega) \cdot \tilde{\mathbf{V}}^*(\omega') \rangle = 12\pi k_B T \frac{\mathrm{Re}\left[\tilde{\zeta}(\omega)\right] \delta(\omega - \omega')}{|\tilde{\zeta}(\omega)|^2 + \omega^2 M_p^2}. \tag{3.63}$$

Inverting the Fourier Transform over ω', at a time $t' = 0$, gives

$$\langle \tilde{\mathbf{V}}(\omega) \cdot \mathbf{V}(0) \rangle = 6k_B T \frac{\mathrm{Re}\left[\tilde{\zeta}(\omega)\right]}{|\tilde{\zeta}(\omega)|^2 + \omega^2 M_p^2}, \tag{3.64}$$

so that

$$\langle \mathbf{V}(t) \cdot \mathbf{V}(0) \rangle = \frac{3k_B T}{2\pi} \int_{-\infty}^{\infty} \frac{e^{i\omega t}}{\tilde{\zeta}(\omega) + i\omega M_p} + \frac{e^{i\omega t}}{\tilde{\zeta}^*(\omega) - i\omega M_p} \, d\omega. \tag{3.65}$$

The two terms are non-zero for $t < 0$ and $t > 0$, respectively, because causality requires $\zeta(t < 0) = 0$, which in turn requires that ζ be analytic in the lower-half plane. The result for $t > 0$ is thus

$$\langle \mathbf{V}(t > 0) \cdot \mathbf{V}(0) \rangle = \frac{3k_B T}{2\pi} \int_{-\infty}^{\infty} \frac{e^{i\omega t}}{\tilde{\zeta}(\omega) + i\omega M_p} d\omega, \qquad (3.66)$$

whereas for $t < 0$, integration over the first term vanishes to leave

$$\langle \mathbf{V}(t < 0) \cdot \mathbf{V}(0) \rangle = \frac{3k_B T}{2\pi} \int_{-\infty}^{\infty} \frac{e^{i\omega t}}{\tilde{\zeta}^*(\omega) + i\omega M_p} d\omega. \qquad (3.67)$$

The $t < 0$ result can can be put in the same form as the $t > 0$ result (eqn 3.66) under the coordinate substitution $\omega \to -\omega'$,

$$\langle \mathbf{V}(t < 0) \cdot \mathbf{V}(0) \rangle = \frac{3k_B T}{2\pi} \int_{-\infty}^{\infty} \frac{e^{i\omega' |t|}}{\tilde{\zeta}(\omega') + i\omega' M_p} d\omega'. \qquad (3.68)$$

In fact, both can be represented via

$$\langle \mathbf{V}(t) \cdot \mathbf{V}(0) \rangle = \frac{3k_B T}{2\pi} \int_{-\infty}^{\infty} \frac{e^{i\omega |t|}}{\tilde{\zeta}(\omega) + i\omega M_p} d\omega, \qquad (3.69)$$

as one might expect from the fact that the VAC is an even function of time. When tracking dislacements in $\mathbb{D}$ dimensions, this expression becomes

$$\langle \mathbf{V}(t) \cdot \mathbf{V}(0) \rangle = \frac{\mathbb{D}k_B T}{2\pi} \int_{-\infty}^{\infty} \frac{e^{i\omega |t|}}{\tilde{\zeta}(\omega) + i\omega M_p} d\omega. \qquad (3.70)$$

There are thus two ways to express the VAC in Fourier space. The *bilateral* Fourier Transform, where the time integration is performed over $-\infty < t < \infty$, is represented by eqn 3.64. Because $\langle \mathbf{V}(t) \cdot \mathbf{V}(0) \rangle$ is even in time, however, the unilateral Fourier Transform (eqn 3.37), which integrates only over positive times $0 \le t \le \infty$, contains identical information, and gives

$$\langle \tilde{\mathbf{V}}(\omega) \cdot \mathbf{V}(0) \rangle_u = \frac{3k_B T}{\tilde{\zeta}(\omega) + i\omega M_p} = \frac{\mathbb{D}k_B T}{\tilde{\zeta}(\omega) + i\omega M_p}. \qquad (3.71)$$

The unilateral Fourier Transform and the Laplace Transform are intimately related via analytic continuation. Indeed, substituting $\omega \to -is$ into eqn 3.71 yields the Laplace Transform-derived analog (eqn

3.83). The bilateral Fourier Transform, on the other hand, can be obtained from the unilateral transform via

$$\langle \tilde{\mathbf{V}}(\omega) \cdot \mathbf{V}(0) \rangle = 2\mathrm{Re}\langle \tilde{\mathbf{V}}(\omega) \cdot \mathbf{V}(0) \rangle_u. \tag{3.72}$$

Finally, the VAC can be related to the MSD using eqn 3.45,

$$\langle \tilde{\mathbf{V}}(\omega) \cdot \mathbf{V}(0) \rangle_u = -\frac{1}{2}\omega^2 \langle \Delta \tilde{\mathbf{r}}^2(\omega) \rangle_u, \tag{3.73}$$

to give the Generalized Einstein Relation (GSER),

$$\langle \Delta \tilde{\mathbf{r}}^2(\omega) \rangle_u = \frac{6k_B T}{(i\omega)^2 (\tilde{\zeta}(\omega) + i\omega M_p)}. \tag{3.74}$$

when displacements in all three dimensions contribute to $\Delta \tilde{r}^2$, or

$$\langle \Delta \tilde{\mathbf{r}}^2(\omega) \rangle_u = \frac{2\mathbb{D}k_B T}{(i\omega)^2 (\tilde{\zeta}(\omega) + i\omega M_p)} \tag{3.75}$$

when displacements are tracked in $\mathbb{D}$ dimensions.

3.3.2 Laplace Transform

As written, eqn 3.55 integrates over times reaching back to $t = -\infty$, introducing problems for the Laplace Transform approach. In principle, doing so gives

$$sM_p\hat{\mathbf{V}}(s) - M_p\mathbf{V}(0) = \hat{\mathbf{f}}_B(s) - \hat{\zeta}(s)\hat{\mathbf{V}}(s) - \int_{-\infty}^{0} \mathscr{L}\{\zeta(t-t')\}\mathbf{V}(t')dt'. \tag{3.76}$$

Because $\mathbf{V}(t' < 0)$ falls outside the realm of the Laplace Transform, one cannot neatly solve for $\hat{\mathbf{V}}(s)$. This issue did not arise in eqn 3.20, because of the instantaneous response of the probe.

A common and appealing approach is to effectively ignore times $t' < 0$ in eqn 3.55,

$$M_p\dot{\mathbf{V}}(t) = \mathbf{f}_B^0(t) - \int_0^t \zeta(t-t')\mathbf{V}(t')dt', \tag{3.77}$$

and then follow the logic of Section 3.2.1. This is problematic though. After all, any probe in thermal equilibrium at $t = 0$ is nonetheless responding (statistically) to the probe's previous velocity history, due to the material's memory, be it inertial or viscoelastic. The Brownian forces $\mathbf{f}_B^0(t)$ in this approach, then, must differ from those in the stationary system (which can not depend on any particular $t = 0$).

In fact, eqn 3.77 can be viewed as an alternate version of eqn 3.55, wherein the Brownian force $\mathbf{f}_B^0(t)$ incudes "memories" of times $t < 0$, above and beyond the stationary Brownian force $\mathbf{f}_B(t)$,

$$\mathbf{f}_B^0(t) = \mathbf{f}_B(t) - \int_{-\infty}^0 \zeta(t-t')\mathbf{V}(t')dt'. \tag{3.78}$$

If $\mathbf{f}_B^0(t)$ in eqn 3.77 had the same statistics as $\mathbf{f}_B(t)$ in the stationary distribution eqn 3.55, then eqn 3.77 would represent a sphere (and material) that was *at rest* for $t < 0$, rather than in equilibrium, and then released to start moving thermally for $t > 0$. As written, $\mathbf{f}_B^0(t)$ is therefore not stationary—it depends on time relative to "initial" time $t = 0$, at which point the system is in equilibrium, and therefore reflects memory of the statistical forces and velocities that preceded the time (arbitrarily) identified as $t = 0$.

Taking the Laplace Transform of (3.77) and using the convolution theorem yields

$$sM_p\hat{\mathbf{V}}(s) - M_p\mathbf{V}(0) = \hat{\mathbf{f}}_B^0(s) - \hat{\zeta}(s)\hat{\mathbf{V}}(s). \tag{3.79}$$

Solving for $\hat{\mathbf{V}}(s)$, taking the scalar product with $\mathbf{V}(0)$, and ensemble averaging gives the velocity correlation function

$$\langle \hat{\mathbf{V}}(s) \cdot \mathbf{V}(0) \rangle = \langle \hat{\mathbf{f}}_B^0(s) \cdot \mathbf{V}(0) \rangle + \frac{M_p\langle \mathbf{V}(0) \cdot \mathbf{V}(0) \rangle}{\hat{\zeta}(s) + sM_p}. \tag{3.80}$$

Irrespective of the viscoelastic properties of the medium, the equipartition theorem requires

$$\frac{1}{2}M_p\langle \mathbf{V}(0) \cdot \mathbf{V}(0) \rangle = \frac{\mathbb{D}}{2}k_B T, \tag{3.81}$$

reflecting the $\mathbb{D}$ translational degrees of freedom, leading to

$$\langle \hat{\mathbf{V}}(s) \cdot \mathbf{V}(0) \rangle = \langle \hat{\mathbf{f}}_B^0(s) \cdot \mathbf{V}(0) \rangle + \frac{\mathbb{D}k_B T}{\hat{\zeta}(s) + sM_p}. \tag{3.82}$$

Finally, for the distribution to be stationary, $\langle \hat{\mathbf{f}}_B^0(s) \cdot \mathbf{V}(0) \rangle$ must vanish, as was assumed in the quasi-steady Newtonian case (eqn 3.10), giving

$$\langle \hat{\mathbf{V}}(s) \cdot \mathbf{V}(0) \rangle = \frac{\mathbb{D}k_B T}{\hat{\zeta}(s) + sM_p}. \tag{3.83}$$

Equation 3.83 can be transformed into its Fourier analog eqn 3.71, via analytic continuation—simply replacing s with $i\omega$. Again, this is to be expected based on the causal nature of $\zeta(t)$, and therefore its analyticity properties. Finally, the velocity correlation function can be related to the mean-squared displacement by invoking eqn 3.43, giving

$$\langle \Delta \hat{r}^2(s) \rangle = \frac{2 \mathbb{D} k_B T}{s^2 \left[\hat{\zeta}(s) + s M_p \right]}. \tag{3.84}$$

Equation 3.84, like eqn 3.74, is called the Generalized Einstein Relation, and is central to passive microrheology. The two expressions are related by analytic continuation, effectively by substiting $s = i\omega$ (Pipkin, 1986).

These equations have the form of eqn 3.44, but the Stokes pseudo-steady-hydrodynamic resistance $\zeta = 6\pi a\eta$ has been replaced by a frequency-dependent memory function. In section 3.4, we will relate $\tilde{\zeta}(\omega)$ to the viscoelastic properties of the surrounding medium, just as ζ is related to the (frequency-independent) viscosity in the more limited case of a particle suspended in a Newtonian fluid.

Before moving on, ponder a key assumption made in this section—equipartition of energy. Equipartition only holds for systems—probe particles and their surrounding materials—that are in thermal equilibrium. This has important implications that we will consider later when discussing the limitations of passive microrheology. In particular, a probe particle in thermal equilibrium with its surroundings *cannot drive the material out of equilibrium*. Passive microrheology is therefore limited to measurements of a material's *linear* rheological response.

Another implication is that the material must not be *driven* by some out-of-equilibrium process—for instance, by swimming bacteria, the action of molecular motors or some other chemical process. Such **active matter**—including living cells—have long been studied using tracer particle methods, but their rheology cannot be measured using passive microrheology alone. For example, Mizuno *et al.* (2007) used a combination of passive and active microrheology methods to study the violation of the fluctuation dissipation theory that occurs when myosin molecular motors perform work on F-actin filament networks, further discussed in Section 7.2.1. The myosin in this case causes relative sliding of the filaments as the protein hydrolyzes ATP. In the absence of ATP, the actin-myosin network is at equilibrium, and the fluctuation dissipation-theorem is restored.

3.4   The Stokes component

In Section 3.3, we derived the Einstein component of the GSER, which relates the (measurable) mean-square displacement of a probe particle to the resistance $\tilde{\zeta}(\omega)$ (or $\hat{\zeta}(s)$ in Laplace space) and inertia $iM_p\omega$ (or $M_p s$) of the probe as it moves in the fluid. The Einstein component thus relates a stochastic, thermally fluctuating quantity to deterministic, mechanical quantities that depend on the probe and the material. To then extract intrinsic material properties requires the probe resistance to be related the linear viscoelastic reponse properties of the material. This step comprises the "Generalized Stokes" component of the GSER.

In the Fourier domain, the frequency-dependent resistance $\tilde{\zeta}(\omega)$ gives the force exerted on the probe by the surrounding material when the probe is forced to move with oscillatory velocity at frequency ω. To actually determine $\tilde{\zeta}(\omega)$ for a material with unknown rheological properties would generally require the equations of motion for the material, in reponse to the oscillating probe. This seems at first to present a conundrum: How can one even write down—much less solve—this mechanics problem, if one does not even know the constitutive equations of the material?

The resolution to this paradox was discussed in Chapter 2. The *Correspondence Principle*, discussed in Section 2.4, demonstrates that the resistance of a spherical probe moving quasi-steadily in an incompressible Newtonian viscous fluid

$$\zeta = 6\pi a\eta, \tag{3.85}$$

yields an identical problem—and solution—in the frequency domain for an incompressible viscoelastic medium, such that

$$\tilde{\zeta}(\omega) = 6\pi a\eta^*(\omega), \tag{3.86}$$

or, by analytic continuation,

$$\hat{\zeta}(s) = 6\pi a\hat{\eta}(s). \tag{3.87}$$

Written in terms of the shear modulus, $G^*(\omega) = i\omega\eta^*(\omega)$, the quasi-steady resistance becomes

$$\tilde{\zeta}(\omega) = 6\pi aG^*(\omega)/i\omega. \tag{3.88}$$

These equations hold in the case where the fluid inertia is negligible. For Newtownian fluids, the oscillatory boundary layer thickness $\lambda_V = \sqrt{2\eta/\rho\omega}$ must be significantly larger than the probe radius a for the quasi-steady Stokes equations to be appropriate. Some tracer particle microrheology experiments, especially those employing light scattering, may approach frequencies where particle and fluid inertia cannot be neglected.

At higher frequencies, the unsteady Newtonian resistance (eqn 2.103),

$$\tilde{\zeta}(\omega) = 6\pi\eta a \left(1 + \frac{a}{\lambda_V} + i\left[\frac{a}{\lambda_V} + \frac{2a^2}{9\lambda_V^2}\right]\right) \qquad (3.89)$$

can be generalized for incompressible, linear viscoelastic materials, by replacing the Newtonian viscosity η with the frequency-dependent complex viscosity $\eta^*(\omega)$ of the linear viscoelastic material, via

$$\tilde{\zeta}(\omega) = 6\pi\eta^*(\omega)a\left(1 + a(1+i)\sqrt{\frac{\rho_f\omega}{2\eta^*(\omega)}}\right) + i\omega M_f, \qquad (3.90)$$

where $M_f = 2\pi\rho_f a^3/3$ is the "added mass" of the surrounding material, which oscillates along with the probe, as discussed in Section 2.5.3. We discuss the effect of inertia in microrheology experiments further in Chapter 5.

3.5 The Generalized Stokes–Einstein Relation (GSER)

Combining the results of Sections 3.3 and 3.4 yields

$$\langle\Delta\tilde{\mathbf{r}}^2(\omega)\rangle_u = \frac{-2\mathbb{D}k_BT}{6\pi\eta^*(\omega)\omega^2 a\left(1 + a(1+i)\sqrt{\frac{\rho_f\omega}{2\eta^*(\omega)}}\right) + i(M_p + M_f)\omega^3}, \qquad (3.91)$$

which expresses the (experimentally-measurable) MSD entirely in terms of material properties (density and complex viscosity), frequency, probe size, and the number $\mathbb{D}$ of dimensions that are tracked and that contribute to the MSD $\langle\Delta\tilde{\mathbf{r}}^2(\omega)\rangle$. Perhaps the most important result for microrheology emerges for frequencies that are low enough for inertia to be negligible, in which case eqn 3.91 reduces to

$$\langle\Delta\tilde{\mathbf{r}}^2(\omega)\rangle_u = \frac{\mathbb{D}k_BT}{3\pi a(i\omega)^2\eta^*(\omega)}, \qquad (3.92)$$

or its analog in Laplace space,

$$\langle\Delta\tilde{\mathbf{r}}^2(s)\rangle = \frac{\mathbb{D}k_BT}{3\pi as^2\tilde{\eta}(s)}. \qquad (3.93)$$

Equations 3.92 and 3.93 are generally called the Generalized Stokes–Einstein Relation, or GSER. Strictly speaking, they hold only at frequencies low enough for inertia to be neglected. Nonetheless, this frequency range is typically broader than would be accessible to macroscopic rheometers, and therefore encompasses many frequencies of interest for soft materials. In short, this limit of the GSER connects the measured MSD of probe particles to the linear viscoelastic spectrum of the surrounding material.

Following the seminal work of Mason and Weitz (1995), eqn 3.93 was sometimes incorrectly treated as fortuitous. Soon after, however, Schnurr *et al.* (1997) essentially invoked the Correspondence Principle to rationalize why the GSER should hold for all (non-inertial) frequencies. Their correction to this sometimes persistent misunderstanding was subtle and was later reiterated clearly by Indei *et al.* (2012*b*).

Alternative forms of the GSER may be derived from eqn 3.92. Relating $\eta^*(\omega)$ to the complex shear modulus $G^*(\omega)$ via

$$G^*(\omega) = i\omega\eta^*(\omega) \tag{3.94}$$

gives an expression that may be solved for $G^*(\omega)$,

$$G^*(\omega) = \frac{\mathbb{D}k_B T}{3\pi a(i\omega)\langle\Delta\tilde{r}^2(\omega)\rangle} \tag{3.95}$$

or alternately,

$$\hat{G}(s) = \frac{\mathbb{D}k_B T}{3\pi as\langle\Delta\hat{r}^2(s)\rangle}, \tag{3.96}$$

where $\hat{G}(s)$ is the Laplace Transform of the memory function $M_r(t)$.

One can not invert these transforms exactly for generic functional forms of $G^*(\omega)$, because transformed functions appear in denominators. These relations may be inverted to give real-time relations, however, by using eqn 1.39,

$$\tilde{\mathcal{J}}(\omega) = \frac{1}{i\omega G^*(\omega)}, \tag{3.97}$$

to relate $G^*(\omega)$ in eqn 3.95 to the transformed creep compliance $\tilde{\mathcal{J}}(\omega)$,

$$\tilde{\mathcal{J}}(\omega) = \frac{3\pi a}{\mathbb{D}k_B T}\langle\Delta\tilde{r}^2(\omega)\rangle. \tag{3.98}$$

This form of the GSER may be immediately inverted (Xu *et al.*, 1998*a*; Mason, 2000), thereby connecting the measured MSD directly to the creep compliance $\mathcal{J}(t)$.

$$\mathcal{J}(t) = \frac{3\pi a}{\mathbb{D}k_B T}\langle\Delta\mathbf{r}^2(t)\rangle, \tag{3.99}$$

where $\mathbb{D}$ is the number of dimensions tracked in the experiment. Recall, the creep compliance contains the same rheological information (*i.e.*, the entire linear viscoelastic spectrum of the material).

Equation 3.99 is important. First, it reveals that it is not necessary to convert passive microrheology data to the frequency domain, despite the widespread adoption of this approach. In fact, such conversions may even introduce numerical artifacts, owing to the limited sampling range in the time domain. The methods of such conversions are discussed in Section 3.8. Second, eqn 3.99 reveals an interesting physical insight—the mean-squared displacement can be understood as a creep experiment. The mean-squared displacement of a probe particle reflects the "strain" that accumulates due to the average thermal stress imposed on the probe particle by the random Brownian force.

Equations 3.95, 3.96, and 3.99 take into account the number of dimensions tracked in a passive microrheology experiment. Measurements methods such as light scattering (Chapter 5) will report this three-dimensional value. Techniques such as multiple particle tracking (Chapter 4) will typically involve analysis of the mean-squared displacement in only one- or two-dimensional projections. It is necessary to alter eqns 3.95, 3.96, and 3.99 by the dimension, $\mathbb{D}$. For instance, data collected as a two-dimensional projection (typical for video microscopy) yields

$$\mathcal{J}(t) = \frac{3\pi a}{2k_B T} \langle \Delta r_{2\mathbb{D}}^2(t) \rangle. \qquad (3.100)$$

3.6 Passive microrheology examples

Having derived the GSER—the fundamental relation that underpins passive microrheology—it is worthwhile to consider several examples of measured probe mean-squared displacements in complex and simple fluids, and how rheological properties may be determined from such data sets. Four examples are shown in Fig. 3.2.

The first example (Fig. 3.2a) shows the diffusing wave spectroscopy data of Cardinaux *et al.* (2002), wherein 0.7 and 1.5 μm diameter particles are dispersed in an aqueous surfactant solution that self-assembles into worm-like micelles (WLM). Entangled micelles form transient physical cross-links, resulting in strong viscoelastic properties. The probe motion at short delay times, far below the material's relaxation time, is sub-diffusive, reflecting the internal dynamics of the network. Over longer times, probe-particle confinement gives rise to an elastic plateau, indicating that the solution behaves like a weak viscoelastic solid. Beyond about 10^{-1} seconds,

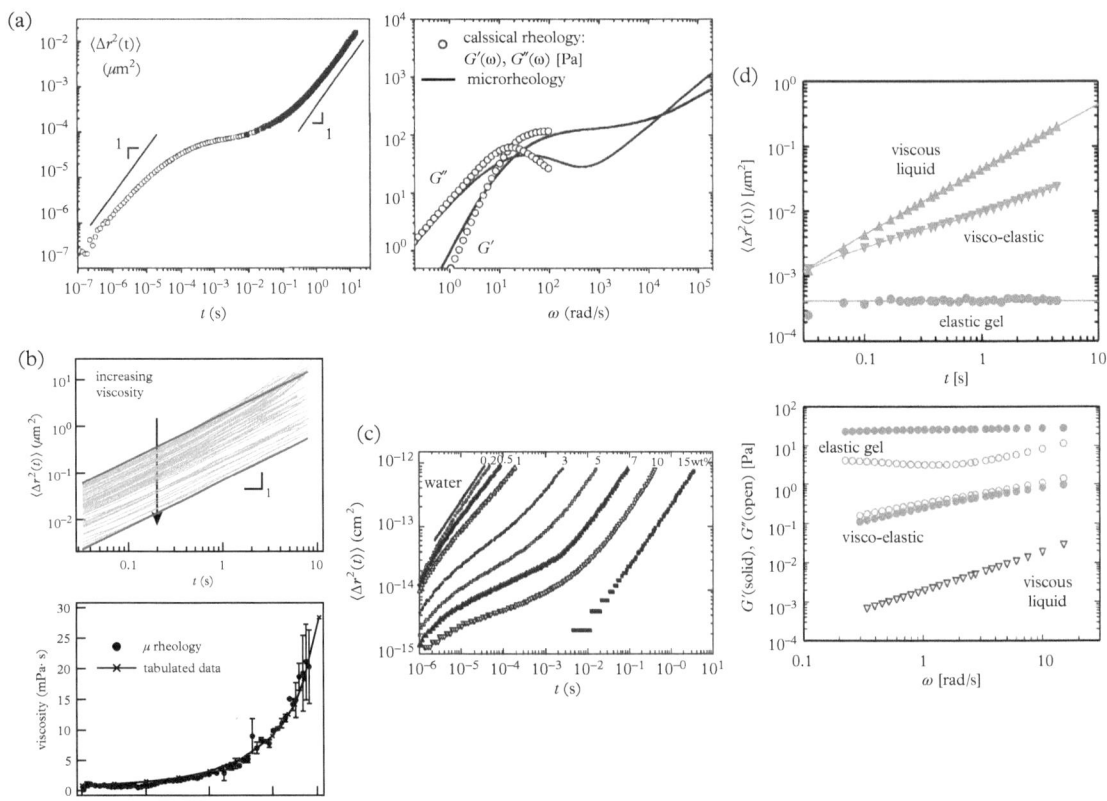

Fig. 3.2 *Examples of tracer particle dynamics in simple and complex fluids. (a) A concentrated aqueous surfactant solution that forms entangled worm-like micelles. The data is converted the to frequency domain and compared with bulk rheology measurements (Cardinaux* et al., *2002). Reprinted with permission from Europhys. Lett., 2002, Number 5, March, http://iopscience.iop.org/journal/0295-5075 (b) A glycerine solution with increasing viscosity. Adapted from Schultz and Furst (2011) with permission from The Royal Society of Chemistry. (c) Microrheology of PEO solutions in water. Reprinted with permission from van Zanten, J. H., Amin, S., & Abdala, A. A. Macromolelcules 37, 3874–80. Copyright (2004) American Chemical Society. (d) Alginate microrheology as the polysaccharide is induced to gel by the addition of calcium chloride. Reprinted with permission from Sato, J. & Breedveld, V. J. Rheol., 50, 1–19 (2006). Copyright (2006), The Society of Rheology.*

the qualitative properties change once again, showing predominantly viscous behavior, as the WLM network relaxes and flows.

Converting the MSD results to frequency-dependent viscoelastic moduli, using methods discussed in Section 3.8, enables direct comparisons to bulk rheology measurements of the same sample, as shown in Fig. 3.2b. Notably, microrheology extends the rheology measurements to considerably higher frequencies—well into the terminal relaxation regime corresponding to the relaxation of individual filaments. Such high-frequency data allows the stiffness of these and other supramolecular assemblies and macromolecules to be characterized, as discussed in Chapter 5.

Figure 3.2b shows the measurements of Schultz and Furst (2011) on simpler samples: Mixtures of glycerine and water for glycerine concentrations up to roughly 75 wt%. Because the mixtures are Newtonian, each MSD is simply a straight line. These measurements were made using particle tracking microrheology (Chapter 4) with microfluidic devices that are used to prepare many samples simultaneously. The viscosities, calculated by the GSER (which reduces to the Stokes–Einstein relation in this case) track the expected viscosity of these fluids.

The third example, Fig. 3.2c, is taken from van Zanten *et al.* (2004), who measured the dynamics of tracer probes in solutions of 333,000 g/mol polyethyele oxide (PEO), again with diffusing wave spectroscopy. As the polymer concentration increases from 0.2 to 10 wt%, the mean-squared displacement initially decreases as the solution viscosity increases. A significant sub-diffusive regime emerges at imtermediate concentrations, ultimately crossing over to normal diffusion.

Our last example (Fig. 3.2d) comes from multiple particle tracking measurements of alginate solutions Sato and Breedveld (2006). Here, the viscous solutions gel as salt is introduced into the sample through a dialysis membrane. At equilibrium, before and after the introduction of the salt, the mean-squared displacement curves are diffusive (viscous liquid) or entirely flat (elastic solid). Transient viscoelastic behavior is captured too, and lies in between the terminal states. In this case, particular particle-tracking microrheology captures the gelation of soft materials, an application we discuss in detail in Chapter 10.

3.6.1   Limiting behavior of the MSD

It is useful to keep in mind the limiting behavior of the mean-squared displacement that results from the GSER. In each example shown in Fig. 3.2, the mean-squared displacement is bound between that of a viscous Newtonian fluid, in which the mean-squared displacement is *linear* with time, and that of an elastic solid, which exhibits a constant, time-independent displacement. These limits will hold for for several model fluids discussed in the next section and are represented by the black curves in Fig. 3.3—one for a fluid with viscosity with a Newtonian viscosity $\eta = 1$ mPa·s and the other for an elastic solid with a modulus $G = 10$ Pa. Note that the logarithmic slope of the MSD, defined by

$$\alpha(t) = \frac{d \ln \langle \Delta r^2(t) \rangle}{d \ln t} \tag{3.101}$$

Fig. 3.3 *Limiting behavior of the MSD scaled by particle radius a and thermal energy $k_B T$ for a Newtonian fluid with viscosity $\eta = 1$ mPa·s and elastic solid with modulus 10 Pa. A Maxwell fluid and the Kelvin-Voigt model are shown as gray lines. Both have a viscosity $\eta = 1$ mPa·s and elastic modulus $G = 1$Pa, giving identical relaxation times $\tau = \eta/G = 10^{-3}$s.*

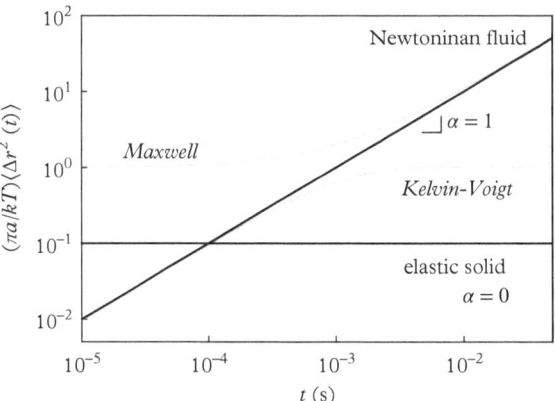

can only have values between

$$0 \leq \alpha \leq 1. \tag{3.102}$$

Logarithmic slopes outside of this range typically indicate a problem has occurred with the passive microrheology measurement, due for instance to statistical noise of the MSD or physical sources of error, such as convection in the sample or vibration. In Section 3.8, we will show that α is related to the loss tangent, $\tan \delta(\omega) = G''(\omega)/G'(\omega)$.

3.7 GSER for model materials

We consider here the moduli and compliances for several models of viscoelastic fluids and solids and the resulting GSER equations—the creep compliance expressed as the mean-squared displacement of tracer particles. Representations of the frequency-dependent moduli and mean-squared displacement are shown in Figs 3.4, 3.5, and 3.6. This discussion should help us interpret passive microrheology results in the time domain of the experiment, rather than relying on a conversion to the frequency domain. Common methods for converting between the two domains are discussed in Section 3.8.

3.7.1 Elastic solid

An incompressible elastic solid with shear modulus $G^*(\omega) = G$, which is a real, constant quantity, and compliance

$$\mathcal{J}(t) = \frac{1}{G} \tag{3.103}$$

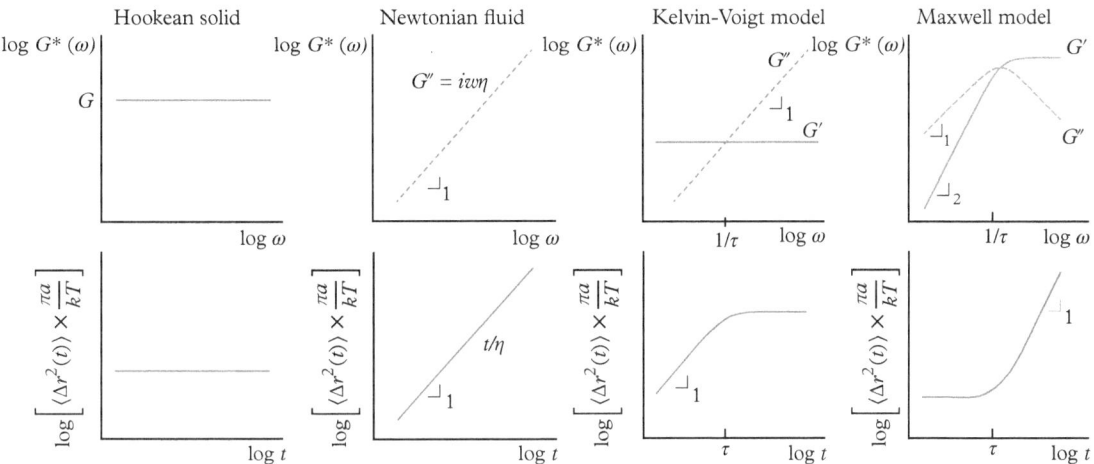

Fig. 3.4 *Schematic representations of the frequency-domain viscoelastic moduli of several model materials and the corresponding scaled mean-squared displacement (or creep compliance) from the GSER. From left to right: A Hookean elastic solid, a Newtonian fluid, the Kelvin–Voigt model (viscoelastic solid), and the Maxwell model (viscoelastic liquid). Both axes of each plot are logarithmic scales.*

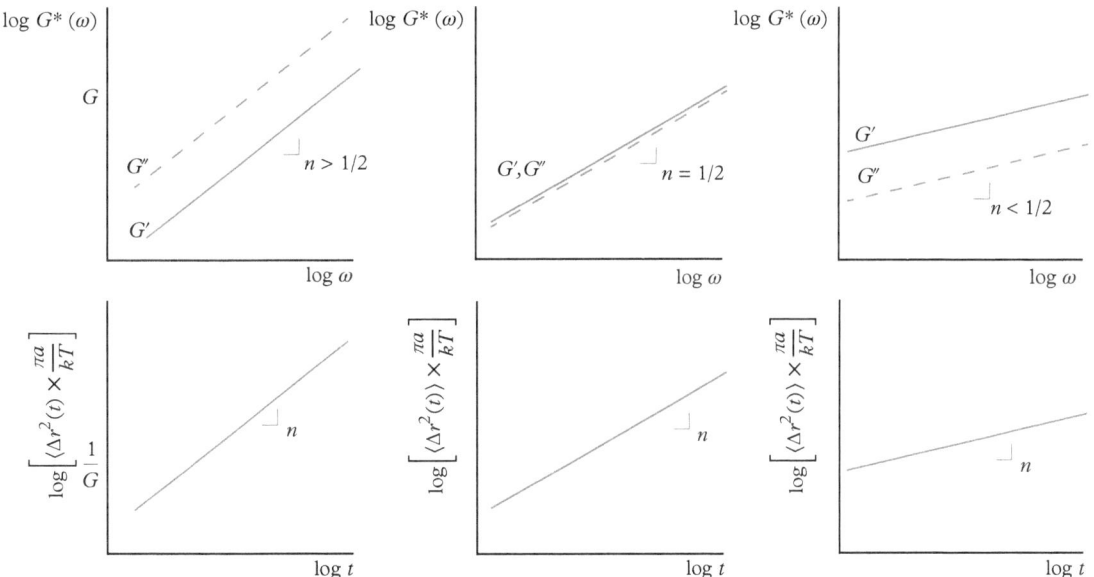

Fig. 3.5 *Schematic representations of the frequency domain viscoelastic moduli of a power-law material and the corresponding scaled mean-squared displacements (or creep compliances) from the GSER for power-law exponents $n > 1/2$, $n = 1/2$, and $n < 1/2$. Both axes of each plot are logarithmic scales.*

will exhibit a mean-squared displacement that is independent of time,

$$\langle \Delta r^2(t) \rangle = \frac{\mathbb{D} k_B T}{3\pi a G}. \tag{3.104}$$

A stiff material, with a large modulus G will have a correspondingly low compliance.

3.7.2 Viscous fluid

A Newtonian-fluid viscosity η has a creep relaxation

$$\mathcal{J}(t) = t/\eta \tag{3.105}$$

and corresponding mean-squared displacement

$$\langle \Delta r^2(t) \rangle = \frac{\mathbb{D} k_B T}{3\pi a\eta} t. \tag{3.106}$$

This is simply the Stokes–Einstein relation, eqn 3.49, with the Stokes resistance $\zeta = 6\pi a\eta$. In the frequency domain, the complex modulus is purely imaginary

$$G^*(\omega) = i\omega\eta. \tag{3.107}$$

Only the loss modulus G'' is non-zero. An increasing viscosity gives rise to a higher loss modulus, but a *lower* compliance.

3.7.3 Kelvin–Voigt model

The Kelvin–Voigt model, with frequency-dependent complex modulus

$$G^*(\omega) = G(1 + i\omega\tau), \tag{3.108}$$

or in Laplace space

$$G(s) = G(1 + s\tau), \tag{3.109}$$

describes a simple viscoelastic solid with constant storage modulus $G'(\omega) = G$, viscosity $G''(\omega) = i\omega\eta$ and characteristic relaxation time $\tau = \eta/G$.

Inverting eqn 3.98 gives

$$\langle \Delta r^2(t) \rangle = \frac{\mathbb{D} k_B T}{3\pi a G}(1 - e^{-t/\tau}). \tag{3.110}$$

The equivalent MSD expression can be found using Eqn 3.99 and the creep compliance of the Kelvin–Voigt model, as shown in Fig. 3.4,

$$\mathcal{J}(t) = \mathcal{J}(1 - e^{-t/\tau}) \tag{3.111}$$

noting that the recoverable compliance is $\mathcal{J} = 1/G$.

3.7.4 Maxwell fluid

The Maxwell fluid represents a simple model for a viscoelastic fluid, with an elastic modulus at short times and viscous relaxation at long times. Its creep compliance is (Ferry, 1980)

$$\mathcal{J}(t) = 1/G + t/\eta, \tag{3.112}$$

where $\mathcal{J} = 1/G$ is the recoverable elastic compliance. The Maxwell fluid has a complex modulus

$$G^*(\omega) = \frac{Gi\omega\tau}{1 + i\omega\tau} \tag{3.113}$$

or

$$\hat{G}(s) = \frac{Gs\tau}{1 + s\tau} \tag{3.114}$$

where

$$\tau = \frac{\eta}{G} \tag{3.115}$$

is the characteristic relaxation time of the Maxwell fluid. Expanding eqn 3.113 by multiplying $(1 - i\omega\tau)/(1 - i\omega\tau)$, we write the storage modulus

$$G'(\omega) = \frac{G\omega^2\tau^2}{1 + \omega^2\tau^2} \tag{3.116}$$

and loss modulus

$$G''(\omega) = \frac{G\omega\tau}{1 + \omega^2\tau^2}. \tag{3.117}$$

Equation 3.99 gives the mean-squared displacement for the Maxwell model,

$$\langle \Delta r^2(t) \rangle = \frac{\mathbb{D}k_B T}{3\pi aG} \left(\frac{t}{\tau} + 1 \right). \tag{3.118}$$

An unusual feature of the Maxwell model at short times is that the mean-squared displacement approaches a constant value given by the elasticity of the material. Of course, a plateau in the MSD below the relaxation time $t < \tau$ is expected, but at short enough times, the probe displacement must go to zero. This apparent contradiction is an artifact of neglecting the fluid inertia in the Stokes equation, and can be corrected by properly accounting for the time-dependent equations of motion (Grimm *et al.*, 2011; Indei *et al.*, 2012*b*). We discuss inertial corrections in Section 5.6.2 when we present high-frequency microrheology with diffusing wave spectroscopy.

3.7.5   Power-law response

A material with a power-law response has a shear relaxation modulus

$$G(t) = Kt^{-n} \tag{3.119}$$

where K, the "consistency" has fractional units Pa·s^n and n is an exponent bounded by $0 < n < 1$. Then the storage and loss moduli scale as $G' \sim G'' \sim \omega^n$ over all frequencies (Winter and Mours, 1997; Jaishankar and McKinley, 2012). Specifically,

$$G^*(\omega) = K\Gamma(1-n)(i\omega)^n \tag{3.120}$$

and

$$G'(\omega) = K\Gamma(1-n)\omega^n \cos\frac{n\pi}{2} \tag{3.121}$$

$$G''(\omega) = K\Gamma(1-n)\omega^n \sin\frac{n\pi}{2} \tag{3.122}$$

where $\Gamma(x)$ is the gamma function. Whether the material response is dominated by viscous or elastic behavior depends on the value of n: For $n > 1/2$ the loss modulus has larger magnitude than the storage modulus. When $n < 1/2$, the storage modulus has larger magnitude than the loss modulus. When $n = 1/2$, neither dominates, meaning that $G' = G''$ over all frequencies.

Power-law rheology arises when a hierarchy of relaxation time scales are established by a microstructure that is self-similar over a wide range of length scales. An example is the fractal structure of an incipient gel at the percolation transition (Winter and Chambon, 1987; Martin *et al.*, 1988; Adolf and Martin, 1990; Winter and Mours, 1997). Power-law rheology also occurs in many complex materials and products, including biofluids, foods, cross-linked polymers, microgels, and hydrogels.

The Laplace Transform of $G(t)$ for the power law material is

$$\hat{\eta}(s) = \int_0^\infty K t^{-n} e^{-st} dt = K \frac{\Gamma(1-n)}{s^{1-n}}, \qquad (3.123)$$

where we note again that the complex viscosity and relaxation modulus are transform pairs, as discussed in Section 1.2.2.[7] The Laplace-transformed creep compliance is then given by

$$\hat{\mathcal{J}}(s) = \frac{1}{K\Gamma(1-n)} \frac{1}{s^{n+1}}, \qquad (3.124)$$

with inverse

$$\mathcal{J}(t) = \frac{t^n}{K\Gamma(1-n)\Gamma(n+1)}. \qquad (3.125)$$

This can be simplified using the relation $\Gamma(1-n)\Gamma(n+1) = n\pi / \sin n\pi$, giving

$$\mathcal{J}(t) = \frac{\sin n\pi}{n\pi K} t^n, \qquad (3.126)$$

from which the GSER determines the mean-squared displacement to be

$$\langle \Delta r^2(t) \rangle = \frac{\mathbb{D} k_B T \sin n\pi}{3n\pi aK} t^n \qquad (3.127)$$

for a probe in a power-law fluid, as shown in Fig. 3.5.

3.7.6   Rouse and Zimm models

The Rouse model is a bead-spring representation of dilute flexible polymers dispersed in a Newtonian solvent (Ferry, 1980; Doi and Edwards, 1986; Rubinstein and Colby, 2003). Beads of radius b are connected by Hookean springs. The beads represent the hydrodynamic drag exerted on the polymer chains, and the stiffness of the springs captures the entropic elasticity of the flexible molecules. The relaxation modulus is a summation over the relaxation of the chain's normal modes

$$G(t) = n k_B T \sum_{p=1}^{N} e^{-t/\tau_p}. \qquad (3.128)$$

Here, n is the number density of molecules and τ_p is the characteristic relaxation time of the pth mode of a chain consisting of N beads,

$$\tau_p = \frac{\zeta b^2 N^2}{6\pi^2 k_B T p^2} \qquad (3.129)$$

[7] The literature often uses $\tilde{G}$ or $\hat{G}$ to denote transforms of the relaxation modulus. We instead use the transformed viscosity $\eta^*(\omega)$ or $\hat{\eta}(s)$, to avoid confusion with the complex modulus $G^*(\omega)$ and its Laplace variant $\hat{G}(s)$.

where the friction coefficient on each bead is $\zeta \approx \eta_s b$ and η_s is the solvent viscosity. The longest relaxation time of the polymer, corresponding to the mode $p = 1$, is

$$\tau_R = \frac{\zeta b^2 N^2}{6\pi^2 k_B T} \tag{3.130}$$

and the shortest time scale of the relaxation is that of the monomer,

$$\tau_0 \approx \frac{\zeta b^2}{k_B T}. \tag{3.131}$$

In the frequency domain, the storage and loss moduli of the Rouse model are

$$G'(\omega) = nk_B T \sum_{p=1}^{N} \frac{\omega^2 \tau_p^2}{1 + \omega^2 \tau_p^2} \tag{3.132}$$

and

$$G''(\omega) = \omega \eta_s + nk_B T \sum_{p=1}^{N} \frac{\omega \tau_p}{1 + \omega^2 \tau_p^2}, \tag{3.133}$$

representing the summation over N distinct relaxation times in the Maxwell model.

As illustrated in Fig. 3.6, Between the frequencies $1/\tau_0 < \omega < 1/\tau_R$ in the terminal regime, the Rouse model moduli scale as

$$G'(\omega) = G''(\omega) - \omega \eta_s \sim \omega^{1/2}. \tag{3.134}$$

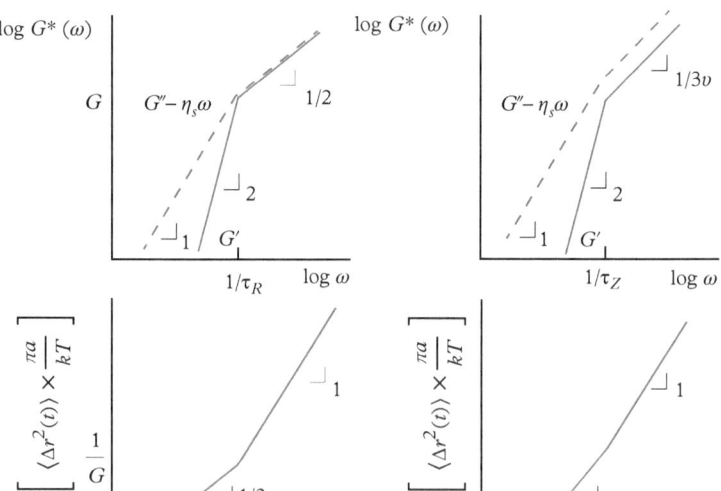

Fig. 3.6 *Schematic representations of the frequency domain viscoelastic moduli of the Rouse and Zimm models of polymer solutions and the corresponding scaled mean-squared displacements (or creep compliances) from the GSER. Both axes of each plot are logarithmic.*

That is, they are equal and have a power-law dependence on frequency of $\omega^{1/2}$—a signature of the "free-draining" hydrodynamics
of the model.[8] Likewise, the creep compliance scales as $t^{1/2}$ at short
times. Thus, the mean-squared displacement of a probe particle in a
passive microrheology experiment will exhibit sub-diffusive motion,

$$\langle \Delta r^2(t) \rangle \sim t^{1/2} \qquad \tau_0 < t < \tau_R. \qquad (3.135)$$

Below $\omega < 1/\tau_R$, $G'(\omega) \sim \omega^2$ and $G''(\omega) \sim \omega$. For $t > \tau_R$,
there is a cross-over from sub-diffusive to diffusive probe dynamics,
$\langle \Delta r^2(t) \rangle \sim t$.

The Zimm model differs from the Rouse model by accounting for
the hydrodynamic interactions between the beads of the bead-spring
model. The polymer is no longer considered "free-draining." This
affects the terminal high-frequency response of the solution, which
takes on a power law form

$$G'(\omega) \sim G''(\omega) - \omega\eta_s \sim \omega^{1/3\nu} \qquad (3.136)$$

where the Flory exponent ν depends on the polymer-solvent interaction. In theta-solvents, $\nu = 1/2$ and the scaling becomes $G'(\omega) \sim$
$G''(\omega) - \omega\eta_s \sim \omega^{2/3}$, while polymers in good solvents exhibit a lower
Flory exponent, $\nu \approx 0.588$, and correspondingly lower exponent
(~ 0.57). The terminal regime will be apparent in the mean-squared
displacement at time scales shorter than the Zimm relaxation time
$\tau_Z \approx \tau_0 N^{3\nu}$,

$$\langle \Delta r^2(t) \rangle \sim t^{1/3\nu} \qquad \tau_0 < t < \tau_Z. \qquad (3.137)$$

3.7.7   Semiflexible polymers

Semiflexible polymers are macromolecules and macromolecular assemblies for which the degree of the polymer backbone rigidity
becomes a significant source of elasticity and dissipation compared
to flexible and rod-like molecules. Comprehensive models for the
rheology of semiflexible polymers have been discussed by Morse
(1998c,a,b), Shankar *et al.* (2002), MacKintosh *et al.* (1995), and
Gittes and MacKintosh (1998). These theories cover a wide range of
conditions, including concentration (ranging from the dilute to tightly
entangled), persistence lengths (from flexible to rigid), and the effect
of cross-linkers.

[8] This is different from the free-draining
limit of the compressibility discussed in
Chapter 2.

The rheology of semiflexible polymers has been especially prominent in the microrheology literature largely due to the number of examples that are found in biological materials, especially the protein filaments and microtubules that dominate cell and tissue mechanics (Dichtl and Sackmann, 2002; Addas *et al.*, 2004), filamentous viruses (Sarmiento-Gomez *et al.*, 2012), and peptide assemblies (Ozbas *et al.*, 2004). A hallmark of semiflexible polymer microrheology is that its high-frequency terminal response takes on the scaling

$$G^*(\omega) \sim \omega^{3/4}, \tag{3.138}$$

which means that the mean-squared displacement scales as

$$\langle \Delta r^2(t) \rangle \sim t^{3/4} \tag{3.139}$$

at short times. This scaling has been measured by microrheology using diffusing wave spectroscopy, magnetic tweezers, and laser tracking microrheology (Amblard *et al.*, 1996; Gittes *et al.*, 1997; Palmer *et al.*, 1998, 1999; Mason *et al.*, 2000). Indeed, this is one direct means of measuring the persistence length l_p of a material composed of semiflexible polymers and supramolecular assemblies, as we discuss in Section 5.6.

3.8 Converting between the time and frequency domains

In Section 3.5, we found that the shear modulus in the frequency domain can be expressed in terms of the Laplace or Fourier Transform of the mean-squared displacement. In experiments such as video microscopy particle tracking (Chapter 4) and light scattering (Chapter 5) the mean-squared displacement is measured directly as a function of real time, at discrete time intervals over a range of times. We noted that the time-domain data can be interpreted directly as a creep measurement, but it is often desirable to express the microrheology in the frequency domain to compare with oscillatory bulk rheology or theory. The widespread use of oscillatory rheology has also made interpretation of rheological measurements in the frequency domain more familiar to many rheologists.

While a direct numerical transform of the time-domain data to the frequency domain is possible, this method is often unreliable, in that it leads to significant truncation errors at the frequency extremes. Others have used polynomial fits of the logarithmic time-domain data

$\ln\langle\Delta r^2(\ln t)\rangle$ and their analytical transforms into the frequency domain (Willenbacher *et al.*, 2007). Drawbacks include the accuracy of the fit, the potential to introduce artifacts into the transformed data, and the possibility that the apparent frequency-dependent moduli do not satisfy the Kramers-Kronig relations (cf. Section 1.2.2). Two other methods for converting the time-domain data to the frequency domain include the power-law approximation and methods of constrained regularization.

3.8.1 Power-law approximation

One common method of calculating the frequency-domain moduli in microrheology is to approximate the MSD at each sampled time t_0 as a power–law function,

$$\langle\Delta r^2(t)\rangle \approx \langle\Delta r^2(t_0)\rangle(t/t_0)^{\alpha(t_0)}, \tag{3.140}$$

where $\alpha(t_0)$ is the logarithmic slope of the mean-squared displacement evaluated at t_0,

$$\alpha(t_0) = \left.\frac{d(\ln\langle\Delta r^2(t)\rangle)}{d(\ln t)}\right|_{t=t_0}. \tag{3.141}$$

The Laplace Transformation of a power-law,

$$\mathcal{L}\{t^p\} = \frac{\Gamma(p+1)}{s^{p+1}} \tag{3.142}$$

where $\Gamma(x)$ is the Gamma function, implies that

$$s\langle\tilde{\Delta}r^2(s)\rangle = \langle\Delta r^2(t_0)\rangle(s_0/s)^{\alpha(t_0)}\Gamma(\alpha+1), \tag{3.143}$$

with $s_0 = 1/t_0$. Equation 3.96 is then recovered, giving

$$G(s_0) = \left.\frac{\mathbb{D}k_B T}{3\pi a\langle\Delta r^2(t_0)\rangle\Gamma[\alpha(t_0)+1]}\right|_{t_0=1/s_0}, \tag{3.144}$$

Here again, $\mathbb{D}$ represents the number of dimensions tracked for the mean-squared displacement. Evaluating eqn 3.144 at each sampled time t_0 in the MSD gives the corresponding relaxation modulus at $s_0 = 1/t_0$ from the value of the MSD and its logarithmic slope, as illustrated in Fig. 3.7. Similarly, the Fourier domain yields the modulus amplitude,

$$|G^*(\omega_0)| = \left.\frac{\mathbb{D}k_B T}{3\pi a\langle\Delta r^2(t_0)\rangle\Gamma[\alpha(t_0)+1]}\right|_{t_0=1/\omega_0}. \tag{3.145}$$

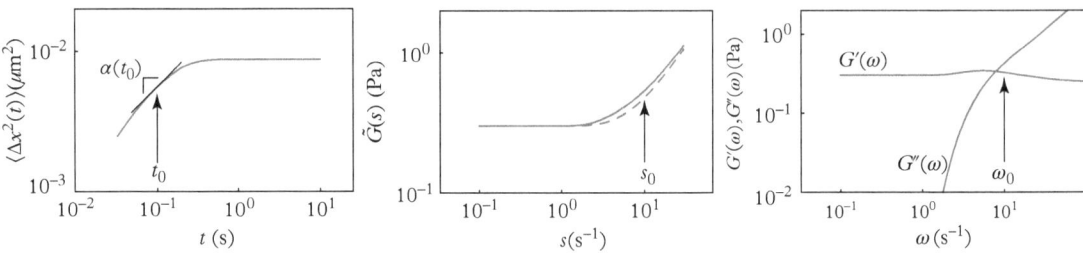

Fig. 3.7 *(Left) MSD for a Voigt model fluid. The MSD and logarithmic slope α are evaluated at t_0. (Middle) The corresponding $\tilde{G}(s)$. The dashed line is Eqn 3.144 neglecting the Gamma function. (Right) The storage (solid line) and loss (dashed line) moduli.*

From this, the storage $G'(\omega)$ and loss $G''(\omega)$ moduli are calculated as

$$G'(\omega) = |G^*(\omega)| \cos(\pi\alpha(\omega)/2) \tag{3.146}$$

$$G''(\omega) = |G^*(\omega)| \sin(\pi\alpha(\omega)/2). \tag{3.147}$$

Alternatively, the loss tangent can be expressed as

$$\tan\delta(\omega) = G''(\omega)/G'(\omega) = \tan[\pi\alpha(\omega)/2]. \tag{3.148}$$

This equation gives the relation between the phase angle and logarithmic slope of the mean-squared displacement,

$$\delta(\omega) = \frac{\pi}{2}\alpha(\omega). \tag{3.149}$$

With α limited to values between 0 and 1, the phase angle is constrained to $0 \le \delta \le \pi/2$. In the limit of an elastic solid, $\alpha = 0$ and $\tan\delta = 0$, and of course the loss tangent diverges for $\alpha = 1$.

The approximate transform to the frequency domain based on the power–law approximation works well when $\langle \Delta r^2(t) \rangle$ is a fairly smooth function of time (on a doubly logarithmic scale). For cases in which the MSD exhibits more curvature, higher-order terms in the power–law expansion can be included (Dasgupta *et al.*, 2002). The chief drawback of the approximate method is the accuracy of the numerical differentiation of the MSD to calculate its logarithmic slope, α. The differentiation mainly affects $\tan\delta$ or G' and G'' and not $G(s)$ or $|G^*(\omega)|$, since the Gamma function is a weak function of its permissible values, $1 \le \alpha + 1 \le 2$, as illustrated by plotting $\Gamma(x)$ for $1 \le x \le 2$ in Fig. 3.8. The calculated values of G' and G'' may also deviate significantly when the cosine and sine functions in eqns 3.146 and 3.147 approach zero. This is clearly evident in Fig. 3.7 for the low-ω behavior of G'', which decays more quickly as $\omega \to 0$ than the $G''(\omega) = \eta\omega$ allowed by the Kelvin-Voigt model.

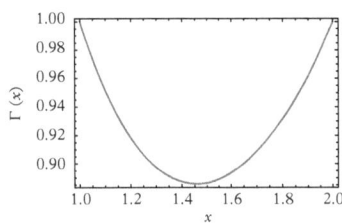

Fig. 3.8 *The Gamma function evaluated over the range of permissible values for the logarithmic MSD slope, $1 \le \alpha + 1 \le 2$.*

3.8.2 Constrained regularization

An alternate to the approximate transform is to calculate $\langle \Delta \tilde{r}^2(s) \rangle = \mathcal{L}\{\langle \Delta r^2(t) \rangle\}$ using a constrained regularization method (Honerkamp and Weese, 1989; Elster *et al.*, 1992; Honerkamp and Weese, 1993; Solomon and Lu, 2001; Lu and Solomon, 2002; Starrs and Bartlett, 2003a). In one version of this approach, the relaxation modulus $G(t)$ is expressed as a summation of N Maxwell model relaxation modes (Ferry, 1980; Honerkamp and Weese, 1989),

$$G(t) = \sum_{i=1}^{N} h_i e^{-t/\tau_i} \tag{3.150}$$

where h_i and τ_i are the amplitude and relaxation time of mode i, respectively.[9] The Laplace Transform of eqn 3.150 is the shear modulus

$$\hat{\eta}(s) = \sum_{i=1}^{N} \frac{h_i s \tau_i}{1 + s \tau_i}. \tag{3.151}$$

The inverse problem is to identify values of h_i and τ_i such that eqn 3.150 is a good description of the time-domain mean-squared displacement.

The number of relaxation times, their values, and their individual weighting is an ill-posed problem—the errors in the inversion are unbounded. Instead, one determines the values of h_i and τ_i using constraints that, for instance, require the relaxation spectrum to be smooth or at least continuous. Lu and Solomon (2002), for instance, use the method of Provencher (1982a) implemented in the program package CONTIN. They suggest that constrained regularization may perform better than the power-law approximation, as previously discussed, when the rheology exhibits a strong frequency dependence. A comparison of the CONTIN and power-law approximation methods using light-scattering microrheology for an associative polymer solution are shown in Figure 3.9. The CONTIN derived data does agree with the low-frequency response of the bulk rheometry data much more closely than the moduli calculated using the power-law approximation.

Another regularization method uses the Tirkhonov regularized fit of the mean-squared displacement (or creep compliance) with a set of N basis functions derived from a Voigt fluid (Mason *et al.*, 2000; Ferry, 1980)

$$\mathcal{J}(t) = \sum_{n=1}^{N} L_n (1 - e^{-t/\tau_n}) \tag{3.152}$$

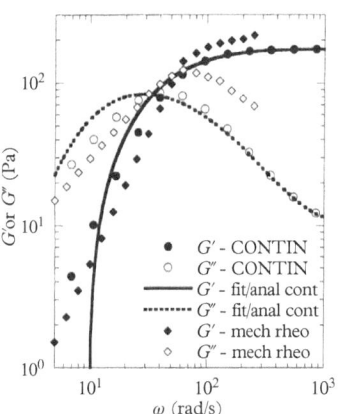

Fig. 3.9 *Microrheology of 4 wt% telechelic associative polymer thickener. Reprinted from* Curr. Opin. Colloid Interface Sci., 6, *Solomon, M. J. & Lu, Q. Rheology and dynamics of particles in viscoelastic media, 430–7, Copyright (2001), with permission from Elsevier.*

[9] This expression can be applied to viscoelastic solids by including a finite static modulus.

or Kelvin-Voigt model (Kloxin and Van Zanten, 2009; Tanner *et al.*, 2011),

$$J(t) = \sum_{n=1}^{N} L_n(1 - e^{-t/\tau_n}) + \frac{t}{\eta_0}. \tag{3.153}$$

The number of terms is kept smaller than the measured data points and τ_N are fixed to be logarithmically spaced. The coefficients L_n are determined subject to the minimization of a function that weights the residual sum of squares between the fitted function and measured values $\hat{J}(t)$ with a "smoothness" constraint,

$$\text{minimize: } [J(t) - \hat{J}(t)]^2 + \lambda \frac{\partial^2 L}{\partial \tau^2}. \tag{3.154}$$

This combination of terms prevents unphysical variations in the values of L_n. The smoothness is determined by the parameter λ and is given by the method implemented by Weese (Weese, 1993; Weese, 1992). Once an appropriate model of the retardation spectrum is found, the complex shear modulus is

$$G^*(\omega) = \left(\sum_{n=1}^{N} \frac{L_n}{1 + i\omega\tau_n} \right)^{-1}. \tag{3.155}$$

An example of the regularized fit of the Voigt fluid basis functions is shown in Fig. 3.10 for entangled semiflexible networks of F-actin. The mean-squared displacement (or creep compliance) is

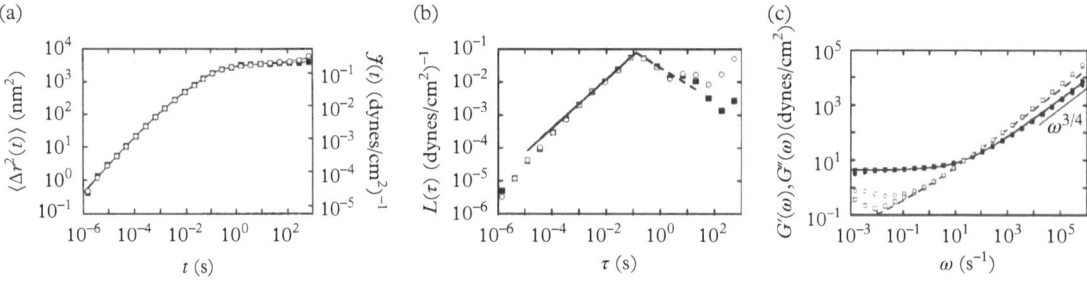

(a)   (b)   (c)

Fig. 3.10 *The microrheology of entangled F-actin solutions. (a) The mean-squared displacement of 1.6 µm diameter probes. (b) The retardation spectrum L(τ). (c) The calculated viscoelastic moduli. The solid points are the storage modulus G′(ω) and the open points are the loss modulus G″(ω). Reprinted with permission from Mason, T. G., Gisler, T., Kroy, K., Frey, E., & Weitz, D. A. J. Rheol., 44, 917–28 (2000). Copyright (2000), The Society of Rheology.*

fit using equation 3.152 with $N = 25$ terms. The resulting retardation spectrum is given in Fig. 3.10b from which the storage and loss moduli are calculated using eqn 3.155. The Tirkhonov regularized fit method has also been used to analyze probe microrheology measurements of micellar solutions of block copolymers (Kloxin and Van Zanten, 2009; Tanner *et al.*, 2011) and associative polymers (Abdala *et al.*, 2015).

3.9   Strengths and limitations of passive microrheology

The reader is no doubt interested in the theory and practice of microrheology due to its many potential applications. The small sample size requirements of microrheology and the ease of its implementation—at minimum requiring simple movies of particle Brownian motion—are among its strengths that will be highlighted throughout the remainder of this book. However, there are several important limits on passive microrheology that can be brought to light based on the discussion in this chapter.

The first and foremost limitation of passive microrheology is that it can *only measure a material's linear rheology*, because probe particles are in thermal equilibrium with the surrounding material. A probe in equilibrium with a material may not drive that material out of equilibrium! As a result, many interesting and technologically important rheological properties are inaccessible to passive microrheology: Yielding, shear thinning, shear thickening, and so on. Such behaviors, which arise when a material is driven strongly out of equilibrium, can only accessed in microrheology by *active* techniques in which the probe motion is driven by non-thermal forces, as discussed in Chapter 7.

Second, passive microrheology is limited to materials with rather weak moduli and low viscosities (or correspondingly large compliances) compared to many bulk-rheology measurements, because it depends on thermal motion as the driving force. The average thermal stress exerted on a particle scales as $\sim k_B T/a^3$. For a one micrometer diameter particle, this is on the order of just 10^{-2} Pa. As we discuss later, this range does not necessarily limit the utility of microrheology; it can still be used to screen whether an elastic gel forms from a precursor viscous fluid, for instance, even if the compliance is too low to be measured quantitatively. Conversely, passive microrheology excels at measuring many weakly-elastic or low-viscosity materials that are otherwise difficult to characterize using bulk rheometry, especially if the material is only available in limited quantities.

3.10   Validity of the GSER

The validity of the GSER depends on how well the assumptions made in its derivation apply to the experiment at hand, the two key ones being: (1) The applicability of the Stokes equation—the probe experiences a continuum mechanical environment—and (2) that the material is at thermal equilibrium, or sufficiently close to thermal equilibrium to constitute a "quasi-equilibrium" in the case of materials undergoing a chemical reaction, such as gelation or degradation. Non-continuum effects are not limited to passive microrheology, but affect all forms of probe microrheology, including active microrheology; they are all based on the Stokes relation relating the force acting on a probe that accompanies the material deformation.

3.10.1   Non-continuum effects

For the Stokes component we can ask whether the material behaves as a continuum on the length scale of a probe particle. This assumption could be violated if the probe size is smaller than the length scales of the material microstructure, as we discussed in Section 2.2. The local microstructure can also be changed by the particle.

A number of ways have been used to experimentally verify the validity of the continuum behavior in passive microrheology. Comparing the calculated moduli to other experiments, including bulk rheology, in overlapping frequency ranges (or by extrapolating one data set to the other) is one approach to testing its validity. Another method is to perform a series of experiments using different probe sizes, and a third is to use particles with different surface chemistries. Finally, the correlated motion of probes, presented in Section 4.11 and called two-point microrheology, can be used to measure the microrheological response on length scales larger than the probe diameter.

Particle size

Measurements made with probes of several diameters should collapse when scaled by the corresponding probe radius

$$\langle \Delta r^2(t) \rangle a = \frac{\mathbb{D} k_B T}{3\pi} \mathcal{J}(t) \tag{3.156}$$

provided the GSER is satisfied. Microrheology measurements of associative polymer solutions using a range of probe particles with diameters from 0.3 to 2.0 μm provide a good illustration of one such breakdown of the GSER (Lu and Solomon, 2002). As we see in

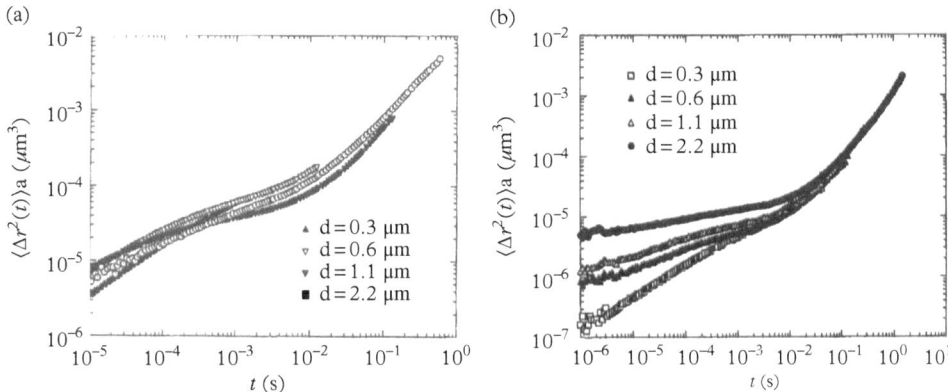

Fig. 3.11 *Microrheology of HEUR associative polymers at (a) low concentration—1 wt%, and (b) higher concentration—4 wt%. Reprinted with permission from Lu, Q. & Solomon, M. J., Phys. Rev. E, 66, 61504 (2002) Copyright (2002) by the American Physical Society.*

Fig. 3.11a, the mean-squared displacements for probes at the lower concentration of polymer collapse when scaled by the particle radius. At a higher-polymer concentration, however, the mean-squared displacements fail to collapse at the shortest times (Fig. 3.11b). The deviation from the GSER behavior is associated with the formation of a growing network of polymers. The results suggest that larger particles are sufficiently entangled in the developing network and exhibit a plateau modulus consistent with Maxwell fluid rheology, but that smaller particles are able to percolate through the presumably inhomogeneous structure. It is also apparent in Fig. 3.11b that beyond some relaxation time of the polymer network, probes of all diameters move as if in a viscous solution, and the proper particle scaling is recovered.

Microrheology experiments by van Zanten *et al.* (2004) for aqueous poly(ethylene oxide) (PEO) solutions demonstrate the proper particle size scaling expected of the GSER. The data are reproduced in Fig. 3.12. Four experiments measuring the Brownian motion of spherical polystyrene tracers in 7 wt% PEO collapse on a single curve when plotted as the creep compliance, $\mathcal{J}(t) = (\pi a/k_B T)\langle \Delta r^2(t)\rangle$, which scales out the particle size. The four probe diameters tested range from 0.195 to 1.55 μm.

Surface chemistry

The surface chemistry of the probes can affect microrheology measurements. Such a dependence usually indicates the depletion,

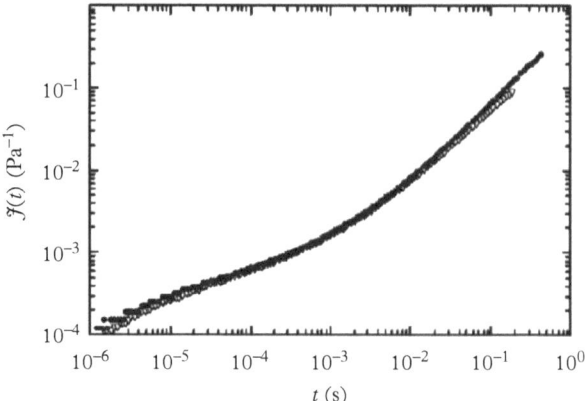

Fig. 3.12 *Microrheology of poly (ethylene oxide) polymers solutions using for probe sizes ranging from 0.195 to 1.55 μm diameter. Adapted with permission from van Zanten, J. H., Amin, S., & Abdala, A. A.* Macromolelcules **37**, *3874–80 (2004). Copyright (2004) American Chemical Society.*

accumulation, or restructuring of the material in the vicinity of the probe.

A good example is the sensitivity of F-actin microrheology on probe chemistry. McGrath *et al.* (2000) used laser tracking microrheology to measure the microrheology of tightly entangled F-actin networks and found that the modulus amplitude and phase angle depended on the surface chemistry of polystyrene probes, as shown in Fig. 3.13. The modulus amplitude had a clear dependence on the capacity of the probes to adsorb F-actin monomer. Also notable is the significant difference in the phase angle, plotted in Fig. 3.13b, across the samples. Probes with the lowest binding capacity have a phase

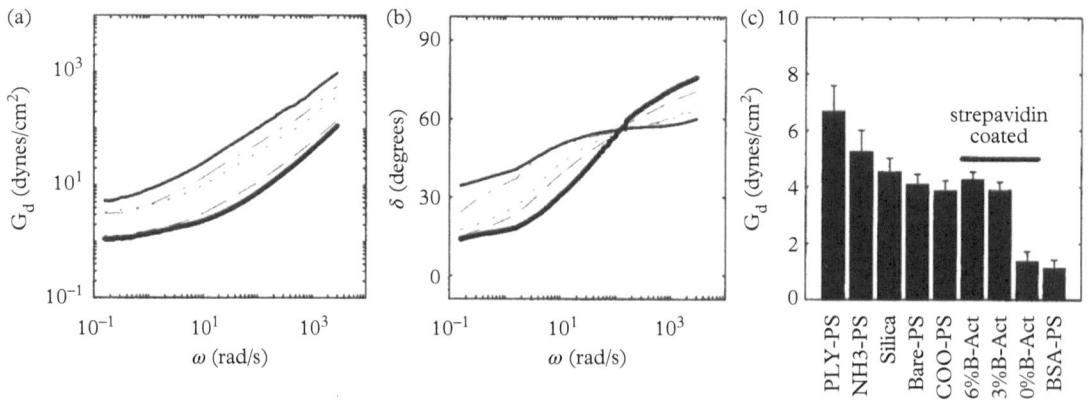

Fig. 3.13 *The dependence of F-actin microrheology on the surface chemistry of the polystyrene probes. Reprinted from* Biophys. J., **79**, *McGrath, J. L., Hartwig, J. H., & Kuo, S. C., The mechanics of F-actin microenvironments depend on the chemistry of probing surfaces, 3258–66, Copyright (2000), with permission from The Biophysical Society.*

angle approaching 90 degrees at high frequency, like a Newtonian fluid, probes with strong interactions to the F-actin produce phase angles consistent with the expected scaling of the semiflexible polymer high-frequency moduli, $\delta \sim 50° \sim \tan^{-1}[(3/4)(\pi/2)]$.

For probes that bind weakly to the F-actin network, the measured rheology is intermediate between that which one expects for the solvent and a tightly entangled semiflexible polymer network. This behavior is an indication that a shell of softer or depleted material has formed around the probes (Levine and Lubensky, 2001), and is discussed in more detail in Section 4.11.4. Similar to depletion, a higher density of material may accumulate near the probe surface, although this will appear as an increase in the effective probe size. Microrheology measurements in this case generally produce the correct frequency response, but with an apparent modulus that is higher than the true modulus. In other cases, discrete contact points between the probes and material may occur, with depleted regions between these. The probes are expected to report the correct frequency dependence of the rheology, but with a lower apparent modulus (Van Citters *et al.*, 2006).

3.10.2   Microrheology without probes?

In the "tracer probe" microrheology that we have been considering, a material is seeded with colloidal particles. The dynamics of these particles are used to measure the microrheology. Most often these probes are particles which are added to the material of interest. However, nothing prevents us from measuring the dynamics of the material itself, for instance if it's a concentrated emulsion or colloidal suspension. Can the rheology be derived from these experiments with the GSER?

Some of the first studies using light-scattering microrheology do measure the dynamics of emulsions and suspensions (Mason and Weitz, 1995; Mason *et al.*, 1997*b*). As we can see for the dynamics of concentrated emulsions in Fig. 3.14a, the droplets exhibit a mean-squared displacement reminiscent of the Kelvin–Voigt model. In fact, the Laplace-Transformed Kelvin–Voigt model (eqn 3.109) is compared to the transformed data by the dashed line Fig. 3.14b, matching the asymptotic solid and viscous limits. The difference between the model and data are represented by the open symbols and fit the power-law function $s^{0.5}$ over about six decades. When the Kelvin–Voigt model and power-law are converted to moduli, they agree well with mechanical rheology. That comparison can be seen in Fig. 3.14c.

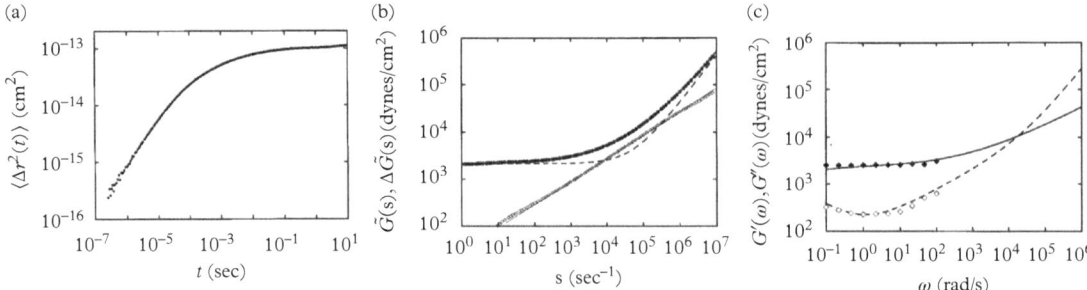

Fig. 3.14 *Dynamics of concentrated-monodisperse emulsion droplets (volume fraction $\phi = 0.65$) measured by diffusing wave spectroscopy. (a) The emulsion droplets' mean-squared displacement and (b) its Laplace Transform (solid symbols). (c) The storage and loss moduli derived from the light scattering are in good agreement with bulk rheology (symbols) in the overlapping frequency range. Adapted with permission from Mason* et al. *(1997b), The Optical Society.*

The agreement between bulk rheology and the rheology derived from the emulsion droplet dynamics seems to violate one of the key assumptions of our derivations in this chapter—that the medium constitutes a continuum on the length scale of the probe particle. In these cases, the probe particles are the material of interest. Earlier studies of hard-sphere suspensions, the Stokes–Einstein relation relating the short-time self-diffusivity D_s^s to the high-frequency viscosity η'_∞,

$$D_s^s(\phi) \overset{?}{\propto} \frac{\mathbb{D}k_B T}{3\pi a \eta'_\infty(\phi)}, \tag{3.157}$$

holds to within experimental accuracy across a wide range of volume fractions (Shikata and Pearson, 1993; Banchio *et al.*, 1999), but fails for dispersions in which particles interact by screened electrostatic interactions (Horn *et al.*, 2000). Such applications of the Stokes–Einstein equation exploring the relation between diffusivity and viscosity have a long history, indeed going back to studies of atomic and molecular fluids, where the *approximate* validity of the Stokes–Einstein formula for molecules was well known (Zwanzig and Bixon, 1970).

So, caution must be exercised when the dynamics of the material are interpreted using the GSER. Still, it can be seen as a potential *index* method of rheology, capable of detecting changes due to curing or gelation, for instance. Such methods have been applied to industrial rheology—the curing of paints and coatings, rheological changes that accompany food processing (yogurt, cheese), cements, and similar rheological changes in consumer-care products (Alexander and Dalgleish, 2007; Moschakis, 2013).

3.11   General limits of operation

The exact range of measurable moduli and time scales for passive microrheology depends on the technique that is used. In the next chapters, we will introduce the methods of multiple particle tracking (Chapter 4) and light scattering microrheology (Chapter 5). There are best practices and nuances for each experiment, but here it is useful to consider a few general limits that apply to any experimental passive microrheology method.

Passive microrheology relies on tracking the motion of particles with respect to time to calculate the mean-squared displacement. The range of time scales and displacements that are accessible with each experimental method define its operating regime. There is a lower time limit set by the MSD acquisition rate and an upper limit determined by the total acquisition time of the ensemble average mean-squared displacement. For example, using video microscopy for particle tracking, the video acquisition frame rate f sets the minimum time between video frames, $\tau_{min} = 1/f$, and thus the shortest lag time for the MSD. In light scattering, the minimum lag time may be as short as tens of nanoseconds—short enough that we may need to take into consideration the particle and fluid inertia.[10] A more common lower limit is $\approx 1\ \mu$s.

3.11.1   Minimum compliance

Consider the accuracy of the particle tracking and the minimum displacement of the probe's movement that can be detected. Let ε be the lower resolution of the position such that the measured mean-squared displacement

$$\langle \Delta \hat{r}^2(\tau) \rangle = \langle \Delta \hat{x}^2(\tau) \rangle + \langle \Delta \hat{y}^2(\tau) \rangle + \langle \Delta \hat{z}^2(\tau) \rangle \qquad (3.158)$$

is given by the "true" mean-squared displacement with a minimum value $2\varepsilon^2$ in each direction by

$$\langle \Delta \hat{x}^2(\tau) \rangle = \frac{kT}{3\pi a} \mathcal{J}(\tau) + 2\varepsilon^2. \qquad (3.159)$$

Then the *minimum compliance* $\mathcal{J}_{min}(\tau)$ must exceed

$$\mathcal{J}(\tau) > \frac{6\pi a \varepsilon^2}{k_B T}. \qquad (3.160)$$

The minimum compliance is independent of the number of dimensions $\mathbb{D}$ of the mean-squared displacement. For light scattering, $\mathbb{D} = 3$,

[10] In light scattering microrheology, the lower time limit τ_{min} is determined under most circumstances by the particle displacement resolution, *i.e.*, the time it takes a particle to diffusive a given length, like 1 nm. The exact value is determined by the scattering geometry, probe scattering properties (size, concentration), and other factors.

while particle tracking is most often performed with a 2D projection of the probe displacement, so $\mathbb{D} = 2$.

We can use eqn 3.160 to identify limits of purely viscous behavior, where $\mathcal{J}(t) = t/\eta$ and purely elastic behavior, where $\mathcal{J}_e = 1/G_e$. Our passive microrheology operating regime is set by the desire to unambiguously distinguish a sample in the two extreme limits of rheological response—a viscous fluid or elastic solid—for $\tau \geq \tau_{min}$. Figure 3.15 illustrates this heuristic with three sets of MSD curves, corresponding to three Newtonian fluids and three purely elastic solids.

In the first case (1) in Fig. 3.15, both MSD curves for the hypothetical viscous fluid and elastic solid are above $\langle \Delta r^2(\tau) \rangle > 2\mathbb{D}\varepsilon^2$ for $\tau > \tau_{min}$. Over all lag times, the fluid can be unambiguously distinguished from the elastic solid. Any sample with complex viscoelastic behavior between these limits could also be measured.

In case (2) of Fig. 3.15, the limit is reached wherein the particle motion in the elastic solid cannot be distinguished from the minimum displacement of the method being used. However, any sample in which the creep compliance is above this line will be measurable. When (3) is reached, however, there is a range of lag times for which the displacement in the limiting viscous behavior falls below $2\mathbb{D}\varepsilon^2$. The measured MSD would be constant, then increase after crossing $2\mathbb{D}\varepsilon^2$. Such a cross-over would be smooth and continuous, and thus the short-time or high-frequency response could be mistaken for elasticity even for a sample that is, in reality, purely viscous (Savin and Doyle, 2005).[11] Indeed, comparing eqn 3.118, the resulting curve resembles the expected MSD for a Maxwell fluid.

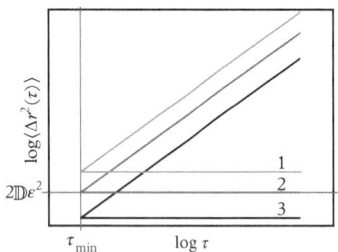

Fig. 3.15 *Limits of the mean-squared displacement.*

Maximum viscosity and shear modulus

The general relation eqn 3.160 can be written for the limiting viscosity of a Newtonian fluid,

$$\eta_{max} = \frac{k_B T \tau_{min}}{6\pi a \varepsilon^2} \tag{3.161}$$

or the shear modulus amplitude,

$$|G^*(\omega)|_{max} \approx \frac{k_B T}{6\pi a \varepsilon^2}. \tag{3.162}$$

Using multiple particle tracking microrheology (Chapter 4) and probe particles with diameter $2a = 1\ \mu$m with a typical particle tracking error of $\varepsilon \approx 10$ nm, the calculated limits above are $|G^*(\omega)| \approx 5$ Pa (or $\mathcal{J}_{min} \approx 0.2\ \mathrm{Pa}^{-1}$) and $\eta_{max} \approx 150$ mPa · s. But eqn 3.160 also gives the extent to which this range of moduli can be changed by selecting different probe particle sizes. Smaller probes can be used

[11] Savin and Doyle's (2005) data are reproduced in Fig. 4.25.

to increase the upper limits of modulus or viscosity, provided that the continuum approximation of the (generalized) Stokes equation is still satisfied, as Cohen and Weihs (2010) nicely demonstrate in microrheology studies of undiluted, viscous honey samples.

Operating diagram

In Fig. 3.16, we show the operating range of microrheology measurements based on eqns 3.161 and 3.162 in terms of the particle mean-squared displacement and time for three passive microrheology experiments: Multiple particle tracking by video microscopy (MPT), light scattering by diffusing wave spectroscopy (DWS), and single particle laser tracking (LT). Because a common probe particle diameter in microrheology measurements is on the order of 1 μm, we use this probe size to calculate the equivalent values of compliance $\mathcal{J}(t)$, viscosity of a Newtonian fluid η, and equilibrium modulus G_0 of an elastic solid. Again, these limits change with probe size and depend on other experimental factors. For instance, DWS microrheology depends on the scattering geometry and probe concentration. See Chapter 5 for a discussion of these and other details.

Each operating range is bound by a practical upper limit of time scale or lower limit of frequency characterized by the maximum MSD lag time τ_{max}. If it was certain that the fluid was Newtonian, one could propose to wait an indefinite time for the particles to move a measurable distance. But in practice, it is usually not feasible to track materials over such long times—small amounts of convection could obscure the particles' diffusive motion, or macroscopic vibration and

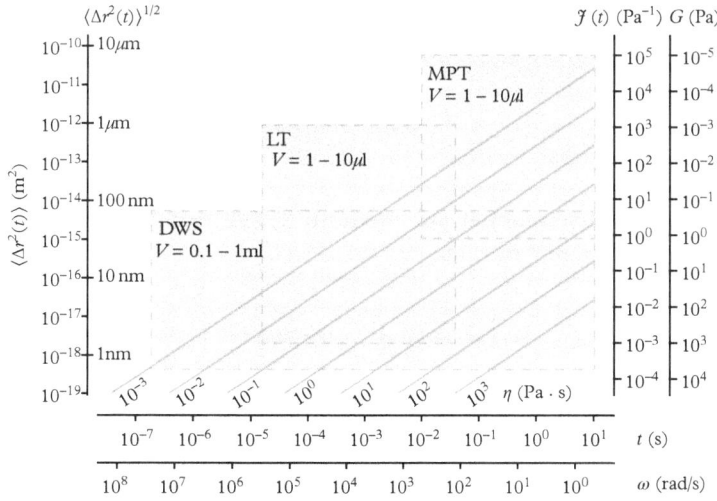

Fig. 3.16 *Operating limits of passive microrheology using multiple particle tracking (MPT), diffusing wave spectroscopy (DWS) and laser tracking (LT). The ranges shown here are calculated for a probe diameter $2a = 1 \mu m$. The limits can be shifted by changing the probe size, provided that this does not violate the continuum assumption of the Stokes equation.*

thermal expansion could contribute to the measured displacement. And there is the overall acquisition time of the measurement to consider; as we discuss later, materials with a rheology that changes with time, during a hydrogelation reaction, for instance, necessitate acquisition times that ensure the ensemble averaged mean-squared displacement approximates a stationary property. One exception is the use of multispeckle imaging discussed in Section 5.7.6.

Typical sample volumes are also indicated in Fig. 3.16. Taken together, the volumes, probe displacements, and time scales identify particular classes of problems amenable to microrheology:

(1) At high compliances, multiple particle tracking is suited to low viscosity samples and the incipient rheology of biomaterial hydrogelators.

(2) At lower compliances and short time scales, diffusing wave spectroscopy can access the terminal relaxation of polymer solutions, networks, and gels. With extended time scales, it can be used to characterize the relaxation time of polymer solutions.

(3) Screening experiments that take advantage of the low volume requirements, and quick mass and heat transfer in samples are.

These and other applications are discussed throughout the text as *application notes* and in Chapter 10.

With the GSER and our understanding of some of the strengths and limitations of passive microrheology, the next two chapters will focus on the experimental methods using microscopy and particle tracking (Chapter 4) and light scattering (Chapter 5).

. . . .

EXERCISES

(3.1) **Particle in an elastic medium.** Consider the solution to the equation of motion, (eqn 3.1)

$$M_p \dot{\mathbf{V}}(t) = \mathbf{f} - \int_{-\infty}^{t} \zeta(t - t') \mathbf{V}(t') dt', \qquad (3.163)$$

in a viscous fluid in the absence of inertia is

$$\mathbf{V} = \mathbf{F}/\zeta, \qquad (3.164)$$

where the resistance to motion is represented by a constant memory function (friction coefficient), $\zeta = 6\pi a\eta$. The velocity is a constant that is proportional to ζ. Since ζ is related to the viscosity η, a higher viscosity means that the particle translates more slowly. Show that the solution to the equation of motion for a purely elastic material is

$$\Delta \mathbf{X} = \mathbf{f}/\kappa, \tag{3.165}$$

where $\kappa = 6\pi a G$.

(3.2) **Green-Kubo formulas.** Various *Green-Kubo formulas* relate deterministic transport coefficients to autocorrelation functions of stochastic quantities. One step in the derivation of the GSER,

$$\langle v(0)\tilde{v}(\omega) \rangle = \frac{-\omega^2}{6} \langle \Delta \tilde{r}^2(\omega) \rangle, \tag{3.166}$$

is follows from one such formula.

(a) Show that eq 3.166 is the Fourier-Laplace Transform of the integral

$$D = (1/3) \int_0^\infty \langle \mathbf{v}(0) \cdot \mathbf{v}(t) \rangle dt, \tag{3.167}$$

where $\langle \Delta r^2(t) \rangle = 6Dt$.

(b) Next, derive eqn 3.167 by starting with the formula for displacement

$$\Delta \mathbf{r}(t) = \int_0^t \mathbf{v}(\tau)d\tau \tag{3.168}$$

by noting that the scalar mean-squared displacement is then

$$\langle \Delta r^2(t) \rangle = \int_0^t \int_0^t \langle \mathbf{v}(\tau_1) \cdot \mathbf{v}(\tau_2) \rangle d\tau_1 d\tau_2. \tag{3.169}$$

(3.3) **Fluctuation-dissipation.** A colloidal particle in water is subject to an impulsive force with magnitude f_0. In this problem, we will consider an energy balance on the particle.

(a) As the particle moves after the impulse is applied, how much work W is done on the particle by the surrounding fluid? What is the rate of work $\dot{W}$ done by the fluid?

(b) The rate of work done by the fluid on the particle represents the rate of energy dissipation, $\dot{W}_{out}$. An energy balance on the particle at equilibrium would yield

$$\dot{W}_{in} + \dot{W}_{out} = 0 \qquad (3.170)$$

(by convention, work done *on* the particle is negative and work done *by* the particle is positive). The impulsive force gives the particle a kinetic energy $\frac{1}{2}mv^2$. Thus, the rate of work done *on* the particle can be estimated as $\dot{W}_{in} = \frac{1}{2}mv^2/\tau$, where τ is the time over which the force f_0 acts on the particle (alternatively, $\dot{W}_{in} = \frac{1}{2}mv^2\delta(t)$, where $\delta(t)$ is the Dirac delta function). At equilibrium, the average kinetic energy should be $\frac{1}{2}kT$ by the equipartition theorem. Show that thermal equilibrium establishes a relationship between the force f_0 and the dissipation of energy via friction. This is (roughly) a statement of the *fluctuation-dissipation theorem*.

(c) Use the velocity autocorrelation and equipartition to show that the magnitude of the Brownian force is given by

$$F = 12\pi a\eta kT. \qquad (3.171)$$

Multiple particle tracking

Multiple particle tracking uses microscopy to measure the displacement of probe particles due to Brownian motion. From the observation of many particle displacements, the mean-squared displacement is calculated and can be interpreted in terms of the Generalized Stokes–Einstein Relation discussed in the last chapter.

Particle tracking has a long history. The earliest descriptions of Brownian motion relied on precise observations, and later quantitative measurements, using light microscopy. Multiple particle tracking microrheology is based on these same principles and tools. In a sense, little has changed since Brown's first reports in the early-nineteenth century, or perhaps more accurately, Jean Perrin's experiments a century later: Microscopy remains an accessible, accurate, and flexible method of measuring the thermal motion of colloidal particles. Another century's developments have brought better optical systems and, of course, modern video and digital image processing technologies, and thus increased the speed and scope with which large data sets may be collected and analysed.

Perrin's seminal treatise on the atomic nature of matter describes particle tracking experiments in detail (Perrin, 1913). Perrin employed a light microscope to view emulsions of gamboge, the resin from evergreen trees of the *Guttiferae* family emulsified by methanol. Perrin prepared emulsions by rubbing the resin under water with his hands, providing the shear necessary to break it into micrometer diameter droplets. Through painstaking measurement of dried samples, these rough emulsions were fractionated to create monodisperse samples for his studies. Perrin and his students Chaudesaigues, Dabrowski, Bjerrum, and Costantin, dispersed gamboge particles in water, then noted their positions at even intervals ranging from five seconds to one minute using a camera lucida and microscope (Figs. 4.1 and 4.2). They took great care to isolate their samples, to maintain constant temperature and prevent evaporation (Bigg, 2011).

With these measurements, Perrin used the Stokes–Einstein relation (eqn 1.3) to determine Avogadro's number (by way of $N_A = k_B/R$). This value agreed with Perrin's independent measurements of N_A based on sedimentation equilibrium measurements in suspensions. The nature and statistics of colloidal diffusion directly tested (and,

Microrheology. Eric M. Furst and Todd M. Squires, Oxford University Press (2017).
© Eric M. Furst and Todd M. Squires. DOI 10.1093/oso/9780199655205.001.0001

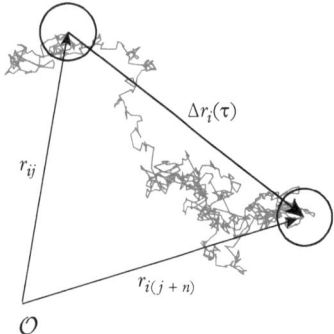

Fig. 4.1 *Particle tracking microrheology measures the displacement of tracer (probe) particles as they move randomly in a fluid or (weak) solid. Each particle i will move a displacement $\Delta r_i(\tau)$ over a time τ between a beginning observation j and a final observation j + n of the particle position. Here the lag time $\tau = n/f$ is given by the number of observations n made at a frequency f. A representative random trajectory is shown connecting the two-particle positions.*

obviously, confirmed) Einstein's molecular theory of Brownian motion. Perrin received the 1926 Nobel Prize in Physics for his work that, in the words of Carl Oseen's award ceremony presentation speech, "put a definite end to the long struggle regarding the real existence of molecules."[1]

Perrin used a microscope to image and track the motion of particles with a known size in a fluid of known viscosity to measure fundamental physical constants. Particle tracking microrheology represents a variation of this experiment, measuring the motion of particles to find the (complex) viscosity of the surrounding medium. Of all experimental techniques for microrheology, particle tracking is likely the simplest to implement, and often yields rich data. Simultaneous tracking of multiple particles improves measurement statistics, and extends the strategy in significant ways. The statistics of different particles may reveal distributions of rheological properties in heterogeneous materials, as discussed in Section 4.10, and cross-correlations between distinct particles encode the rheology of the material that lies between them (so-called two-point microrheology, discussed in Section 4.11). Thus, multiple particle tracking enables rheology to be measured on different length scales and positions to map out heterogeneities in the rheological properties in the sample.

4.1   Video microscopy

The most common particle tracking tools today are a light microscope, a video camera, and a computer. Image frames of monodisperse particles, usually a polymer latex dispersed in the medium of interest, are taken with the camera, stored by the computer, and later processed using a program that implements a particle tracking algorithm. The result is a trajectory of each particle, not unlike those shown in Fig. 4.2, from which the particles' mean-squared displacements $(\langle \Delta \mathbf{r}^2(t) \rangle)$ and other statistics may be determined. As discussed in Chapter 3, the rheology of complex materials may be determined quantitatively from these measurements, with multiple successful examples shown in Fig. 3.2.

In the following sections, we describe this process in detail. Later sections concern the practice of multiple particle tracking, as well as the accuracy and associated errors of locating each particle position in every image frame.

The role of the video microscopy apparatus is to record the particle positions at known time intervals. An example experimental setup is illustrated in Fig. 4.3, which shows an inverted microscope equipped with both a halogen lamp for bright field microscopy and a mercury arc lamp for epifluorescence microscopy. The microscope should be secured against sources of vibration, which include air handlers, pumps, lab refrigerators, and people. Pneumatic isolation tables are commonly used to dampen vibration (Crocker and Hoffman, 2007).

Images are recorded with a video camera and saved on a computer for later analysis. The method of storing images and transferring them to a computer depends on the camera that is used. Many scientific grade high-speed cameras store images in an internal memory buffer before transferring them directly to the acquisition computer, through an ethernet connection, for instance. Less expensive video cameras that output an analog or digital signal directly require additional equipment to store images and important them into the computer for analysis. A computer may store images directly to the hard drive using a frame grabber. Some investigators prefer to store images on a video recording device such as a digital video (DV) recorder to archive the experiment before transferring the data to the computer.

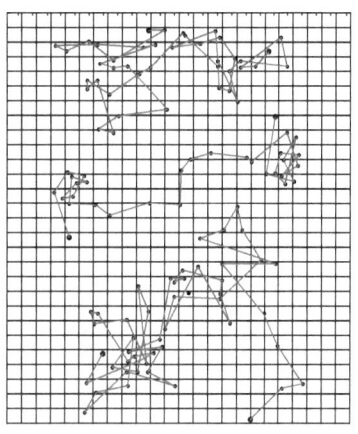

Fig. 4.2 *Three particle trajectories from Perrin's book (Perrin, 1913). How strange these must have seemed to those who were used to thinking of the motion a body in terms of a smooth, continuous trajectory! The position of a single particle was noted here at 30 second intervals. The method required two experimenters: One to note the particle location and the other to call out even time intervals from a chronograph. Each square is 3.2 μm.*

While bright field images can be used for particle tracking, fluorescence microscopy provides a high contrast and a good signal-to-noise ratio (SNR). The chief disadvantages of fluorescence imaging are photobleaching of the tracer particles over time and the background fluorescence produced by particles out of the focal plane, which contributes to the noise when determining the locations of the particles. Figure 4.4 shows a fluorescence microscopy image of 1 μm diameter polystyrene latex particles dispersed in water. The brightest particles are those closest to the focal plane, while particles above and below the focus appear as a concentric ring or rings.

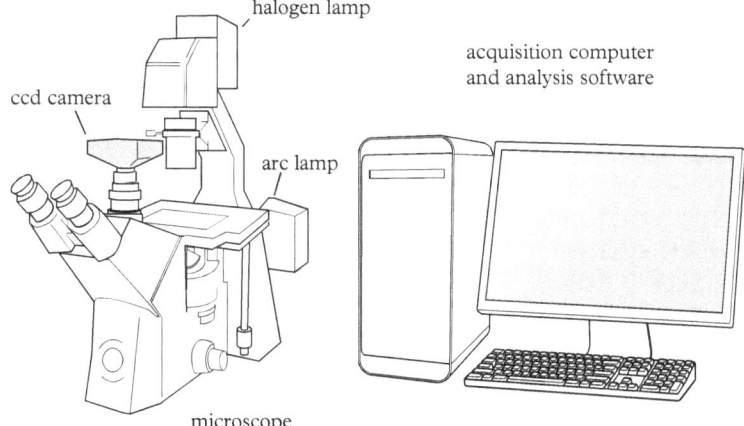

halogen lamp

acquisition computer and analysis software

ccd camera

arc lamp

microscope

Fig. 4.3 *Video microscopy equipment consists of a light microscope, a camera, and a computer for data acquisition and analysis.*

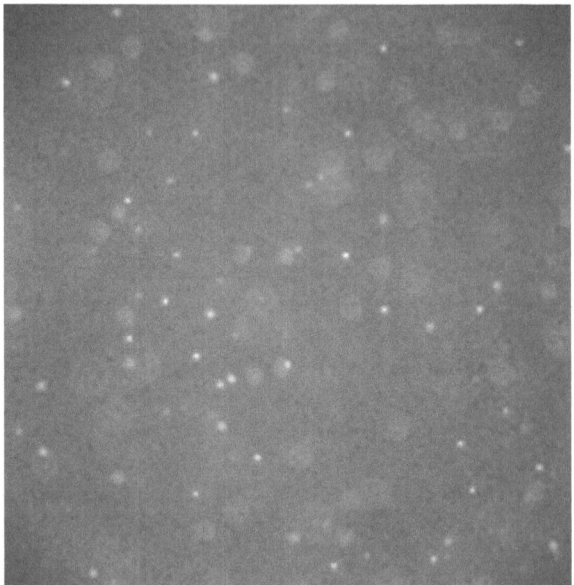

Fig. 4.4 *Fluorescence image of 1.06 μm diameter polystyrene particles, 100× total magnification. The image is 574×574 pixels.*

4.1.1   Video camera

Video microscopy uses an electronic solid-state chip camera with a pixelated array of detectors, such as a charge-coupled device (CCD), intensified CCD (ICCD), electron-multiplied CCD (EM-CCD) or so-called CMOS device based on an active-pixel sensor. CCD detectors are more common in scientific video cameras, and while CMOS detectors are typically found in consumer digital cameras, they are becoming more widely used in scientific applications.

Like any optical detector, the camera sensor converts light power to electrical current. The basis of operation is similar among detector types. In a CCD, light strikes a capacitor array, which causes each capacitor to accumulate charge. The charge is proportional to the incident light intensity. Each pixel accumulates photoelectrons during an *integration time,*[2] σ_t. The accumulated charge in each capacitor is determined in the read-out process, a sequence in which the charge in each capacitor is transferred to its neighbor by a control circuit. The last capacitor in the array transfers its charge to an amplifier, which converts it to a voltage. Thus, the array is converted to a sequence of voltages (analog output), which can then be converted to a digital value and stored in memory. The precision of the analog-to-digital conversion, the number of bits used to represent the intensity, is typically between 8 and 14. A camera with an 8-bit depth has the ability to discriminate 255 intensities, while one with a 14-bit depth can discriminate 16384.

[2] Also referred to as exposure time or shutter time.

Particle tracking is reasonably forgiving with respect to the choice of video camera. Plenty of experiments have been performed with relatively inexpensive NTSC standard CCD cameras with 8-bit depth. These cameras typically output an analog signal, which needs to be acquired by the computer through a *frame grabber* and software. Scientific grade CCDs offer advantages such as greater control over the imaging region of interest, integration times, pixel binning, and acquisition (frame) rates. Many of the latter cameras store images in onboard memory. Cameras that use cooled CCDs or intensifiers are generally unnecessary for particle tracking when the probe particles are sufficiently bright.

4.1.2   Image file types

When saving files it is best to use a file format that does *not* use a lossy compression algorithm. The portable network graphics (PNG) format is a good choice, and TIFF files are also acceptable, as long as they do not use a lossy JPEG compression. Since native TIFF files are not compressed, using them will lead to large image files. A single 16-bit megapixel image is 2 megabytes.[3] One ten second movie at 100 fps requires 1 gigabyte of storage.

4.1.3   Imaging basics

Particle tracking starts with obtaining the best images possible. The image in Fig. 4.4 is 574×574 pixels. It was cropped from a larger image from a megapixel (1024×1024) camera. The image has 8-bit depth, so the grayscale pixels are represented by integer values between 0 and 255. Notice that the image has a non-uniform background intensity, *e.g.*, with darkened corners. This must be corrected before particle tracking. The image has a good dynamic range that uses a wide extent of the possible 8-bit values. A histogram of the pixel intensity is shown in Fig. 4.5. Only two pixels reach the maximum value 255, and the rest are distributed with intensities well below this. The brightest pixels associated with in-focus particles are in the range of 200-55. A slightly lower intensity could be used to avoid saturation, when multiple pixels are at the highest-intensity value of the camera. It is best to detect saturated pixels at the beginning of an experiment and to eliminate them by adjusting the incident light intensity or exposure time.

In a fluorescence image, particles that are closest to the focal plane appear as bright disks with an approximately Gaussian intensity distribution. Particles above and below the focal plane will appear as rings with a central spot, typical of the Airy-disk pattern in the

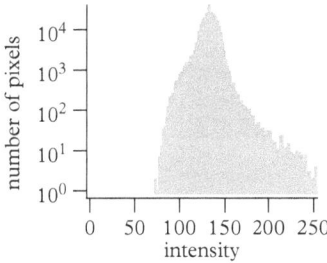

Fig. 4.5 *Sample fluorescence image and image histogram.*

---

[3] While many high-grade cameras output a 12- or 14-bit depth image, each pixel is stored as two bytes (16 bits).

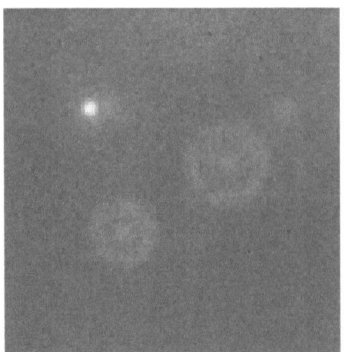

Fig. 4.6 *Images of one particle close to the paraxial-focal plane and three nearby particles that are out of focus.*

diffraction limit, as shown in Fig. 4.6. These images represent a convolution of the "true" particle image (the spatial distribution of light emitting points in the sample) with the *point spread function* (PSF) of the imaging system. The PSF represents an intensity distribution resulting from a point source of light imaged through the microscope. A microscope image is the convolution between the spatial distribution of light emission (or reflection) from the sample and the PSF.

The maximum lateral-resolving power d of a microscope is determined by the diffraction limit (Born and Wolf, 1999)

$$d_l = \frac{0.61\lambda_0}{\text{NA}} \tag{4.1}$$

where NA is the numerical aperture of the objective.[4] For a high-quality water immersion microscope objective with NA = 1.2 and fluorescence emission at the vacuum wavelength λ_0 = 520 nm (the peak emission for fluorescein isothiocyanate, FITC) the lateral resolution is at best approximately half the wavelength of the light, which is a substantial fraction of the physical particle size. Non-immersion objectives are limited to NA values below 1; the lateral resolution for a 40× plan-apochromat objective with NA = 0.7 is 450 nm. The rings that appear as the particles move in and out of the focal plane must remain symmetric. Rings that are pinched or skewed are an indication that the illumination or imaging system is misaligned or suffering from other aberrations. These must be corrected before performing particle tracking experiments.

The axial (or longitudinal) resolving power of a microscope is also important in particle tracking microrheology. The axial resolution determines the depth of field, the distance between the nearest and farthest object planes that are simultaneously in focus. Since many particle tracking microrheology experiments investigate samples where the particle mobility is relatively large (*e.g.*, they aren't trapped strongly in a fixed position), particles are typically free to move in and out of the focal plane. This movement affects the statistics of particle tracking by shortening the particle trajectories, and also introduces bias in heterogeneous materials, a topic we will return to in Section 4.10.2.

Like the lateral resolution, the depth of field is also determined by the numerical aperture of the objective. For high numerical apertures, the depth of field is (Pawley, 2006)

$$d_i = \frac{\lambda_0 n}{\text{NA}^2}. \tag{4.2}$$

The 63× NA 1.2 water immersion objective and sample refractive index n = 1.33 used to produce the image in Fig. 4.4 has a depth of

[4] We assume that the emission from the fluorophores is *incoherent*. Imaging with *coherent* sources results in a slightly lower-resolving power. See Born and Wolf (1999) for a more complete discussion.

field of about $d_i = 500$ nm for fluorescence at $\lambda_0 = 520$ nm. However, depending on the particle tracking algorithm, particles that would be considered out-of-focus using eqn 4.2 may be tracked as well.[5]

The choice of microscope objective will determine the efficiency of light collection from the sample, the depth of field, and the magnification. An additional and important consideration is the working distance of the objective, which determines the maximum distance of the image plane from the sample boundary. The image plane must be sufficiently far from the boundary to minimize hydrodynamic interactions with the no-slip surface, as discussed in Sections 2.6.4 and 4.3.1.

4.2   Image quality

A number of factors influence the quality of video microscopy data, including:

(1)   exposure time and frame rate of the camera;

(2)   detection noise;

(3)   brightness of the particles and electronic gain of the detector.

In this section, we examine each of these aspects.

4.2.1   Frame rate and exposure time

Video cameras acquire images at a *frame rate*, such as 100 frames per second (fps). The frame rate is the time between complete images. It ultimately limits the shortest lag times of particle tracking data $\tau_{min} = 1/f$. The frame rate is chosen such that the exposure time σ_t is at most one tenth as long as the time between frames, $\sigma_t \leq 0.1\tau_{min}$. This ensures that the particles do not move too much during the image acquisition. The minimum exposure time will depend on the sensitivity of the camera, the intensity of fluorescence emission of the particles, and the overall tolerance to noise.

The frame rate of NTSC compatible CCD cameras (known as the RS-170 standard) is fixed at 30 fps. Analog cameras are common and can be recorded directly to video tape or DVD for later retrieval.[6]

4.2.2   Detection noise

Noise is inherent to electronic imaging systems due to both the quantization of light as photons and the electronics that carry minute

[5] The human eye can accommodate a wide range of focus. More particles may be visible when viewed through the microscope oculars than are captured by the thin fixed plane of a CCD sensor.

[6] The NTSC standard produces 480i video—480 interlaced vertical lines. The horizontal resolution is typically 640 pixels and the pixels have an aspect ratio of 4:3. PAL video standard used in Europe and parts of Asia are 525i at 25 Hz.

photoelectron charges and convert them into the information we see as a pixel value in an image.

Photons are absorbed by the detector to create photoelectrons.[7] The instantaneous conversion rate $\alpha(t)$ in photoelectrons per time is proportional to the instantaneous incident power $W(t)$ at the detector (Mertz, 2010),

$$\alpha(t) = \frac{Q(\lambda)}{h\nu} W(t) \tag{4.3}$$

where $Q(\nu)$ is the detector *quantum efficiency* at the wavelength $\lambda = c/\nu$ and h is Planck's constant.[8]

While eqn 4.3 is straightforward in that the instantaneous photocurrent is proportional to the instantaneous power, consider that photons arrive at random intervals, and thus $W(t)$ is stochastic. Therefore, the number of photoelectrons generated during the detector integration time varies. This fluctuation is the *shot noise*. If n is the measured number of photoelectron conversions over the detector integration time σ, then the mean value is simply $\langle n \rangle = \sigma \langle \alpha \rangle_\sigma$ photoelectrons per second,[9] but the normalized variance is (Mertz, 2010)

$$\frac{\langle n^2 \rangle - \langle n \rangle^2}{\langle n \rangle^2} = \frac{1}{\langle n \rangle} + \frac{\langle S^2 \rangle_\sigma - \langle S \rangle_\sigma^2}{\langle S \rangle_\sigma^2} \tag{4.4}$$

where $\langle S \rangle_\sigma$ is the time-averaged source intensity, which is also assumed to fluctuate. There are two contributions to the normalized variance of n in eqn 4.4—one from the fluctuations of the source, and the first term on the right side of the equation, which comes from the quantum noise. We see that the shot noise scales inversely with the mean photoelectron current. Shot noise will limit the signal-to-noise ratio of an image at low-light intensity levels or short exposure times. In general, the SNR is the inverse of the normalized variance. In the absence of source fluctuations, eqn 4.4 reveals the SNR to be proportional to the photoelectron conversion rate. The SNR may therefore be improved by increasing the light intensity, integration time, or imaging system aperture.

In most cases, shot noise will be the limiting source of noise of the camera. Nonetheless, there are several other forms of noise in video imaging that should also be kept in mind.

One additional source of noise is the *dark current* of the detector. Dark current is thermionic in origin for semiconductor-based detectors—the thermal energy at the detector active area produces a background current even in the absence of light. For this reason, some sensitive cameras used for low-light intensity conditions or for short exposure times employ a cooled detector. Similarly, there is the *Johnson noise* of the camera electronics due to the finite

---

[7] This is the semi-classical description of light; it propagates as waves, but is detected as particles.

[8] Typical efficiencies for CCD cameras lie between 60–80% in the visible spectrum.

[9] The average conversion rate is $\langle \alpha \rangle_\sigma = \frac{1}{\sigma} \int_t^{t+\sigma} \alpha(x) \, dx$.

temperature of the charge carriers. Johnson noise is a manifestation of fluctuation-dissipation, which relates spontaneous fluctuations in current to the resistance (or more accurately, the impedance), analogous to the relation between the Brownian force and viscous dissipation (Nyquist, 1928). Finally, *readout noise* arises in the photoelectron digitization process. For CCD cameras with slow readout frequencies (< 1 MHz), typical readout noises are small. However, for fast CCD cameras (readout frequencies ≥ 10 MHz), readout noise can dominate shot noise. Janesick *et al.* (1987) provide robust estimation techniques for different sources of noise in CCD sensors.

4.2.3   Image signal-to-noise ratio

In the previous section, we considered the sources of noise in an imaging system that affects the signal-to-noise (SNR) ratio of an acquired image. In this section, we will discuss methods of quantifying the image SNR for video microscopy data. Such measures are important for evaluating and optimizing the image quality of a multiple particle tracking experiment.

In image processing, it is common to characterize image SNR in decibels (dB) as

$$\text{SNR (dB)} = 10 \log \left(\frac{\sigma_{\text{image}}}{\sigma_{\text{noise}}} \right) \tag{4.5}$$

where σ_{image} and σ_{image} are standard deviations of the image and noise pixel intensities, respectively (Russ, 2011). The noise is measured by calculating the variation between multiple still images. For multiple particle tracking experiments, this can be achieved by first preparing a sample in which particles are immobile and imaged under similar imaging conditions (particle size, volume fraction, illumination intensity, focal plane, *etc.*). For instance, in one standard practice, particles are trapped in a sufficiently strong gel (*e.g.*, polyacrylamide) such that their motion is arrested within the precision of the tracking.

The image frames of bright particles on a (mostly) dark background in multiple particle tracking microrheology suggest other criteria for the SNR that can be applied to any particle tracking video. Savin and Doyle (2005) for instance, apply the theory of Rose (1948) to calculate the signal as the difference between the local brightness of a particle and the average brightness of the background relative to the fluctuations of the background.[10] Written in terms of the average intensity of the particles $\langle I_S \rangle$ (the signal) and the average intensity of the background $\langle I_N \rangle$ (the noise),

$$\text{SNR} = \frac{\langle I_S \rangle - \langle I_N \rangle}{\sigma_N}. \tag{4.6}$$

[10] This expression, which is related to the reciprocal of the coefficient of variation of the image, is sometimes referred to as a contrast-to-noise ratio.

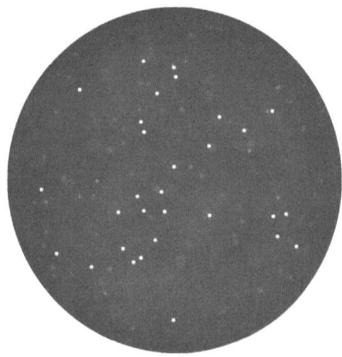

Fig. 4.7 *An image in which the signal particles in the paraxial focus, have been masked out (indicated by white regions) to calculate the background noise. The vignetted area of the image has also been masked.*

where $\sigma_N = (\langle I_N^2 \rangle - \langle I_N \rangle^2)^{1/2}$ is the standard deviation of I_N. The numerator in eqn 4.6, $\Delta I_S = \langle I_S \rangle - \langle I_N \rangle$, is the excess expected intensity of the signal over the background. The noise I_N is the intensity in regions of an image that exclude the in-focus particles, and the calculation consists of applying a mask over the particles and averaging the background noise, as in Fig. 4.7.

Figure 4.8 shows a comparison of two images that illustrate the SNR calculation. Both images are 1.06 μm fluorescent particles at 63× magnification using a NA 1.2 water immersion objective. The black region surrounding the fluorescence image is caused by vignetting and is excluded from the analysis, although it can be used to distinguish sources of noise generated by the acquisition system, such as the readout and dark-current noise previously discussed, from shot noise. In the left-most image, particles are clearly distinguishable from the darker background due to the longer camera integration time, 0.9 ms. Using a 16-bit range of pixel values, the average signal of the bright particle image is $\langle I_S \rangle = 3.9 \times 10^4$, the average noise is $\langle I_N \rangle = 1.8 \times 10^4$, and the standard deviation of the noise is $\sigma_N = 1.9 \times 10^3$. The SNR is 11 by eqn 4.6. The image on the right, in which particles are much dimmer, is taken using an exposure of 10 μs. The average signal is $\langle I_S \rangle = 2 \times 10^4$, and the average and standard deviation of the noise are $\langle I_N \rangle = 1.5 \times 10^4$ and $\sigma_N = 1 \times 10^3$,

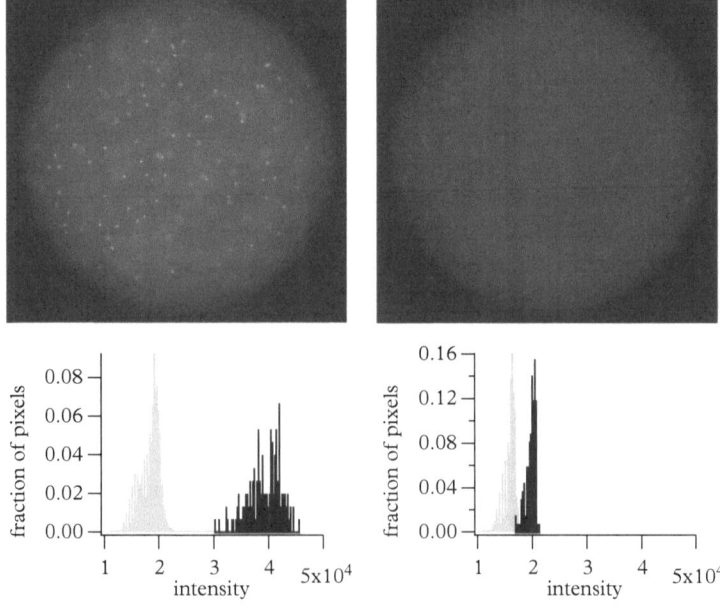

Fig. 4.8 *Images of 1.06 μm fluorescent particles at 63× magnification using a NA 1.2 water immersion objective. The left image is taken with a 0.9 ms exposure, and the right image with 10 μs exposure. Below the images, normalized histograms of the background noise (light gray) are compared with the signal histogram (dark gray).*

giving SNR = 4. A minimum desirable signal-to-noise ratio SNR ≥ 5 is often cited (Burgess, 1999).

Histograms of the signal and noise show the relative separation that determines the SNR in both images. Despite the clear visual differences between the images, the histograms demonstrate that the signal is robustly distinguishable from noise even under less than optimal imaging conditions; however, as we discuss in Section 4.8, the SNR will affect the overall tracking accuracy.

What are the choices a microrheologist can make to yield the best particle tracking results? Answering this question is a matter of optimizing several competing and often antagonistic contributions that are governed by the system magnification, brightness of the tracer particles, and noise of the camera and electronics systems, among other factors (Crocker and Hoffman, 2007). Multiple particle tracking must balance a field of view wide enough to image on the order of a hundred particles, while maintaining a sufficient resolution such that each particle image covers a reasonable number of pixels. The magnification of the optical system affects both the field of view and the light reaching the detector camera. The condenser numerical aperture also affects the brightness and contrast of tracers in an image.

Figure 4.9 are images of 1.06 μm diameter latex particles at three magnifications, 32, 64, and 100×. At the lowest magnification, the particle image is no more than three pixels across, so the sub-pixel resolution of the tracking methods we will see in Section 4.4 will be less accurate. Better images are obtained with a 40× NA 0.75 objective and 63× NA 1.2 water-immersion objective, which yields particle images that are ten to 20 pixels across. At higher magnification, the large particle image is offset by a smaller field of view, and hence fewer visible particles to track. The magnification also affects the SNR, which is to a first approximation proportional to the inverse square of the magnification,

$$\text{SNR} \sim M_T^{-2}. \tag{4.7}$$

Higher magnifications lead to lower signal-to-noise ratios because the light collected by the objective is spread out over more pixels. This

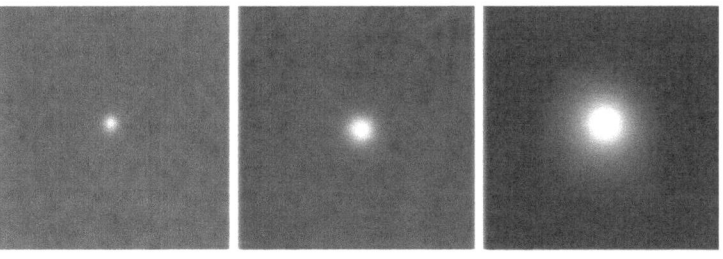

Fig. 4.9 *Images of the same 1.06 μm diameter particles. Each image is 40×40 pixels at three total magnifications: (left) 32×, (middle) 64×, and (right) 100×. The images were taken using a 0.9 ms integration time.*

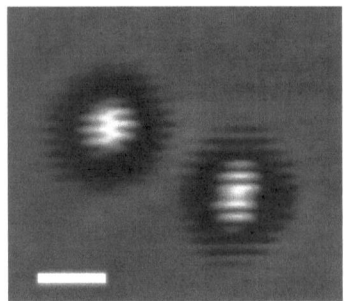

Fig. 4.10 *An interlaced image of 1μm diameter particles. The particles' movement between the acquisition of the odd and even fields is apparent by their displacement. The scale bar is 1 μm. Reprinted from* J. Colloid Interface Sci., *179,* Crocker, J. C. & Grier, D. G. Methods of digital video microscopy for colloidal studies, pp. 298–310, Copyright 1996, with permission from Elsevier.

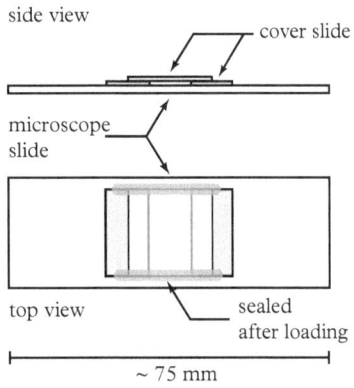

Fig. 4.11 *A typical multiple particle tracking sample cell constructed from a glass microscope slide and glass cover slides.*

---

[11] Interlacing was introduced to reduce flicker in broadcast television.

decrease can be compensated somewhat with a higher illumination level or by using a higher numerical aperture objective.

4.2.4   Other image artifacts

When the integration time of a CCD detector is sufficiently long, photoelectron charges collected in the sensor's electronic "bins" in the brightest part of the image will overflow to neighboring bins. This results in *blooming* of the image. The incident light intensity or exposure time should be reduced to avoid this condition.

Finally, some cameras produce *interlaced* images. Each video frame consists of two *fields* consisting of the odd and even lines of the image. For the RS-170 standard,[11] the fields are acquired 1/60th of a second apart. Since particles can move during the time between the fields are scanned, the two fields in the image will be offset and have a "wavy" appearance that results in poor tracking. Interlaced image frames should be *de-interlaced* by using only one of the fields before they are processed further. An example of an interlaced image is shown for 1 μm diameter particles in Fig. 4.10.

4.3   Particle tracking samples

Before moving on to particle tracking, we discuss the basics of sample preparation for video microscopy. For simple sample cells, there are a few things to consider. Sample chambers can be built by hand with microscope slides and cover slips or using pre-made glass capillaries. These sample chambers should be meticulously cleaned beforehand. Microfluidic devices provide additional capabilities for sample preparation and manipulation, as discussed in "Microfluidics."

Sample chambers

A schematic of a typical particle tracking sample is shown in Fig. 4.11. The chamber is constructed using a glass microscope slide and glass coverslips, both as a cover or spacer. The glass pieces are glued together with a UV-curing epoxy. The cover slide thickness, typically No. 1.5 (0.16–0.19 mm thick) are reasonable for minimizing probe interactions with the wall, and the final channel dimensions lead to volumes in the order of 15 μl. Spacers with similar dimensions include strips of plastic parafin film, which is less durable, but often adequate. The top cover slide or capillary wall should have a thickness that is compatible with the microscope objective; many require the glass to be 0.13–0.17 mm thick (No. 1 cover slips), for instance.

Other objectives can be adjusted to accommodate variations in cover slip thickness with a correction collar.

Introducing a sample into chambers or capillaries is usually straightforward, since capillary action is normally sufficient to draw in the sample, assuming that the samples wet the chamber adequately. The sample chamber should produce minimal optical distortion. If the material of interest is a viscoelastic solid, such as a weak gel, then it is desirable to form this gel *in situ* so as not to disrupt the microstructure by the strong shear rate as it wicks into the chamber. Of course, sufficiently solid samples or highly viscous fluids may not flow into the sample chamber quickly or at all.

The three most important considerations for MPT samples are:

(1) effectively sealing the sample,
(2) the sample dimensions,
(3) the probe particle concentration.

After introducing the sample, the chamber must be sealed, especially to eliminate potential sources of convection. Epoxies, mixtures of beeswax and parafin, and even nail polish (color optional) are common. Sealants that set quickly should be chosen, to avoid sample contamination due to sealant/sample mixing. Leaky samples are difficult to detect, except by the drift that they inevitably introduce in particle tracking data (discussed in Section 4.8.4). Slowly leaking samples may eventually dry, but over a period of days. Finally, care should be taken to avoid introducing air pockets in the sealed sample. Interfaces readily adsorb macromolecules, proteins, and particles, and their expansion, contraction, or even wandering by buoyancy, can generate additional data drift or disrupt the sample integrity.

Microfluidics

Beyond the simple sample chambers previously described, the small sample volumes required for multiple particle tracking microrheology are especially amenable to interfacing with microfluidic devices. The small dimensions enable rapid heat exchange or mass transfer, and thus sample conditions can be changed quickly to monitor their effects, or merely increase the throughput of the rheology measurements. Several examples of microfluidic microrheology systems are shown in Fig. 4.12.

Figure 4.12a shows the microfluidic dialysis cell of Sato and Breedveld (2006), which exploits the rapid mass transfer times in microfluidic dimensions. The device consists of a quiescent, constant volume sample chamber overlaid with a rigid, porous membrane. On the other side of the membrane, a chamber connected to an external

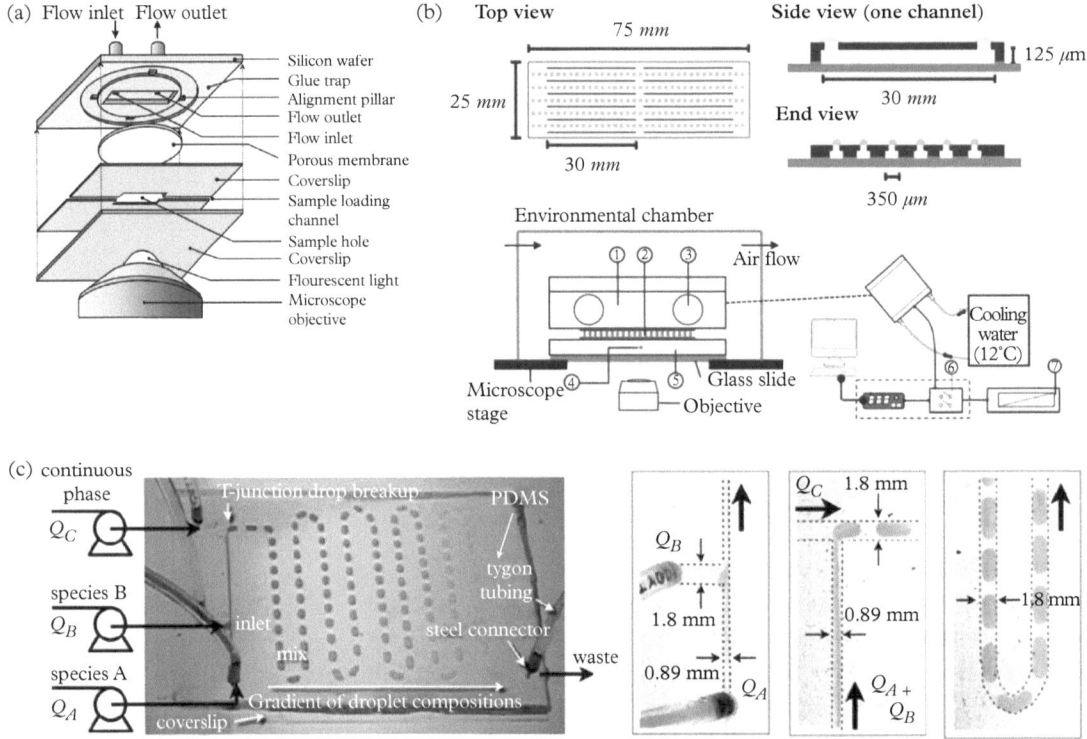

Fig. 4.12 *Examples of microfluidic devices used as sample environments for particle tracking microrheology experiments. (a) A sample cell that incorporates a rigid porous membrane for rapid solvent exchange. Reprinted with permission from Sato, J. & Breedveld, V., J. Rheol., 50, 1–19 (2006). Copyright 2006, The Society of Rheology. (b) The temperature of multiple microrheology samples in microfluidic channels is controlled by a Peltier stage. Reprinted from Josephson, L. L., Galush, W. J., & Furst, E. M. Biomicrofluidics, 10, 43503 2016b with the permission of AIP Publishing. (c) A microfluidic T-junction device is used to make multiple microrheology samples spanning a range of compositions. Reproduced from Schultz and Furst (2011) with permission from The Royal Society of Chemistry.*

flow loop is designed to rapidly exchange solvents in the sample. The dialysis membrane was used to study the effect of ionic strength on the rheology of sodium alginate and sodium polystyrene sulfonate (NaPSS).

Similarly, small samples give rise to fast heat exchange and rapid temperature equilibration. With multiple samples on a single microscope slide, either by custom-fabricated sample channels or by simply attaching several glass capillaries, the temperature dependence of several samples can be measured simultaneously (Josephson *et al.*, 2016*b*). One such device is shown in Fig. 4.12b, which uses a

Peltier stage to control the temperature of several samples. Finally, microfluidics can be used to prepare the samples themselves. The device shown in Fig. 4.12c, generates tens to hundreds of microliter-volume samples in the form of aqueous droplets in a non-aqueous carrier fluid using a microfluidic T-junction (Schultz and Furst, 2011). By controlling the inlet composition to the T-junction, each sample can be made with a unique composition.

4.3.1 Sample dimensions

Particle tracking samples can be small—their dimensions often contain volumes that are microliters to tens of microliters. To reliably measure material properties, the probe motion should be free of significant hydrodynamic interactions with the sample walls.

In Section 2.6.4 we discussed the hydrodynamic coupling between a probe of radius a and a wall separated by a distance l, revealing corrections to the probe mobility of order a/l. More cumbersome calculations are required to compute the hindered mobility b_l of a sphere at the mid-plane between two solid, no-slip walls, translating parallel to those walls, as shown in Fig. 4.13 (Faxén, 1922; Happel and Brenner, 1983)

$$\frac{b_l}{b_0} = 1 - 1.004 \left(\frac{a}{l}\right) + 0.418 \left(\frac{a}{l}\right)^3 + 0.21 \left(\frac{a}{l}\right)^4 - 0.169 \left(\frac{a}{l}\right)^5, \quad (4.8)$$

where l is the distance between the particle and the channel walls, (*i.e.*, half the height of the channel). Swan and Brady (2010) calculate the full translational and rotational mobility of spheres for arbitrary distances between two walls. The channel height (or minimum dimension in any direction) relative to the probe radius should be $l/a \sim 10^2$ in order to ensure the confined hydrodynamic mobility of a probe agrees with its "unbounded medium" limit to within 1%. Probes of 1 μm diameter require a total sample thickness of order 100 μm.

4.3.2 Probe concentration

The greater the number of probe particles that are tracked, the better the statistics of the experiment. At the same time, however, care must be taken to limit the total particle volume fraction to avoid compromising particle tracking accuracy or material integrity. As evident in Fig. 4.4, out-of-plane particles contribute to the background fluorescence, and hence background noise of the image.

When imaging a suspension of particles with volume fraction ϕ, some number N_p will therefore appear in focus within the imaging

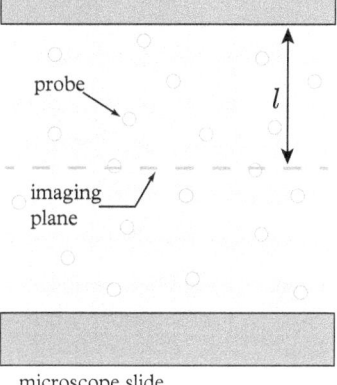

cover slide

probe

imaging plane

microscope slide

Fig. 4.13 *Distance of the imaging plane to the sample boundaries l determines the degree that the mobility of particles will be hindered by hydrodynamic interactions with the walls.*

volume $V_f = Ad_i$ defined by field of view area A and depth of field d_i. The effective focal depth d_i for particle tracking experiments, given by eqn 4.2, depends on the wavelength of light, the magnification, and the numerical aperture of the imaging system. One can therefore derive a relation,

$$\phi = \frac{N_p V_p}{V_f} = \frac{4N_p \pi a^3}{3d_i A}, \tag{4.9}$$

that enables ϕ to be selected in order to achieve a desired number N_p of particles to be tracked. Typical values involve 1 μm diameter probes, $d_i \approx 1\mu$m, and $A \approx 10^4 \mu$m^2, suggesting that probe particles should be dispersed in the sample with typical volume fractions $\phi \approx 1-5 \times 10^{-3}$, or about 0.1–0.5 vol%. Stock colloidal solutions must be diluted, as they typically contain 1-10% particles (by mass or volume). The particles in this case should be on average 10–20 μm apart in the focal plane based on an average separation between probes of approximately $\sim (V_p/\phi)^{1/3}$. Eq. 4.9 suggests that 100 nm diameter particles should be diluted another 1000-fold.

4.4 Particle tracking

Tracking particles requires their location to be determined in each frame, and for particle identities to be connected between frames. Many methods have been developed to find features in a digital image. For example, thresholding operations can be applied to images, producing binary images that highlight points of chosen brightness. Open source software (*e.g.*, ImageJ) easily locates features in thresholded images and reports their centroid-weighted position.

Crocker and Grier (1996) presented the particle-tracking algorithm most commonly implemented in microrheology. The algorithm consists of three basic steps: (i) Locate the brightest pixel of each particle; (ii) refine each particle's location to sub-pixel accuracy by calculating the image centroid, using the brightest pixel as the initial guess; and (iii) link particle locations in adjacent image frames, to generate time trajectories for each particle.

Each image frame is typically preprocessed to remove noise prior to tracking, so we start by discussing these image filtering processes.

4.4.1 Image filtering

Regardless of the particle tracking routine to be used, most video microscopy images benefit from some form preprocessing to reduce the effects of image noise. Typically, two bandpass filters are applied to

reduce high- and low-spatial frequency noise. Low frequencies are filtered to remove gradients in the background illumination caused by
uneven illumination or a nonuniform illumination source. High spatial frequencies are filtered to remove the pixel-to-pixel noise from
cameras and shot noise at low illumination levels.

Contrast gradients in the image complicate the process of identifying potential particles, but such variations can easily be subtracted
using a boxcar average due to the small image size and wide separation of the particle features. The boxcar average is taken over a region
$2w + 1$, where w is an integer larger than the single sphere's image
radius in pixels, but smaller than the spacing between particles. Each
pixel is given a new value

$$I_w(x, y) = \frac{1}{(2w + 1)^2} \sum_{i,j=-w}^{w} I(x + i, y + j). \qquad (4.10)$$

A second filter reduces the random noise of individual pixels by
averaging the value of nearby pixels. A Gaussian average for each
pixel is suitable, with a half-width $\xi = 1$ pixel over the same region w,

$$I_\xi(x, y) = \frac{\sum_{i,j=-w}^{w} I(x + i, y + j) \exp\left(-\frac{i^2 + j^2}{4\xi^2}\right)}{\left[\sum_{i=-w}^{w} \exp\left(-i^2 / 4\xi^2\right)\right]^2}. \qquad (4.11)$$

Since both high- and low-pass filters are performed over the same
region w, they are typically calculated simultaneously. The output image, Fig. 4.14, shows the result. The image appears quite dim, since
it now has a range of values between 1 and 51.2. The brightness of
the images increases as w increases. The brightness is given by the
normalization constant

$$K_0 = \frac{1}{B} \left\{ \sum_{i=-w}^{w} \exp\left[-(i^2 / 2\xi^2)\right] \right\}^2 - \left[B/(2w + 1)^2\right] \qquad (4.12)$$

where $B = \left[\sum_{i=-w}^{w} \exp\left(-i^2 / 4\xi^2\right)\right]^2$.

Fig. 4.14 *Image detail after applying the bandpass filters. The left-most
image is the original. The second image uses the entire dynamic range
of the 8-bit color map, while the
right-most has been rescaled for better
contrast.*

Having filtered the background intensity variations and pixel-to-pixel noise, we now proceed to particle identification and location.

4.4.2   Locating the brightest pixels

A magnified image of a single particle is shown in Fig. 4.15. Let $I_n(x, y)$ represent the intensity of the pixels associated with the nth particle's image; which is not an "exact" image of the particle, but a convolution of the particle's fluorescence emission (or scattering) and the point spread function of the imaging system. Additionally, $I_n(x, y)$ has likely been filtered to remove low- and high-frequency noise.

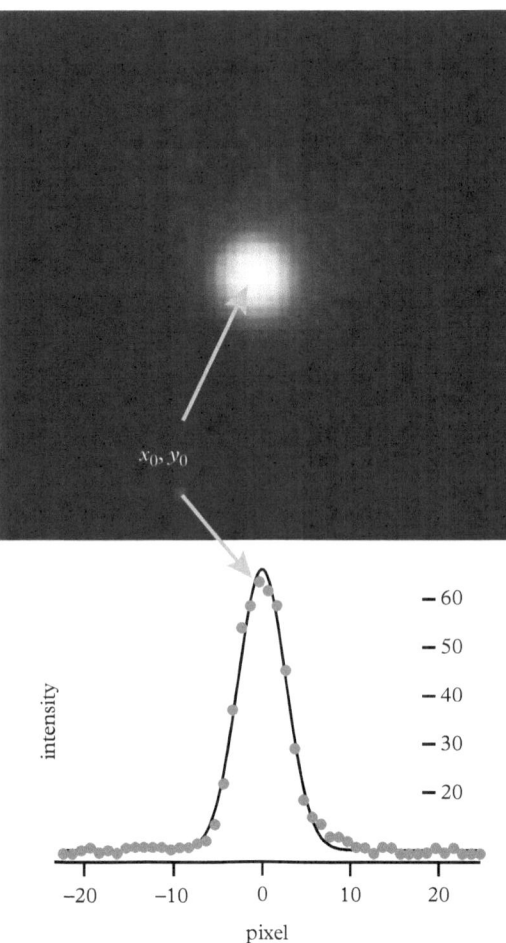

Fig. 4.15 *A magnified, unfiltered image of a single fluorescent particle. The brightest pixel is indicated at* (x_0, y_0). *Below the image is the intensity along the x-axis.*

The brightest pixel within $I_n(x, y)$, serves as an initial estimate for the location (x_0, y_0) of the nth particle. The locally brightest pixels in the image are those with the highest values within w pixels, where w is approximately the radius of the particle image and smaller than the average distance between particles. From this point, most tracking routines only accept the brightest 30–40% of these pixels as particle candidates.

Referring again to the magnified particle image in Fig. 4.15, the brightest pixel is close to the image center. Moreover, the image I_n of this n^{th} particle is well-approimated by a two-dimensional Gaussian centered at a position $\mathbf{r}_i$,

$$I_n(\mathbf{r}) = I_0 \exp\left(-\frac{|\mathbf{r} - \mathbf{r}_i|^2}{s^2}\right). \tag{4.13}$$

The width of the intensity distribution in Fig. 4.15 is $s = 3.81 \pm 0.07$ pixels. Notice that the brightest pixel isn't quite at the center of the fitted Gaussian distribution, but is instead located to the left by about 0.3 pixels. This offset reflects the fact that the center of the intensity *distribution* can be located more accurately than the brightest pixel— which is at best accurate to within a half a pixel. Consequently, particle locations can be determined with sub-pixel resolution.

4.4.3 Refining the initial location estimates

The intensity-weighted centroid surrounding the brightest pixel serves as a refinement for the particle location and provides *sub-pixel* resolution. For the nth particle, this correction is calculated as

$$\begin{pmatrix} \epsilon_x \\ \epsilon_y \end{pmatrix} = \frac{1}{m_0} \sum_{i^2 + j^2 \leq w^2} \begin{pmatrix} i \\ j \end{pmatrix} I_n(x_0 + i, y_0 + j) \tag{4.14}$$

where

$$m_0 = \sum_{i^2 + j^2 \leq w^2} I_n(x_0 + i, y_0 + j) \tag{4.15}$$

is the integrated brightness of the particle image. The refined particle location is

$$(x, y)_n = (x_0 + \epsilon_x, y_0 + \epsilon_y)_n. \tag{4.16}$$

The output of a particle location routine will produce a list of all of the particles identified in an image. Table 4.1 shows sample output

Table 4.1 *Sample location output.*

x	y	m_0	m_2
127.1	330.1	2074.2	8.5
164.1	293.4	2142.2	8.6
175.7	247.1	1304.5	8.6
194.5	483.6	1683.0	10.0
209.9	306.0	2305.2	12.6
213.3	72.4	2243.8	12.6

from a particle location routine. In addition to the x- and y-positions, a number of properties may also be reported, such as the brightness of each particle m_0 given by eqn 4.15, and the second moment (its radius of gyration squared),

$$m_2 = \frac{1}{m_0} \sum_{i^2+j^2 \le w^2} (i^2 + j^2) I_n(x + i, y + j). \qquad (4.17)$$

The values of m_0 and m_2 are helpful for distinguishing true particles from particle aggregates or other types of misidentification (Crocker and Grier, 1996), since these parameters tend to cluster systematically, as shown in Fig. 4.16. The exact values of m_0 and m_2 will depend on the microscopy method and how the image of a particle changes as it moves in and out of focus. It is also possible to create a reference calibration that allows the position orthogonal to the focal plane to be calculated from the moments (Crocker and Grier, 1996).

An important check of the centroid refinement algorithm is to plot a histogram of the particle location corrections produced by the centroid correction. This is accomplished by calculating the x- and y-position values *modulo* 1—that is, the fraction of a pixel. A histogram of these remainders should be flat, which ensures there is no bias in the centroid correction. A common failure is to have peaks in the histogram near 0 and 1 and a dip at 0.5. This pattern appears when the specified feature size w is too small, causing the x- and y-coordinates to round off to the nearest integer value. Figure 4.17 shows the centroid remainders for 10^4 particle positions. The top graph is relatively flat. In this case, the feature size was $w = 11$ pixels. Using a feature size $w = 7$ pixels in the same images yields biased centroid corrections.

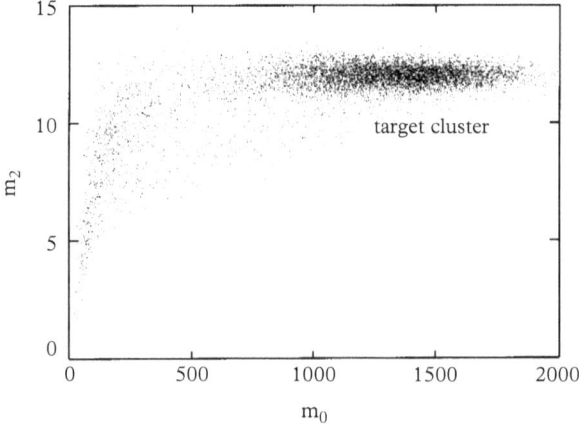

Fig. 4.16 *A target cluster identified by plotting m_0 and m_2 can be used to discriminate single particles from aggregates or image artifacts. These values were generated from 15,000 images of 0.65 μm diameter colloids. Reprinted from* J. Colloid Interface Sci., *179, Crocker, J. C. & Grier, D. G. Methods of digital video microscopy for colloidal studies, pp. 298–310, Copyright 1996, with permission from Elsevier.*

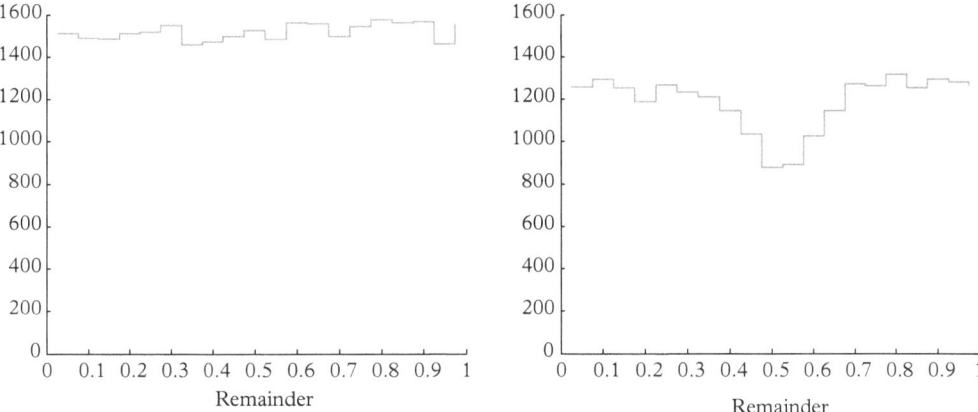

Fig. 4.17 *Histogram of x- and y-position remainders for 10^4 particle positions after the centroid refinement. The left and right plots were generated from the same image set using particle feature sizes of 11 and 7 pixels, respectively. The left histogram is flat, while bias in the right histogram indicates biased centroid corrections.*

4.5 Linking trajectories

After locating all particles in each frame of a video, a particle tracking algorithm must identify the same particle in neighboring frames and connect these into particle trajectories. The linking is typically accomplished by only considering the proximity of particles in neighboring frames, since particles cannot normally be distinguished from each other. Thus, trajectory linking relies on the probability of finding a particle in a subsequent frame in the vicinity of its previous location. Trajectory linking is also independent of how the particles are identified and located in each frame. Particle tracking provides direct observations of the displacements that make up the random walk, so this is a good point to consider their statistics. Before discussing the linking method, we will first consider the probability distribution function for a diffusing particle.

4.5.1 Van Hove correlation function

Section 3.2 discussed the random Brownian motion of tracer particles in complex fluids and soft solids. The stochastic thermal forces lead to each particle to execute a random trajectory, or a *random walk*.

The probability that a particle at **r** is displaced by a distance $\Delta\mathbf{r}$ during a time t is written

$$G_s(\Delta\mathbf{r}, t) = \langle \delta(\mathbf{r} - \Delta\mathbf{r}_j) \rangle. \qquad (4.18)$$

The delta function in eqn 4.18 sorts members of the ensemble, giving those with a displacement $\Delta r_j \equiv r_j(t) - r_j(0)$ in the vicinity of r a value of 1. Members of the ensemble with greater or lesser displacement than Δr are given the value zero. Thus, $G_s(\Delta r_j, t)d^3r$ is the probability that particle j will move in the vicinity r within time t, and is known as the *Van Hove self space-time–correlation function* (Van Hove, 1954).[12]

Since we have divided the total particle trajectories into intervals of the lag time τ, we expect the displacement of the particles to vary, in some general way, from one interval to the next. With a sufficiently large sample of displacements, the central limit theorem states that the probability of a particle displacement $r d^3 r$ is expected to be Gaussian; the Van Hove function may then be written

$$G_s(\Delta \mathbf{r}, t) = \left(\frac{2\pi}{3} \langle \Delta r^2(t) \rangle \right)^{-\frac{3}{2}} \exp \left(\frac{-3\Delta r^2}{2\langle \Delta r^2(t) \rangle} \right). \tag{4.19}$$

Although each particle executes a three-dimensional trajectory, particle tracking data is typically collected as a two-dimensional projection. Moreover, it is common to analyze the motion along each axis separately as one-dimensional functions of the displacement. With some video microscopy systems, one-dimensional Van Hove functions are calculated when the pixel dimensions are rectangular, thus requiring separate pixel-to-displacement calibrations along the vertical and horizontal directions. An example is when only one of the fields of an interlaced camera image is used (Valentine *et al.*, 2001).

In the $\mathbb{D}$-dimensional case, the Van Hove correlation function is written as

$$G_s(\Delta r, t) = \left(\frac{2\pi}{\mathbb{D}} \langle \Delta r^2(t) \rangle \right)^{-\frac{\mathbb{D}}{2}} \exp \left(\frac{-\mathbb{D}\Delta r^2}{2\langle \Delta r^2(t) \rangle} \right). \tag{4.20}$$

$G_s(\Delta r, t)dr$ represents the probability of locating a particle within dr of the displacement Δr at time t, and $\langle \Delta r^2(t) \rangle$ is the $\mathbb{D}$-dimensional projection of the mean-squared displacement—the particle displacement tracked along just $\mathbb{D}$ cartesian axes.

An example of the Van Hove correlation is shown in Fig. 4.18 for $\mathbb{D} = 1$. When plotted on a semi-log axes, the Van Hove function appears parabolic. Non-Gaussian statistics, and hence a non-parabolic shape to the curve, can indicate microheterogeneity of the sample or errors in the particle tracking. $N_{\Delta x}$ is the number of displacements of magnitude Δx observed, and is therefore equal to $N_{\text{tot}} G_s(\Delta x, t)d(\Delta x)$. From the fitted Van Hove functions, we can determine the corresponding mean-squared displacements.

Equation 4.19 makes no assumptions with respect to the mean-squared displacement at a lag time t. It applies to any complex fluid

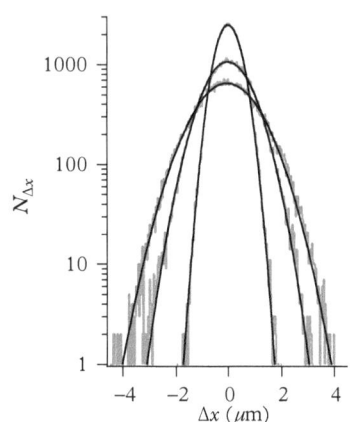

Fig. 4.18 *Van Hove plots for 1.063 ± 0.01 μm diameter polystyrene probe particles in 10 wt% glycerin solution at lag times 0.3, 1, and 2 s. The histogram bin size is 30 nm. The temperature is 20.8° C.*

---

[12] Named for Léon Van Hove, the Belgian physicist and director at CERN who used it to model neutron scattering. Sometimes it is abbreviated as "VHSSTCF" or "VHSTCF."

for which the Generalized Stokes–Einstein Relation is valid, since the probability distribution is merely a consequence of many independent observations of the dynamics that are driven by fluctuations that arise from the medium over a sum of uncorrelated regions. Hence, the Brownian forces are Gaussian in nature (Höfling and Franosch, 2013). The evolution of the mean-squared displacement, $\langle \Delta r^2(t) \rangle$ can be described by the generalized Langevin equation introduced in Section 3.5. However, eqn 4.19 does assume that the displacement is *isotropic* in three-dimensional space, and thus precludes materials with anisotropic rheology and the displacements of *anisotropic probes*.

We illustrate this limitation by considering the conservation of the particle probability in a slightly more restricted model—the uncorrelated diffusion of particles in a purely Newtonian medium. Any change in time in the number of particles in the region $\mathbf{r}d^3R$ would be balanced by the flux into and out of this volume,

$$\frac{\partial}{\partial t} G_s(\mathbf{r}, t) = -\nabla \cdot \mathbf{j} \tag{4.21}$$

where $\mathbf{j}$ is the probability flux given by

$$\mathbf{j} = -\mathbf{D} \cdot \nabla G_s \tag{4.22}$$

where the diffusivity tensor is written in terms of the grand-mobility tensor (eqn 2.141), $\mathbf{D} = k_B T \mathbf{b}_{ij}$. Then,

$$\frac{\partial}{\partial t} G_s(\mathbf{r}, t) = \nabla \cdot \mathbf{D} \cdot \nabla G_s. \tag{4.23}$$

Thus, only for the isotropic probe (and material), does this expression simplify to

$$\frac{\partial}{\partial t} G_s(\mathbf{r}, t) = D\nabla^2 G_s, \tag{4.24}$$

for which eqn 4.19 is a solution with $\langle \Delta r^2 \rangle = 6Dt$,

$$G_s(\mathbf{r}, t) = (4\pi Dt)^{-3/2} \exp\left(-r^2/4Dt\right). \tag{4.25}$$

4.5.2 Random walks

The derivation of the Van Hove correlation function is related to the statistics of a random walk. We will derive the random walk model here, then return to it when we discuss the interpretation of particle-tracking experiments.

First, consider a particle that moves with equal probability along a coordinate axis x. Steps are taken at even time intervals Δt over a fixed length l. The probability of moving in the positive x direction is p, and equal to the probability of moving in the negative direction q, such that

$$p = q \tag{4.26}$$

and

$$q = 1 - p. \tag{4.27}$$

Thus, the probabilities of moving forward or backward are $p = q = 1/2$.

After taking n steps, a particle has moved a certain number of steps in the positive direction, and the rest in the negative direction. Let k be the number of forward steps and $n - k$ be the number of backward steps. The probability that the particle has moved k steps in the forward direction out of n total steps follows the binomial distribution,[13]

$$P(k; n) = \frac{1}{2^n} \frac{n!}{k!(n-k)!}. \tag{4.28}$$

It is the probability of the particle taking k steps forward and $n - k$ steps backward,

$$\frac{1}{2^n} = \frac{1}{2^k} \frac{1}{2^{(n-k)}}, \tag{4.29}$$

while recognizing that there are

$$\binom{n}{k} = \frac{n!}{k!(n-k)!} \tag{4.30}$$

combinations of taking k steps available to our "random walker."

With eqn 4.28, we can take the first and second moments of the displacement. That is, we can ask what is the average position of the particle and the mean-squared displacement. The first moment shows that after n steps the average displacement is zero,

$$\langle \Delta x(n) \rangle = \langle x(n) - x(0) \rangle = 0. \tag{4.31}$$

On average there is an equal probability the walker will have moved to the left or right, but the second moment is non-zero,

$$\langle \Delta x^2(n) \rangle = \langle (x(n) - x(0))^2 \rangle = n l^2. \tag{4.32}$$

Since $n = t/\Delta t$, we can write

$$\langle \Delta x^2(t) \rangle = (l^2/\Delta t)t, \tag{4.33}$$

where $l^2/\Delta t$ has units of the diffusivity, and thus, $2D = l^2/\Delta t$, which not only recovers the Einstein relation, but identifies l as the mean-free

---

[13] This is equivalent of course to the probability of having k "heads" out of a total n flips of a coin.

path length of the random walk over the time Δt. With a sufficiently large number of steps n, the binomial distribution of eqn 4.28 can be expressed as a Gaussian distribution by writing the equivalent finite difference equation for eqn 4.24 (Ogunnaike, 2009). The solution is identical (in three dimensions) to eqn 4.25.

A random walk is a Markovian process, which is by definition memory-less. Each random walk step is independent of the previous step. This approximation is fine in a Newtonian fluid, but for a tracer particle moving in a complex, non-Newtonian material, there is memory between the steps that is unaccounted for by the Smolu-chowski equation 4.24 or the discrete random walk model we have just discussed.[14] If we were to write the time Fourier Transform of eqn 4.24 with a *frequency (time)-dependent* diffusivity given by the GSER $\tilde{D}(\omega) = k_B T/6\pi a \tilde{\eta}(\omega)$,

$$i\omega \tilde{G}_s(\mathbf{r}, \omega) = \tilde{D}(\omega)\nabla^2 \tilde{G}_s \tag{4.34}$$

we quickly recognize that the inverse transform must be an integral that represents the correlation between a time-dependent diffusivity and the Laplacian of G_s. The process is non-Markovian. Nonetheless, such simplified random walk models are useful for illustrating statistical issues that arise in tracer particle measurements, especially in the analysis heterogeneous materials. The reader is encouraged to use them to help understand the statistics of random particle motion.[15] Indeed, we'll return to this subject again in Section 4.10 when we discuss the microrheology of heterogeneous materials.

4.5.3 Application to trajectories

For particles tracked in the focal plane, the two-dimensional projection of the Van Hove correlation function is

$$G_s(\mathbf{r}, t) = \left(\pi \langle \Delta r^2(t)\rangle_{\mathbb{D}=2}\right)^{-1} \exp\left(\frac{-r^2}{\langle \Delta r^2(t)\rangle_{\mathbb{D}=2}}\right), \tag{4.35}$$

or in terms of the particle diffusivity $4Dt = \langle \Delta r^2(t)\rangle_{\mathbb{D}=2}$,

$$G_s(\mathbf{r}, t) = (4\pi Dt)^{-1} \exp(-r^2/4Dt). \tag{4.36}$$

We will now use eqn 4.36 to calculate the probability that two particles identified in separate frames comprise a trajectory in time of the same particle. Extending this by calculating the probability for *all* particles, we find the most likely identification of all particles between frames.

---

[14] The material may exhibit a relaxation time scale after which the process resumes, when normalized by this longest relaxation time scale, is Markovian.

[15] See Exercise 4.2.

Based on the Van Hove space-time correlation function, the probability that a single particle diffuses a distance δ in the time between two frames τ is

$$P(\delta, \tau) = (4\pi D\tau)^{-1} \exp(-\delta^2/4D\tau). \tag{4.37}$$

While eqn 4.37 strictly applies only to non-interacting particles in a viscous solvent, it is a reasonable approximation provided that the displacements of particles between frames are small. The probability distribution for N identical particles corresponding to the displacements δ_i of each particle is the product

$$P(\{\delta_i\}, \tau) = \prod_{i=1}^{N} P(\delta_i, \tau) = (4\pi D\tau)^N \exp\left(-\sum_{i=1}^{N} \frac{\delta_i}{4D\tau}\right). \tag{4.38}$$

Table 4.2 *Sample particle tracking output. The position data is reported in pixels (with sub-pixel accuracy). The time step is the frame number.*

x	y	frame	id
614.5	284.4	0	0
614.7	284.3	1	0
614.9	284.4	2	0
614.6	284.1	3	0
525.1	283.4	0	1
525.2	283.4	1	1
525.2	283.2	2	1
525.2	283.3	4	1
525.3	283.6	5	1
753.0	495.0	0	2
752.6	494.8	1	2
753.0	494.9	2	2
753.2	494.9	3	2
753.1	494.7	4	2
753.0	494.8	5	2
⋮	⋮	⋮	⋮

The correct identification of particles will maximize $P(\{\delta_i\}, \tau)$, or equivalently, minimize the total mean-squared displacement between frames, $\sum_{i=1}^{N} \delta_i^2$ (Crocker and Grier, 1996).

Since it is computationally inefficient to calculate the probabilities considering the displacement between all particle pairs in the two movie frames, which would scale as $\mathcal{O}(N!)$ calculations, a cut-off distance l, typically a fraction of the inter-particle spacing, is imposed; thus, a much smaller subset of particles within this distance are considered as candidates in the calculation of the mean-squared displacement between frames. In cases when no candidate particle exists in the next frame within a distance l, the maximum value is assigned. In this way, the tracking algorithm can account for particles that move out of the focal plane. Either the routine will try to match these with particles in subsequent frames, providing a "memory" function for the tracking, or it will end a trajectory and start a new one, regardless if it is the same or different particles. Such trajectory truncation can have consequences for the interpretation of particle tracking data, particularly in heterogeneous materials.

Table 4.2 is representative data that is reported from particle-tracking software. Each particle tracked is given a unique number (id), and the x and y position are reported and the time step (or *frame* number). The program may also report other quantities, such as the brightness, radius of gyration, and eccentricity of the tracked objects. These values can be used to discriminate which particles to include in an analysis—tracked features with a large eccentricity or brightness may represent particle dimers or small aggregates, and can be excluded. Note that the first time step of a particle may not correspond to the first frame, nor is the trajectory likely to last as long as the total number of frames collected. In Table 4.2, particle 0 is tracked for four frames, while particles 1 and 2 are tracked out to frame 5. Unless particles are strongly trapped in a gel or the viscosity of the medium is high, particles will tend to move into and out of the focal plane.

Microrheology experiments require dilute concentrations of probe particles. Particles will be separated on average by $d = (4\pi a^3/3\phi)^{1/3}$ (a length of about 10 μm for 1 μm diameter probes dispersed at a volume fraction $\phi = 10^{-3}$). Besseling *et al.* (2009) estimate that the trajectory algorithm due to Crocker and Grier (1996) previously described will perform well when the particle displacement between frames is no more than about half of this average separation, $\delta_i \approx 0.5d$. For most particle-tracking microrheology applications, this is more than sufficient. However, particles can also approach much closer than d. Two closely-spaced particles can potentially swap positions, causing an error in both trajectories that would be perceived as "hops" or "jumps" in the position (Besseling *et al.*, 2009). Such tracking errors, if they occur with sufficient frequency, should be apparent by non-Gaussian behavior of the Van Hove function. However, as we discuss later, non-Guassian statistics can also be a result of microheterogeneity in the material.

If a particle moves out of the focal plane and returns some time later, it will be given a different identifier and be treated as a separate particle. In some tracking routines it is possible to specify the number of frames which a single particle can be "missed" and still retain its identifying number. The sample data in Table 4.2 shows that particle 1 is not tracked in frame 3. Its location is missing, but the particle is tracked again in frames 4 and 5. The potential advantage of allowing missed frames is that longer trajectories, and hence, longer time scales of the rheology can be captured. Missed frames, however, introduce greater complexity in the analysis routines.

4.6   Analysis of particle tracking

For reasons we will see, inferring the properties of a sample from a visual analysis of individual trajectories is risky. Trajectories are merely observations of a particle position at regular time intervals, with straight lines drawn to connect these points. Perrin recognized the abstract and somewhat artificial nature of this representation. He noted that the trajectories of particles exhibited "prodigious entanglement,"

Fig. 4.19 *Three probe particle trajectories for particles diffusing in a viscous fluid (top), a viscoelastic fluid (middle) and a viscoelastic solid (bottom).*

and if shorter time intervals were used, "each straight segment would be realized as a curve as complex as the greater one" (Bigg, 2011).[16] Compare Perrin's trajectories in Fig. 4.2 to the three trajectories in Fig. 4.19.

The three test trajectories in Fig. 4.19 are taken from three samples. The first trajectory is from a probe particle diffusing in a Newtonian fluid with viscosity close to water. The second is taken from a gelling material near the liquid-to-solid transition, and the third trajectory was measured in a hydrogel. The last trajectory certainly exhibits more confinement than the other two, but by disregarding the time dependence of the displacement, it is not possible to distinguish the gel from a more viscous fluid. Likewise, despite the different material states of the first two trajectories, visually, it would be difficult to discern these properties from single particle trajectories. Distinguishing material structure and rheology, whether heterogeneity in a single sample, or the differences between samples, requires a proper statistical analysis of individual trajectories and their averages.

After obtaining the positions of the particles in each movie frame, the statistics of the the particle motion is analysed to yield the mean-squared displacement, and hence the microrheology via the Generalized Stokes–Einstein Relation. Here, we will first discuss the statistics of random walks and their probability functions.

4.6.1   Mean-squared displacement

We calculate the mean-squared displacement by calculating the displacement of each particle over the desired lag time τ. The lag time ranges between the time between individual frames, given by the frame rate of the acquisition f, and the total length of the collected video T,

$$1/f \leq \tau \leq T \tag{4.39}$$

although the longest time for any single trajectory will likely be less than T due to trajectory truncation as particles move into and out of the focal plane. The number of frames corresponding to each lag time is $n = \tau f$.

The first step in our analysis is to collect the displacements of each trajectory out of N_{traj} trajectories. The position data from the particle tracking software in Table 4.2 is of the form

$$r_{ij} = (x_{ij}, y_{ij}) \tag{4.40}$$

where i is the particle number and j is the frame number. Starting with the first lag time corresponding to the displacement of particles

[16] We recognize now that there is indeed a lower limit to this division: The distance the particle moves during the initial ballistic trajectory on the decay time scale of the velocity autocorrelation function.

between individual frames, $\tau_1 = 1/f$, the first three displacements for particle "1" are:

$$\Delta r_{10}(\tau_1) = (\Delta x_{10} = x_{11} - x_{10}, \Delta y_{10} = y_{11} - y_{10})$$
$$\Delta r_{11}(\tau_1) = (\Delta x_{11} = x_{12} - x_{11}, \Delta y_{11} = y_{12} - y_{11}) \qquad (4.41)$$
$$\Delta r_{12}(\tau_1) = (\Delta x_{12} = x_{13} - x_{12}, \Delta y_{12} = y_{13} - y_{12}).$$

Likewise, displacements calculated for the lag time $2\tau_1$ are

$$\Delta r_{10}(2\tau_1) = (\Delta x_{20} = x_{12} - x_{10}, \Delta y_{10} = y_{12} - y_{10})$$
$$\Delta r_{11}(2\tau_1) = (\Delta x_{21} = x_{14} - x_{12}, \Delta y_{11} = y_{14} - y_{12}) \qquad (4.42)$$
$$\Delta r_{12}(2\tau_1) = (\Delta x_{22} = x_{16} - x_{14}, \Delta y_{12} = y_{16} - y_{14}).$$

and in general may be written

$$\Delta r_{ij}(n\tau_1) = (\Delta x_{ij} = x_{i(j+n)} - x_{ij}, \Delta y_{ij} = y_{i(j+n)} - y_{ij}). \qquad (4.43)$$

for $n = \tau f$. These displacements are calculated over non-overlapping intervals to generate statistically independent observations. Here we assumed that the particle position is only resolved in a two-dimensional focal plane.

Consider the examples of particle displacements in Fig. 4.20 that were calculated with the previous equations for 1 μm probe particles in a 10wt% glycerol solution. Each series of displacements is

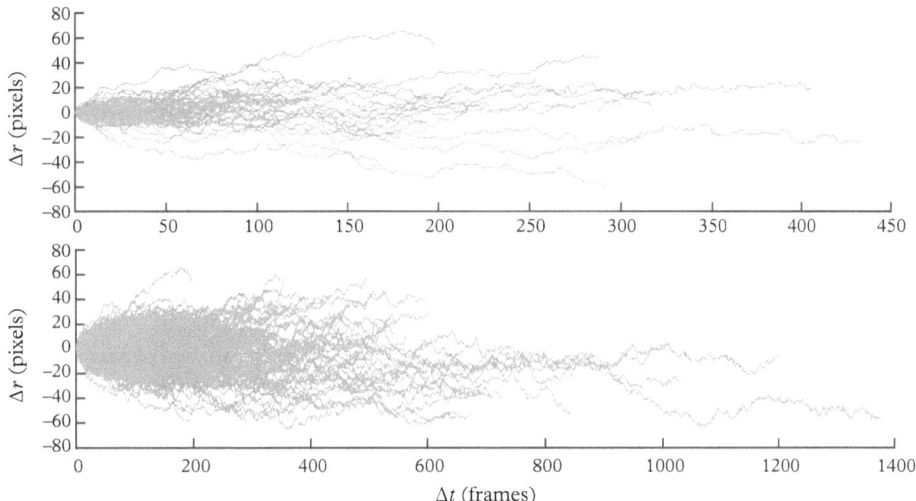

Fig. 4.20 *The displacements of 1.06 ± 0.01 μm diameter polystyrene probe particles in 10 wt% glycerin for (top) the first 200 trajectories and (bottom) all 1502 trajectories.*

a one-dimensional random walk trajectory. The displacements are clearly distributed around $\Delta r_{ij} = 0$. Taking a cross section of this data at any particular frame (or corresponding lag time) yields the Van Hove distributions shown in Fig. 4.18. The second moment of each distribution is the mean-squared displacement at that lag time.

For each lag time there are at most

$$N_j = T_j/\tau - 1 \tag{4.44}$$

statistically independent displacements from the ith particle trajectory of duration T_j. Naturally, lag times τ must be increments of the acquisition time $1/f$. The *time averaged* mean-squared displacement calculated from the gathered values of Δr_{ij} for the the ith particle is

$$\langle \Delta r_i^2(\tau) \rangle_t = \sum_{j=0}^{N_j} \left(\Delta x_{ij}^2 + \Delta y_{ij}^2 \right) \tag{4.45}$$

which may also be expressed as the time average of the one-dimensional x and y displacements,

$$\langle \Delta r_i^2(\tau) \rangle_t = \langle \Delta x_i^2(\tau) \rangle_t + \langle \Delta y_i^2(\tau) \rangle_t \tag{4.46}$$

where

$$\langle \Delta x_i^2(\tau) \rangle_t = \frac{1}{N_j} \sum_{j=0}^{N_j} \Delta x_{ij}^2, \quad \langle \Delta y_i^2(\tau) \rangle_t = \frac{1}{N_j} \sum_{j=0}^{N_j} \Delta y_{ij}^2. \tag{4.47}$$

The ensemble averaged mean-squared displacement is an average over all individual trajectories

$$\langle \Delta r^2(\tau) \rangle = \frac{1}{N_{\text{traj}}} \sum_{i=1}^{N_{\text{traj}}} \langle \Delta r_i^2(\tau) \rangle_t \tag{4.48}$$

at each lag time τ. An example mean-squared displacement for micrometer-diameter latex particles dispersed in water is shown in Fig. 4.21.

Also shown in Fig. 4.21 are the number of observations N that are included in the average at each lag time. At the shortest lag time, one video frame in the 400 frame movie, there are approximately 20,000 displacements, corresponding to about 50 particles on average visible in the imaging volume. The number of observations decreases nearly linearly with increasing lag time due to our requirement of sampling the particle positions in non-overlapping time intervals. At later lag times, the number of displacements decreases more quickly due to

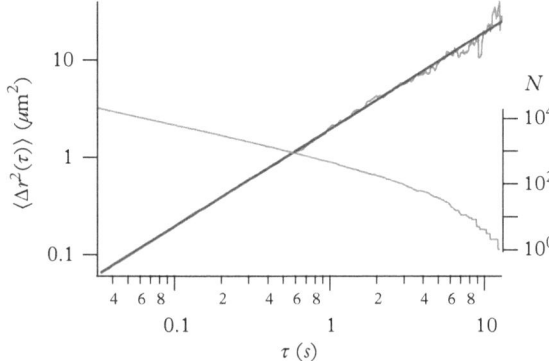

Fig. 4.21 *The ensemble-averaged mean-squared displacement of 1.06 μm diameter particles in water. A line is fitted to the data with a slope of 1 and a magnitude that gives the diffusivity 1.96 μm²/s. The right axis shows the number of independent observations at each lag time, which is offset to cross the mean-squared displacement curve at N = 1000. The image data is acquired at 30 frames per second for 400 frames. The longest recorded trajectory is 389 frames.*

trajectory truncation because particles leave the imaging volume. The curve demonstrates the decreasing statistical sampling of the average with increasing lag time, an effect that has important implications for statistical bias when studying heterogeneous materials, as we will see later.

In conjunction with the decreased number of observations N at each lag time, the mean-squared displacement curve exhibits greater variation as the lag time increases. The standard error of the Gaussian variance is $\sim 2\langle \Delta x^2 \rangle/\sqrt{N}$. Thus, to maintain an error of the mean-squared displacement to within 5% requires $N = 1600$ independent observations.

4.7 Non-Gaussian parameter

In Section 4.5.1 we discussed the Van Hove correlation function. Higher-order moments of the displacement can be calculated over $\mathcal{N}$ particles,

$$\langle \Delta r^k(t) \rangle = \frac{1}{\mathcal{N}} \sum_i [r_i(t) - r_0]^k. \tag{4.49}$$

A characteristic feature of the normal distribution is that the third moment and higher are zero. The departure of the Van Hove correlation from Gaussian behavior at any lag time can be quantified in three-dimensions by the ratio of successive moments (Rahman, 1964),

$$\alpha_n(t) = \frac{\langle \Delta r^{2n} \rangle}{C_n \langle \Delta r^n \rangle^n} - 1, \tag{4.50}$$

where $n = 1, 2, 3, \ldots$ and the coefficient C_n is

$$C_n = 1 \cdot 3 \cdot 5 \cdot 7 \cdots (2n+1)/3^n. \tag{4.51}$$

The **excess kurtosis** α_2, or non-Gaussian parameter, is the standardized fourth centralized moment,

$$\alpha_2 = \frac{\langle \Delta r^4 \rangle}{3 \langle \Delta r^2 \rangle^2} - 1 \qquad (4.52)$$

written here for a displacement distribution in one-dimension. It is commonly used in statistical analysis and reflects tailedness and peakedness in a distribution. For a Gaussian distribution, $\alpha_2 = 0$.

The test statistic Z_{α_2} is used to determine how far the non-Gaussian parameter must be from the value $\alpha_2 = 0$ to be considered non-zero within a statistical confidence interval, and is defined as

$$Z_{\alpha_2} = \alpha_2 / \sigma_{\alpha_2}, \qquad (4.53)$$

where

$$\sigma_{\alpha_2} = \sqrt{\frac{24 N (N-1)^2}{(N-3)(N-2)(N+3)(N+5)}} \qquad (4.54)$$

is the standard error of kurtosis (SEK) and N is the number of observed probe displacements at lag time τ. The critical value of Z_{α_2} is 1.96 at 0.05 significance level; when $|Z_{\alpha_2}| > 1.96$, the excess kurtosis is considered to be significantly different from zero.

4.8   Tracking accuracy and error

While so far we have assumed that the tracking algorithm provides positions of each particle that are exact, this is clearly impossible. The inherent spatial quantization of the camera's pixel array, combined with the choice of magnification and image SNR will affect the position information obtained with particle tracking. It is critical that errors are properly accounted for and, where possible, minimized. Otherwise, these errors will significantly alter the physical interpretation of the particle tracking data (Cheezum *et al.*, 2001; Thompson *et al.*, 2002).

There are two sources of potential error in particle tracking with video microscopy: *static* and *dynamic* error. Static error is the inherent inaccuracy of locating the position of a particle within a spatial resolution ε. Dynamic error occurs due to the finite acquisition time of the camera, during which particles continue to move. The position of the particle determined in an image frame is actually an average of its excursions during the exposure time. We discuss both separately, then look at their combined effect on particle tracking data.

When tracking a particle, the *apparent* position measured $\hat{x}(t)$ at time t, is

$$\hat{x}(t) = x(t) + \chi \qquad (4.55)$$

where $x(t)$ is the true tracer position and χ is a stationary random offset with mean $\langle \chi \rangle = 0$ and variance $\langle \chi^2 \rangle = \varepsilon^2$. Jean Perrin also described the error associated with his measurements, which was dominated by the uncertainty of the particle position, what we now call static error (Perrin, 1909).

4.8.1   Static error

Static error is an inherent characteristic of particle tracking data due to the pixelated detectors used in video microscopy. We introduced a centroid calculation in Section 4.4.3 that provided sub-pixel resolution. The accuracy of the centroid calculation depends on the image SNR and the choice of feature size w relative to the characteristic half-width of the particle image, s.

The effect noise has on multiple particle tracking can be understood by considering the images shown in Fig. 4.22. Both are of 1.06 μm diameter fluorescent polystyrene particles imaged at 63× magnification. The left image was taken with a 0.9 ms integration time, while

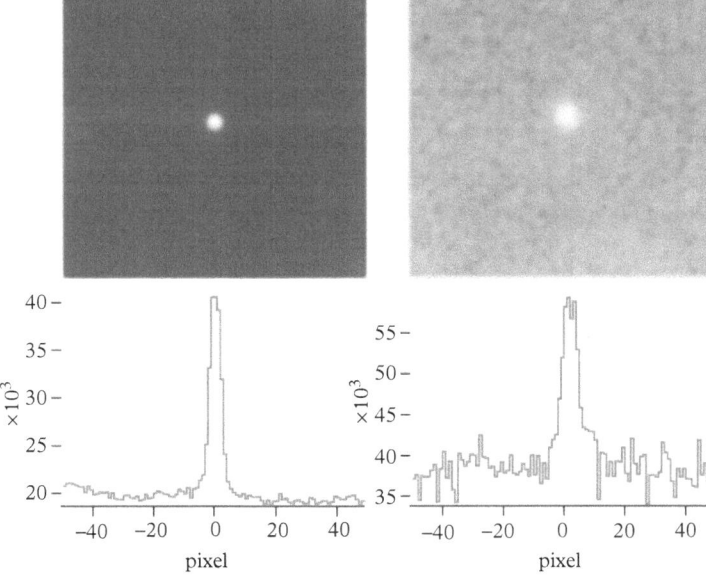

Fig. 4.22 *Two images of 1.06 μm fluorescent particles at 63× magnification using a NA 1.2 water immersion objective. The left image is taken with a 0.9 ms exposure, and the right image with 10 μs exposure. The corresponding intensity profiles taken through the centers of the particles are plotted in Fig. 4.23.*

the image on the right had a 10 μs exposure. The effect of the short exposure time on the SNR is apparent by the significantly lower contrast of the second image. Assuming that the particle image is modeled as a Gaussian function of brightness (eqn 4.13), the measurement error of each pixel caused by noise contributes

$$\varepsilon_{\text{noise}} \approx \left(\frac{l_N}{2\pi^{1/2}} \right) \left(\frac{(\langle I_N^2 \rangle - \langle I_N \rangle^2)^{1/2}}{\langle I_N \rangle} \right) \frac{2w^2/s^2}{1 - e^{-2w^2/s^2}} \qquad (4.56)$$

to the error in the particle location, where l_N is the noise-correlation which is typically assumed to be one pixel (Savin and Doyle, 2005; Crocker and Grier, 1996). The error due to noise grows larger with the feature size w used in the centroid calculation, suggesting that smaller values improve the tracking resolution. However, as w decreases, a second contribution to the static error arises due to clipping at the edges of the particle in the centroid estimate. For a particle offset by ϵ from the pixel grid, the error is

$$\varepsilon_{\text{clip}} \approx \epsilon \left(\frac{2w^2}{s^2} \right) \frac{e^{-2w^2/s^2}}{1 - e^{-2w^2/s^2}}. \qquad (4.57)$$

This error grows as w becomes small relative to the particle image half-width s. The combined total error due to eqns 4.56 and 4.57 is

$$\varepsilon = \left(\varepsilon_{\text{noise}}^2 + \varepsilon_{\text{clip}}^2 \right)^{1/2}. \qquad (4.58)$$

The combined effect of noise and clipping leads to an optimum value of w and an estimate of the maximum tracking accuracy due to static error. Shown in Fig. 4.23 Crocker and Grier (1996) plotted the estimated static error for imaging conditions in which the noise is $(\langle I_N^2 \rangle - \langle I_N \rangle^2)^{1/2}/\langle I_N \rangle = 0.02$ and the particle image half-widths are 6 and 2 pixels. For each condition, there exists an optimum w slightly larger than the image half-width. For smaller particle images, the static error increases quickly beyond the optimum value, but larger particle image sizes show more tolerance to noise, allowing a wider range of values for w to be chosen. These results were validated by Monte Carlo simulations.

It is also possible to estimate the maximum resolution of the particle tracking, approximately 0.05 pixel in each direction for the noise and half-widths in Fig. 4.23. Savin and Doyle (2005) report similar values of the tracking resolution as a function of the SNR, as shown in Fig. 4.24. Savin and Doyle use an estimate of the SNR which excludes out-of-focus particles, leading to higher-SNR values than are calculated for the images in Fig. 4.8, but the values for ε are typical. Under

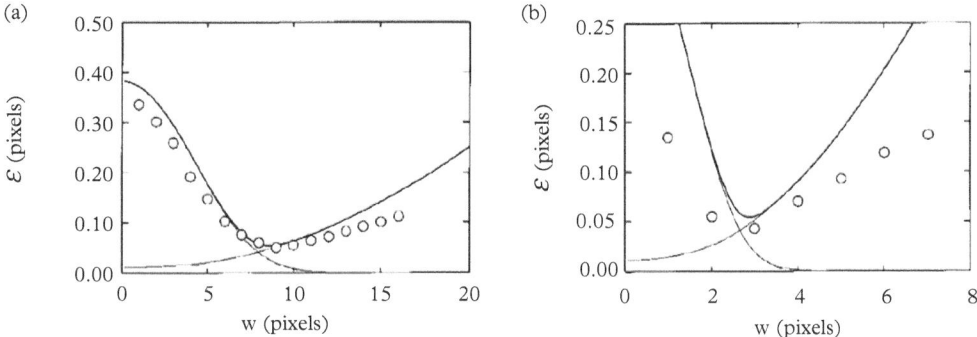

Fig. 4.23 *The total static error estimated by eqns 4.56 and 4.57 with noise $(\langle I_N^2 \rangle - \langle I_N \rangle^2)^{1/2}/\langle I_N \rangle = 0.02$ for particle image half-widths (a) s = 6 pixels and (b) s = 2 pixels. Reprinted from* J. Colloid Interface Sci., *179, Crocker, J. C. & Grier, D. G. Methods of digital video microscopy for colloidal studies, 179, pp. 298–310, Copyright 1996, with permission from Elsevier.*

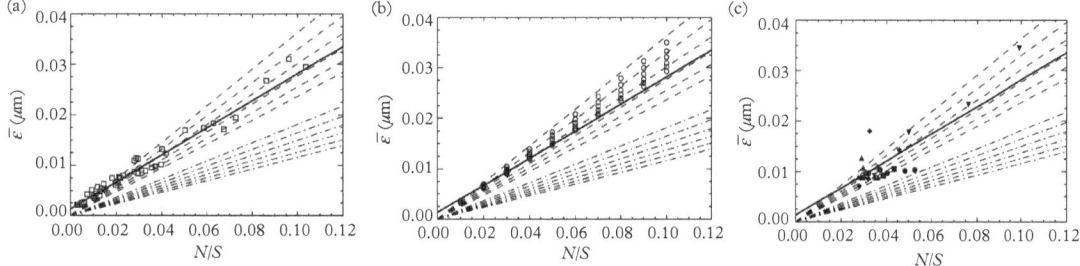

Fig. 4.24 *Estimates of static error as a function of the reciprocal image SNR for (a) immobilized probe particles, (b) simulated particle images, and (c) mobile particles. The lines are eqn 4.56 for values of the particle width s ranging from 4–5 pixels and a fixed feature width w = 7. Reprinted from* Biophys. J., *88, Savin, T. & Doyle, P. S., Static and Dynamic Errors in Particle Tracking Microrheology, pp. 623–38, Copyright 2005, with permission from The Biophysical Society.*

reasonably good imaging conditions, particle tracking is typically accurate to approximately $\varepsilon \sim 10$ nm. Particle tracking can be improved to resolutions of a few nanometers using techniques described by Crocker and Hoffman (2007), including the use of high-intensity filtered illuminator and low-noise, non-interlaced cameras.

Static error is often estimated using similar imaging conditions as those employed in the experiment, but fixing the particles in a strong gel, such as poly-acrylamide. Using a similar concentration of particles, imaging conditions—the illumination intensity and focal plane—ensures that the signal-to-noise will be similar to the microrheology samples of interest. However, static error will vary from sample-to-sample, and so calibrations like these must be used carefully, especially if the error is to be subtracted from the measured mean-squared displacement. Imaging particles fixed to the coverslip will generally produce an *underestimate* of the static tracking error due to the uniformity of the particle images from smaller aberrations close to the sample boundary and a lower contribution of out-of-plane light reaching the detector under fluorescence imaging conditions.

4.8.2   Dynamic error

The acquisition time of a video microscopy image means that the observed tracer position is actually an average of the true particle position over the integration time (Savin and Doyle, 2005),

$$\bar{x}(t, \sigma) = \frac{1}{\sigma} \int_0^\sigma x(t - \xi) d\xi. \tag{4.59}$$

Thus, the particle movement on time scales less than σ cannot be resolved. Dynamic error in a trajectory is illustrated in Fig. 4.26. The original and exact trajectory is generated by a Brownian dynamics simulation. The trajectory taken at every 50th time step is shown in Fig. 4.26b overlaid with the original trajectory. This exact sampled trajectory is next compared in Fig. 4.26c to the trajectory calculated by averaging the position of the previous 20 time steps, simulating the averaging process during the exposure time of a camera.

4.8.3   Tracking error and the MSD

Static and dynamic tracking errors have important consequences in microrheology, especially when calculating the mean-squared displacement and corresponding rheology from the Generalized Stokes–Einstein Relation. The influence of static error especially can be mistaken for rheological behavior. Moreover, the effects of static and dynamic error in particle tracking can be subtle because the two contributions can cancel each other.

We start with the simple case of a Newtonian fluid. The apparent one-dimensional mean-squared displacement for diffusing probe particles is (Savin and Doyle, 2005)

$$\langle \Delta \hat{x}^2(t) \rangle = 2D(t - \sigma_t) + 2\varepsilon^2. \tag{4.60}$$

From this expression, we see that dynamic error will cause the mean-squared displacement to curve downward as the lag time t decreases toward the exposure time σ_t. The result is an apparent, and unphysical, super-diffusive motion of the probes. Static error, on the other

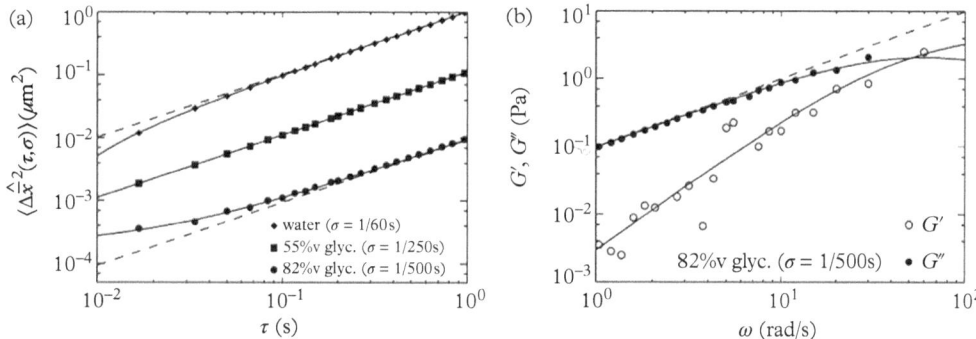

Fig. 4.25 *Dynamic and static error are apparent in these multiple particle tracking measurements of New-tonian glycerine solutions. The influence of dynamic error was controlled by adjusting the acquisition time of the camera. Reprinted from* Biophys. J., **88**, Savin, T., & Doyle, P. S., Static and Dynamic Errors *in Particle Tracking Microrheology, pp. 623–38, Copyright 2005, with permission from The Biophysical Society.*

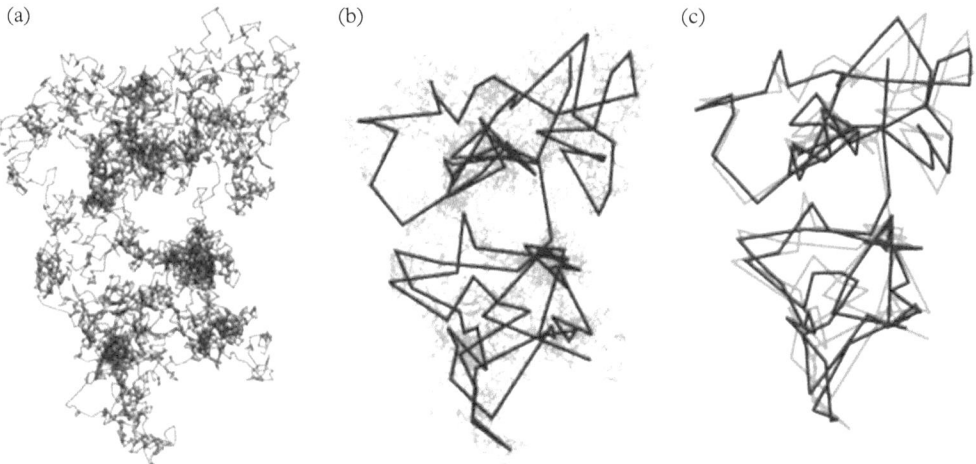

Fig. 4.26 *A simulation of the dynamic tracking error due to particle motion during the exposure time. (A) A simulated trajectory is (B) sampled at every 50th time step. (C) The sampled trajectory is compared with a trajectory that averages the previous 20 time steps. Reprinted from* Biophys. J., **88**, Savin, T. & Doyle, P. S., Static and Dynamic Errors in Particle Tracking Microrheology, pp. 623–38, Copyright 2005, with permission from The Biophysical Society.*

hand, produces a plateau value of the mean-squared displacement at a value $2\varepsilon^2$ as the lag time decreases. Under some conditions, these errors may appear to cancel.

The signatures of static and dynamic error are nicely illustrated in Fig. 4.25 from Savin and Doyle (2005). Multiple particle tracking was used to measure the mean-squared displacement of probes in Newtonian solutions of water and glycerine. The dynamic error is controlled by changing the shutter time. The mean-squared displacement for the lowest-viscosity solution (water) is well above the static error of the measurement, but the relatively long exposure time (1/60 sec) leads to an apparent super-diffusive motion at the shortest lag times, as expected by eqn 4.60. At the highest glycerine content and shortest exposure time, the static error produces a plateau-like feature at short lag times. Intermediate to these conditions, the static and dynamic error are "antagonistic," but they do not cancel each other in the sense that one corrects for the other. They merely satisfy the condition $2\varepsilon^2 - 2D\sigma_t/3 = 0$. The authors use the approximate form of the Generalized Stokes–Einstein Relation to calculate the frequency-dependent storage and loss moduli, to show that the mild upturn of the mean-squared displacement produces an apparent elastic behavior at high frequencies, much like a Maxwell fluid.

Static error can also be apparent in the excess kurtosis, α_2. The experiment cannot distinguish displacements within a spatial resolution $|\varepsilon|$. All displacements Δx smaller than this value are distributed evenly by the particle tracking algorithm (see Fig. 4.17). This particle displacement probability distribution can then be written (in one dimension) as

$$P(\Delta x) = \begin{cases} A\delta(\Delta x) & |\Delta x| \leq \varepsilon \\ B\exp\left(\frac{-(\Delta x)^2}{2D\tau}\right) & |\Delta x| > \varepsilon. \end{cases} \tag{4.61}$$

Calculating the second and fourth moments using the distribution yields, for small values of ε, a non-Gaussian parameter

$$\alpha_2 \approx \sqrt{\frac{8}{9\pi}}(D\tau)^{-3/2}\epsilon^3 + \mathcal{O}(\varepsilon^4). \tag{4.62}$$

From eqn 4.62 we see that the excess kurtosis should decrease with lag time as $\tau^{-3/2}$. This is confirmed by experimental data shown in Fig. 4.27 for several Newtonian solutions of increasing viscosity.

In practice, it is easier to avoid dynamic error than static error by using an exposure time that is significantly less (by at least a factor of 10) than the frame acquisition rate of the camera. Of course, this can lead to a lower signal-to-noise, and thus larger static error, so there is some compromise.

4.8.4   Convective drift and vibration

Another source of error in microrheology experiments arises from convective drift in the sample. These can be due to simple leaks in the

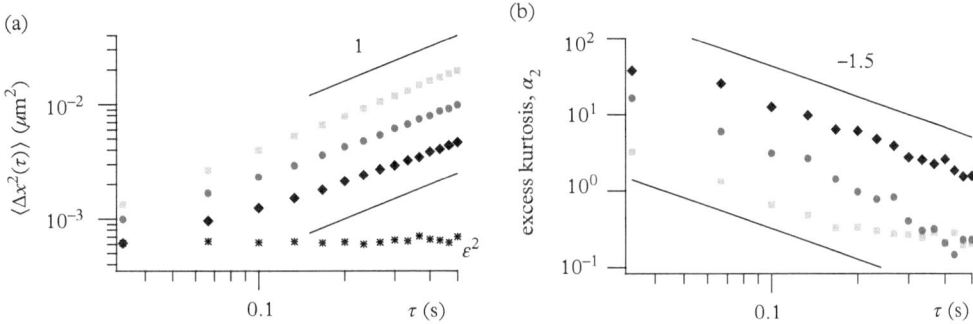

Fig. 4.27 *The (a) 1D mean-squared displacement and (b) excess kurtosis for 1 µm diameter particles in Newtonian solutions of increasing viscosity at short lag times.*

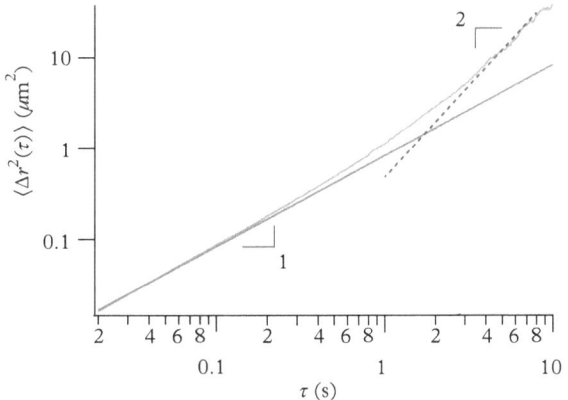

Fig. 4.28 *The ensemble-averaged mean-squared displacement of 1.06 µm diameter particles in water subject to drift. The solid line is the expected MSD. The dashed line has a logarithmic slope of 2, consistent with the displacement $\Delta r = vt$.*

sample chamber or occur from mass convection during changes in a sample, like degradation.

In Fig. 4.28, the mean-squared displacement for 1 µm particles in a Newtonian fluid (water) at about 22°C is plotted. As expected, the initial logarithmic slope is 1, but after a lag time of approximately $\tau = 1$s, the MSD crosses over and reaches a terminal slope of 2. This limiting behavior at long times is expected for particles with a drift velocity v, in which case the mean-squared displacement is

$$\langle \Delta r^2(\tau) \rangle = (vt)^2. \tag{4.63}$$

In the figure, the drift velocity is approximately $v = 0.69$ µm/s, just over one particle radius per second. Notice how the drift is dominated by the thermal motion of the probes at short lag times. Depending on the material properties, experimental window of mean-squared displacement delay times, and magnitude of the drift, a range of logarithmic slopes in the mean-squared displacement could be observed, from $1 \leq n \leq 2$. Data exhibiting these slopes should always be examined with a discerning eye.

In practice, drift can be difficult to distinguish while taking video microscopy data. Quickly scanning back and forth through the captured frames might make it more obvious. Often, however, the magnitude of drift is so small to be nearly undetectable except by particle tracking, with its exquisite sensitivity to probe motion. Good sample preparation methods should be used to ensure that samples are free of convection, where possible. If drift in a sample has occurred, the collective motion of probes has been used to characterize the drift velocity and *subtract* it from the measured mean-squared displacement (Mason *et al.*, 1997*a*).

Vibration due to mechanical (ventilation systems, pumps, people walking, elevators) and acoustic sources (talking, equipment, air handling) is ubiquitous in the laboratory environment and is easily detected in a sensitive microrheology experiment. The use of a vibration isolation table or platform is generally required to achieve acceptable results. Crocker and Hoffman (2007) estimate that a normal microscope placed on a laboratory bench or table will exhibit vibration amplitudes on the order of 10–100 nm. The use of a pneumatic vibration isolation table can reduce the vibration by an order of magnitude. Care must also be taken to reduce conduits of vibration, such as electrical cords and pump tubing if a cooling system is used.

Several mean-squared displacement curves of tracer particles in Newtonian fluids subjected to varying degrees of vibration are shown in Fig. 4.29. Probes dispersed in water are relatively insensitive to the vibration—their displacements at these lag times are large enough to mask the mechanical noise. For higher-viscosity sucrose solutions, however, the vibration becomes apparent as an oscillation in the

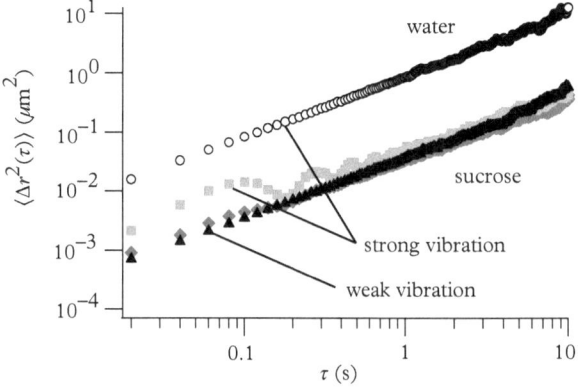

Fig. 4.29 *The ensemble-averaged mean-squared displacement of 1.06 μm diameter particles in water and a sucrose solution subject to varying magnitudes of vibration. The strong vibration is apparent in high-viscosity sucrose, but not the lower-viscosity water sample.*

mean-squared displacement. The source of the vibration in this case was a water bath circulator pump. The gray squares are the data with the pump operating and the black triangles represent data taken with the pump off. The frequency of the vibration is about 5.4 Hz and the amplitude is on the order of 100 nm. The vibration is most apparent at short lag times, when the amplitude of the motion is comparable to the probe displacement.

4.9 Operating regimes of particle tracking

In video microscopy, the reciprocal of the video frame frequency sets the shortest time between video frames, τ, while the static tracking error sets the minimum mean-squared displacement (or compliance), and is given by eqn 3.160.

Static error becomes more important for viscous fluids. At times short enough that the MSD approaches ϵ^2, it exhibits an apparent plateau. Fig. 4.30 plots the one-dimensional MSD given by eqn 4.60 as a function of lag time τ, for fluids with viscosities ranging from 1 mPa · s to 10 Pa · s. To avoid this plateau, the compliance should be greater than $\mathcal{J}_{\min}$ for all lag times $\tau > \tau_{\min}$. This is shown by the dashed line in Fig. 4.30, and corresponds to $\eta < k_B T \tau_{\min}/6\pi a \epsilon^2$.

For probe particles with diameter $2a = 1\,\mu$m and a typical particle tracking error of $\epsilon \approx 10$ nm, the limits calculated above give $G_e^{\max} \approx 5$ Pa (or $\mathcal{J}_e^{\min} \approx 0.2\,\mathrm{Pa}^{-1}$) and $\eta^{\max} \approx 150$ mPa · s.

Finally, there is an upper practical limit on the MSD lag times, $\tau_{\max}$. This longest lag time is somewhat arbitrary. If it was certain that the fluid was Newtonian, one could wait an indefinite time for the particles to move a sufficient distance to track. But in practice, it is usually

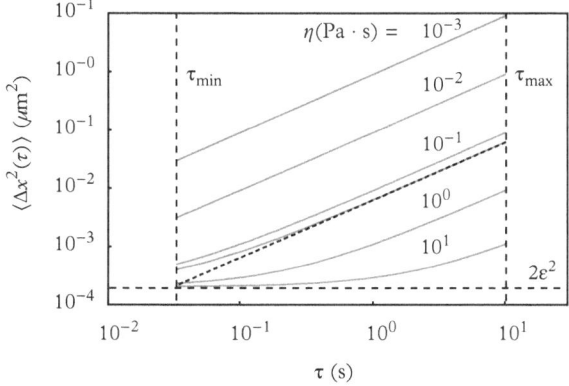

Fig. 4.30 *Calculated mean-squared displacements for 1 μm diameter probe particles in fluids of increasing viscosity from 10^{-3} to 10 Pa · s. The operating regime for multiple particle tracking microrheology is defined by the shortest and longest lag times and the particle tracking error. Solid curves are the apparent MSD, while the dashed line for $\eta = 0.146$ Pa · s is the true MSD.*

not feasible to perform particle tracking in materials over such long times. Moreover, it is important not to mistake the curvature caused by particle tracking errors for the rheology of the sample. Another concern for more compliant materials is the degradation of the particle tracking statistics due to movement of the particles out of the focal plane, which truncates trajectories (Savin and Doyle, 2007a). Finally, there is the overall acquisition time of the measurement to consider; as we discuss later in Chapter 10, changes in the material rheology with time, during a hydrogelation reaction, for instance, necessitate shorter acquisition times.

Given this operating regime, particle tracking microrheology has been most frequently used to characterize the rheology of materials with low viscosities and elastic moduli. Among these are protein biopolymers, such as those of the cytoskeleton, including F-actin and intermediate filaments (Apgar *et al.*, 2000; Gardel *et al.*, 2003; Wong *et al.*, 2004; Valentine *et al.*, 2004; Liu *et al.*, 2006). Hydrogelators derived from polymers and biopolymers (Yamaguchi *et al.*, 2007; Schultz *et al.*, 2009a; Schultz *et al.*, 2012a), self-assembling peptides (Xu *et al.* 2005; Zimenkov *et al.*, 2006; Savin and Doyle, 2007a; Larsen and Furst, 2008; Larsen *et al.*, 2009; Corrigan and Donald, 2010; Corrigan and Donald, 2009b; Corrigan and Donald, 2009a), protein assemblies (Mulyasasmita *et al.*, 2011) and solutions (Tu and Breedveld, 2005; Josephson *et al.*, 2016b; Josephson *et al.*, 2016a) have also been studied.

4.10   Heterogeneous materials

Microrheology data in multiple particle tracking microrheology is collected across a field of view in the sample. An individual particle response therefore tracks the rheology of an isolated region in the sample—the particle *microenvironment*—whose size is set approximately by the particle radius. An obvious advantage of multiple-particle tracking microrheology is its potential to measure spatial variations in a material's rheology (Fig. 4.31). In principle, the rheology of a material can be mapped throughout a sample by analyzing the motion of particles as a function of their position.

In practice, measuring material heterogeneity in a microrheology experiment requires care. The perceived variation in rheology must be distinguishable from the inherent statistical variation of the random trajectories of Brownian motion. And of course, one must also take care that the fundamental assumptions of microrheology are met. Foremost among our assumptions is the validity of the Stokes equation. Each probe particle must experience a microenvironment that is consistent with the continuum approximation of the equations of

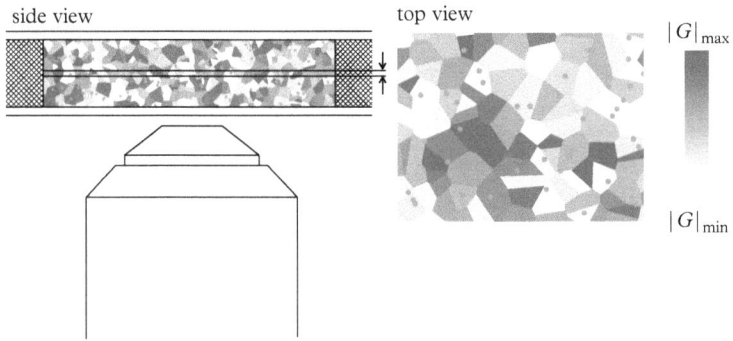

Fig. 4.31 *Probe particles distributed throughout a material the variations in the local rheology. Here, different shades represent different values of the modulus amplitude. Adapted with permission from Savin and Doyle (2007b). Copyrighted by the American Physical Society.*

motion in Chapter 2 if data are to be interpreted as a local rheological property.

In a heterogeneous material, which contains domains with distinct rheological responses, we expect probe particles to form sub-populations. Each of these sub-populations exhibits particle trajectories that are determined by the local rheology. But can one compare these trajectories? We saw earlier in Fig. 4.19 that some differences between trajectories are immediately obvious by visual inspection, but our perception (and their interpretation) can be deceived by their inherently random nature.

The answer lies in comparing the particle trajectory statistics. For example, if we consider probe particles suspended in a material with two distinct equilibrium moduli, $G_1 > G_2$—perhaps formed by spatially patterning a hydrogel using photochemistry as in Savin and Doyle (2007b)—then each *sub-population* of particles will exhibit a different distribution of displacements, given by the Van Hove correlation function, eqn 4.19. Provided one obtains enough displacement measurements, it should be possible to distinguish members of these two populations by their Van Hove correlation functions.

For example, Fig. 4.32 shows the Van Hove correlation for two hypothetical probe populations: One dispersed in a region with $G_1 = 1$ Pa and the other in a region with a modulus ten-times lower, $G_2 = 0.1$ Pa. Here, the mean-squared displacement is independent of time $\langle \Delta r^2 \rangle_{\mathbb{D}=1} = k_B T / 3\pi a G$, leading to the Van Hove correlation

(a)

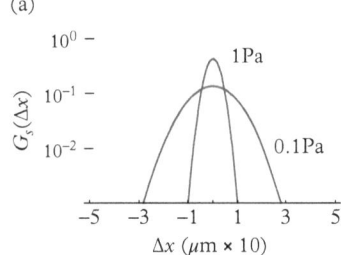

(b)

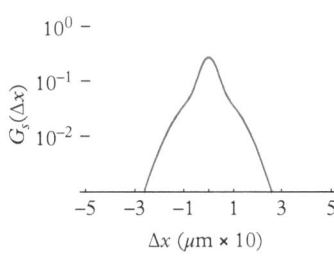

Fig. 4.32 *(a) Calculated Van Hove functions for probe particles in elastic microenvironments with moduli 1 and 0.1 Pa. (b) The Van Hove correlation is non-Gaussian when multiple microenvironments are present.*

$$G_s(\Delta x) = \left(\frac{\pi k_B T}{3\pi a G} \right)^{-1/2} \exp\left(\frac{-3\pi a G \Delta x^2}{k_B T} \right). \tag{4.64}$$

The corresponding Van Hove functions are clearly different. We could distinguish these two populations easily. Moreover, if the probes were evenly divided between the two regions, the Van Hove function for the entire sample would be non-Gaussian (Fig. 4.32b). Again, $\mathbb{D}$-dimensional displacements would have an analogous distribution function

$$G_s(\Delta x) = \left(\frac{\mathbb{D}\pi k_B T}{3\pi aG}\right)^{-\mathbb{D}/2} \exp\left(\frac{-3\pi aG\Delta x^2}{\mathbb{D}k_B T}\right). \tag{4.65}$$

4.10.1   f-test method

Valentine *et al.* (2001) developed a statistical test to distinguish different probe microenvironments arising from material heterogeneity. A stunning example of non-Gaussian probe statistics is shown in Fig. 4.33, and comes from their work on tracking probe particles in an agarose gel.

Agarose forms a network of pores on similar length scales as the probe particles, so the GSER is invalid—the material is inhomogeneous on the length scale of the probes. Although it is tempting to interpret the agarose particle tracking measurements in terms of "local" compliance, viscosity, or viscoelastic moduli, the breakdown in the continuum approximation means this interpretation is incorrect. However, probe particles moving in pores obviously exhibit strong or weak confinement, and hence a spatial variation in microenvironment that can be useful to characterize in its own respect. Caggioni *et al.* (2007), for instance, use multiple particle tracking to follow the effects of processing on the microstructure of gellan gums. The smooth

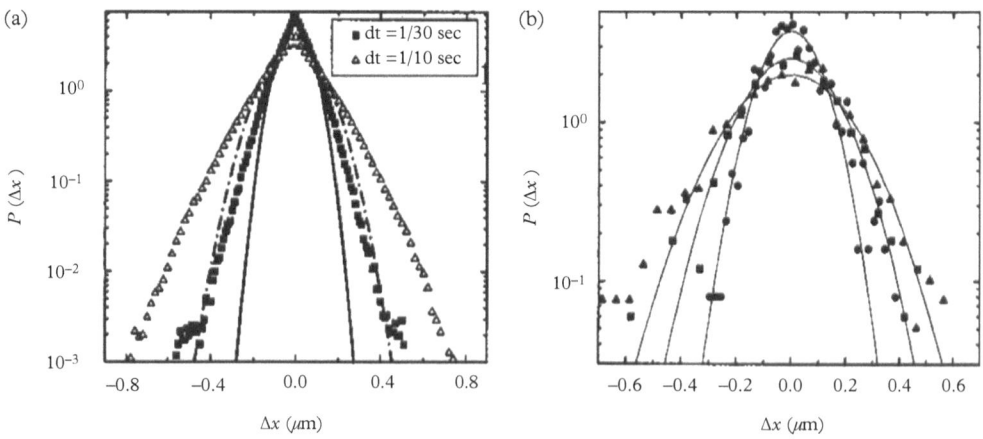

Fig. 4.33 *(a) The ensemble-averaged Van Hove correlation functions for particles diffusing in a micro-heterogeneous agarose gel is non-Gaussian. (b) Single particle Van Hove correlations taken at the lag time 0.1 s are statistically different. Reprinted with permission from Valentine, M. T. et al., Phys. Rev. E,* **64**, *61506 (2001) Copyright 2001 by the American Physical Society.*

curves in Fig. 4.33a compared to our calculated Van Hove functions with just two distinct populations (Fig. 4.32) suggest a wide range of microenvironments, reflecting a broad distribution of pore sizes in the agarose gels.

Because the Van Hove correlation function provides a statistical description of probe motion, Valentine *et al.* (2001) differentiate between sub-populations of probes on the basis of the f-statistic, (Ogunnaike, 2009)

$$f = \frac{\sigma_1^2/n_1}{\sigma_2^2/n_2} \tag{4.66}$$

which compares the variances of any random, mutually-independent, normally distributed populations with standard deviations σ_1 and σ_2 and sizes n_1 and n_2. Two probes can be distinguished—identified as belonging to two distinct sub-populations or microenvironments—by calculating the Van Hove correlation for each particle for non-overlapping time segments, then comparing the variances to the null hypothesis

$$H_0 : \sigma_1^2 = \sigma_2^2. \tag{4.67}$$

If the Van Hove correlations represent independent distributions, then the f-statistic follows the $F(v_1, v_2)$ distribution, where $v_1 = n_1 - 1$ and $v_2 = n_2 - 1$ are the degrees of freedom. From this criteria, we can establish the relation of σ_1 and σ_2 with the following tests:

$$H_a : \sigma_1^2 < \sigma_2^2 \ \ f < F_{1-\alpha}(v_1, v_2)$$
$$H_a : \sigma_1^2 > \sigma_2^2 \ \ f > F_{\alpha}(v_1, v_2)$$
$$H_a : \sigma_1^2 \neq \sigma_2^2 \ \ f < F_{1-\alpha/2}(v_1, v_2)$$
$$\text{or} \ \ f > F_{\alpha/2}(v_1, v_2) \tag{4.68}$$

where α is the significance level, which is typically specified at 5%, or $\alpha = 0.05$.

We require independent measurements of the displacement of single probes, which become n_1 and n_2. Obviously, this means that the chosen lag time for this comparison and trajectory length for the probe provide a sufficient number of non-overlapping observations of displacement at a specified lag time. The standard deviations are found by fitting eqn 4.20 to the displacement data.

Examples of single particle Van Hove correlations are shown in Fig. 4.33b. The data are taken from the same agarose gels used to calculate the ensemble average Van Hove correlations in Fig. 4.33a. Instead of three distinct lag times, three separate particles are shown at the same lag time 0.1s. To within a reasonable approximation, the individual curves are Gaussian, and are shown with respective curve fits. The F-test can now be used to distinguish probe microenvironments and even cluster particles with similar, statistically indistinguishable, Van Hove correlations (Valentine *et al.*, 2001).

The statistical test described overcomes a significant drawback when comparing the mean-squared displacement of single particles, which tend to suffer from under-sampled statistics. Compare, for instance, the single particle mean-squared displacements for glycerol and agarose, shown in Fig. 4.34. From these plots, it looks as though both glycerol and agarose exhibit heterogeneity, despite the fact that we know the viscous solution is a homogeneous Newtonian fluid. This *apparent* heterogeneity in glycerol is no more than a consequence of the limited statistics of probe motion at longer lag times. Both the ensemble average Van Hove correlations and representative distributions for single particles shown in Fig. 4.35 confirm this suspicion. The ensemble average is clearly Guassian at two distinct lag times. The distributions for two individual probes, plotted in Fig. 4.35b are identical, confirming that the movement of individual probes in the glycerol is the same to within the significance of the f-test.

Non-Gaussian Van Hove correlation functions are not always an indicator of rheological heterogeneity. Transient dynamical events such as "hopping," also known as Levy flights, can also give rise to non-Gaussian distributions. Hopping has been observed by particles comprising a colloidal glass (Weeks *et al.*, 2000) and in microrheology measurements of entangled F-actin networks (Wong *et al.*, 2004; Valentine *et al.*, 2001). The latter may reflect the transient relaxation of entanglements between filaments. These sources of heterogeneity fall broadly into the category of non-continuum effects and a breakdown of the Stokes equation.

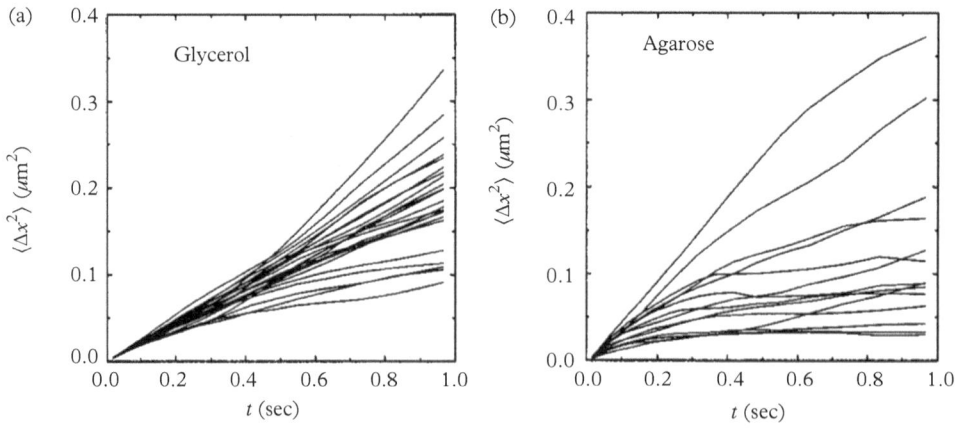

Fig. 4.34 *Single particle mean-squared displacements for probe particles dispersed in (a) a homogeneous glycerol solution and (b) a heterogeneous agarose gel. Reprinted with permission from Valentine, M. T. et al., Phys. Rev. E,* **64**, *61506 (2001) Copyright 2001 by the American Physical Society.*

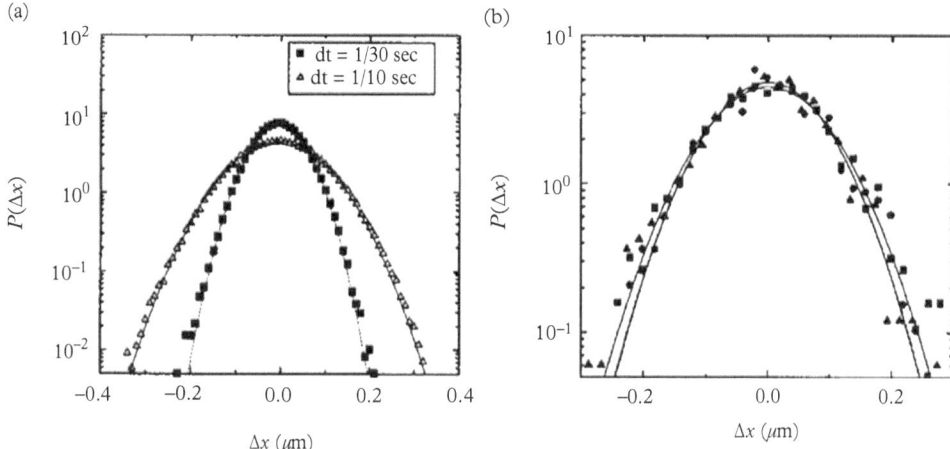

Fig. 4.35 *(a) The ensemble-averaged Van Hove correlation functions for particles diffusing in a homogeneous Newtonian solution of glycerol and water. (b) Single particle Van Hove correlations taken at the lag time 0.1 s are statistically identical. Reprinted with permission from Valentine, M. T. et al., Phys. Rev. E, **64**, 61506 (2001) Copyright 2001 by the American Physical Society.*

4.10.2 Global measures of heterogeneity

The method of characterizing heterogeneity discussed in the previous section recognized the statistical nature of the measurement and used this to distinguish probes in regions of different rheology throughout a material. Such information can then be used to map the heterogeneity of a sample—grouping regions of similar microenvironments to generate a measure of local rheology. However, the f-test does not provide an overall quantitative measure of heterogeneity, something that is of interest in industrial rheology for understanding the role and effects of processing on microstructure, for instance (Caggioni *et al.*, 2007).

Based on this need, Savin and Doyle (2007*b*) introduced a method for producing quantitative measures of spatial heterogeneity in multiple particle tracking experiments. The approach addresses an important statistical bias in tracking data of heterogeneous materials: Standard tracking algorithms produce data that is significantly skewed by more mobile particles, which enter and leave the imaging plane frequently and create many short trajectories as part of the ensemble average mean-squared displacement. *Unbiased estimators* of the ensemble average mean-squared displacement $M_1(\tau)$ and ensemble-average variance $M_2(\tau)$ can be generated by weighting each trajectory i by a factor w_i proportional to its duration T_i,

$$M_1(\tau) = \sum_i w_i \langle \Delta r^2(\tau) \rangle_i \qquad (4.69)$$

$$M_2(\tau) = \sum_i w_i \left[\langle \Delta r^2(\tau) \rangle_i - M_1(\tau) \right]^2. \qquad (4.70)$$

The set of weighting factors w_i is normalized. From $M_1(\tau)$ and $M_2(\tau)$, the *heterogeneity ratio,* HR is defined as

$$HR = \frac{M_2(\tau)}{M_1^2(\tau)}. \qquad (4.71)$$

Two examples of heterogeneity characterized by HR are shown in Fig. 4.36. The first shows the development of heterogeneity in 1 wt% Laponite at a lag time $\tau = 0.33$s (Rich *et al.,* 2011*b*). Structure develops as the dispersion ages over a waiting time t_w. The heterogeneity is more complicated than the development of regions with different viscoelastic moduli because the HR depends on the probe size—a clear indication of a breakdown of the continuum limit. The second example (Fig. 4.36b) is measurements of a fluorenylmethoxycarbonyl-tyrosine hydrogel that is induced to gel by a decrease in pH generated by the slow-hydrolysis kinetics of glucono-δ-lactone (Aufderhorst-Roberts *et al.,* 2012). There is a small but steady increase in heterogeneity up to the gel point, after which HR reaches a plateau.

The heterogeneity ratio HR discussed in this section is similar to the excess kurtosis α_2 (Section 4.7), which has also been used to

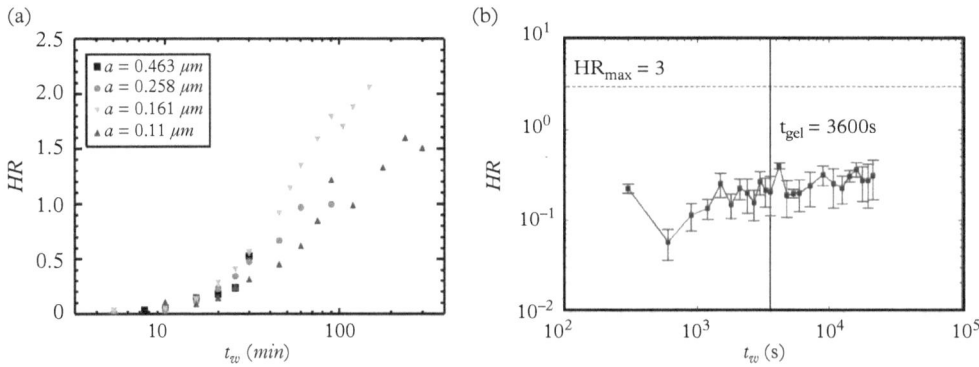

Fig. 4.36 *Heterogeneity ratios (HR) plotted for (a) a Laponite suspension as it ages with waiting time* t_w. *Reprinted with permission from Rich, J. P., McKinley, G. H., & Doyle, P. S. J. Rheol., 55, 273 (2011). Copyright 2011, The Society of Rheology. (b) HR in a hydrogel as it forms. Reproduced from Aufderhorst-Roberts* et al. *(2012) with permission from The Royal Society of Chemistry.*

characterize heterogeneity in microrheology studies (Oppong *et al.*, 2008). Remember, however, that HR defined by eqn 4.71 corrects for the peculiar statistical biases of multiple particle tracking microrheology, while the calculation of α_2 does not.

4.11 Two-point microrheology

Two-point microrheology analyzes the correlated motion of probe particle pairs. In Section 2.6.2, we computed the entrainment of particles due to a force on another particle. Similarly, a displaced particle in an elastic solid disturbs the positions of nearby particles by the strain field it generates. Normally, we consider the interactions between probes to contribute negligibly to their displacement, given that they are dilute and widely-separated. In fact, particles are coupled due to the hydrodynamic interactions we depict in Fig. 4.37, mediated by the intervening material. Crucially, this coupling depends on the rheology of the material, in addition to the distance between probes.

Two-point microrheology extracts the rheology of the medium between two probes on the basis of measured cross-correlations between particles. Two-point measurements are therefore less sensitive to the material structure immediately surrounding each probes—and thus to the probe surface chemistry and probe-material interactions—than conventional "single-point" measurements. The same trajectory data collected for single-point experiments is used to calculate two-point correlations, but as we will see, the weak interactions require a significantly greater number of measurements.

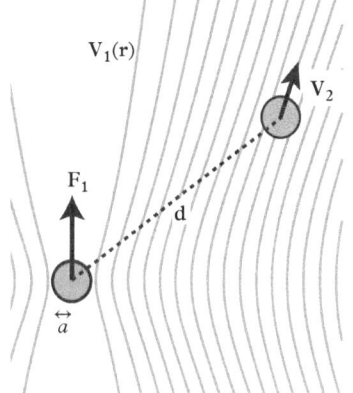

Fig. 4.37 *Two-point microrheology measures the correlation of probe displacements due to their hydrodynamic interactions. Here, the velocity field v_1 of one particle forces the motion of a second, nearby particle, V_2.*

4.11.1 Two-point GSER

To start, we will derive the generalized Stokes–Einstein relation for a two-point microrheology experiment. The Langevin equations for two spheres in a quiescent fluid are

$$M_p \dot{\mathbf{V}}_1 = \mathbf{f}_1^R - \int_{-\infty}^{t} \left[\boldsymbol{\zeta}_{11}(t-t') \cdot \mathbf{V}_1(t') + \boldsymbol{\zeta}_{12}(t-t') \cdot \mathbf{V}_2(t') \right] dt' \quad (4.72)$$

$$M_p \dot{\mathbf{V}}_2 = \mathbf{f}_2^R - \int_{-\infty}^{t} \left[\boldsymbol{\zeta}_{21}(t-t') \cdot \mathbf{V}_1(t') + \boldsymbol{\zeta}_{22}(t-t') \cdot \mathbf{V}_2(t') \right] dt', \quad (4.73)$$

Laplace Transforming gives

$$sM_p \hat{\mathbf{V}}_1(s) - M_p \mathbf{V}_1(0) = \hat{\mathbf{f}}_1^R - \hat{\boldsymbol{\zeta}}_{11}(s) \cdot \hat{\mathbf{V}}_1(s) - \hat{\boldsymbol{\zeta}}_{12}(s) \cdot \hat{\mathbf{V}}_2(s) \quad (4.74)$$

$$sM_p \hat{\mathbf{V}}_2(s) - M_p \mathbf{V}_2(0) = \hat{\mathbf{f}}_2^R - \hat{\boldsymbol{\zeta}}_{21}(s) \cdot \hat{\mathbf{V}}_1(s) - \hat{\boldsymbol{\zeta}}_{22}(s) \cdot \hat{\mathbf{V}}_2(s). \quad (4.75)$$

Written in matrix form, the transformed equations of motion are

$$sM_p\hat{\mathbf{V}}(s) - M_p\mathbf{V}(0) = \hat{\mathbf{f}}^R - \hat{\boldsymbol{\zeta}}(s) \cdot \hat{\mathbf{V}}(s) \tag{4.76}$$

where $\hat{\mathbf{V}}(s)$ is a two-particle velocity vector, $\hat{\boldsymbol{\zeta}}(s)$ is the two-particle resistance tensor, and $\hat{\mathbf{f}}^R$ is the two-particle stochastic force vector. Solving for $\hat{\mathbf{V}}(s)$ gives

$$\hat{\mathbf{V}}(s) = (sM_p\boldsymbol{\delta} + \hat{\boldsymbol{\zeta}})^{-1} \cdot M_p\mathbf{V}(0) + (sM_p\boldsymbol{\delta} + \hat{\boldsymbol{\zeta}})^{-1} \cdot \hat{\mathbf{f}}^R. \tag{4.77}$$

To simplify the analysis, we assume low enough frequencies that inertia is irrelevant, so that $M_p s$ is much smaller than any diagonal element of $\boldsymbol{\zeta}$, which are of order $6\pi\eta a$. Neglecting these terms simplifies the matrix inversion:

$$(\hat{\boldsymbol{\zeta}} + sM_p\boldsymbol{\delta})^{-1} \approx \hat{\boldsymbol{\zeta}}^{-1} \equiv \hat{\mathbf{b}}(s), \tag{4.78}$$

to give

$$\hat{\mathbf{V}}(s) = M_p\hat{\mathbf{b}}(s) \cdot \mathbf{V}(0) + \hat{\mathbf{b}}(s) \cdot \hat{\mathbf{f}}^R. \tag{4.79}$$

Now, we will use index notation to write,

$$\hat{V}_i(s) = M_p\hat{b}_{ik}(s)V_k(0) + \hat{b}_{il}(s)\hat{f}_l^R. \tag{4.80}$$

Following the same methods of Chapter 3, we multiply by $V_j(0)$ and ensemble average,

$$\langle \hat{V}_i(s)V_j(0)\rangle = \hat{b}_{ik}(s)M_p\langle V_k(0)V_j(0)\rangle + \hat{b}_{il}(s)\langle \hat{f}_l^R(s)V_j(0)\rangle. \tag{4.81}$$

Again, the stochastic thermal force is uncorrelated with particle velocity,

$$\langle \hat{f}_l^R(s)V_j(0)\rangle = 0 \tag{4.82}$$

and equipartition gives

$$\langle M_p V_k(0)V_j(0)\rangle = k_B T\delta_{kj} \tag{4.83}$$

so that ultimately

$$\langle \hat{V}_i(s)V_j(0)\rangle = k_B T\hat{b}_{ij}(s). \tag{4.84}$$

The VAC can be related to the MSD in the standard way,

$$\langle \hat{V}_i(s) V_j(0) \rangle = \frac{1}{2} s^2 \langle \Delta \hat{R}_i \Delta \hat{R}_j \rangle (s) \tag{4.85}$$

where

$$\langle \Delta \hat{R}_i \Delta \hat{R}_j \rangle (s) = \mathscr{L} \left\{ \langle \Delta R_i \Delta R_j \rangle (t) \right\}. \tag{4.86}$$

The two-point mean-squared displacement may thus be expressed as

$$\langle \Delta \hat{R}_i \Delta \hat{R}_j \rangle (s) = \frac{2 k_B T}{s^2} \hat{b}_{ij}. \tag{4.87}$$

Analytic continuation connects the Laplace and Fourier Transforms, giving

$$\langle \Delta \tilde{R}_i \Delta \tilde{R}_j \rangle (\omega) = \frac{2 k_B T}{(i\omega)^2} \tilde{b}_{ij}. \tag{4.88}$$

Remember that $\tilde{b}_{ij}$ refers to the grand, multi-particle mobility tensor, so that i and j refer to both particle number and displacement direction. Therefore, one can measure the displacement of particle 1 in the x-direction from time t to $t + \tau$, multiply that by the displacement of particle 2 in the y-direction over the same interval, and average this product over all times t to obtain the two-point mean-squared displacement $\langle \Delta R_i(\tau) \Delta R_j(\tau) \rangle$. Fourier transforming over the lag time τ gives the left-hand side of eqn 4.88, which is therefore equal to a constant times the ij-component of the multiparticle mobility tensor, b_{ij}.

Of particular interest to microrheology are the coupling components of the mobility tensor, which we computed in eqn 2.143 to be

$$\tilde{b}_{i \neq j}(\omega) = \frac{1}{8\pi \tilde{\eta}^*(\omega) d} \left(\delta + \hat{d}\hat{d} \right), \tag{4.89}$$

or

$$\tilde{b}_{i \neq j}(\omega) = \frac{i\omega}{8\pi G^*(\omega) d} \left(\delta + \hat{d}\hat{d} \right), \tag{4.90}$$

where

$$\hat{d} = \frac{\mathbf{r}_2 - \mathbf{r}_1}{|\mathbf{r}_2 - \mathbf{r}_1|} \tag{4.91}$$

is the unit vector between the particle centers. Note we have used the Correspondence Principle to generalize the (Newtonian) computation for LVE materials via $\eta \to G^*(\omega)/(i\omega)$.

For a pair of particles (1 and 2), separated by $\mathbf{d}$, we thus find for displacements along the line-of-centers

$$\langle \Delta \tilde{R}_1^{\|} \Delta \tilde{R}_2^{\|} \rangle(\omega) = \frac{2k_B T}{4\pi (i\omega) G^*(\omega) d} \tag{4.92}$$

and displacements perpendicular to the particles' line-of-centers

$$\langle \Delta \tilde{R}_1^{\perp} \Delta \tilde{R}_2^{\perp} \rangle(\omega) = \frac{2k_B T}{8\pi (i\omega) G^*(\omega) d}. \tag{4.93}$$

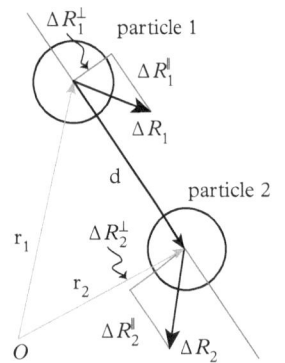

Fig. 4.38 *The displacement projections of two particles over a lag time* t, ΔR_1 *and* ΔR_2, *that are used to calculate two-point microrheology.*

These displacement projections are illustrated in Fig. 4.38.

Using the relation between creep compliance and shear modulus, eqn 1.39, the time-domain displacement correlations are

$$\langle \Delta R_1^{\|} \Delta R_2^{\|} \rangle(t) = \frac{k_B T}{2\pi \mathcal{J}(t) d} \tag{4.94}$$

and

$$\langle \Delta R_1^{\perp} \Delta R_2^{\perp} \rangle(t) = \frac{k_B T}{4\pi \mathcal{J}(t) d}. \tag{4.95}$$

Because $\langle \Delta R_1^{\|} \Delta R_2^{\|} \rangle$ is a factor of two greater than $\langle \Delta R_1^{\perp} \Delta R_2^{\perp} \rangle$, the probes' motion along their line of centers is typically analyzed in a two-point experiment.

Most significant here is the fact that neither particle's radius or mass appears anywhere in eqns 4.92–4.95. This represents a clear advantage of two-point microrheology: The cross-correlated fluctuations between distinct particles are dominated by the response of the material that lies between the particles, through which the stress propagates. The size and shape of either particle plays a small role in the cross-correlated fluctuations of two particles, which are dominated by the total force on one particle, and the local velocity experienced by the other particle. To the leading order, neither of these quantities depend on the size or shape of either probe, so long as the probes are small compared with their separation distance. By contrast, the self-mobility of a probe depends significantly on its own size and shape, as well as on the material properties in its immediate vicinity. If a probe affects the microstructure of the material around it, for example, its self-mobility will be sensitive to this perturbation. The coupling mobility, on the other hand, arises due to the stress propagated through the space between the probes, and is therefore much less sensitive to the immediate environment around each probe.

The *cross-correlated* fluctuations of two particles $\langle \Delta R_1(\tau) \Delta R_2(\tau) \rangle$ depends on the distance between the particles (which can be measured accurately) and the material properties $\eta^*(\omega)$, which one wants to measure. They do not depend strongly on the probe size (which always has some polydispersity, and is difficult to measure with much accuracy using standard optical microscopy), nor on the local material environment (which may be affected by probe chemistry). Two-point microrheology is thus well-suited to measure the homogeneous rheology of materials. Differences between the one-point and two-point measurements encode information about probe-material interactions, or length scale-dependent rheology, or material heterogeneities.

4.11.2   Data requirements of two-point microrheology

A downside to two-point microrheology is that the cross-correlated fluctuations are much weaker than the self fluctuations by an amount on the order of $\sim 2a/d$. Significantly greater statistics are thus required to obtain a clear measurement of the rheology. This requirement can be onerous in cases where the particle mobility is fast, given that the calculation requires the presence of both particles in the imaging plane in two frames a lag time t apart and may preclude samples with a time-dependent rheology, such as gelators, unless the gelation kinetics are slow. Crocker and Hoffman (2007) estimate that, under best conditions, approximately $(d/2a)^2$ more realizations of $\langle \Delta R_1(\tau) \Delta R_2(\tau) \rangle$ are required than the corresponding single point mean-squared displacement measurement, $\langle \Delta R^2(t) \rangle$. Since $d/2a \sim 10$, two-point microrheology (conservatively) requires about two orders of magnitude more measurements to achieve the same level of statistical certainty as a one-point measurement.

4.11.3   Two-point experiments

Crocker *et al.* (2000) introduced and validated two-point microrheology in a series of experiments that employed both Newtonian and viscoelastic materials. The two-point response of particles in a glycerol / water mixture served as a validation experiment. Fig. 4.39 shows $D_{rr} \equiv \langle \Delta R_1^{\|}(t) \Delta R_2^{\|}(t) \rangle$ as a triple-logarithmic plot in both time and particle separation. Mean-squared displacement indeed decays as $\sim 1/d$ and grows as $\sim t$. Defining a "distinct" mean-squared displacement[17]

$$\langle \Delta r^2(t) \rangle_D = \frac{2d}{a} \langle \Delta R_1^{\|}(t) \Delta R_2^{\|}(t) \rangle, \tag{4.96}$$

[17] The term *distinct* refers to a correlation of a particle at time zero to the position of a *different* particle at a position **r** some time later, in contrast to a *self* correlation, like the Van Hove function.

allows two-point results to be compared directly with one-point measurements calculated using the same image data (Fig. 4.39b). As expected, one-point and two-point measurements give the same values for the Newtonian fluid. Both the one- and two-point mean-squared

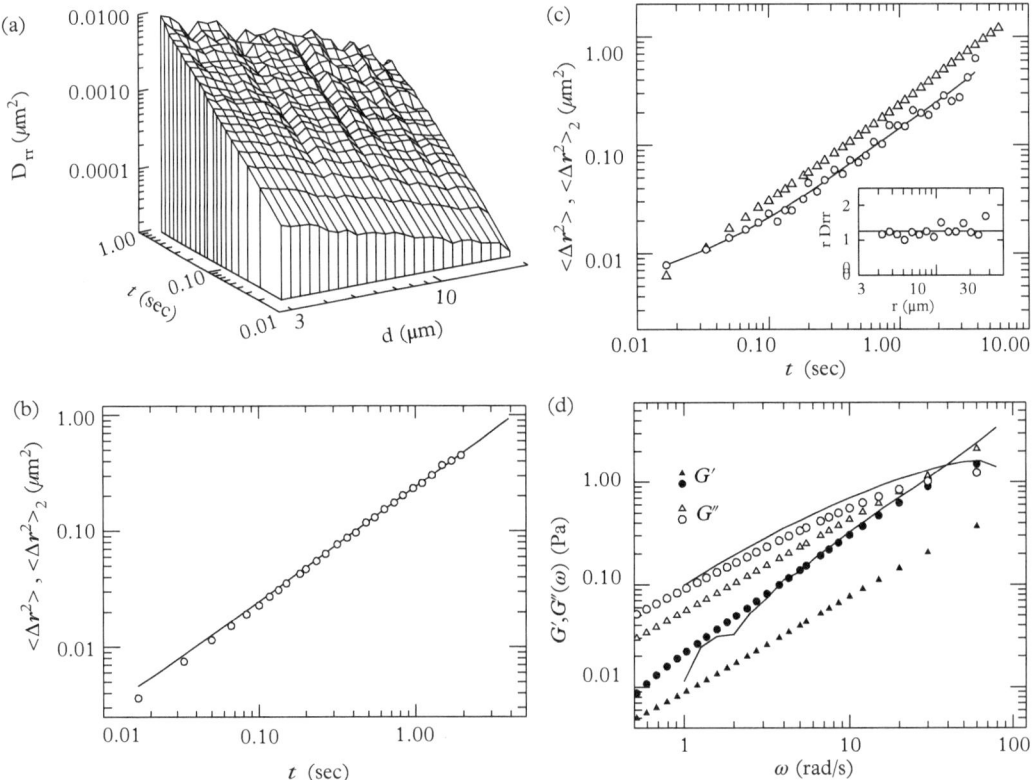

Fig. 4.39 *A two-point microrheology experiments in aqueous glycerol solutions and guar gels. (a) The two-point displacement correlation function of 0.47 μm diameter particles in a glycerol / water solution as a function of time and particle separation. (b) A comparison between one-point (solid line) and two-point (symbols) measurements. (c) Two-point displacements of a guar sample. (d) The calculated moduli from one-point and two-point microrheology compared with bulk rheometry. Reprinted figure with permission from Crocker, J. C. et al., Phys. Rev. Lett., 85, 888–91 (2000). Copyright 2000 by the American Physical Society.*

displacements should be equal to

$$\frac{1}{\pi\, a \mathcal{F}(t)} = \frac{t}{\pi\, a \eta},\tag{4.97}$$

expressing the one-point mean-squared displacement in three-dimensions by multiplying it by 3.

Subsequent measurements of a guar sample illustrate two-point microrheology's unique capabilities to measure rheology on length scales greater than the probe dimensions. Guar, a natural polysaccharide, exhibits significant heterogeneity on the micrometer scale. Locally, probe particles experience different degrees of cross-linking and polymer concentrations. A comparison of the self and distinct mean-squared displacements (Fig. 4.39c) and corresponding shear moduli (Fig. 4.39d) show that the one-point and two-point measurements give different results. The one-point mean-squared displacement is higher, and moduli computed from the one-point data using the GSER do not agree with bulk rheology. By contrast, the moduli calculated from the two-point analysis match bulk rheology measurements.

The guar sample results are a strong indication that local structure affects the one-point microrheology, but that the longer-length scales probed by two-point microrheology "average out" inhomogeneities, as they only reflect shear stresses propagated across the separation between particles. Similarly, bulk rheology deforms samples over macroscopic dimensions.

Two-point microrheology provides one remedy when one-point microrheology "fails" due to probe-material interactions or inhomogeneity in a sample (which violates the Stokes criterion of one-point microrheology). When either of these limitations are not an issue, one-point measurements are far less cumbersome. The two-point measurements described by Crocker *et al.* (2000) required ten minutes of recorded data at 60 frames per second. As we will see in the next section, the combination of one- and two-point microrheology can also provide new rheological information that would otherwise be inaccessible to bulk rheology—the structure and rheology of material near the interface of probe particles. Thus, when one-point "fails," it can provide new information about a material.

4.11.4   Shell model

While one-point and two-point microrheology analyses should yield identical rheological properties for homogeneous materials that satisfy

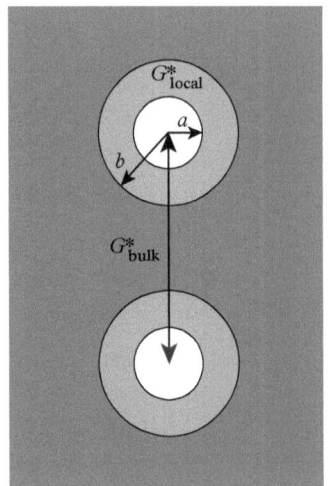

Fig. 4.40 *The shell model of two-point microrheology.*

the Stokes criteria, one- and two-point measurements will almost certainly differ if the material is heterogeneous.

Levine and Lubensky (2000) and (2001) introduced a model for inhomogeneous materials in two-point microrheology, illustrated in Fig. 4.40, wherein a "shell" of material with moduli G^*_{local} and thickness $(b - a)$ surrounds each probe (of radius a), whereas the material between the probes has modulus G^*_{bulk} (Levine and Lubensky, 2000; Levine and Lubensky, 2001). Such shell models have been used to study the two-point response in inhomogeneous materials and can be experimentally realized when strong depletion of a polymer occurs in the vicinity of probes, especially in solutions of semiflexible polymers, such as DNA (Chen *et al.*, 2003) or F-actin (Chae and Furst, 2005; Huh and Furst, 2006).

The apparent moduli that would be measured using one-point and two-point analyses of the system shown in Fig. 4.40 can be computed, giving

$$\frac{G^*_2(\omega)}{G^*_1(\omega)} = \frac{4\beta^6\kappa'^2 - 9\beta^5\kappa\kappa' + 10\beta^3\kappa\kappa' - 9\beta\kappa'^2 - 15\beta\kappa' + 2\kappa\kappa''}{2[\kappa'' - 2\beta^5\kappa']}$$

(4.98)

where $\beta = a/b$, $\kappa = G^*_{\text{bulk}}(\omega)/G^*_{\text{local}}(\omega)$, $\kappa' = \kappa - 1$ and $\kappa'' = 3 + 2\kappa$. Equation 4.98 applies strictly in the limit of incompressible materials, which is often a reliable assumption; Levine and Lubensky (2001) account for compressibility as well, in addition to rheological variations that are smoother than this discrete shell. Notably, the apparent moduli predicted in eqn 4.98 depend on both the material rheology and shell geometry. As we will show, $G^*_2(\omega) \approx G^*_{\text{bulk}}$, but $G^*_1(\omega) \neq G^*_{\text{local}}(\omega)$. The single-point response depends on the viscoelastic properties of both the local and bulk materials.

One-point response

Some analysis of the one-point calculation illustrates the range of effects that arise even with a simple shell. We consider a particle of radius a with a thin shell of radius $b = a(1 + \epsilon)$ around it, within which the viscosity is η_{local}, in a viscous liquid of viscosity $\eta_{\text{bulk}} = \kappa\eta_{\text{local}}$. We solve the Stokes equations in both regions ($a < r < a(1 + \epsilon)$) and $(a(1 + \epsilon) < r)$, with (1) no-slip boundary conditions on the sphere, so that the inner-fluid velocity matches the uniform velocity on the sphere V; (2) a decaying far-field velocity; (3) matching velocities at the shell boundary, and (4) matching normal and tangential stresses at the shell boundary.

With the methods of Section 2.5.2, the inner (shell) and outer-velocity fields have stream functions

$$\psi_i = \frac{(A + Br^2 + Cr^3 + Dr^5)V\sin^2\theta}{2r} \tag{4.99}$$

$$\psi_o = \frac{(A_o + B_or^2)V\sin^2\theta}{2r}. \tag{4.100}$$

In this case, the drag on the sphere is given by

$$F = 4\pi\eta BV, \tag{4.101}$$

so finding the coefficient B gives the drag.

The self-resistance of the sphere, normalized by what one would expect for a homogeneous fluid of viscosity η_s, is given by

$$\frac{\zeta_T}{6\pi\eta_2 a} = 1 - \epsilon(\kappa - 1)\frac{N(\epsilon,\kappa)}{P(\epsilon,\kappa)} \tag{4.102}$$

where N and P are polynomials of the shell thickness ϵ,

$$\begin{aligned} N(\epsilon,\kappa) =& 10 + 30\epsilon + 60\epsilon^2 + 15\epsilon^3(4+\kappa) \\ &+ 15\epsilon^4(2+\kappa) + 2\epsilon^5(3+2\kappa) \end{aligned} \tag{4.103}$$

and

$$\begin{aligned} P(\epsilon,\kappa) =& 10 + 30\epsilon(1+\kappa) + 30\epsilon^2(2+3\kappa) \\ &+ 20\epsilon^3(3+7\kappa) + 15\epsilon^4(2+7\kappa+\kappa^2) \\ &+ 3\epsilon^5(2+13\kappa+5\kappa^2) + 2\epsilon^6\kappa(3+2\kappa). \end{aligned} \tag{4.104}$$

More important than the detailed functional forms are the various limits that follow. If the shell is infinitely thin (so that $\epsilon = 0$), then the self-resistance reduces to

$$\zeta_T(\epsilon = 0) = 6\pi\eta_{\text{bulk}}a \tag{4.105}$$

as expected. When the shell and bulk media have the same rheology ($\kappa = 1$),

$$\zeta_T(\kappa = 1) = 6\pi\eta_{\text{bulk}}a \tag{4.106}$$

as well. If the shell viscosity were much higher than the outside viscosity ($\kappa \to 0$), then

$$\zeta_T(\kappa \to 0) \to 6\pi\eta_{\text{bulk}}a(1 + \epsilon), \tag{4.107}$$

as though the shell were simply part of a larger sphere, with total radius $a(1 + \epsilon)$.

Some care is required in looking at thin shells and low-viscosity ($\kappa \gg 1$) shells, since we have products of small and large parameters. We keep the highest-order κ term for each power of ϵ in the numerator and denominator,

$$\frac{\zeta_T}{6\pi\eta_2 a} =$$
$$1 - \epsilon\kappa\,\frac{10 + 30\epsilon + 60\epsilon^2 + 15\epsilon^3\kappa + 15\epsilon^4\kappa + 4\epsilon^5\kappa}{10 + 30\epsilon\kappa + 90\epsilon^2\kappa + 140\epsilon^3\kappa + 15\epsilon^4\kappa^2 + 15\epsilon^5\kappa^2 + 4\epsilon^6\kappa^2}. \tag{4.108}$$

then take the thin shell limit ($\epsilon \ll 1$), drop ϵ^2 compared with ϵ, but make no assumptions regarding the size of $\epsilon\kappa$, giving

$$\frac{\zeta_T}{6\pi\eta_2 a} = 1 - \epsilon\kappa\,\frac{2 + 3\epsilon^3\kappa}{2 + 6\epsilon\kappa + 3\epsilon^4\kappa^2} = \frac{2 + 4\epsilon\kappa}{2 + 6\epsilon\kappa + 3\epsilon^4\kappa^2}. \tag{4.109}$$

Three regimes appear. If $\kappa\epsilon \ll 1$, corresponding to a relative shell thickness ϵ that is smaller than the relative viscosity of the shell κ^{-1}, the self-resistance of the sphere becomes

$$\frac{\zeta_T(\epsilon\kappa \ll 1)}{6\pi\eta_{\text{bulk}}a} = 1 - \epsilon\kappa, \tag{4.110}$$

which is approximately equal to the result without any shell, but with a shell that reduces the resistance slightly.

If the relative-shell viscosity were a bit smaller, so that $1 \ll \epsilon\kappa \ll 1/\epsilon^2$, then

$$\frac{\zeta_T(1 \ll \epsilon\kappa \ll \epsilon^{-2})}{6\pi\eta_{\text{bulk}}a} \sim \frac{2}{3}, \tag{4.111}$$

so that the thin shell acts like a "slip" boundary condition for the sphere,

$$\zeta_T(1 \ll \epsilon\kappa \ll \epsilon^{-2}) \sim 4\pi\eta_{\text{bulk}}a. \tag{4.112}$$

The resistance of the one-point probe continues to be dominated by the outer fluid.

Finally, if the shell viscosity η_{local} is extremely small – $\kappa\epsilon^3 \gg 1$, then the self-resistance becomes

$$\frac{\zeta_T}{6\pi\,\eta_{\text{bulk}}a} \sim \frac{4}{3\epsilon^3\kappa}, \tag{4.113}$$

which corresponds to

$$\zeta_T \sim \frac{8\pi\,\eta_{\text{local}}a}{\epsilon^3}. \tag{4.114}$$

In this limit, the drag is completely insensitive to the viscosity of the outer fluid, but instead behaves like a probe moving in a fluid of (shell) viscosity η, confined within a rigid spherical cavity of radius $a(1+\epsilon)$.

So we have expressions for the self-resistance, from which we can get the self-mobility by simple inversion.

Two-point response

The two-point response requires the coupling mobility to be computed, as discussed in Section 2.6.2. Although the shell model would appear to provide significant complications for hydrodyanamic-interaction calculations, the coupling mobility is almost completely unaffected by the properties of the shell. We first note that the drag computed by integrating the stress around a sphere within the outer fluid is

$$F = 4\pi\,\eta_{\text{bulk}}B_o V, \tag{4.115}$$

which must be equal to the drag on the probe as computed in eqn 4.101. The far-field flow velocity around sphere 1 is thus given by

$$\psi_o \to \frac{B_o r V \sin^2\theta}{2} = \frac{r\sin^2\theta}{8\pi\,\eta_{\text{bulk}}}F, \tag{4.116}$$

which is simply the Stokeslet flow due to the point force F in a medium of viscsosity η_{bulk}. Sphere 2—along with its shell—then simply translates with this far-field velocity, to leading order. The coupling mobility is thus simply given by eqn 2.143, in a medium of viscosity η_{bulk}.

To summarize how local "shells" affect one-point and two-point measurements, the apparent viscosity reported by one-point measurements of probe self-diffusion depend upon the thickness and viscosity of the shell, according to

$$\eta_1^{\mathrm{app}} = \begin{cases} \eta_{bulk}(1 + \epsilon) & \text{if } \kappa \ll 1; \\ \eta_{bulk}(1 - \epsilon\kappa) & \text{if } \kappa \gg 1 \text{ and } \epsilon\kappa \ll 1; \\ \frac{2}{3}\eta_{bulk} & \text{if } \kappa \gg 1 \text{ and } 1 \ll \epsilon\kappa \ll \epsilon^{-2}; \\ \frac{4}{3\epsilon^3}\eta_{bulk} & \text{if } \kappa \gg 1 \text{ and } \epsilon^3\kappa \gg 1. \end{cases} \tag{4.117}$$

By contrast,

$$\eta_2^{\mathrm{app}} \approx \eta_{bulk} \tag{4.118}$$

when the viscosity is inferred from the coupling mobility (*i.e.*, two-point). These results can be immediately generalized for viscoelastic medium with modulus G^*_{bulk} using the Correspondence Principle.

Application note: Rheological microscopy

Discrepancies between one- and two-point microrheological measurements encode structural information about the material and its interaction with the probes. Measuring both one- and two-point responses can thus yield the bulk material rheology, local rheology, and extent of the perturbed zone around each particle—more information than possible from a bulk rheology experiment alone. In an early demonstration of the principle, Chen *et al.* (2003) used the shell model to characterize the depletion of polymer around probes dispersed in semidilute DNA solutions between 0.03 and 0.4 mg/ml. The lower concentration is approximately the overlap concentration of the DNA, which has a contour length of 16.5 μm and a persistence length of 50 nm. Fig. 4.41 shows their microrheology results, which we will now discuss.

In DNA microrheology studies, the apparent moduli derived from one- and two-point measurements do not agree (Fig. 4.41a). The moduli calculated from the one-point mean-squared displacement depend on the probe-particle diameter, which varies between 0.46 and 2.0 μm. Chen *et al.* (2003) assumed that the two-point moduli reported the true bulk-rheology of the material, and modeled a depleted layer as a shell of undetermined thickness $\Delta = b - a$ that consists of pure solvent (so that $G^*_{\mathrm{local}} = i\omega\eta_s$). They found that apparent one-point moduli collapsed onto the two-point measurements when Δ was used as a fitting parameter, using eqn 4.98, shown in Fig. 4.41b. The best-fit values for Δ were consistent with the mesh size ξ of the DNA solutions (Fig. 4.42), reinforcing an interpretation that a depletion of

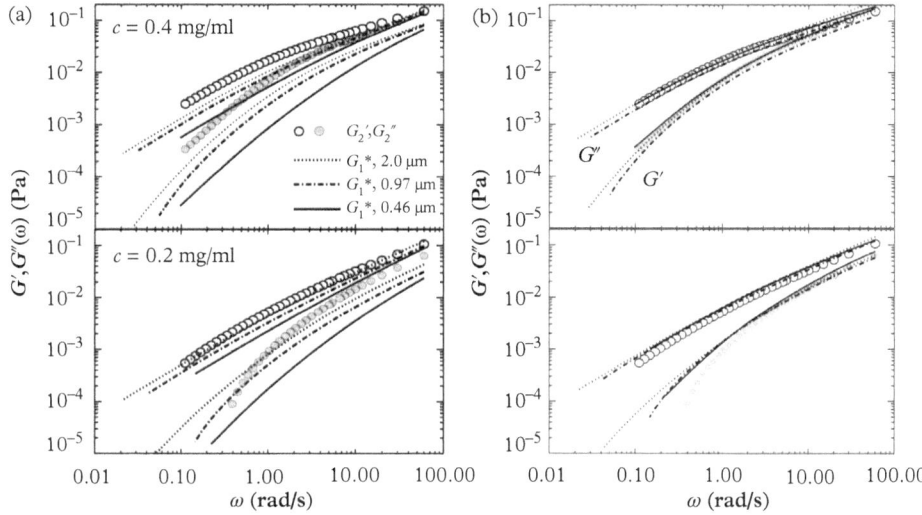

Fig. 4.41 *One- and two-point microrheology of 0.4 mg/ml λ-DNA solutions. (a) The apparent one- and two-point moduli. (b) The data are collapsed using the shell model. Reprinted figure with permission from Chen, D. T. et al., Phys. Rev. Lett., 90, 108301 (2003). Copyright 2003 by the American Physical Society.*

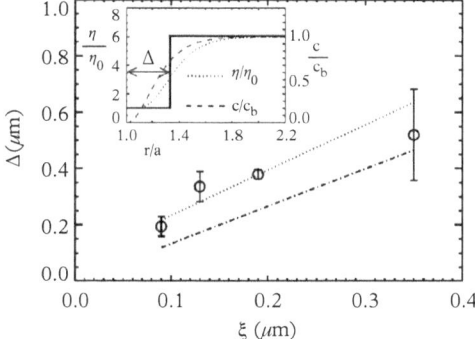

Fig. 4.42 *Using the shell model, the shell thickness Δ is inferred from one- and two-point microrheology and compared to the mesh size ξ of entangled DNA solutions. Reprinted figure with permission from Chen, D. T. et al., Phys. Rev. Lett., 90, 108301 (2003). Copyright 2003 by the American Physical Society.*

polymer segments near the probe surface has a strong effect on the one-point mean-squared displacement. Hence, the combination of one- and two-point microrheology provides an understanding of the fluid structure near the probe surface. Analogous experiments have been performed in entangled solutions of F-actin, where depletion occurs when the probe particle surface chemistry prevents adsorption of the actin protein (Chae and Furst, 2005; Huh and Furst, 2006).

....

EXERCISES

(4.1) **Perrin's data.** Jean Perrin used microscopy to track the Brownian motion of monodisperse emulsion particles. The particles were composed of a tree sap, which solidifies. Careful fractionation was used to produce particles with radius $a = 0.367 \mu$m. The position of the particles was noted every 30 seconds. Perrin constructed a plot of the particle positions, examining every step with reference to the previous step. Thus, this is a probability distribution of the displacement of a particle after 30 seconds. The plot is shown in Fig. 4.43.

(a) Use Perrin's data to plot the probability distribution of displacement. This is the *Van Hove self space–time–correlation function*.

(b) Plot the distribution of angles that particles make. Compare this to what you think you should observe.

(c) Can you estimate the viscosity from this calculation?

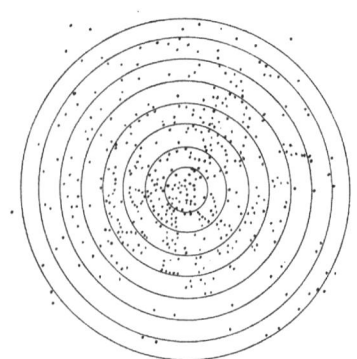

Fig. 4.43 *Perrin's data for the displacement of $a = 0.367 \mu m$ particles in water at 30 second intervals. Each circle represents fractions of the root mean-squared distance $e = 7.84 \mu m$. The circles in the plot are $\frac{e}{4}, \frac{2e}{4}, \frac{3e}{4}$, etc. (Perrin, 1913)*

(4.2) **Statistics of a random walk.** Write a program that generates a 1D random walk. The algorithm should be similar to our derivation of the random walk in Section 4.5.2. Define a time step τ such that at each time point, the particle will step at the length δ along the axis. The direction the particle will step should be determined randomly with equal probability of moving left or right. The output of your program should be the particle position for each time step. Use the program to develop a sense of the statistics of random walks.

(a) Plot a sample "trajectory" (a plot of position versus time) for one run for 1500 time steps.

(b) Plot more individual trajectories (at least 100). How do these compare to the trajectories from experimental data shown in Fig. 4.20?

(c) Plot the mean-squared displacement of *each particle* and the ensemble-averaged mean-squared displacement as a function of lag time. How do these compare?

(d) Plot the non-Gaussian parameter at several lag times. How does this change with lag time?

(e) Implement a time average in addition to the ensemble average for the MSD by breaking trajectories into shorter segments and compare these averages to the ensemble and single-particle MSDs.

(4.3) **Probe polydispersity.** The coefficient of variation of a probe particle used in a particle tracking experiment is C.V. = 2%. If the probes are dispersed in a Newtonian fluid, what is the minimum precision for which the viscosity can be determined? Calculate the the Van Hove correlation and non-Gaussian parameter for several logarithmically-spaced lag times between 10^{-3} and 1 s.

(4.4) **Probe concentration.** Probe particles are dispersed in water at a volume fraction $\phi = 0.001$ and imaged using video microscopy at a rate of 30 frames per second. Calculate the average interparticle distance.

(a) How does the average particle separation compare with the distance the particles are expected to move between frames?

(b) Multiple particle tracking is performed using 1 μm diameter fluorescent probe particles. Using a high-numerical microscope objective, the effective depth of field is about 1 μm. There are approximately 30 particles being tracked in any given frame. What is your estimate of the volume fraction of the probes?

(4.5) **Shell model.** The shell model discussed in Section 4.11.4 is similar to the motion of a sphere passing through the center of a spherical container containing a fluid of viscosity η. This "concentric spheres" problem is discussed by Happel and Brenner (1983) who show that the mobility of the sphere is changed by a factor K

$$\zeta = 6\pi a \eta K \tag{4.119}$$

where

$$K = \frac{1 + \frac{3}{2}\beta^5}{1 - \frac{3}{2}\beta + \frac{3}{2}\beta^5 - \beta^6} \tag{4.120}$$

for a fluid sphere of vanishing viscosity (*i.e.*, a bubble) and

$$K = \frac{1 - \beta^5}{1 - \frac{9}{4}\beta + \frac{5}{2}\beta^3 - \frac{9}{4}\beta^5 + \beta^6} \tag{4.121}$$

for a sphere of infinite viscosity. Here, $\beta = a/b$, the ratio of the radius of the particle to that of the spherical container. In this exercise, show that eqn 4.98 gives identical results in these limits.

5

Light scattering microrheology

Dynamic light scattering (DLS) is a well-established method for measuring the motion of colloids, proteins, and macromolecules. Light scattering has several advantages for microrheology, especially given the availability of commercial instruments, the relatively large sample volumes that average over many probes, and the sensitivity of the measurement to small particle displacements, which can extend the range of length and time scales beyond those typically accessed by the methods of multiple particle tracking (discussed in the previous chapter) and bulk rheology.

Like multiple particle tracking experiments (Chapter 4), light scattering experiments are a form of passive microrheology, as they measure the thermally-fluctuating displacements of probe particles dispersed within the medium of interest, via fluctuations in the interference patterns of scattered light. Examples of studies of colloid motion in complex fluids abound, but are not always interpreted in terms of microrheology (Lin and Phillies, 1984). There is a deep literature on the subject, stretching back to the development of the laser and dynamic light scattering. Readers are referred to classic texts, such as those by Berne and Pecora (2000) and Chu (1991), as well as more recent treatments by Brown (1993) and Pusey (2002).

Two light scattering methods are typically used in microrheology experiments. The first is the traditional DLS experiment. The second, diffusing wave spectroscopy (DWS), was developed more recently, and takes advantage of multiple scattering. Both techniques are described here, but any dynamic light scattering technique, *e.g.*, evanescent wave dynamic light scattering or multi-color DLS, can be adapted for microrheology measurements. DWS is easily incorporated into older DLS instruments with a few optics and detectors, and several commercial systems are now available.

We begin this chapter with a brief discussion of time–correlation functions. We will then review the basic principles of light scattering and derive the working equations for DLS and DWS microrheology. This lays a foundation for discussing the operating range of the experiment and several applications, including high-frequency

Microrheology. Eric M. Furst and Todd M. Squires, Oxford University Press (2017).
© Eric M. Furst and Todd M. Squires. DOI 10.1093/oso/9780199655205.001.0001

microrheology, measurements of non-ergodic samples like gels, and "broad-band" experiments capable of measuring rheology over several decades of time scales.

5.1 Time–correlation functions

Time-correlation functions are important for describing random signals and noise. Two examples in microrheology are the velocity autocorrelation function, which we introduced in Chapter 3 in the context of Brownian motion and used to derive the Generalized Stokes–Einstein Relation, and the intensity autocorrelation function, which is measured in the dynamic light scattering methods discussed in this chapter.

The time-averaged, time autocorrelation function of a stochastic function A is

$$C_{AA}(t; t_0) = \lim_{T \to \infty} \frac{1}{T} \int_{t_0}^{t_0+T} A(t')A(t'+t)dt', \qquad (5.1)$$

where t_0 is the initial time, T is the total time over which data is taken and the correlation function is calculated, and t is the *delay time, shift time*, or *lag time* of the correlation. As the integration time T becomes large, we expect the time-averaged, time autocorrelation function to become independent of the initial time t_0.

The ensemble average, denoted $\langle A(t_0)A(t_0+t)\rangle$, represents a different averaging process than eqn 5.1. Instead, the ensemble average is computed by (i) making many measurements ($N \gg 1$) of the same system, (ii) computing the product of $A_j(t_0)$ measured at time t_0 with the value $A_j(t_0 + t)$ measured a lag time t after t_0, for each and every individual measurement ($1 < j < N$); and (iii) computing the average:

$$\langle A(t_0)A(t_0 + t)\rangle = \lim_{N \to \infty} \frac{1}{N} \sum_{j=1}^{N} A_j(t_0)A_j(t_0 + t). \qquad (5.2)$$

Unlike the time-averaged, time–correlation function (eqn 5.1), the ensemble-averaged autocorrelation function may depend on the initial time t_0, depending on how the experiment was prepared.

If a system is *stationary*—meaning that its statistics do not change in time—then the ensemble average does not depend on t_0, so that

$$\langle A(t_0)A(t_0+t)\rangle = \langle A(0)A(t)\rangle = \langle A(-t)A(0)\rangle = \lim_{N \to \infty} \frac{1}{N} \sum_{j=1}^{N} A_j(0)A_j(t).$$

$$(5.3)$$

Ensemble-averaged measurements on a system in thermodynamic equilibrium will not depend on t_0, since any time is statistically equivalent to all others once equilibrium has been achieved.

In *ergodic* systems, the time-average and ensemble-average correlation functions are equal. In the microrheology context, where A might represent a probe velocity, ergodic systems are ones in which the statistical behavior of a single probe particle, measured over all times, is identical to the statistical average of individual probes measured in a large number of identical experiments. In other words, an individual probe, if left long enough in an ergodic system, would sample all possible conditions with the same statistical probability as would be observed over short times with many individual probes, prepared in many identical systems. For ergodic systems, then,

$$C_{AA}(t) = \langle A(0)A(t) \rangle = \lim_{T \to \infty} \frac{1}{T} \int_0^T A(t')A(t' + t)\,dt'. \tag{5.4}$$

Systems may be non-ergodic for several reasons. Highly-heterogeneous materials may show non-ergodic properties: Probes may require extremely long—even infinite—times to explore all different configurations, whereas measurements on many such samples would immediately reveal a spectrum of mechanical responses. In other cases, non-ergodicity accompanies kinetic arrest or slowing down of the dynamics, which is a hallmark of gels and glasses. Since the time average time–correlation function is no longer equal to the ensemble average in this case, special care must be taken when measuring non-ergodic samples. Methods for analyzing light scattering from non-ergodic samples are discussed later in this chapter.

Time-average autocorrelation functions are used in microrheology experiments—most commonly in dynamic light scattering (photon correlation spectroscopy) experiments discussed in this chapter, but passive microrheology often effectively combines time- and ensemble-averaging.

The autocorrelation function has an upper bound given by its value at zero lag time,

$$\langle |A(0)|^2 \rangle \geq \langle A(0)A(t) \rangle, \tag{5.5}$$

whereas at long-lag times, stochastic functions become uncorrelated,

$$C_{AA}(t \to \infty) \to |\langle A \rangle|^2, \tag{5.6}$$

so that the long-time value is simply the average value squared.

The Wiener–Kintchine theorem states that the Fourier Transform of the time–correlation function of a stationary process is its spectral density,

$$I_{AA}(\omega) = \langle |A(\omega)|^2 \rangle = \int_{-\infty}^{\infty} \langle A(0)A(t) \rangle e^{-i\omega t} dt. \qquad (5.7)$$

While we typically employ Langevin equations in this book—which effectively average over the fast (thermal) degrees of freedom—this approach may be generalized (*e.g.*, using the Fokker–Planck equation approach) to expicitly resolve momentum degrees of freedom as well. Such approaches give rise to ensemble average time–correlation functions given by (Berne and Pecora, 2000)

$$C_{AA}(t) \equiv \int A(\Gamma_0)A(\Gamma_t) \left(\frac{e^{-\mathscr{H}(\Gamma_0)/k_B T}}{Q} \right) d\Gamma_0 \qquad (5.8)$$

where $\mathscr{H}$ is the Hamiltonian of the system, defined as the kinetic and potential energy over the $6N$ canonical positions q and momenta p of an N particle system,

$$\mathscr{H}(\{q, p\}) = \text{K.E.} + \text{P.E.} \qquad (5.9)$$

Γ_t represents a point in phase space at time t (the $3N$ positions and $3N$ momenta of all particles at that instant) and Q is the partition function,

$$Q \equiv \int \rho(\Gamma) d\Gamma. \qquad (5.10)$$

Because the probability that a system occupies a point in phase space is given by

$$\rho(\Gamma) d\Gamma = \frac{e^{-\mathscr{H}(\Gamma)/k_B T}}{Q}, \qquad (5.11)$$

we can rewrite the ensemble average time–correlation function as

$$C_{AA}(t) \equiv \int A(\Gamma_0)A(\Gamma_t)\rho(\Gamma_0)d\Gamma_0. \qquad (5.12)$$

Hence, any dynamical variable A of such systems depends on the phase-space trajectories Γ_t in time.

The trajectory of the system in phase space coordinates is governed by Hamilton's equations of motion, in which the position and momenta of particle i are

$$\dot{q_i} = \frac{\partial \mathcal{H}}{\partial p_i} \tag{5.13}$$

and

$$\dot{p_i} = -\frac{\partial \mathcal{H}}{\partial q_i}, \tag{5.14}$$

respectively. In principle, one may solve the equations of motion to find Γ_t as a function of the initial value Γ_0 and time t, so that

$$C_{AA}(t) \equiv \int A(\Gamma_0)A(\Gamma_0, t)\rho(\Gamma_0)d\Gamma_0. \tag{5.15}$$

Alternately, the probability $\mathcal{P}(\Gamma_0, \Gamma_t, t)$ of finding the system at Γ_0 at time zero and Γ_t at time t, if it is known, can be used to evaluate the ensemble-averaged time–correlation function,

$$C_{AA}(t) \equiv \int \mathcal{P}(\Gamma_0, \Gamma_t, t)A(\Gamma_0)A(\Gamma_0, t)d\Gamma_0 d\Gamma_t. \tag{5.16}$$

An example of one such probability distribution function is the Van Hove space-time correlation function, Eqn 4.19.

5.2 Light scattering

In light scattering, coherent laser light with the incident wavevector $\mathbf{k}_i$ illuminates a sample and scatters. A detector in the far-field collects the scattered light with wavevector $\mathbf{k}_s$ at an angle θ relative to the incident beam. An example of the geometry is shown in Fig. 5.1a.

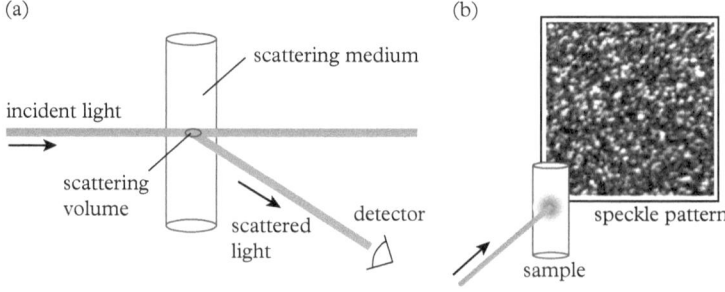

(a)

scattering medium

incident light

scattering volume

scattered light

detector

(b)

speckle pattern

sample

Fig. 5.1 *(a) A typical scattering experiment. (b) A speckle pattern is the random constructive and destructive interference of scattered light from a sample.*

The total scattered field E at the detector is a superposition of the field from each scatterer in the *scattering volume*—the overlap of the volume illuminated by the incident beam and volume visible to the detector—resulting in

$$E(t) = \sum_{i=1}^{N} E_i \exp\left[i\mathbf{q} \cdot \mathbf{r}_i(t)\right] \tag{5.17}$$

where $\mathbf{q}$ is the scattering wavevector $\mathbf{q} = \mathbf{k}_s - \mathbf{k}_i$ and $\mathbf{r}_i(t)$ is the position of the ith particle. The magnitude of the scattering wavevector is

$$q = |\mathbf{q}| = (4\pi n/\lambda_v) \sin \theta/2 \tag{5.18}$$

for light with a vacuum wavelength of λ_v in a scattering medium with refractive index n. In inelastic scattering, for which negligible momentum is exchanged with the scatterer, $|\mathbf{k}_s| = |\mathbf{k}_i|$, the distribution of scattering particle positions leads to phase differences in the electric field reaching the detector $\Delta\Phi$, leading to constructive and destructive interference. Scattering also depends on the light's polarization. In Fig. 5.2, the light is *s*-polarized, and hence its electric field would be oriented perpendicular to the figure plane. It is shown in the plane to highlight the phase difference between to scatterers that contribute to the total electric field at the detector.

   With light from many single scatterers, a random diffraction pattern of scattered light, or speckle pattern, will form. This pattern is illustrated in Fig. 5.1b. The interference pattern of light and dark spots is generated by light arriving from the scattering volume, an overlapping region of the laser passing through the sample and viewed by the detector. Each speckle changes intensity as scatterers move and the phase lag between their respective scattered fields change. An average intensity pattern with scattering angle or wavevector follows from the distribution of particles, given by the structure factor $S(q)$, and the particle form factor $P(q)$,

$$I(q) \sim P(q)S(q). \tag{5.19}$$

The structure factor is defined as

$$S(q) = \frac{1}{N} \sum_{i,j=1}^{N} \langle \exp[i\mathbf{q} \cdot (\mathbf{r}_i - \mathbf{r}_j)]\rangle. \tag{5.20}$$

It describes the scattered interference pattern from the ensemble average of N point sources located at positions $\mathbf{r}_i$ throughout space based

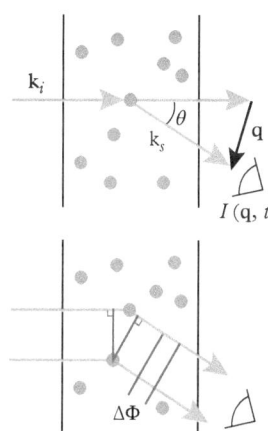

Fig. 5.2 *In inelastic single scattering, the intensity of scattered light from the incident wavevector* $\mathbf{k}_i$ *is measured at a detector. Viewed here from above, the angle of the detector relative to the incident wavevector (or laser source) defines the scattering vector* $\mathbf{q} = \mathbf{k}_s - \mathbf{k}_i$. *The observed intensity reflects the constructive and destructive interference of the scattered electric fields emanating from all of the scatterers in the* scattering volume. *The electric field in this diagram would be orthogonal to the page, but is shown in the plane to highlight the changes in the phase* $\Delta\Phi$ *between two scatterers.*

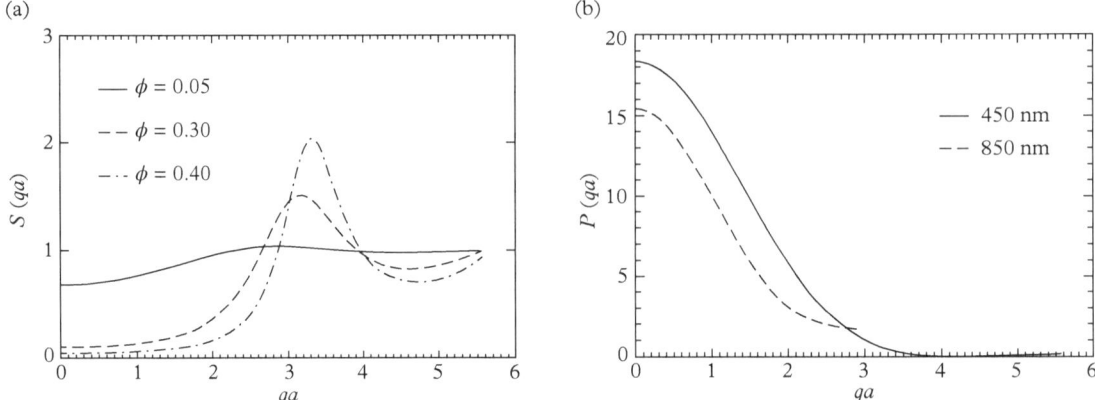

(a)   (b)

Fig. 5.3 *The structure factor $S(q)$ of hard spheres and the scattering form factor $P(q)$ from Mie theory plotted as a function of the dimensionless wavevector aq. Reprinted figure with permission from Kaplan, P. D., Dinsmore, A. D., Yodh, A. G., Pine, D. J. Phys. Rev. E, 50, 4827–35 (1994). Copyright 1994 by the American Physical Society.*

on their relative positions with respect to the scattering wavevector $\mathbf{q}$, and is derived from eqn 5.17. The form factor describes the interference of the scattered electric fields from different volume elements within single particles. Examples of $S(q)$ and $P(q)$ are shown in Fig. 5.3 from Kaplan *et al.* (1994). Here, the structure factor is calculated using the Percus-Yevick closure to the Ornstein-Zernicke integral equation (McQuarrie, 2000). As we see in Fig. 5.3b, larger particles scatter more strongly in the forward direction (smaller scattering vectors $\mathbf{q}$ or scattering angles θ) relative to the incident light.

Returning to the speckle pattern, each speckle represents a single coherence area of the scattered light. Incoherent point scatterers distributed over an area A_s will generate a field that appears to be correlated over an area

$$A_c = \frac{z^2}{k_0^2 A_s} \tag{5.21}$$

where z is the distance from the scatterers to the detector and $k_0 = 2\pi/\lambda$ (Mertz, 2010). The area A_s can also be related to the angle subtended by the detector aperture Ω, leading to $A_c \sim \lambda^2/\Omega$ (Berne and Pecora, 2000). In light scattering, better signal-to-noise is achieved when the detector collects light from only one coherence area. With a classical pinhole-detection scheme, this limit can be achieved using an aperture somewhat smaller then A_c. With a modern detection scheme, a single coherent mode can be selected using an appropriate single-mode fiber optic (Gisler *et al.*, 1995).

The experiment described here is *homodyne* light scattering—all light received at the detector has been scattered from the sample. An alternative *heterodyne* experiment is sometimes used in light scattering, in which the scattered light is collected along with unscattered light. In this case, the detected intensity reflects the interference of the scattered and unscattered fields.

5.3 Dynamic light scattering

If the particles scattering light were fixed and not moving in time, then the speckle pattern in Fig. 5.1b would not change. As particles move, the random scattering pattern evolves when the interfering waves of the scattered light change phase. If we measure the intensity of a single speckle, it should fluctuate randomly about some mean value $\langle I \rangle$, as depicted in Fig. 5.4. The variance of the intensity $\langle I^2 \rangle - \langle I \rangle^2$, a measure of the magnitude of the intensity fluctuation, is also shown as the standard deviation from the mean, $\sqrt{\langle I^2 \rangle - \langle I \rangle^2}$. Detecting and measuring such fluctuations is the basis of the dynamic light scattering experiment, since the time scale on which the intensity of the speckles fluctuate is related to the motion of the scatterers.

Our first step towards modeling the dynamic light scattering experiment is to relate the fluctuating light intensity to the motion of the scatterers. We start with the normalized autocorrelation of the scattered *electric field*

$$g_{(1)}(t) = \frac{\langle E(t)E^*(0)\rangle}{\langle |E(t)|^2 \rangle} \tag{5.22}$$

where E^* denotes the complex conjugate and the brackets indicate the ensemble average, which is equivalent to the time average in ergodic

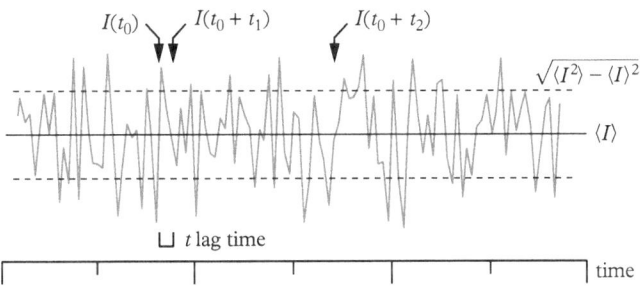

Fig. 5.4 *A representation of the random light-intensity fluctuations of a speckle.*

systems. Note also that eqn 5.22 assumes the process is stationary. Combining eqns 5.17 and 5.22 yields

$$g_{(1)}(t) = \frac{\sum_{i=1}^{N} \sum_{j=1}^{N} \langle E_i E_j^* \exp\{i\mathbf{q} \cdot [\mathbf{r}_i(0) - \mathbf{r}_j(t)]\}\rangle}{\sum_{i=1}^{N} \sum_{j=1}^{N} \langle E_i E_j^* \rangle}. \tag{5.23}$$

In the case of non-interacting scatterers, such as those in dilute suspensions or tracer experiments, the cross-terms $i \neq j$ in eqn 5.23 vanish, leading to

$$g_{(1)}(t) = \langle \exp[i\mathbf{q} \cdot \Delta\mathbf{r}(t)] \rangle \tag{5.24}$$

where $\Delta\mathbf{r}(t) \equiv \mathbf{r}(t) - \mathbf{r}(0)$ which is also referred to as the *self-intermediate scattering function*, $F_s(\mathbf{q}, t)$.

The self-intermediate scattering function is related to the Van Hove correlation function introduced in Section 4.5.1. Recall that the Van Hove function $G_s(\mathbf{R}, t)d^3R$ is the probability that particle i will move in the vicinity R within time t. Taking the spatial Fourier Transform[1] of $G_s(\mathbf{R}, t)$,

$$\mathscr{F}\{G_s(\mathbf{R}, t)\} = \iiint \langle \delta(\mathbf{R} - [\mathbf{r}_i(t) - \mathbf{r}_i(0)])\rangle e^{i\mathbf{q}\cdot\mathbf{R}} d^3R \tag{5.25}$$

yields

$$\langle \exp i\mathbf{q} \cdot [\mathbf{r}_j(t) - \mathbf{r}_j(0)] \rangle \tag{5.26}$$

after commuting the integral with the ensemble average and applying the sifting property of the delta function. This result confirms that $F_s(\mathbf{q}, t)$ is simply the Fourier Transform of the Van Hove correlation function,

$$F_s(\mathbf{q}, t) = \iiint G_s(\mathbf{R}, t)e^{i\mathbf{q}\cdot\mathbf{R}} d^3R. \tag{5.27}$$

Now we can derive the self-intermediate scattering function based on the statistics of our particle trajectories. Dividing time into small intervals Δt we expect the displacement of the particles to vary, in some general way, from one interval to the next. With a sufficiently large sample of displacements, the central-limit theorem may be invoked, implying that the probability of a particle displacement $\mathbf{R}d^3R$ is expected to be Gaussian, and the Van Hove function can be written

$$G_s(\mathbf{R}, t) = \left(\frac{2\pi}{3}\langle \Delta r^2(t)\rangle\right)^{-\frac{3}{2}} \exp\left(\frac{-3R^2}{2\langle \Delta r^2(t)\rangle}\right). \tag{5.28}$$

[1] See Appendix A.1.2 for a short discussion of the spatial Fourier Transform. Note that the sign convention of the spatial transform common in the scattering literature is different than the time-frequency domain Fourier Transform convention.

The Fourier Transform $\mathscr{F}\{G_s(\mathbf{R}, t)\}$ yields

$$g_{(1)}(\mathbf{q}, t) = \exp\left(-q^2 \langle \Delta r^2(t) \rangle / 6\right). \qquad (5.29)$$

Notice that the field correlation function decays appreciably as particles move distances on the order of the length $\lambda/2\pi n$, or in other words, on the order of the reciprocal scattering vector, q^{-1}. If the particles are suspended in a viscous Newtonian fluid, where the mean-squared displacement is given by the Stokes–Einstein relation, $\langle \Delta r^2(t) \rangle = 6Dt = (k_B T / \pi a\eta)t$, then the field correlation function decays as an exponential function of time,

$$g_{(1)}(t) = \exp\left[-t/\tau\right] \qquad (5.30)$$

where $\tau = 6\pi a\eta / q^2 k_B T$.

5.3.1   Light intensity and the Siegert relation

So far, we've derived the *field* autocorrelation function of scattered light (eqn 5.29). Light detectors measure the *intensity*, $I = E^*E$. Fluctuations in the scattered light intensity are characterized by the normalized intensity autocorrelation function,

$$g_{(2)}(t) = \frac{\langle I(t_0)I(t_0 + t) \rangle}{\langle I \rangle^2}. \qquad (5.31)$$

The quantity $\langle I \rangle$ is the average intensity, sometimes referred to as the *baseline* intensity. The random fluctuating intensity of a speckle represented in Fig. 5.4 may have short durations—tens of nanoseconds to microseconds, depending on the size of the probe particles and surrounding material rheology. Conversely, the fluctuations, and resulting correlation function, may stretch out to seconds in a high-viscosity or viscoelastic medium.

The correlation function is the average product of the intensity $I(t_0)$ with the intensity a specified lag time later, $I(t_0 + t)$. Since the process is assumed to be stationary, the reference time t_0 is arbitrary (eqn 5.4). At short lag times the correlation is the mean-squared intensity, $\langle I(t_0)I(t_0 + t) \rangle_{t \to 0} = \langle I^2 \rangle$. At a small lag time, indicated by t_1, the intensity would not have changed much from its measurement at time t_0. These intensities would be strongly correlated. In contrast, measurements of the intensity at two widely-separated times, indicated by the separation between t_0 and t_2 in Fig. 5.4 would be uncorrelated, and the intensity correlation function becomes the product of the average intensity squared, $\langle I(t_0)I(t_0 + t_2) \rangle \to \langle I \rangle^2$.

The normalized intensity and field autocorrelations functions are related by the **Siegert relation**

$$g_{(2)}(t) = 1 + |g_{(1)}(t)|^2, \tag{5.32}$$

which is a fairly general expression that holds for fields obeying quasi-stationary circular Gaussian statistics. Often an empirical pre-factor β is included in the Siegert relation to account for the short-time "intercept" value, or **dynamical contrast**, of the correlation function,

$$g_{(2)}(t) = 1 + \beta |g_{(1)}(t)|^2. \tag{5.33}$$

This contrast factor, which has a value $\beta < 1$, depends on the speckle size relative to the detector area and the presence of noise in the intensity measurement. As we already noted, optical mixing from multiple speckles degrades the dynamical contrast. The use of collimated single-mode fiber optic detectors instead of arrangements of pinholes is now common for maximizing the dynamical contrast of the intensity measurement.

In Fig. 5.5, we plot the calculated intensity correlation function for 1 μm diameter particles in water and a scattering vector $q = 2.3 \times 10^7 \mathrm{m}^{-1}$. The latter corresponds, for instance, to a laser vacuum wavelength 514.5 nm and a scattering angle of 90 degrees. With $\beta = 1$, the normalized intensity correlation function is a monotonically decreasing function from the value $g_{(2)} = 2$ at short times to $g_{(2)} = 1$ at long times. Since the latter represents the square of the average intensity $\langle I \rangle^2$, it is common to subtract this baseline and plot $g_{(2)}(t) - 1$. The intercept value as the delay time $t \to 0$ then becomes

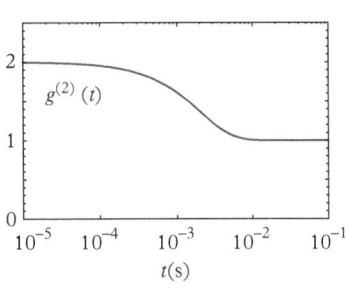

Fig. 5.5 *The calculated intensity correlation function using Eqns 5.29 and 5.33 for 1 μm diameter probes in water and $\beta = 1$. The scattering vector is $q = 2.3 \times 10^7 m^{-1}$.*

$$g_{(2)}(0) - 1 = \frac{\langle I^2 \rangle - \langle I \rangle^2}{\langle I \rangle^2} \tag{5.34}$$

which is the normalized variance of the light's fluctuations. Many commercial light scattering instruments will report the normalized intensity correlation function $g_{(2)}(t) - 1$.

The correlation function plotted in Fig. 5.5 decays on a time scale of the particles' diffusion over the length $q^{-1} \approx 43$ nm. This is much shorter than the time scale for a 1 μm particle to diffuse its own radius,

$$t = \frac{a^2}{D_0} = \frac{6\pi a^3 \eta}{k_B T}, \tag{5.35}$$

which is approximately 0.5 s for a 1 μm diameter particle in water. Here, the wavelength and scattering angle determines the characteristic length scale over which particle motion is measured. As the particles are displaced a few multiples of the inverse scattering vector $\langle \Delta r^2 \rangle^{1/2} \sim q^{-1}$, the correlation function decays completely. At short lag times, the particles must move a sufficient fraction of q^{-1} to produce intensity changes of the scattered speckles. For the current example, $g_{(2)}(t) - 1 = 0.95$ at a lag time $t \approx 0.1$ ms.

5.3.2 Microrheology with DLS

Equations 4.19 and 5.33 provide a means to perform microrheology using dynamic light scattering. If the intensity fluctuations of scattered light from tracer particles are measured, then the mean-squared displacement can be calculated and interpreted by the Generalized Stokes–Einstein Relation. The mean-squared displacement is calculated by inverting equation 5.29 after substituting into the Siegert relation, eqn 5.33,

$$\langle \Delta r^2(t) \rangle = \frac{3}{q^2} \left(\ln[g_{(2)}(0) - 1] - \ln[g_{(2)}(t) - 1] \right) \tag{5.36}$$

assuming that $g_{(2)}(0) - 1 \approx \beta$. Each point in a mean-squared displacement is calculated by subtracting the natural logarithm of the normalized intensity correlation function from its initial "intercept" value. By changing the scattering angle, the length scales (and time scales) of probe motion can be changed.

Equations 5.29 and 5.36 are only strictly valid in the limit that the probe dynamics obey a Gaussian displacement distribution, as one would expect for large probe particles in a material—the Stokes continuum limit discussed in Chapter 2. This relation, as well as the equivalent equations for the other light scattering methods discussed later in this chapter, almost certainly will not hold as the probe particle shrinks to a size commensurate with the material structure.

Dasgupta *et al.* (2002) report measurements of complex fluid microrheology using DLS. Their data is shown in Fig. 5.6a for 0.96 μm diameter carboxylate-modified polystyrene probes in a 4 wt%, 900 kDa poly(ethylene oxide) polymer solution. Two scattering angles, $\theta = 20°$ and $90°$, are used to probe different length scales of probe motion, which extends the data over a larger range of values of the mean-squared displacement (and time). The DLS data are compared to measurements at much smaller displacements (and lag times) using diffusing wave spectroscopy (DWS), which we discuss starting in Section 5.4. The microrheology results are in good agreement with

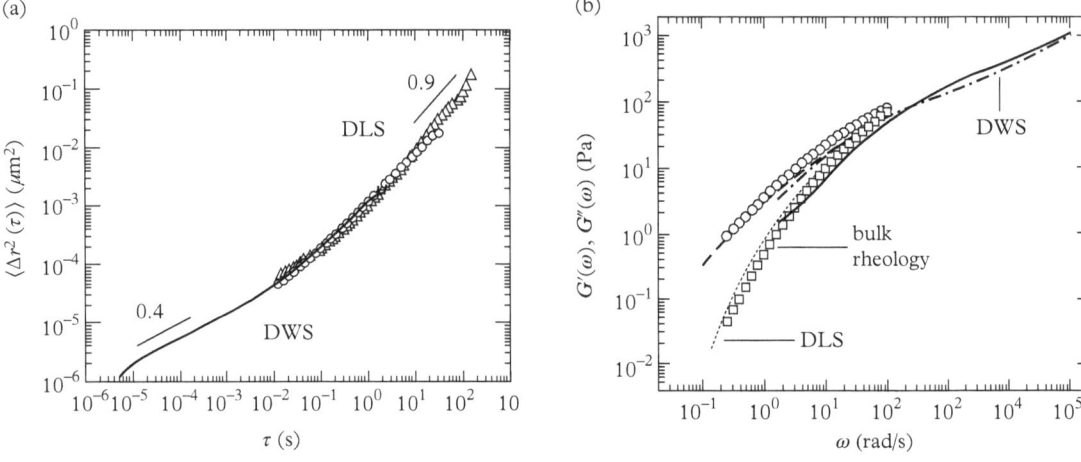

Fig. 5.6 *Light scattering microrheology of 4 wt%, 900 kDa poly(ethylene oxide) using 0.96 μm diameter carboxylate-modified polystyrene probes. (a) A comparison the MSD obtained using DLS (triangles θ = 20° and circles 90°) and diffusing wave spectroscopy (DWS, solid line). (b) The microrheology-derived moduli (lines) for DLS and DWS compared with bulk rheology (symbols). Adapted from Dasgupta* et al. *(2002).*

bulk-rheology measurements of the polymer solutions (Fig. 5.6b) using a strain-controlled rheometer and double-walled concentric cylinder geometry.

Application note: Viscosity of protein solutions

For simple dispersions of macromolecules or proteins, which are expected to exhibit Newtonian behavior, DLS microrheology can be an efficient and high-throughput means to measure the solution viscosity. Newer light scattering instruments are available that take advantage of multi-well plates which hold up to hundreds of samples and small sample volumes ($\sim 35\,\mu l$), making it possible to screen many concentrations or solution conditions.

In one example, He *et al.* (2010) used DLS probe microrheology to measure the viscosity of monoclonal antibody protein solutions with increasing concentration. Here, the authors used a CONTIN analysis (Provencher, 1982*b*) to discern scattering of the particles from that of the protein. The results, shown in Fig. 5.7, show two narrowly distributed particle sizes corresponding to the protein and particles. For sufficiently concentrated protein, the apparent radius of the protein is not the true hydrodynamic radius, since strong interactions (both thermodynamic and hydrodynamic) are present. Likewise, the apparent probe radius is given by assuming the buffer viscosity. With

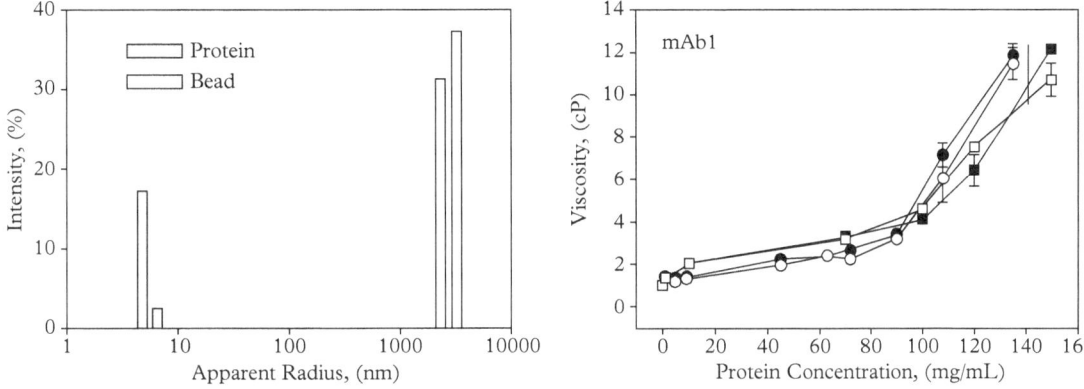

Fig. 5.7 *The viscosities of monoclonal antibody solutions measured using DLS microrheology. Adapted from.
Reprinted Anal.* Biochem., **399**, *He, F. et al., High-throughput dynamic light scattering method for measuring
viscosity of concentrated protein solutions, 141–3, Copyright 2010, with permission from Elsevier.*

a known particle radius a, measured prior by DLS, the true viscosity of the protein solution is just $\eta = \eta_s(a_{\text{apparent}}/a)$. The viscosities measured by DLS are in good agreement with those made using a microcapillary device.

A potential problem to be aware of when measuring protein rheology with tracer particle microrheology is the colloidal stability of the probes in the protein solution. Strong adsorption of the protein onto particles like polymer latex and silica can lead to bridging and destabilization (see Fig. 1.18). Probe aggregation in a DLS sample isn't always immediately obvious, and the effectively higher-hydrodynamic size of clustering particles could be mistaken for an increase in viscosity. The narrow distribution of the apparent probe radii in Fig. 5.7 confirms the stability of the tracers in this example.

Another way to detect clustering of probes is through a change in the angular distribution of the scattering intensity, *e.g.*, by making measurements at two or more scattering angles. At least one of the angles must be chosen sufficiently small to capture q values well below the first minimum $q_{\text{min}} \approx 4.5/a$ of the particle form factor $P(q)$.

5.3.3 Scattering from the material under test

The application we examined points out a signficant complication of DLS microrheology: Scattering is detected from both the *material* and the tracer particles. For relatively dilute proteins and polymers, the total correlation function is a weighted sum over all species i,

$$g_{(2)}(t) - 1 = \frac{\sum_i \rho_i I_i^2 \exp(-q^2 D_i t)}{\sum_i \rho_i I_i^2} \qquad (5.37)$$

each with a characteristic scattering intensity I_i and number density $\rho_i = (\frac{4}{3}\pi R_i^3)^{-1}$ (Berne and Pecora, 2000; Russel *et al.*, 1989). For Rayleigh scatterers of radius R_i with refractive index n_i in a solvent of index n_s,

$$I_i = \frac{4\pi R_i^3 (n_i^2 - n_s^2)}{n_i^2 + 2n_s^2}. \qquad (5.38)$$

Rayleigh scattering is isotropic and therefore independent of the scattering vector q. The scattering intensity from particles with dimensions that approach the wavelength of light is given in terms of the scattering form factor $I_i = P(q)$, discussed further in Section 5.4.6.

Differentiating the scattering by molecular solutes like glycerol or sugar dissolved in water from dispersed tracers is not a great problem—probe scattering will dominate the intensity fluctuations at the measurable correlation times. But concentrated macromolecules, surfactants, and proteins may contribute significantly to the scattered light intensity. In these cases, the self-diffusivity of the probe particles and the collective dynamics of the material of interest may not allow the contribution of the tracers to be separated easily (Brown and Smart, 1997). Unwanted scattering from the material can be masked by selecting a sufficiently high probe particle concentration. However, this may then lead to perturbations of the measured signal, now due multiple light scattering. The latter can be actively suppressed by DLS cross correlation techniques (discussed in Section, 5.3.4) or reduced by shortening the optical beam path in the cell.

5.3.4 Suppressing multiple scattering

Before we discuss dynamic light scattering in the high multiple scattering (DWS) limit, it is worth pointing out that there are several dynamic light scattering methods that have been designed to *suppress* multiple scattering (Phillies, 1981; Schätzel, 1991). These, too, may be used for light scattering microrheology in a manner identical with DLS. Such methods include modulated 3D cross-correlation light scattering (Block and Scheffold, 2010) and two-color dynamic light scattering (Drewel *et al.*, 1990; Segrè *et al.*, 2005). Similar to DLS microrheology, a significant complication is that the medium will likely contribute to the total scattering intensity, requiring careful analysis to separate the signal from the motion of probes from any signal emanating from the dynamics of the material under test (Xue *et al.*, 1992*a*; Joosten *et al.*, 1990).

5.4 Diffusing wave spectroscopy

Multiple scattering can be thought of as successive single scattering events. In dynamic light scattering, the standard practice is to avoid multiple scattering because the scattering angles, and thus the length scales over which motion is probed, cannot be inferred from the angle between the source and detector. Methods to suppress multiple scattering in DLS, such as two-color light scattering (Segrè *et al.*, 2005) and "3D cross-correlation DLS" (Block and Scheffold, 2010), rely on extracting the signal from singly-scattered light and suppressing signal from multiply-scattered light. These are effective methods if a sufficient amount of singly scattered light still penetrates through a sample, as in the case for mildly turbid media.

Diffusing wave spectroscopy (DWS) is unique in that it solves the problem of unknown scattering angles in turbid, multiple scattering samples by treating the ensemble of scattering angles statistically (Maret and Wolf, 1987; Pine *et al.*, 1988; Fraden and Maret, 1990; Weitz *et al.*, 1992; Weitz and Pine, 1993). For the statistical analysis to be accurate, a large number of scattering events are required, and thus the DWS method applies to highly-turbid samples—samples that transmit no singly-scattered light at all. In this limit, photons can be thought of as taking random walks through the sample, hence the name "diffusing wave," as illustrated in Fig. 5.8. Here, we present an overview of the derivation of the DWS working equations. These will help us understand the strengths and limitations of the measurement as it applies to microrheology. We follow with a discussion of the two primary scattering geometries that are used, transmission and backscattering.

5.4.1 Multiple scattering

Light scattering in the high multiple scattering limit assumes that photons are scattered numerous times. High multiple scattering gives materials opacity, like the sample shown in the image of Fig. 5.9. On average, the distance between scattering events is $l = (\rho\sigma)^{-1}$, where

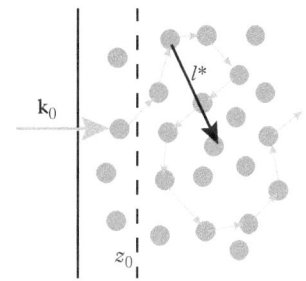

Fig. 5.8 *An illustration of multiple scattering. Many scattering events randomize the photon path over a photon mean-free path length l^*.*

Fig. 5.9 *An image of a typical DWS sample. This 1 cm wide cuvette with a 4 mm path length contains 1 μm diameter polystyrene probe particles dispersed at 1% volume fraction. A high degree of multiple scattering gives rise to the milky white appearance.*

$\rho = \phi/(4\pi a^3/3)$ is the number-density of scatterers and σ is the scattering cross-section. This scattering mean-free path is written in terms of the particle scattering form factor $P(q)$ (Ishimaru, 1990)

$$l = \left(\rho\frac{2\pi}{k_0^4}\int_0^{2k_0} P(q)q\,dq\right)^{-1}.\qquad(5.39)$$

Depending on the scattering characteristics of the probes, several scattering events may be required to randomize the direction of a photon. Probe particles with diameters similar to the wavelength of the scattered light scatter predominantly in the forward direction, for instance. The longer-length scale required to randomize the photon direction is the **photon mean-free path**, l^*. It is the random walk step of a photon through the turbid medium. In the following sections we discuss the light transport model in greater detail, and l^* in particular in Section 5.4.6. Here, our first task is to calculate the field autocorrelation function for multiply scattered light.

In the multiple scattering regime, the magnitude of the scattered field E_p after N_p scattering events is

$$E_p(t) = E_p \exp\left[i\Phi(t)\right]\qquad(5.40)$$

where $\Phi(t)$ is the phase change of the light after each scattering event defined as

$$\Phi(t) = \sum_{i=0}^{N_p}\mathbf{k}_i\cdot[\mathbf{r}_{i+1}(t)-\mathbf{r}_i(t)]\qquad(5.41)$$

In eqn 5.41, $\mathbf{k}_i$ is the wavevector for light scattered between particles i and $i+1$, $\mathbf{r}_0$ is the position of the source, $\mathbf{r}_{N+1}$ is the position of the detector, and $\mathbf{k}_0$ and $\mathbf{k}_{N+1}$ are the incident and detected wavevectors, respectively. Summing eqn 5.41 over all paths and combining with eqn 5.17 yields

$$g_{(1)}(t) = \sum_p\langle I_p\rangle\langle\exp\left[-i\Delta\Phi_p(t)\right]\rangle\qquad(5.42)$$

where $\Delta\Phi_p(t) = \Phi_p(t)-\Phi_p(0)$. Equation 5.42 includes only uncorrelated photon paths (Scheffold and Maret, 1998). Noting that $\Delta\Phi_p(t)$ is a cumulative function of the motion of many scatterers, $N_p \gg 1$, this phase factor can be represented as a Gaussian random variable by the Central Limit Theorem, allowing us to write

$$\langle\exp\left[-i\Delta\Phi_p(t)\right]\rangle = \exp\left[-\langle\Delta\Phi_p^2(t)\rangle/2\right].\qquad(5.43)$$

To the leading order, $\Delta\Phi_p(t)$ is written as

$$\Delta\Phi_p(t) = \sum_{i=1}^{N} \mathbf{q}_i \cdot \Delta\mathbf{r}_i(t), \qquad (5.44)$$

where $\mathbf{q}_i$ is the scattering wavevector $\mathbf{q}_i = \mathbf{k}_i(0) - \mathbf{k}_{i-1}(0)$ with magnitude $q = 2k_0 \sin(\theta/2)$. The magnitude of the wavevector appears in a number of our calculations, so we will note it explicitly here:

$$k_0 = \frac{2\pi n}{\lambda_v} \qquad (5.45)$$

where again λ_v is the vacuum wavelength of the laser light and n is the refractive index of the scattering medium.

Assuming the independence of successive phase factors and the independence of the scattering vector $\mathbf{q}_i$ from the displacement vector $\Delta\mathbf{r}_i$, Weitz and Pine (1993) show that $\langle\Delta\Phi_p^2(t)\rangle$ can be written as

$$\langle\Delta\Phi_p^2(t)\rangle = \frac{1}{3}N_p\langle q^2\rangle\langle\Delta r^2(t)\rangle \qquad (5.46)$$

noting that $\langle q^2\rangle$ is weighted over by the single-particle form factor, and can be expressed as $\langle q^2\rangle = 2k_0^2 l/l^*$. For long photon paths where $N_p \gg 1$, the total number of scattering events through the sample is given by $N_p = c\tau/l$, where c is the speed of light in the medium and τ is the time a photon takes to traverse the path p. The final expression for the phase factor is

$$\langle\Delta\Phi_p^2(t)\rangle = \frac{c}{3l^*}k_0^2\langle\Delta r^2(t)\rangle\tau. \qquad (5.47)$$

Using eqn 5.47, the final expression for the field autocorrelation in diffusing wave spectroscopy is

$$g_{(1)}(\mathbf{r}, t) = \int_0^\infty \mathcal{P}(\mathbf{r}, \tau) \exp[-(c/3l^*)k_0^2\langle\Delta r^2(t)\rangle\tau]d\tau, \qquad (5.48)$$

where the sum in eqn 5.42 has been rewritten as an integral weighted by the probability $\mathcal{P}(\mathbf{r}, \tau)$ that a diffusing photon will arrive at position $\mathbf{r}$ at time τ. From eqn 5.48, we see that the autocorrelation function decays exponentially, similar to a DLS experiment, with the difference that multiple scattering events occurring sequentially are accounted for by the path length. The total correlation function is a sum, represented by the integral, of all of the photon paths through the sample.

Written in terms of the distribution of photon path lengths $\mathcal{P}(s)$ with the change of variables $s = c/\tau$, eqn 5.48 becomes[2]

$$g_{(1)}(t) = \int_0^\infty \mathcal{P}(s) \exp[-(s/3l^*)k_0^2\langle\Delta r^2(t)\rangle]ds. \qquad (5.49)$$

A path is composed of s/l^* random walk steps, each of which contributes, on average, $\sim \exp(-k_0^2\langle\Delta r^2(t)\rangle/3)$ to the decay of the autocorrelation function (Weitz and Pine, 1993). Thus, shorter path lengths require the constituent particles along the path to move further in order to induce the same phase change and the same decay of the correlation function. Longer path lengths require smaller motion of the particles. Since different scattering geometries lead to different path-length distributions, the scattering geometry can be used to tailor the range of displacements, and hence time scales, probed by DWS microrheology.

As we will see, eqn 5.49 is a useful form of the field-correlation function. We will use it in Section 5.4.2 to derive the correlation function for transmission and backscattering DWS experiments.

Scattering geometry affects the length scales of probe motion detected by DWS.

5.4.2   Diffusive-light transport

In order to use eqn 5.48, we must know the distribution of path-lengths photons take through a sample. This can be thought of as a "time of flight" experiment: When a pulse of light strikes the sample, the measured intensity at the detector should "stretch" depending on the random journey each photon takes through the material. The normal approach in a DWS experiment is to model the photon transport in order to describe the path-length distribution. Treating multiple-light scattering by only the scattered intensities is an approximation, but generally a good one, as long as the scattering-medium dimensions exceed the photon mean-free path l^* (Ishimaru, 1990). The phase correlations between the scattered waves are ignored. Higher order correlations caused by crossing scattering paths only become significant if the light is strongly confined, for instance, by introducing a small cylindrical pinhole in the sample (Scheffold and Maret, 1998).

Using the classical transport theory, the diffusion equation governs the path-length distribution

[2] Take care not to confuse the particle-scattering form factor $P(q)$ with the photon path-length distribution $\mathcal{P}(s)$.

$$\frac{\partial U(\mathbf{r}, \tau)}{\partial \tau} = D_l\nabla^2 U(\mathbf{r}, \tau) \qquad (5.50)$$

where $D_l = c/3l^*$ is the diffusivity of light.[3] The diffusive probability is found from

$$P(\mathbf{r}, \tau) = \frac{U(\mathbf{r}, \tau)}{\int_0^\infty U(\mathbf{r}, \tau) d\tau} \qquad (5.51)$$

which is the fraction of photons received at the detector, located at $\mathbf{r}$ at time τ. Equation 5.50 constitutes a boundary-value problem typical of diffusive transport problems, such as those that occur in heat or mass transfer. Analogous solutions to heat transfer problems are particularly useful for a variety of DWS geometries (Carslaw and Jaeger, 1986).

It is possible to specify several boundary conditions (Ishimaru, 1990). Using the "zero net-flux" boundary condition at the sample walls (Pine *et al.*, 1990)

$$U + \frac{2}{3}l^*\mathbf{n} \cdot \nabla U = 0, \qquad (5.52)$$

where $\mathbf{n}$ is the outward normal vector, ensures that there is no flux of *diffusing* photons entering the sample. The initial condition is

$$U(z, t = 0) = U_0 \delta(z - z_0, t). \qquad (5.53)$$

The incident light enters the sample "ballistically" to a position $z_0 \sim l^*$ within the sample, then proceeds to diffuse (Pine *et al.*, 1990). This boundary condition can also be interpreted as a source of diffusing photons at $z = z_0$. For micrometer diameter probes, which are comparable in size to the light wavelength, one typically finds $z_0/l^* = 1.13$. However, for smaller particles, z_0 is sensitive to polarization (MacKintosh *et al.*, 1989; Rojas-Ochoa *et al.*, 2004).

Representative photon paths are shown in Fig. 5.10 for backscattering and transmission from a point source. The path distribution for backscattering is dominated by many short paths around the point source. There are a few paths out of several hundred that reach longer lengths. Transmission requires significantly longer path lengths. Histograms of the path lengths are shown in Fig. 5.11 and again highlight the short paths that dominate $P(s)$ for backscattering.

5.4.3   Transmission geometry

In the transmission geometry, the sample is illuminated by a plane wave source and transmitted light is detected on the opposite side. The photon diffusion is one-dimensional, mimicking the steady-state transmission of heat due to an instantaneous planar source (Weitz and Pine, 1993).

---

[3] Compare with eqn 4.33.

$z = L$

$z = 0$

Fig. 5.10 *Representative diffusive-light paths for (left) backscattered light and (right) transmitted light from a point source. Most trajectories in backscattering are short. Paths for transmitted light are longer. Path length affects the range of length scales of probe motion measured, and hence the range of time scales of the microrheology experiment. The illustration here is for a thick sample, $L = 150l^*$, to highlight the photon random walks.*

Fig. 5.11 *Simulated path-length distributions $\mathcal{P}(s)$ for backscattering (solid line) and transmission (dotted line) in a sample with thickness $L = 15l^*$.*

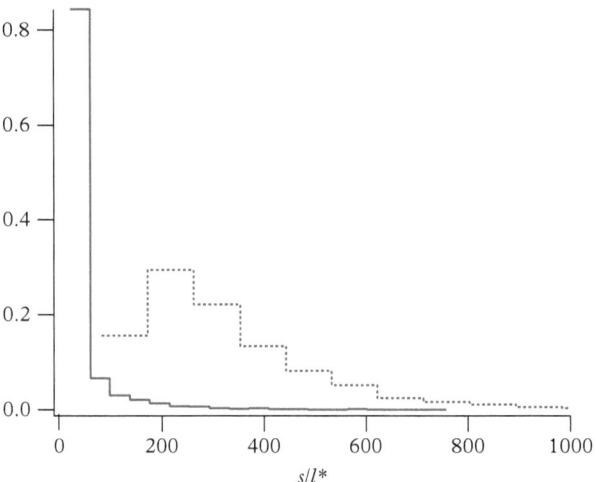

Using the diffusion eqn 5.50 with the boundary conditions

$$U - \frac{2}{3}l^* \frac{dU}{dz} = 0 \qquad z = 0 \tag{5.54}$$

$$U + \frac{2}{3}l^* \frac{dU}{dz} = 0 \qquad z = L \tag{5.55}$$

and eqn 5.48, the field autocorrelation function for the transmission geometry with a plane wave illumination is

$$g_{(1)}(t) = \frac{\frac{L/l^*+4/3}{z_0/l^*+2/3}\left(\sinh\left[\frac{z_0}{l^*}R(t)\right] + \frac{2}{3}R(t)\cosh\left[\frac{z_0}{l^*}R(t)\right]\right)}{\left(1 + \frac{4}{9}R^2(t)\right)\sinh\left[\frac{L}{l^*}R(t)\right] + \frac{4}{3}R(t)\cosh\left[\frac{L}{l^*}R(t)\right]} \tag{5.56}$$

with

$$R(t) \equiv \sqrt{k_0^2\langle\Delta r^2(t)\rangle}, \tag{5.57}$$

which can be interpreted as a root mean-squared displacement scaled by the light wavevector, $k_0 = 2\pi/\lambda$.

In the transmission geometry, the path-length distribution is determined by the sample thickness L, such that the number of *random walk* steps is $n_c \sim (L/l^*)^2$ and the path lengths are distributed around $s \sim n_c l^* \sim L^2/l^*$. For example, in Fig. 5.11, the path-length distribution peaks around $s/l^* \sim 200$ for the sample thickness $L = 15l^*$. By changing the sample thickness, longer or shorter path lengths can be selected to alter the range over which the probe particle displacement is measured. We will come back to this point when we discuss the operating regime of DWS microrheology.

A second convenient transmission geometry is a point source on axis with the detector. The correlation function in this case is (Pine et al., 1990)

$$g_{(1)}(t) = \frac{\int_{(L/l^*)R(t)}^{\infty}\left[A(y)\sinh y + e^{-y(1-z_0/L)}\right]dy}{\int_0^{\infty}\left[A(y)\sinh y + e^{-y(1-z_0/L)}\right]dy} \tag{5.58}$$

where $\epsilon = 2l^*/3L$,

$$(L/l^*)R(t) = (L/l^*)\sqrt{k_0^2\langle\Delta r^2(t)\rangle}, \tag{5.59}$$

and

$$A(y) = \frac{(\epsilon y - 1)\left[\epsilon y e^{-yz_0/L} + (\sinh y + \epsilon y \cosh y)e^{-y(1-z_0/L)}\right]}{(\sinh y + \epsilon y \cosh y)^2 - (\epsilon y)^2}. \tag{5.60}$$

The calculated field correlation functions for plane wave and point source transmission are shown in Fig. 5.12 for 1 μm diameter particles in water. There is little difference between the two, but $g_{(1)}(t)$ for the plane-wave transmission is expected to decay a bit faster. The plane wave illumination leads to a slightly stronger weighting of longer diffusive photon path lengths.

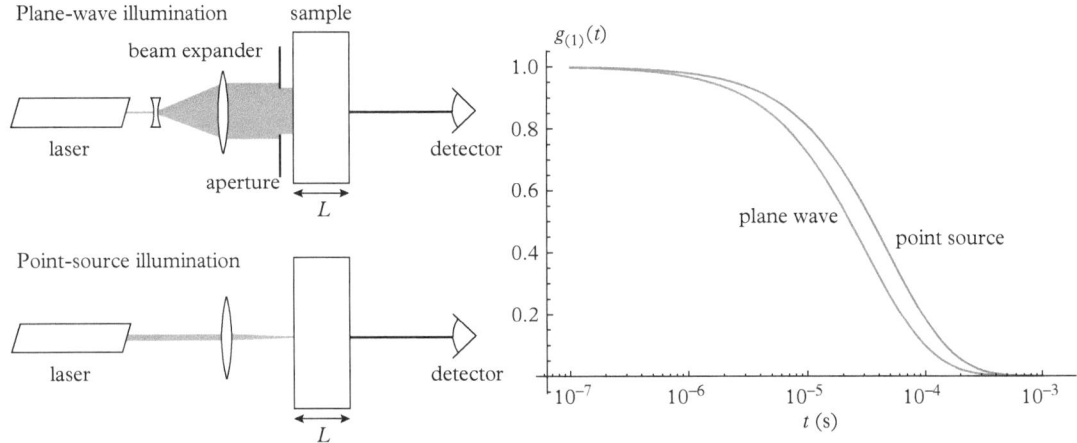

Fig. 5.12 *DWS transmission geometry with two possible illumination schemes: An incident plane wave (approximated by an expanded Gaussian beam) and a point source. The field correlation functions for both geometries are plotted for 1 μm diameter particles diffusing in water with $l^* = 260\,\mu m$ and $L = 5\,mm$.*

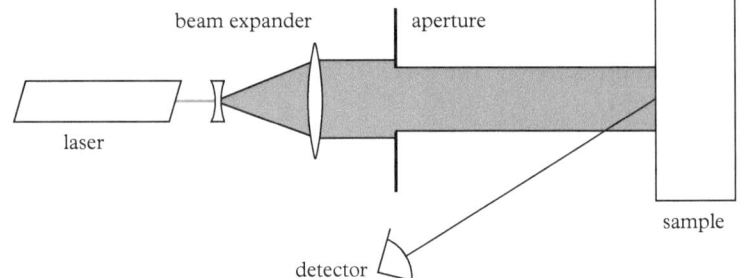

Fig. 5.13 *DWS backscattering geometry with a plane wave illumination source.*

5.4.4 Backscattering geometry

Plane-wave backscattering DWS uses the same sample geometry as plane-wave transmission, as in Fig. 5.13, but the scattered light is collected near the center of the illuminated area. Backscattering measures longer-length scales of probe motion due to the presence of shorter paths in the path-length distribution, $\mathcal{P}(s)$. Some photons enter then exit the sample quickly relative to transmission. The path-length distribution scales as $\mathcal{P}(s) \sim s^{-3/2}$ (Cardinaux *et al.*, 2002).

The field autocorrelation function for plane-wave backscattering is

$$g_{(1)}(t) \propto \frac{\sinh\left[R(t)\left(\frac{L}{l^*} - \frac{z_0}{l^*}\right)\right] + \frac{2}{3}R(t)\cosh\left[R(t)\left(\frac{L}{l^*} - \frac{z_0}{l^*}\right)\right]}{\left(1 + \frac{4}{9}R^2(t)\right)\sinh\left[\frac{L}{l^*}R(t)\right] + \frac{4}{3}R(t)\cosh\left[\frac{L}{l^*}R(t)\right]}$$

(5.61)

again, with $R(t) \equiv [k_0^2 \langle \Delta r^2(t) \rangle]^{1/2}$. The field correlation function described asymptotes at short delay times to the value

$$g_{(1)}(t \rightarrow 0) = \frac{L + \frac{2}{3}l^* - z_0}{L + \frac{4}{3}l^*}, \qquad (5.62)$$

which becomes $g_{(1)}(t \rightarrow 0) \approx 1$ as $L \gg l^*$ (and z_0). The reciprocal of eqn 5.62 can be used as the proportionality constant to properly normalize eqn 5.61. For a sample that is sufficiently thick to be considered "semi-infinite" such that $L/l^* \gg 1$, the cumbersome (but accurate) eqn 5.61 simplifies to

$$g_{(1)}(t) = \frac{\exp\left[-\frac{z_0}{l^*}\sqrt{k_0^2 \langle \Delta r^2(t) \rangle}\right]}{1 + \frac{2}{3}\sqrt{k_0^2 \langle \Delta r^2(t) \rangle}}. \qquad (5.63)$$

Because $z_0 \sim l^*$, there is only a weak dependence on the scattering mean-free path length under these conditions.

The path-length distribution in backscattering is broader than transmission, and many shorter paths contribute to the correlation function. This broad distribution has the effect of increasing the probe displacement necessary to achieve a similar decay of the autocorrelation function when compared to transmission experiments. In Newtonian fluids, for which $\langle \Delta r^2(t) \rangle = 6Dt$ with $D = k_B T/6\pi a\eta$, the breadth of the path-length distribution is apparent by the stretched exponential form of the resulting field-correlation equation,

$$g_{(1)}(t) = \frac{\exp\left[-\frac{z_0}{l^*}\sqrt{6t/\tau}\right]}{1 + \frac{2}{3}\sqrt{6t/\tau}}, \qquad (5.64)$$

where $\tau = 1/k_0^2 D$.

5.4.5   Comparison of transmission and backscattering

The effect of scattering geometry can be seen by comparing the intensity-correlation functions in Fig. 5.14 for 1.02 μm diameter polystyrene particles in water. The sample dimension is $L = 4$ mm in the transmission direction. The probes are dispersed at a volume fraction $\phi = 0.01$ and temperature 24°C and their dynamics measured

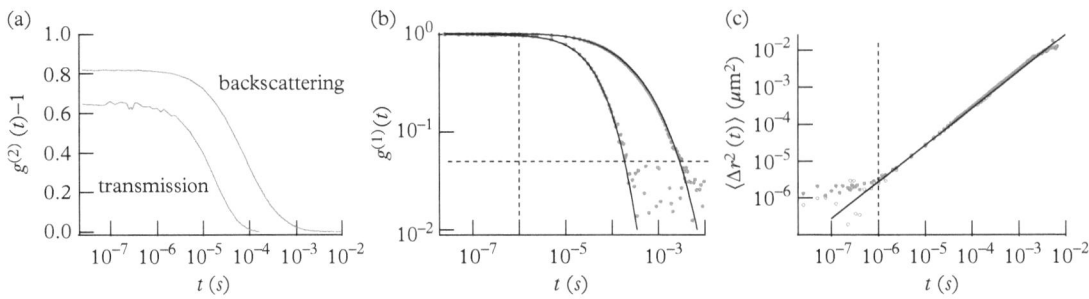

Fig. 5.14 *(a) DWS normalized intensity autocorrelation functions for plane-wave transmission and backscattering geometries. The correlation intercept values are 0.65 and 0.82, respectively and the sample thickness is 4 mm. (b) The corresponding field autocorrelation functions calculated with the Siegert relation. The horizontal dashed line is $g_{(1)} = 0.05$ and the vertical dashed line marks $t = 1$ μs. (c) The mean-squared displacement calculated from the measured correlation function. The solid line through the data has slope 1.*

using both the plane-wave illumination transmission and backscattering geometries. Note that the intercept value of $g_{(2)}(t) - 1$ is different for the two geometries here, which is largely due to differences in the coherence factor β.

For the transmission geometry, $g_{(2)}(t)$ decays within approximately 20 μs, while the correlation function for the backscattering geometry goes to almost 4 ms. The time scales of the measurements are clearer when the field correlation function is plotted, 5.14b, which is calculated using $\beta = 0.65$ and 0.82 for the transmission and backscattering experiments, respectively. The horizontal dashed line indicates a value of $g_{(2)} = 0.05$, where the signal-to-noise diminishes. The total span of time scales probed is different, too, reflecting the range of probe motion captured by DWS. The transmission correlation function decays over about two decades in time—the correlation reaches the noise floor at about 0.2 ms. The backscattering geometry covers a wider range of delay times—over three decades—and becomes noisy at about 3 ms. This extension reflects the broad path-length distribution in backscattering compared to the more narrow distribution in transmission.

In Fig. 5.14b the DWS field correlation functions for plane-wave transmission (eqn 5.56) and the "full" backscattering equation (eqn 5.61) show excellent agreement with the measured correlation functions. Both represent measurements of probes moving in water, so $\langle \Delta r^2(t) \rangle = 6Dt = (k_B T / \pi a \eta)t$. This allows us to fit for the scattering length $l^* = 303 \mu m$, in reasonable agreement with the expected value, which is discussed in the next section. The backscattering also requires the ballistic-length parameter z_0.

Plotting the corresponding mean-squared displacement provides another sense of the time and length scales probed by DWS, especially when $\langle \Delta r^2(t) \rangle$ is compared to the correlation functions. First, notice the small length scales of probe motion, which ranges between $10^{-6} \mu m^2$ and to $5 \times 10^{-4} \mu m^2$ for transmission and $10^{-2} \mu m^2$ for backscattering. These extraordinarily small displacements are perhaps better appreciated by their root mean-squared values—1 nm at the earliest times and 20 and 100 nm at later times for transmission and backscattering, respectively. The largest measured displacement is only a tenth of the particle diameter. As a consequence, the time scales of probe motion are short compared to other microrheology techniques. In terms of frequency, these DWS experiments can be used to measure rheology over the approximate range of $10^3 - 10^6$ Hz.

5.4.6 Photon mean-free path

The photon mean-free path l^* is the length over which photon transport is randomized in a multiple scattering medium. Knowing the value of l^* is important because it affects the path-length distribution and the length- and time scales probed by DWS. It's also necessary to verify that the diffuse light transport model is valid, such that $L \geq 2l^*$ when $k_0 a > 3$, for instance (Kaplan *et al.*, 1994). According to Mie scattering calculations and experiments, 1 μm diameter polystyrene spheres in water at a volume fraction $\phi = 0.01$, have a photon transport mean-free path $l^* \approx 260$ μm (Lu and Solomon, 2002).

The photon mean-free path l^* is proportional to the weighted average over the particle scattering form factor $P(q)$ and structure factor $S(q)$ (Pine *et al.*, 1988; Pine *et al.*, 1990; Fraden and Maret, 1990; Kaplan *et al.*, 1994; Rojas-Ochoa *et al.*, 2002)

$$l^* = k_0^{-6} \left(\pi \rho \int_0^{2k_0} P(q)S(q)q^3 dq \right)^{-1}. \tag{5.65}$$

where ρ is the number density of scatterers. Thus, changes in the spatial arrangement and interactions of scatterers alter the transmission of diffuse light through the sample.

Given the dilute concentration of tracer particles in microrheology samples and the desire to minimize direct interactions between probes, the structure factor becomes $S(q) \approx 1$, and eqn 5.65 will be governed almost solely by the particle form factor $P(q)$. As we saw earlier in Section 5.2, larger particles, for which $k_0 a \geq 1$, tend to scatter light in the forward direction, while smaller particles scatter light

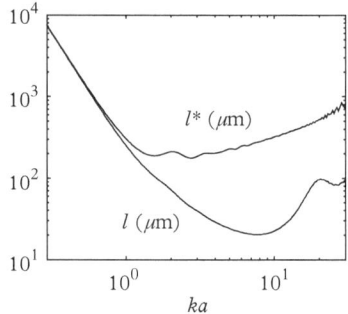

Fig. 5.15 *The photon mean-free path-length l* as a function of particle-size a scaled by the scattering wavevector k = 2π/λ for a volume fraction φ = 0.01.*

Table 5.1 *Photon mean free path-length values calculated based on Mie theory for particles with diameter D in water (Lu and Solomon, 2002).*

D (μm)	ϕ	l^* (μm)
0.6	1.0%	220
	2.0%	110
1.1	1.0%	290
	2.0%	140
2.2	1.0%	450
	2.0%	220

more isotropically. Over the range of particle sizes of interest to microrheology, l^* becomes larger as the particle size increases. That is, more scattering events are necessary to randomize the direction of the photon path. When $k_0 a \ll 1$, scattering is isotropic but weak, and l^* (as well as l) becomes quite large.

In Fig. 5.15, l^* is plotted as a function of scattering particle radius a scaled by the scattering wavevector k for a constant volume fraction $\phi = 0.01$, a concentration used in many DWS microrheology studies. For small particles, which scatter more or less isotropically, the photon mean-free path tracks with the average distance between scattering events, l. For bigger particles, l^* becomes significantly larger than the average distance between scattering events. A summary of l^* values from calculated form Mie theory is given in Table 5.1 from Lu and Solomon (2002).

Characterizing the photon mean free path l*

In DWS microrheology, the photon mean free path value is determined independently by measuring the diffuse light intensity that is transmitted through the sample. The transmitted intensity through a non-adsorbing, multiple scattering slab of thickness L is proportional to (Sheng, 1990)

$$T \sim \frac{l^*/L}{1 + 4l^*/3L}. \tag{5.66}$$

The mean free path is determined by measuring the transmitted light passing through the same material in sample holders, typically cuvettes, with different path lengths.

An alternate and often-used approach is to measure the transmittance of a sample with known l^* and use this to determine the l^* value of for an unknown sample. A common reference sample is an aqueous solution of micrometer diameter polystyrene particles at a volume fraction $\phi = 0.01$. The appropriate DWS correlation function, eqn 5.56, 5.58, or 5.61 is used in a non-linear least-squares fit to find the value of l^*, since all other parameters should be known, including the particle size, fluid viscosity, and temperature. Using the transmittance T_{ref} and photon mean-free path l^*_{ref} of the reference sample, the unknown value of a sample may be calculated (Kaplan *et al.*, 1994; Dasgupta *et al.*, 2002)

$$l^* = \frac{T}{T_{\text{ref}} + \frac{4l^*_{\text{ref}}}{3L}(T_{\text{ref}} - T)} l^*_{\text{ref}} \tag{5.67}$$

assuming identical thicknesses L of the sample and reference.

5.4.7 Light absorption

Light absorption is potentially a significant problem in DWS microrheology. Aside from attenuating the light scattering signal and possibly introducing heating and thermal convection in the sample, absorption alters the photon path-length distribution. Fortuitously, many polymeric, biomacromolecular, and surfactant solutions exhibit negligible absorption in the visible region. If some absorption is present, it can be accounted for in the analysis using the methods described in this section.

Recall that the Beer-Lambert law states that the attenuation due to absorption is

$$T = \exp(-L\alpha) \tag{5.68}$$

where $T = I/I_0$ is the transmittance through a sample, a ratio of the incident intensity I_0 and measured intensity I, α is the extinction coefficient, and L is the path length through the sample.

The effect of absorption on the autocorrelation function is to change the path-length distribution $\mathcal{P}(s)$ by biasing it to shorter-photon paths. By eqn 5.68, longer photon paths are attenuated by $\mathcal{P}(s)\exp(-s/l_a)$, where $l_a = \alpha^{-1}$ is the characteristic length a photon travels before it is absorbed. As a result of the new path-length distribution, the probe motion measured in DWS is shifted towards longer length scales. The transmission and backscattering field autocorrelation equations remain the same (eqns 5.56, 5.58, and 5.61, respectively) with the term $R(t)$ now including the absorption length,

$$R(t) = (k_0^2 \langle \Delta r^2(t) \rangle + 3l^*/l_a)^{1/2}. \tag{5.69}$$

The effect is similar to the effect of laser coherence, which we discuss in Section 5.5.2, with the exception that the intercept of the correlation function does not change. Incorporating eqn 5.69 in the respective field correlation function, one should rescale the result such that $g_{(1)}(t) \to 1$ as $t \to 0$. Sarmiento-Gomez *et al.* (2014) provide the full expression for the normalized correlation function in a plane-wave transmission geometry,

$$g_{(1)}(t) = \frac{(1 + \frac{4}{9}\eta^2)\sinh\left(\frac{L}{l^*}\eta\right) + \frac{4}{3}\eta\cosh\left(\frac{L}{l^*}\eta\right)}{\sinh(\frac{z_0}{l^*}\eta) + \frac{2}{3}\eta\cosh(\frac{z_0}{l^*}\eta)}$$

$$\times \frac{\sinh\left[\frac{z_0}{l^*}R(t)\right] + \frac{2}{3}R(t)\cosh\left[\frac{z_0}{l^*}R(t)\right]}{\left(1 + \frac{4}{9}R^2(t)\right)\sinh\left[\frac{L}{l^*}R(t)\right] + \frac{4}{3}R(t)\cosh\left[\frac{L}{l^*}R(t)\right]} \tag{5.70}$$

where $\eta = \sqrt{3l^*/l_a}$ and $R(t)$ is again given by eqn 5.69.

Using the approach described, it is possible to perform accurate microrheology experiments in *moderately* absorbing materials, provided $l_a/l^* \gg 1$. Due to the change in the detected length scales of probe motion, the operating regimes of the experiment will shift.

Measuring sample absorption

For materials that absorb light, l_a should be measured using the transmittance through samples *without* probe particles. Assuming only weak scattering, the path length is determined by the sample thickness L by the Beer–Lambert equation we have looked at. If multiple scattering is present, then it results in a path-length distribution and eqn 5.68 does not apply; the transmittance through a sample of thickness L when $L > l_a > l^*$ is instead given by

$$T(L) = \frac{\gamma l^*/L_a}{\sinh(L/L_a)} \tag{5.71}$$

where $L_a = \sqrt{l^* l_a/3}$ and $\gamma \approx 5/3$ (Genack, 1990). This equation is useful if l^* is measured or calculated independently of the transmission. Conversely, absorption must be accounted for if the transmission through a sample is being used to characterize the mean-free scattering length l^*. For strong absorption such that $L \gg l^*$ and $L/L_a \gg 1$,

$$T(L) = (2\gamma l^*/L_a) \exp(-L/L_a). \tag{5.72}$$

5.4.8 Mean-squared displacement

Once the field correlation function has been measured and the photon transport mean-free path characterized, the mean-squared displacement of a sample can be calculated using eqns 5.56, 5.58, and 5.61. Calculating the mean-squared displacement requires inverting these expressions using an iterative root-seeking method, such as a Newton–Raphson method.

In Fig. 5.16 we show MSD curves measured for aqueous poly(ethylene oxide) (PEO, $M_w = 333\,000$ g/mol) solutions that nicely captures the length and time scales typical of DWS microrheology. The probe particles are 0.966 μm diameter polystyrene particles at a volume fraction $\phi = 0.01$. The root mean-squared probe displacement for a point source in a transmission geometry ranges between 3 nm and 100 nm. Two sample lengths, $L = 2$ mm and 10 mm, are used to obtain an extended range of probe motion in this geometry. The combination of scattering geometries to extend

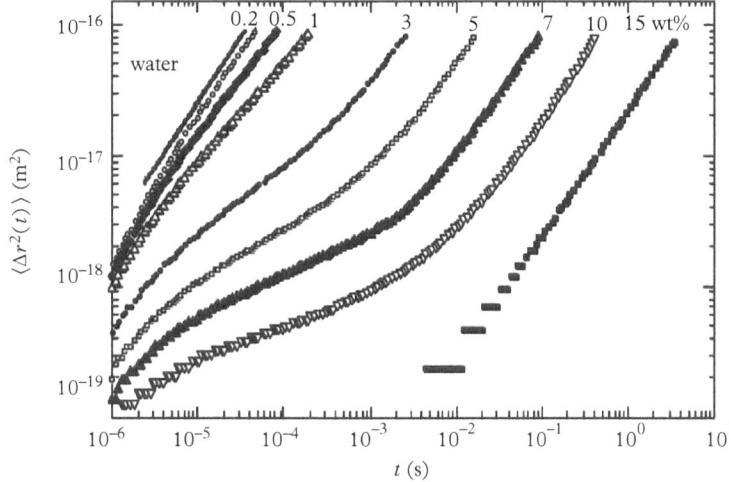

Fig. **5.16** *Examples of DWS microrheology for polystyrene probe particles (0.966 μm diameter) dispersed in 330 kDa molecular weight poly(ethylene oxide) solutions between 0.2 and 15 wt%. Reprinted with permission from van Zanten, J. H., Amin, S., & Abdala, A. A.,* Macromolelcules, *37, 3874–80 (2004). Copyright 2004 American Chemical Society.*

the operating range of DWS microrheology is discussed more in Section 5.4.9.

Given the short *distances* of probe motion measured by DWS, one common misconception is to mistake this length scale for the characteristic length scale of the sample in which the rheology is being measured. Certainly, light scattering is a one-point microrheology method, and as we saw in Section 4.11.4, it is therefore sensitive to local variations in structure (and rheology) near the probe-material interface. Nonetheless, by Stokes' equation, the region of material deformed by the probe's motion is still on the order of the probe size. The Stokes solution applies when the shear wave is $\delta \gg a$.

Comparison to bulk rheology

We've considered several examples of light scattering microrheology that agree with bulk rheology, at least in the overlapping regimes of frequency and modulus. These include measurements of poly(ethylene oxide) polymer solutions—Dasgupta *et al.* (2002) (Fig. 5.6) and van Zanten *et al.* (2004) (Fig. 5.18, to be discussed shortly)—and surfactant solutions.

With respect to the latter material, Willenbacher *et al.* (2007) performed extensive comparisons of microrheology measurements for cetylpyridinium chloride (CpCl) and sodium salicylate (NaSal) over an extended range of frequencies to shear rheology, squeeze flows, and measurements using torsional resonators. The data are plotted in Fig. 5.17. Similar to Fig. 5.18, the experiments employ two DWS geometries to cover an extended operating range. Other examples where

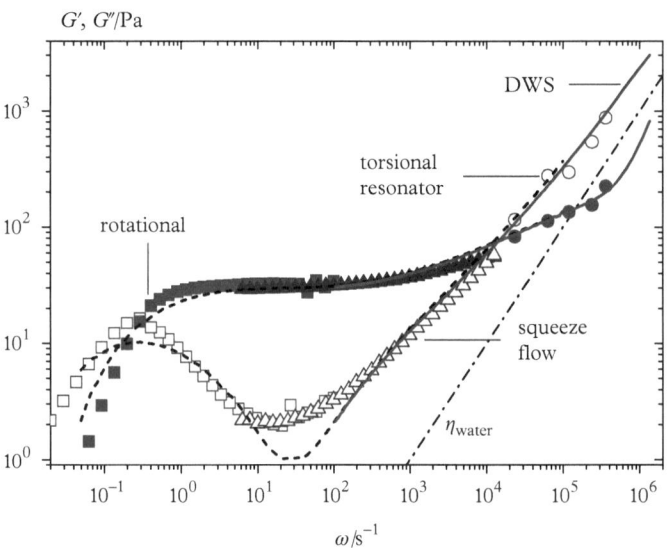

Fig. 5.17 *DWS microrheology of CpCl-NaSal surfactant solutions (solid and dashed lines) compared to rotational bulk rheology (squares), squeeze flow (triangles) and torsional resonator (circles). Reprinted with permission from Willenbacher, N. et al., Phys. Rev. Lett.* **99**, *68302 (2007). Copyright 2007 by the American Physical Society.*

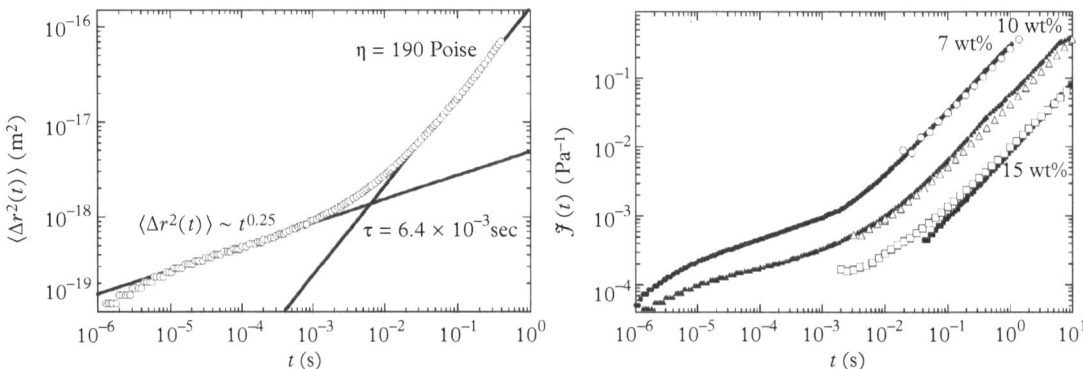

Fig. 5.18 *Characterization of the relaxation time of an entangled solution of PEO using DWS microrheology is shown on the left. On the right is a comparison between creep experiments on a rheometer (open symbols) and DWS-based tracer microrheology. Reprinted with permission from van Zanten, J. H., Amin, S., & Abdala, A. A., Macromolelcules* **37**, *3874–80 (2004). Copyright 2004 American Chemical Society.*

good agreement is observed between micro and macro methods include those of Oelschlaeger *et al.* (2010).

In the practice of DWS microrheology, it sometimes occurs that comparisons between bulk and microrheology (in the frequency domain) give quantitative differences in magnitude of the measured moduli, but that the frequency dependence generally tracks well for both techniques. Thus, one measurement is simply shifted in magnitude versus the other, usually by a factor of two or less. Such discrepancies can arise from a few sources, but weak probe aggregation (or poor dispersion) and uncertainty in the value of l^* are likely causes, at least when the microrheology reports *higher* values than bulk measurements. In one case, Cardinaux *et al.* (2002) report DWS microrheology of worm-like micellar solutions that is 3/2 *lower* than corresponding bulk measurements.

Application note: Relaxation time of polymer solutions

An unique application of DWS microrheology is to measure the longest relaxation time of entangled polymer solutions (van Zanten *et al.*, 2004). At low concentrations, the weak rheological signature and short time scales of the relaxation are difficult to capture by bulk rheology, yet they play an important role in governing elastic instabilities in micro-scale flows of particular importance to enhanced oil recovery, environmental remediation, and other microfluidic processes (Clarke *et al.*, 2015; Casanellas *et al.*, 2016).

An example examining the cross-over from sub-diffusive probe dynamics to diffusive probe dynamics is shown in Fig. 5.18 for aqueous PEO solutions—a subset of the data set presented in Fig. 5.16. The data shown in Fig. 5.16 were measured using a transmission DWS geometry for two sample cuvettes with path lengths of 2 and 10 mm. The combined geometries enable a wider range of time and length scales of probe motion to be measured—the smaller cuvette and correspondingly shorter path-length distribution provides mean-squared displacement data for larger values of the probe displacement (and longer time scales), while the longer cuvette length returns results for smaller displacements and shorter times. Many of the resulting composite mean-squared displacement curves nicely capture the complex-probe motion as it passes through the polymer solutions's relaxation.

At 10 wt% PEO, the relaxation time is just 6.4 ms. Experiments at 7 and 15 wt% are also plotted along with creep experiments using a mechanical rheometer. The long-time terminal regimes of both experiments are in good agreement, but mechanical rheometry has difficulty resolving the cross-over from the initial power-law relaxation of the fluid, and hence, cannot resolve the relaxation time.

5.4.9 Operating regime

Within the array of passive microrheology methods, DWS microrheology is unique in its ability to detect exquisitely small displacements of probe particles. This characteristic is important. It means that short *time scales* are also captured, enabling the high-frequency response of materials to be measured. We discuss several applications of high-frequency DWS microrheology in Section 5.6. Recalling the general operating conditions of passive microrheology, eqn 3.160, the detection of small displacement, also extends passive microrheology to significantly smaller compliances.

The length and time scales of probe motion captured by DWS can be adjusted primarily by the choice of the scattering geometry and probe concentration, both of which affect the light transport (the scattering mean-free path length) and the corresponding *path-length distribution*. Changing the probe size also affects light transport in addition to the mobility of the particles.

The characteristic length scales probed by the transmission geometry is (Pine *et al.*, 1990)

$$l_T = l^*/k_0 L \qquad (5.73)$$

while the backscattering geometry probes length scales on the order of

$$l_B = \frac{1}{k_0 (z_0/l^* + 2/3)} \qquad (5.74)$$

$$\text{or } l_B \approx 3/5k_0 \text{ for } z_0 \approx l^*.$$

Thus, while the limiting tracer particle motion measured by the transmission geometry can be controlled by changing the thickness of the sample L, the backscattering geometry limit is determined mainly by the wavelength of the laser. The PEO microrheology data reported in Fig. 5.16 uses two path-lengths, $L = 2$ and 10 mm, to extend the range of the mean-squared displacement. Because $L/l^* \gg 1$, comparing eqns 5.73 and 5.74 reinforces the degree of the separation of length scales between the two geometries.

In Fig. 5.19 we summarize the operating regime of DWS microrheology. The normalized correlation function values $g_{(2)}(t)/\beta - 1 = 0.95$ and $g_{(2)}(t)/\beta - 1 = 0.05$ calculated by eqn 5.56 (plane wave transmission) and the Siegert relation are used to estimate the range of the mean-squared displacement. The calculations are representative for a probe diameter 1 μm dispersed at volume fraction $\phi = 0.01$, for which $l^* = 260$ μm. For a sample length $L = 10$ mm, the root mean-squared displacement $\langle \Delta r^2(t) \rangle^{1/2}$ ranges between about 0.6 nm and 5 nm. Using a thinner sample, $L = 1$ mm, shifts the detected probe displacement to the range 5–50 nm and shows how varying sample thicknesses can be used to increase the operating range of DWS microrheology (see Fig. 5.20). The time limits here are somewhat arbitrary. Correlators can typically reach as low as 5–25 ns delay times and calculate a

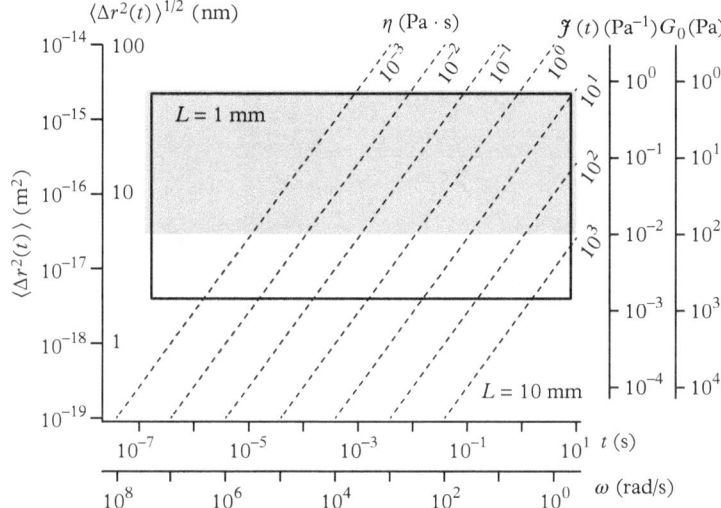

Fig. 5.19 *A DWS microrheology operating diagram. The filled boxes indicate plane-wave transmission experiments with sample thicknesses L = 10 mm (light gray) and L = 1 mm (dark gray). The black unfilled box is calculated for plane-wave backscattering. The compliance $\mathcal{J}(t)$, viscosity η, and equilibrium modulus G_0 values are representative for a probe diameter 1 μm dispersed at volume fraction $\phi = 0.01$, for which $l^* = 260\ \mu m$.*

correlation to an hour (although this makes the integration times quite long). Camera-based methods, discussed in Section 5.7.6, can extend to even longer times. What is most important to remember is that the time scale of the experiment will be set by the probe motion relative to the detectable displacement limits.

The corresponding compliances are also shown. Significantly, the smallest compliances comfortably reach $\mathcal{J}(t) \sim 10^{-3}\ \text{Pa}^{-1}$ or lower. Gels with moduli upwards of several hundred Pascals can be measured, which is significantly higher than the materials accessible to particle tracking microrheology. Of course, since $\mathcal{J}(t) = (\pi a/k_B T)\langle \Delta r^2(t) \rangle$, using a smaller probe particle will further decrease the minimum compliance, increasing the modulus or viscosity that can be measured. A complication, however, is that the light transport will change accordingly through the photon mean-free path length l^* and alter the length scales of the measurement.

Lastly, we've also plotted the limits of particle motion for a plane-wave backscattering experiment using the same limits of the correlation function and $L = 10$ mm. In this case, the root mean-squared displacement ranges between 2 and 50 nm. The expected range of length and time scales is shown by the black box in Fig. 5.19.

5.5 Light scattering experiment

A light scattering apparatus (Fig. 5.20) consists of a laser source, a light detector, and a high-speed digital correlator. DLS requires

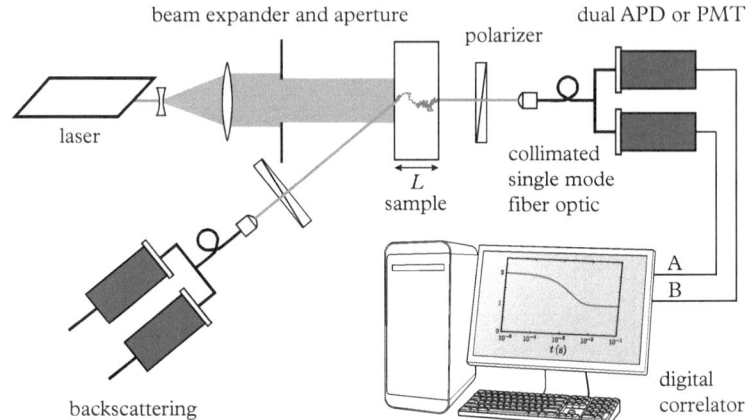

Fig. 5.20 *A schematic of a DWS instrument showing the major components: Laser, detector optics, and digital corrlator.*

a precise alignment of the illumination and detector with the same scattering volume, often with a goniometer capable of sampling a continuous range of scattering angles, or with one or more detectors at fixed angles. Such instruments are generally acquired as a commercial package, and many compact, capable devices are available on the market. DWS can accommodate less stringent alignment requirements, and while several robust commercial instruments are available, it is also feasible to construct a system or add simple DWS capabilities to existing DLS instruments.

In this section, we discuss the choice and operation of lasers for dynamic light scattering, detection schemes and electronics, single-to-noise and measurement error, and the importance of the correlation baseline and intercept in the interpretation of light scattering microrheology. Since the development of coherent sources (lasers), light scattering has had a rich and long history. Excellent references that discuss the design and operation of light scattering instruments include those by Chu (1991) and Brown (1993).

5.5.1   Light scattering samples

Light scattering samples are typically prepared in cuvettes or ampules. Square cuvettes may be plastic, glass, or quartz. A commercial DLS instruments will often require a geometry so that the scattering angle can be known precisely from the position of the laser and detector. These constraints are relaxed for DWS, and preparation of DWS samples is less demanding—disposable plastic cuvettes available for spectrophotometry are often used. An image of a sample is shown in Fig. 5.9. For DWS, the fact that cuvettes come in a variety of path lengths means that the transmission geometry can be tailored easily to measure a range of length scales of probe motion (see Section 5.4.9 and the preceding application note).

One of the unique strengths of passive microrheology is that the sample does not have to be manipulated by an external mechanical disturbance. Using light scattering to measure probe motion further frees us to prepare samples in environments that are difficult to achieve on a microscope or in a mechanical rheometer. An example we explore in the next application note are sample environments that reach high pressures and temperatures.

Application note: High-pressure microrheology

The laser source and detector do not have to be in close proximity to the sample in a light scattering experiment. Thus, an experimentalist is free to use unique sample environments, including high-pressure and temperature cells (some capable of reaching 8000 bar), or even a material's storage container.

Reliable high-pressure scattering cells have been designed for light and neutron scattering experiments, including DLS (Lesemann *et al.*, 2003; Kermis *et al.*, 2004; Meier *et al.*, 2008). A common configuration uses flat sapphire windows mounted in a high-pressure steel body (Lentz, 1969; Claesson *et al.*, 1970; Kirby and McHugh, 1997; DiNoia *et al.*, 2000; Kermis *et al.*, 2004). One example is shown in Fig. 5.21. Multiple windows are desirable for DLS to allow

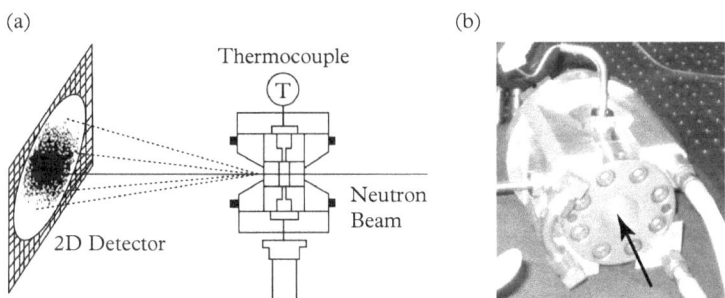

Fig. 5.21 *(a) Diagram of a high-pressure cell used for neutron and light scattering. Adapted with permission from DiNoia, T. P., Kirby, C. F., Van Zanten, J. H., & McHugh, M., Macromolecules **33**, 6321–9 (2000). Copyright (2000) American Chemical Society. (b) A cell constructed with a heating jacket used for DWS microrheology (Kloxin, 2006). The arrow shows the direction of the incident light, which enters the sample through a sapphire window. Image courtesy of C. J. Kloxin.*

measurements at several scattering angles (Richards and Fisch, 1994), but scattering cells for DWS can be simpler—only two windows are necessary for transmission and one for backscattering.

Using DWS and a high-pressure scattering cell, Kloxin and van Zanten (2010) measured the microrheology of PEO-PPO-PEO triblock copolymer solutions in deuterium oxide (D_2O) from atmospheric pressure to 207 MPa (2070 bar, or about 3×10^4 psi) over a range of temperatures from 35 to 75°C. Their data, reproduced in Fig. 5.22, shows a transition at each temperature from a viscoelastic material, in which the probe motion is sub-diffusive, to a viscous, Newtonian fluid as the sample pressure increases. The transition is associated with a traversal across the phase boundaries of the solution, from "sticky micelles" at low pressure to free surfactant solutions at high pressure. The initial rheology at atmospheric pressure depends on the starting phase. The measurements not only produce rheological information, but also provide a clear measurement of the phase diagram.

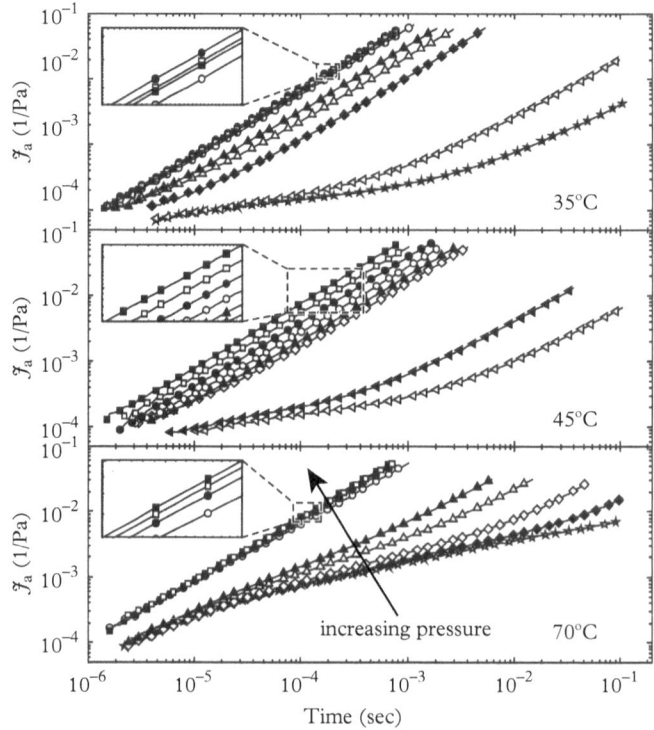

Fig. 5.22 *Microrheology of a 25 wt% PEO-PPO-PEO triblock co-polymer in D_2O for three isotherms, increasing from atmospheric pressure to 207 MPa. Adapted with permission from Kloxin, C. J. & van Zanten, J. H., Macromolelcules, 43, 2084–7 (2010). Copyright (2010) American Chemical Society.*

5.5.2 Laser

Lasers used in light scattering vary in power, wavelength, and type. Gas ion lasers, especially helium neon (HeNe) and argon ion (Ar⁺), were long prized for their power and stability, but solid state and semiconductor lasers have improved in stability, power, and coherence in recent years, and now provide some of the best values.

Power and wavelength

The power of a light scattering laser varies from tens to hundreds of milliwatts. Laser power in a DWS experiment is a chief concern, since expanding the beam for plane-wave geometries and the diffusive nature of light cuts down the light intensity propagating through the sample significantly (consider, for instance, the transmittance equation, 5.66).

Along with laser power, another consideration is the laser wavelength. Most visible wavelengths will do, but ultimately the wavelength should be compatible with the sample, avoiding absorption by the material or probes. Aside from truncating diffusive photon paths in a DWS experiment (see Section 5.4.7), absorption may heat the sample and cause thermal convection, an unwanted source of probe motion. Probes such as polystyrene and silica are generally good choices for the most frequently-encountered laser wavelengths. Gas lasers used in light scattering include argon ion (Ar⁺), which typically lases at a vacuum wavelength λ = 488 or 514.5 nm, and helium neon lasers (HeNe, λ = 632.8 nm). Solid-state lasers include larger garnet lasers equipped with a frequency doubling crystal ("doubled YAG," λ = 532 nm). More recently, diode lasers with suitable coherence lengths (discussed next) and lifetimes are becoming more common. These lasers, their vacuum wavelength, and vacuum wavevector $k_0 = 2\pi/\lambda$ are summarized in Table 5.2.

Table 5.2 *Several common laser sources used in light scattering.*

type	λ (nm)	$k_0/10^7$ (m⁻¹)
Ar⁺	488	1.288
	514.5	1.221
HeNe	632.8	0.9929
Doubled YAG	532	1.181

Coherence length

A laser is a coherent light source. Ideally, if light emitted from the laser was monochromatic and in phase, then the propagating wavefront of the beam would have a perfectly well-defined oscillation over its entire spatial and temporal extent. In contrast, incoherent sources arise from the superposition of emissions from a large number of atoms radiating independently, at different frequencies and phases.

As we have seen, the interference of coherent rays traveling different paths through a sample in either single or multiple scattering is what makes DLS and DWS possible. But lasers are not perfectly coherent sources. Here we discuss a few concepts related especially to their temporal coherence, which is important to consider when selecting a laser for light scattering and evaluating the correlation functions obtained from an experiment. Readers are directed to more thorough treatments for a deeper understanding of coherence and statistical optics such as Born and Wolf (1999) and Mertz (2010).

The temporal coherence of a laser can be characterized by an autocorrelation function

$$G(\tau) = \langle E^*(t)E(t+\tau)\rangle. \tag{5.75}$$

By the Wiener-Kintchine theorem (eqn 5.7), this *temporal coherence function* is a Fourier Transform pair of the power-spectral density,

$$S(\nu) = \mathscr{F}\{G(t)\} \tag{5.76}$$

also referred to as the spectral density, or simply, spectrum. Thus, if the light were perfectly monochromatic, with a spectral density represented by a delta function, the temporal coherence function would be a constant—perfectly correlated over all time. But laser light has some *spectral width* or *linewidth*, $\Delta\nu_c$ about its frequency $\nu = c/\lambda$, that characterizes its deviation from monochromaticity or degree of partial coherence. There are several definitions of the spectral width, but a convenient one is

$$\Delta\nu_c = \frac{\left(\int_0^\infty S(\nu)d\nu\right)^2}{\int_0^\infty S^2(\nu)d\nu}. \tag{5.77}$$

Other definitions include the full-width at half the maximum value of $S(\nu)$, $\Delta\nu_{\text{FWHM}}$, which depends on the functional form of $S(\nu)$—whether it is characterized as a rectangular, Lorentzian, or Gaussian function, for example.

The finite width of the spectral density means that the temporal coherence of the beam is limited, and $G(\tau)$ will exhibit a finite decay on a time scale $\tau_c = 1/\Delta\nu_c$. Thus, over the longitudinal length of the beam, at any instant in time, the phase is decorrelated over a length

$$l_c = c/\Delta\nu_c = c\tau_c. \tag{5.78}$$

This length is the beam's *coherence length*. A *single-mode* laser might have a spectral width $\Delta\nu_c \sim 10^6$ Hz, which gives a coherence length of $l_c \sim 300$ m. Other lasers with a wider spectrum, $\Delta\nu_c \sim 10^9$ Hz, have a smaller coherence length, $l_c \sim 30$ cm.[4]

[4] Given that $\nu \sim 5 \times 10^{14}\, s^{-1}$, for visible light this spread is still a tiny fraction of the laser frequency!

The effect of coherence length is most pronounced for DWS experiments because the path lengths in multiple scattering can become quite long. Paths longer than $s > l_c$ will lose their coherence and not contribute to the correlation function. The loss of coherence has an effect much like light absorption by shifting the measurement sensitivity to larger probe particle displacements. Similar to the treatment of attenuated photon paths by absorption, we rewrite the scaled root mean-squared displacement $R(t)$ in the field-correlation functions to include the truncation,

$$R(t) = (k_0^2 \langle \Delta r^2(t) \rangle + 3l^*/l_c)^{1/2}. \tag{5.79}$$

Transmission DWS experiments are affected the most, due to the significant weighting of long diffusive paths, which are on the order of the sample thickness L^2/l^*.

In Fig. 5.23, we show the expected effect on the intensity correlation function $g_{(2)}(t)$ in a plane-wave transmission DWS experiment (eqn 5.56 and the Siegert relation, eqn 5.33) as the coherence length decreases from infinity (an ideal, perfectly coherent beam) to a value $l_c = 70$ cm. The loss of coherence over longer diffusive paths lowers the intercept of the correlation function, since the detector now receives partially coherent light. The effect is more pronounced for samples of greater thicknesses.

5.5.3 Detectors

Detectors in a light scattering apparatus are designed to measure low intensity light sources with a fast response. An older but

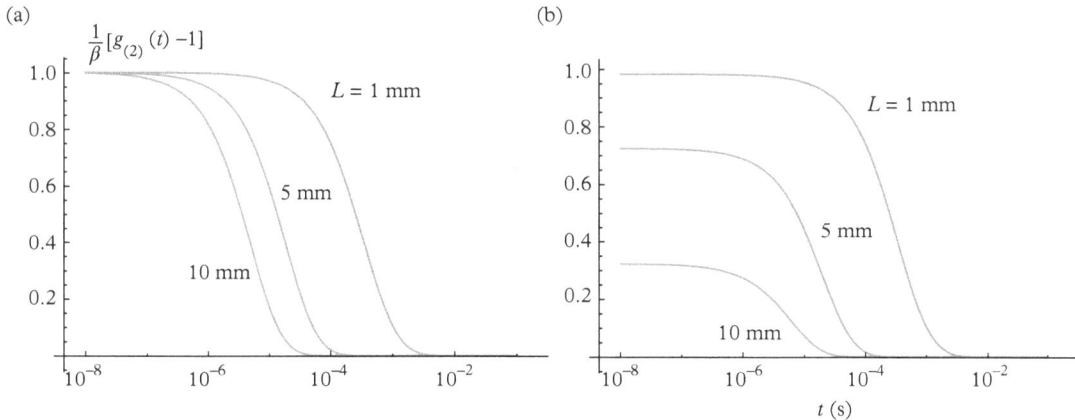

Fig. 5.23 *Calculated plane-wave transmission DWS intensity correlation functions when the laser-coherence length is (a) infinite ($l_c = \infty$), and (b) $l_c = 0.7$ m. The particles are 1 μm diameter in water with $l^* = 260$ μm.*

common detector is a photomultiplier tube (PMT). Contemporary instruments have moved towards the use of solid state detectors—avalanche photodiodes (APD). Both APDs and PMTs amplify the photocurrent generated by a single photoelectron. With a high-speed correlator, these detectors enable measurements of the intensity correlation on time scales as short as tens of nanoseconds (100 MHz bandwidth).

PMTs and APDs are subject to noise—spurious dark counts, afterpulsing, and other sources that are described in more detail later (Brown and Smart, 1997). The noise levels are typically 100 counts-per-second or less. To reach the shortest-time scales, PMTs and APDs are used in pairs and the cross-correlation between the detectors is calculated by the correlator. The cross-correlation reduces the artifacts and noise introduced by each detector, since these signals are uncorrelated. With recent improvements in the spurious noise of APDs, cross-correlation may be unnecessary, which reduces the cost of the instrument.

Detectors are coupled directly to single-mode fiber optics, while in the past pin-hole apertures would be used to define the scattering angle and volume (Rička, 1993). The fiber optic collects light from the sample using a collimating lens matched to the numerical aperture of the fiber. Light can be launched through a second fiber and the lens assembly for precise alignment of the detection point—centering it on the sample and on axis with the illuminating beam. In DWS experiments, a cross-polarizer is typically placed in front of the detector optics to select depolarized light scattered from the sample. Depolarization ensures that the light has been multiply scattered, and is especially important for eliminating singly-scattered light in a backscattering geometry.

5.5.4 Signal-to-noise and measurement error

The signal-to-noise ratio of light scattering microrheology is determined by the statistics of the intensity correlation function. There are two primary contributions to the signal-to-noise: The noise of the light intensity measurement, and the number of correlation samples comprising the average.

The photons that enter the detector do so with a Poisson distribution. The statistical error in the measurement is therefore $1/\sqrt{N}$, where N is the number of photons. One way to increase the signal-to-noise is to increase the intensity of the incident light. If the detected photon count reaches on the order of 10^5 counts per second (often reported by correlators in kilocounts per second, or kcps) then the expected error should be on the order of 3×10^{-3} or 0.3%. An error of 1% corresponds to about 10 kcps.

At low light intensities, the signal-to-noise is degraded further by the dark counts of the photodetector. Like the dark signal of a CCD camera discussed in Chapter 4, these noise sources are most often caused by thermally-excited electrons generating a photocurrent that is independent of the absorption of light. These are referred to as thermionic sources. Next, the amplification process is itself noisy. In a PMT, dark current is generated by field emission and afterpulsing, and even by background radioactivity and cosmic rays. The noise factor F of a photodetector is (Mertz, 2010)

$$F^2 = \frac{\sigma_{Im}^2}{\langle M \rangle^2 \sigma_{Iq}^2} \tag{5.80}$$

where $\langle M \rangle$ is the average detector amplification gain, σ_{Iq}^2 is the variance of the photocurrent that includes contributions of shot noise and dark current, and σ_{Im}^2 is the variance of the gain fluctuations. For a PMT, which has a gain $M \approx 10^6 - 10^7$, the noise factor is $F \approx 2$. An APD has a typical gain $M \approx 10 - 1000$ with a noise factor that scales as $F \approx M^{0.15}$, placing its noise factor in the same range as a PMT.

After considering the detector and amplification noise, the next most significant contribution of error to the correlation function comes from the number of samples included in the average of the time–correlation function. Because correlators calculate a time-average, the total measurement time must be many times the longest delay time of the correlation, t_{max}. This measurement time is also referred to as the correlation sampling time. Again, the measurement uncertainty will scale as $1/\sqrt{n}$, where $n = T/2t_{max}$ is the number of independent measurements over the integration or averaging time T. A rule of thumb is $n \approx 1000$. An estimate of the correlation function error that is sometimes used is (Berne and Pecora, 2000)

$$\Delta_T(t) = \pm \frac{1}{\sqrt{n}} [1 - g_{(2)}(t)]. \tag{5.81}$$

For fast processes, in which the correlation function decays quickly, an accurate correlation function can be measured in minutes because of the short integration time that is required. For example, the correlation function for micrometer-diameter probe particles dispersed in a low viscosity fluid, such as water, decays on the order of ~ 1 ms (see Fig. 5.14). More slowly-relaxing systems, or techniques that use a second material or device to force the correlation function to decay,[5] sometimes have relaxation times of seconds to tens of seconds. Such conditions will require integration times on the order of an hour or more.

[5] See the discussion of nonergodic samples in Section 5.7.

5.5.5   Correlator

The job of the correlator hardware is to count input pulses n from the detector over regularly spaced sampling intervals Δt, delay the counts for some lag time $\tau = k\Delta t$, multiply the current input counts by the delayed counts, and finally accumulate the results in an average, giving the photon correlation function (Schätzel, 1993)

$$\langle n_j n_{j+k} \rangle = \frac{1}{N} \sum_{i=1}^{N} n_j n_{j+k}. \tag{5.82}$$

The photon correlation function is equivalent to the time-averaged intensity correlation,

$$\langle n_j n_{j+k} \rangle = \delta_{k0} \langle \mu_j \rangle_\mu + \langle \mu_j \mu_{j+k} \rangle_\mu \tag{5.83}$$

where μ_j is the time-integrated intensity value for a detector with quantum efficiency $\mathcal{Q}$,

$$\mu_j = \int_{(j-1)\Delta t}^{j\Delta t} \mathcal{Q}I(t)\,dt \tag{5.84}$$

and δ_{k0} vanishes for non-zero k.

   The data rate of the correlator is typically on the order of 100 MHz for correlation delay times as low as several nanoseconds to tens of nanoseconds. All of the calculations required to compute the photon correlation function occur in real-time, to shorten the measurement and improve its statistical accuracy. The noise performance, accuracy, and design architecture of correlators is beyond our scope, and generally are transparent to the average user. A key feature of a modern digital correlator is its large dynamic range: Current hardware is capable of computing correlations spanning 10 ns to an hour. A few additional concepts are useful to consider, such as the normalization of the correlation and other aspects of data processing, which is discussed next.

Correlation intercept and baseline

In Section 5.4.8, we showed that by numerically inverting the field-correlation function (combined with the Siegert relation), the mean-squared displacement of tracer probes can be derived from experimental data. Properly calculating $\langle \Delta r^2(t) \rangle$ requires accurate values of the correlation intercept and baseline. Here, we discuss the effect of intercept and baseline on microrheology measurements. We will see that poor calculations of the intercept and baseline produce erroneous and possibly misleading microrheology results.

Let's start by considering the correlation functions calculated during a microrheology experiment. Three correlation functions for DWS backscattering using 1 μm diameter probe particles in water, 25% glycerol, and a 7500 ppm solution of PEO are shown in Fig. 5.24a. These curves are the raw calculations output by the correlator $\langle I(t_0)I(t_0 + t)\rangle$ in units of *(counts-per-second)*^{2}. The correlation values depend on the scattering intensity of each experiment. In this particular series of measurements, the scattering intensity is higher for the tracer probes in water and about the same for the PEO and glycerol samples. By dividing each correlation by its baseline $\langle I\rangle^2$, the correlation functions are normalized to dimensionless intensity correlations, $g_{(2)}(t)$, which are shown in Fig. 5.24b. Normalizing the correlation function results in a partial cancellation of the inherent count-rate noise (Schätzel, 1993).

The baseline intensity $\langle I\rangle$ can be measured or calculated. For a measured baseline, several "extended" or "monitor" correlator channels at lag times greater than the last specified time are reserved used to average the intensity. A calculated baseline is the total number of pulses received by the correlator squared and divided by the number of samples.

The correlations in Fig. 5.24b have been normalized by the calculated baselines (solid lines) or the measured baselines (dotted lines). In the majority of measurements, the two agree within less than 1%

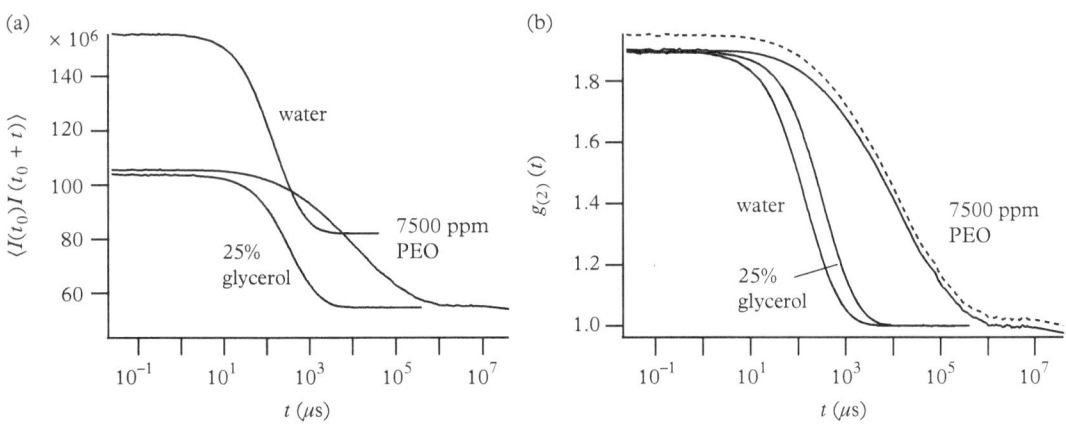

Fig. 5.24 *(a) Plane wave backscattering correlation functions (b) and normalized correlation functions for 1 μm diameter probe particles in water, 25% glycerol, and a mixture of 4 MDa PEO and 25 wt% glycerol. Dotted lines are correlation functions normalized by the measured baseline.*

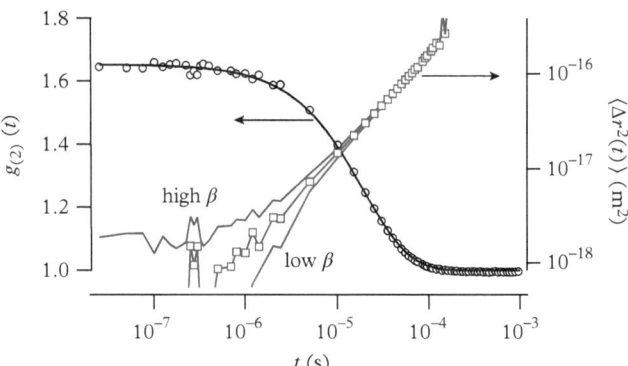

Fig. 5.25 *The intensity correlation function and mean-squared displacement for 1.05 μm diameter probes in water. The square symbols are the mean-squared displacement with an intercept value $\beta = 0.653$. Two additional mean-squared displacement curves are calculated using a high-intercept value, $\beta = 0.69$ and a lower one, $\beta = 0.62$.*

difference, but for the 7500 ppm PEO solution, the measured baseline is significantly lower. The lower value is due to the long lag times selected for the baseline averaging (> 10 s) and subsequent poor averaging statistics. However, this issue is not uncommon for complex fluids, where relaxation in a material can persist for seconds, as in this example.

In Fig. 5.25 we illustrate the sensitivity of the mean-squared displacement to the value of β using the data presented first in Fig. 5.14 for 1 μm diameter probes in water. As was noted earlier, a best fit of the plane-wave transmission correlation function shown here gives an intercept value of $\beta = 0.653$, and the mean-squared displacement calculated by eqn 5.56, indicated by the line of connected symbols. If instead a value $\beta = 0.69$ is used, the calculated mean-squared displacement exhibits an upwards curvature. Likewise, a lower intercept value, $\beta = 0.62$ produces a mean-squared displacement that has a downwards curvature.

The example we have seen is a relatively straightforward illustration due to the simple rheology of a viscous Newtonian fluid. We can analytically calculate the correlation function by the plane-wave transmission equation (eqn 5.56) and $\langle \Delta r^2(t) \rangle = k_B T / \pi a \eta$. Normally, however, one would select a series of short correlation times, which stretch below the times for which significant motion of the probes is detected, to establish the intercept value.

If for some reason there is error in the baseline intensity used to normalize the correlation function, then additional error will be introduced into the mean-squared displacement. Consider, for instance, the mean-squared displacement of probes calculated for the correlation functions in Fig. 5.24. A difference between the measured and calculated baseline for the PEO solutions leads to a false plateau at long times in the mean-squared displacement, illustrated in Fig. 5.26.

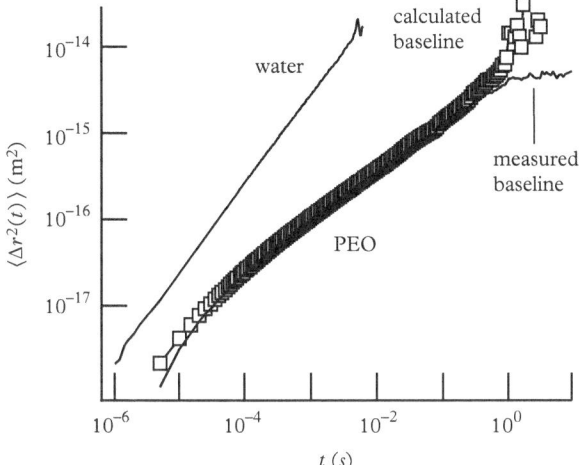

Fig. 5.26 *Mean-squared displacement for 1 μm diameter probe particles in water and a mixture of 7500 ppm 4MDa PEO and 25 wt% glycerol. The symbols are data using the calculated baseline, while the solid lines are calculated using the measured baseline.*

5.6 High-frequency rheology

Here we discuss several examples of DWS microrheology. We will focus on measurements of high-frequency rheology. These examples will motivate us to examine the effect of *inertia* in microrheology.

As we've seen in the derivations of the DWS correlation functions, DWS-based microrheology is sensitive to short displacements of the probe particles, in several instances down to several nanometers—remarkably small given a typical particle size on the order of a micrometer. Probes typically take short times to move such small distances. The GSER relates measurements of the mean-squared displacement at these short time scales to the medium viscoelastic properties, and thus probes the short-time (or equivalent high-frequency) behavior often far outside the operating range of conventional mechanical rheometers (Willenbacher and Oelschlaeger, 2007).

DWS microrheology provides an important and unique means for characterizing the dynamic mechanical properties of solutions of soft materials. At such short time scales, the storage and dissipation of mechanical energy by the material is dominated by the relaxation modes dictated by the internal dynamics of the structure. For polymers, this regime provides a direct rheological measurement of stiffness, molecular architecture, or solvent–polymer interactions. In entangled or cross-linked networks, the single polymer mechanical response may be obtained from the macroscopic shear modulus of the network (Gittes and MacKintosh, 1998). While an extended frequency range can often be obtained with mechanical rheometry using methods such as time-temperature superposition, DWS microrheology is especially useful for polymer, surfactant, and protein solutions that would not withstand such treatments.

In the following section, we provide some examples of measurements in the high-frequency regime and applications to measuring the stiffness of surfactant assemblies. We also discuss the effects of inertia in microrheology measurements, which can become significant for high-frequency microrheology measurements and applications.

5.6.1   High-frequency DWS examples

Microrheology characteristic of a Rouse polymer has been measured using DWS for semi-dilute poly(ethylene oxide) (PEO) solutions, shown in Fig. 5.27 (Gisler and Weitz, 1998). The high-frequency scaling of the storage and loss moduli, $G'(\omega) \sim G''(\omega) \sim \omega^{1/2}$ that clearly emerges for $\omega > 1000$ rad/s reflects the strong hydrodynamic screening between segments in the polymer (Doi and Edwards, 1986). As the frequency decreases below the longest relaxation time, the dynamic-moduli cross-over to the expected scaling behavior for the storage $G'(\omega) \sim \omega^2$ and loss $G''(\omega) \sim \omega$ moduli.

Solutions of the protein F-actin have been widely studied as a model semiflexible polymer (Mason *et al.*, 2000; Palmer *et al.*, 1999; Le Goff *et al.*, 2002; Huh and Furst, 2006). The high-frequency

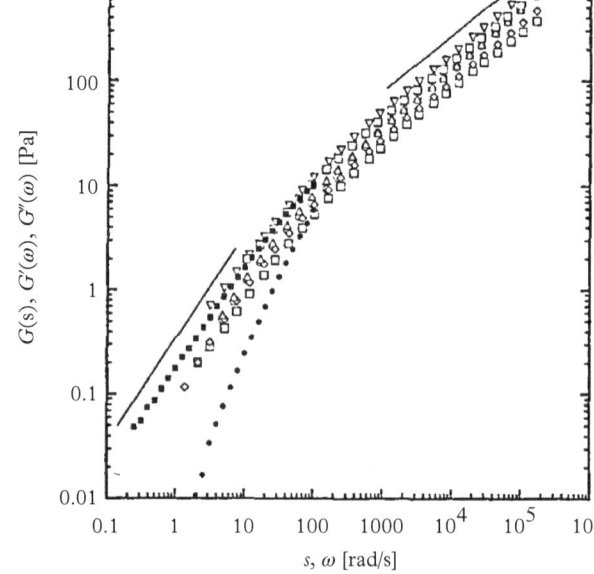

Fig. 5.27 *DWS microrheology (open symbols) and mechanical rheometry (closed symbols) of poly(ethylene oxide) solutions (molecular weight 9×10^5, concentration $c = 2.2$ mg/ml). DWS is performed with two probe sizes, 0.52 μ and 6.3 μm diameter polystyrene. Reprinted from* Curr. Opin. Colloid Interface Sci., *3, Gisler, T. & Weitz, D. A., Tracer microrheology in complex fluids, 586–92, Copyright (1998), with permission from Elsevier.*

rheology of F-actin and other semiflexible polymers is given in terms of the relaxation modulus by Morse (1998c),

$$G(t) \sim \frac{\rho k_B T}{l_p} \left(\frac{k_B T}{\zeta_\perp l_p^3} t \right)^{-3/4} \tag{5.85}$$

where ρ is the polymer length per unit volume, l_p is the polymer persistence length, η is the background solvent viscosity, and $\zeta_\perp \approx 4\pi\eta/\ln(0.6L/d)$ is the lateral drag coefficient per unit length for filaments of length L and diameter d. Using the relation between $G(t)$ and $\mathcal{J}(t)$, the corresponding high-frequency compliance is

$$\mathcal{J}(t) \sim \frac{2\sqrt{2}l_p}{3\pi\rho k_B T} \left(\frac{k_B T}{\zeta_\perp l_p^3} t \right)^{3/4}. \tag{5.86}$$

From eqn 5.86, we see that the characteristic short-time scaling of the mean-squared displacement is $\langle \Delta r^2(t) \rangle \sim \mathcal{J}(t) \sim t^{3/4}$. In the corresponding frequency domain, the shear moduli scale as $G^*(\omega) \sim \omega^{3/4}$.

The high-frequency scaling of semiflexible polymers is nicely demonstrated in data from DWS measurements of F-actin by Mason *et al.* (2000), which are shown in Fig. 5.28 for both the time and frequency domains. The origin of the three-fourths scaling exponent lies in the bending mechanics of the filaments, not the entropic stiffness and hydrodynamic interactions between polymer segments that gives rise to the Rouse or Zimm dynamical scalings (Doi and Edwards, 1986). As we discuss next, DWS can be used to measure the rigidity of dilute, entangled, and cross-linked suspensions of semiflexible polymers.

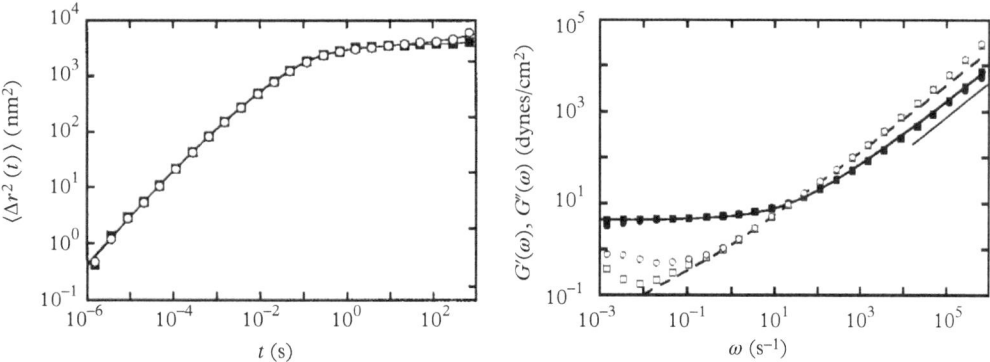

Fig. 5.28 *Mean-squared displacement and moduli of weakly cross-linked F-actin (Mason et al., 2000).*

Application note: Rigidity of semiflexible polymers

The operating range of DWS microrheology has an important application: It can be used to measure the mechanical rigidity of macromolecular protein, and supramolecular assemblies, such as suspended carbon nanotubes, F-actin, and rod-like viruses. Worm-like micellar solutions of self-assembled surfactant molecules have provided especially good demonstrations of this capability (Galvan-Miyoshi *et al.*, 2008; Sarmiento-Gomez *et al.*, 2010; Oelschlaeger *et al.*, 2010; Oelschlaeger *et al.*, 2013).

The individual mechanics of suspended semiflexible filaments can be characterized by DWS because the high-frequency response is dominated by the bending modes and its viscous dissipation to the surrounding solvent, as given by eqns 5.87 and 5.86, while contributions due to entanglements and cross-links are much weaker. Writing the corresponding shear modulus in the high-frequency regime (Morse, 1998*a*; Morse, 1998*b*; Gittes and MacKintosh, 1998),

$$G^*(\omega) \approx \frac{1}{15}\rho\kappa l_p(-2i\zeta_\perp/\kappa)^{3/4}\omega^{3/4} + i\omega\eta \qquad (5.87)$$

where $\kappa = l_p k_B T$ the bending stiffness, we find that when the concentration and geometry of the filaments is known, the only free parameter is the persistence length, l_p, which alone determines the magnitude of the modulus. The rigidity can therefore be found by fitting the high-frequency response.

An example for a worm-like micellar solution of cetylpyridinium chloride (CpCl) and sodium salicylate (NaSal) is shown in Fig. 5.29. It has the characteristic $\omega^{3/4}$ scaling. With this approach, several studies have focused specifically on the micelle flexibility (Oelschlaeger and Willenbacher, 2012; Oelschlaeger *et al.*, 2013) including the effect of counter-ion binding (Oelschlaeger *et al.*, 2010). Another interesting example is from Oelschlaeger *et al.* (2009), who investigated the temperature dependence of the surfactant solution rheology. The data, plotted in Fig. 5.30 exhibit a shift towards higher frequencies corresponding to an increase in the micelle scission with increasing temperature from 20°C to 40°C. Despite this change, the terminal high-frequency response is the nearly the same for all of the measurements, indicating that, on the molecular scale, the micelle mechanics haven't changed significantly.

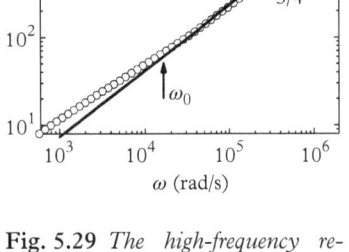

Fig. 5.29 *The high-frequency response of a 100 mM CpCl and 60 mM NaSal worm-like micellar solution is fit by eqn 5.87 to find the persistence length* $l_p = 29 \pm 3$ *nm at 20° C. Reprinted figure with permission from Willenbacher, N. et al., Phys. Rev. Lett. 99, 68302 (2007). Copyright (2007) by the American Physical Society.*

5.6.2 Inertia in microrheology

In the analysis of many of the experiments in the previous section, inertial effects must be properly taken into account. Our derivation

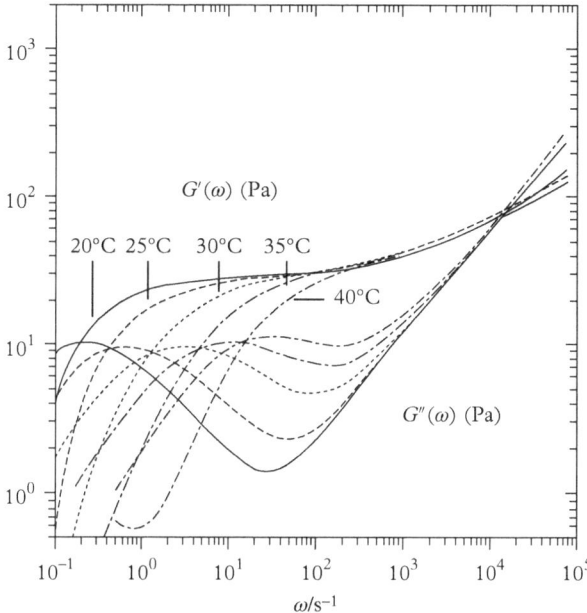

Fig. 5.30 *DWS microrheology me-asurements of an aqueous solution of 100 mM CpCl/60 mM NaSal as a function of temperature over the temperature range 20 to 40°C. Reprinted with permission from Oelschlaeger, C., Schopferer, M., Scheffold, F., & Willenbacher, N., Langmuir 25, 716–23 (2009). Copyright (2009) American Chemical Society.*

of the GSER in Chapter 3 yielded the following expression in the frequency domain for the Generalized Einstein Relation:

$$\langle \Delta \tilde{r}^2(\omega) \rangle = \frac{6 k_B T}{(i\omega)^2 \left[\tilde{\zeta}(\omega) + i\omega M_p \right]}. \tag{5.88}$$

We neglected the inertia of the particle $i\omega M_p$ by reasoning that the probe mass $M_p = \frac{4}{3}\pi a^3 \rho_p$ was sufficiently small that the elastic and viscous resistance $\tilde{\zeta}(\omega)$ overwhelmed the inertial resistance. With many of the microrheology experiments we have considered until now, this limit is fine. However, at the short times of probe motion that are often captured in diffusing wave spectroscopy microrheology (*e.g.* $\sim 10^{-6}$ s) inertia can become significant. Viscoelastic functions (the compliance, complex viscosity, or equivalent shear moduli) cannot be naively calculated from the mean-squared displacement without accounting for it.

Inertia plays a role both the Generalized Einstein Relation as we just saw and the Generalized Stokes equation, the relation between the material resistance and bulk modulus, that together make up the GSER. With respect to the Stokes equation, its time-independence no longer applies. We found in Section 2.5.3 that the shear wave must have sufficient time to propagate from the bead surface. Solving the full time-dependent equation, we arrived at eqn 2.103 for the particle resistance,

$$\tilde{\zeta}(\omega) = 6\pi a\eta^*(\omega)\left[1 + \frac{a}{\lambda_V} + i\left(\frac{a}{\lambda_V} + \frac{2a^2}{9\lambda_V^2}\right)\right] \qquad (5.89)$$

with

$$\lambda_V = \left(\frac{2\eta^*(\omega)}{\rho\omega}\right)^{1/2}. \qquad (5.90)$$

From eqn 5.90 we infer that there are *two* length scales relevant to inertia that emerge in a viscoelastic material: One that represents the penetration depth of the shear wave into the medium from the bead surface and a second that is the wavelength of the damped oscillation of the shear wave (Indei et al., 2012a).[6]

Taken together, eqns 5.88 and 5.90 constitute the full GSER with particle and medium inertia. In the high-frequency limit such that $\lambda_V \ll a$, the mean-squared displacement becomes

$$\langle\Delta\tilde{r}^2(\omega)\rangle \approx \frac{6k_BT}{(i\omega)^2\,[i\omega M^*]} \qquad (5.91)$$

with the effective mass given by $M^* = M_p + M_s = M_p + \frac{2}{3}\pi a^3\rho_f$. The inverse transform yields

$$\langle\Delta r^2(t)\rangle \approx \frac{6k_BT}{M^*}t^2. \qquad (5.92)$$

The probe motion on these time scales is "ballistic" and independent of the medium rheology (Weitz et al., 1989; Lukić et al., 2005).

An example of mean-squared displacement data with inertia is shown in Fig. 5.31 for two- and three-micrometer diameter particles

[6] See also Cordoba et al. (2012) and Indei et al. (2012b) for related discussions, including the relation between the inertial length scales.

Fig. 5.31 *The mean-squared displacement of melamine resin microbeads with radii $a = 0.94$ or $1.47\ \mu m$ and density $\rho_p = 1570$ kg/m³ in water at $T = 21\,°C$. Reprinted (abstract/excerpt/figure) with permission from Domínguez-García, P. et al., Phys. Rev. E **90**, 060301 (2014). Copyright (2014) by the American Physical Society.*

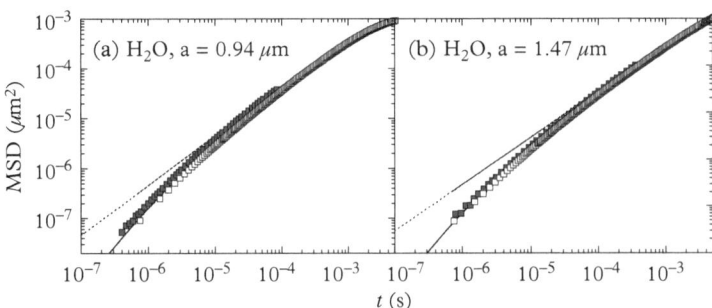

in water. The data is in good agreement with the GSER including inertia given by eqns 5.88 and 5.90 for an incompressible Newtonian fluid.[7] The dashed lines in the figure are the calculated MSDs when the probe inertia $(i\omega M_p)$ is included but the fluid inertia is neglected.

In the range of frequencies between the limits of a fully developed flow $\lambda_V \gg a$, for which the "inertialess" GSER applies (eqn 3.91), and one which is dominated by inertia, $\lambda_V \ll a$ (eqn 5.92), it is possible to separate the inertial and medium contributions to the mean-squared displacement. A straightforward approach is to rewrite eqns 5.88 and 5.90, solving for the shear modulus, which Indei *et al.* (2012*a*) do to find

$$G^*(\omega) = \frac{k_B T}{\pi a i \omega \langle \Delta \tilde{r}^2(\omega) \rangle}$$
$$+ \frac{a^2 \omega^2}{2} \left\{ \left[\rho^2 + \frac{2\rho}{3\pi a^3} \left(\frac{6 k_B T}{(i\omega)^3 \langle \Delta \tilde{r}^2(\omega) \rangle} - M^* \right) \right]^{\frac{1}{2}} - \rho \right\}$$
$$+ \frac{M^* \omega^2}{6\pi a}. \tag{5.93}$$

The first term in the equation 5.93 is the GSER in the absence of inertia. With the mean-squared displacement in the inertial regime, one can now solve for the modulus.

An example of an inertial correction is shown in Fig. 5.32 for particles in a worm-like micellar solution. Taking the mean-squared displacement and applying the GSER without inertial corrections results in a loss modulus $G''(\omega)$ that appears to curve upwards with a frequency scaling that is higher than the expected $G'' \sim \omega^{3/4}$ behavior, and even exceeds a logarithmic slope of one for frequencies $\omega/2\pi > 10^6$ Hz. The loss modulus calculated using eqn 5.93 exhibits the expected scaling over nearly the entire frequency range. The storage modulus, however, appears anomalous in both cases. Nonetheless, like the examples with worm-like micellar solutions discussed, the loss modulus can be used to calculate the bending rigidity of the self-assembled molecular structures.

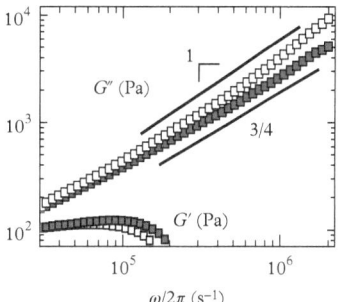

Fig. 5.32 *Moduli calculated without inertial correction (white symbols) and using eqn 5.93 (solid symbols) for probe radius a = 0.94 μm in viscoelastic surfactant solution. Replotted from data from Dominguez-Garcia et al. (2014).*

5.7 Gels and other nonergodic samples

Ergodicity requires that a system sample is able to explore enough configurations in time to be representative of the full ensemble of configurations. One of the first consequences of nonergodicity is that the Siegert relation, eqn 5.32, is invalid—it only applies when the scattered field has a mean value of zero (Joosten *et al.*, 1990). Instead, we need to distinguish *ensemble-averaged* and *time-averaged* quantities.

[7] Hinch (1975) provides the solution in the form of the velocity-autocorrelation function. One of the interesting consequences that emerges is that the probe should oscillate when its density relative to the medium exceeds $\rho_p > (5/8)\rho$, a behavior investigated further by Indei *et al.* (2012*a*).

In light scattering microrheology, this means that the scattering particles in the scattering volume must generate enough configurations that the random speckle pattern becomes completely decorrelated. Simply stated, the bright regions of the speckle pattern shown in Fig. 5.2 would become dark, and the dark regions light—on average, the intensity fluctuations of each speckle would have identical mean values and variances (magnitude).

In nonergodic systems, the time-average (eqn 5.1) and ensemble-average (eqn 5.8) correlation functions are not equivalent. This occurs because the scatterers cannot explore a sufficiently large number of configurations in the integration time of the experiment, if ever. A speckle no longer produces all possible intensity values, but instead varies around its own average. Reflecting on Fig. 5.2 in this chapter, in an ergodic system, the particles in the scattering volume can reasonably sample its entire configuration space and generate all possible intensity values of the scattered field in the form of a speckle intensity. In a nonergodic sample, the scatterer positions, and hence their configurations, are limited. The scattered field and speckle intensity, vary about some average value, but are not representative of all potential configurations.

In microrheology, viscoelastic solids such as polymer gels (Dasgupta and Weitz, 2005), are common nonergodic samples. Fig. 5.33 shows an example of several individual time–correlation functions of a nonergodic sample taken from Scheffold *et al.* (2001). These time–correlation functions of the same sample show different intercept and long-time values of the correlation function, $(g_{(2)}(\infty) - 1)$. Besides being essentially irreproducible, there is also the risk that

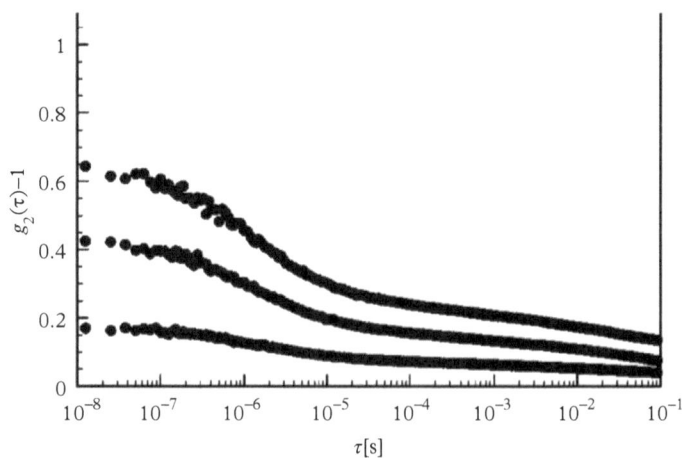

Fig. 5.33 *Light scattering from a nonergodic sample. Reprinted figure with permission from Scheffold, F., Skipetrov, S. E., Romer, S., & Schurtenberger, P., Phys. Rev. E, 63, 61404 (2001). Copyright (2001) by the American Physical Society.*

such measurements would be interpreted as a characteristic of locally heterogeneous dynamics in the sample.

One solution to the problem of measuring the time–correlation function of nonergodic systems is to correct a single-time average correlation function by measuring the average scattered intensity in many static light scattering experiments. This is the original idea introduced by Pusey and Van Megen (1989). Taken together, the static intensity values give the true ensemble average light intensity $\langle I \rangle_E$, which is then used to correct the time average correlation by a nonergodicity parameter $Y = \langle I \rangle_E / \langle I \rangle_T$, where $\langle I \rangle_T$ is the average intensity of the time correlation. An alternative is to measure an ensemble of time-averaged correlation functions, which is straightforward, but tedious. In addition to Pusey's methods, which apply to both DLS and DWS scattering methods, several "optical mixing" techniques have also been introduced, specifically for DWS measurements. Finally, it is possible to collect a representative ensemble using array detectors such as a CCD camera, termed multispeckle detection. We discuss these methods in Section 5.7.1.

5.7.1   Simple model of nonergodicity

Let's examine a simple model of nonergodicity. Pusey and Van Megen (1989) provide a nice illustration by considering the light scattering of non-interacting scattering particles that are restricted to random fixed positions $\{R_j\}$ by weak harmonic forces. The model is similar to what we would expect for a DLS experiment with probe particles dispersed in a viscoelastic gel modeled by a Voigt fluid (Joosten *et al.*, 1990).

The mean-squared displacement of the harmonically bound particles is,

$$\langle \Delta r^2(t) \rangle = 2\langle \Delta r_j^2 \rangle \left[1 - \exp\left(-\frac{D_{0j}\tau}{\langle \Delta r_j^2 \rangle} \right) \right] \qquad (5.94)$$

where D_{0j} is the effective diffusivity of the jth particle and $\langle \Delta r_j^2 \rangle$ is its mean-squared displacement. Compare this expression to the MSD of the Voigt fluid, eqn 3.110. For identical particles in identical microenvironments, such that $\langle \Delta r_j^2 \rangle = \langle \Delta r^2 \rangle$ and $D_{0j} = D_0$, the field correlation function for single scattering eqn 5.29 gives

$$g_{(1)}(q, t) = \exp\left\{ -q^2 \langle \Delta r^2 \rangle \left[1 - \exp\left(-\frac{D_0 t}{\langle \Delta r^2 \rangle} \right) \right] \right\}. \qquad (5.95)$$

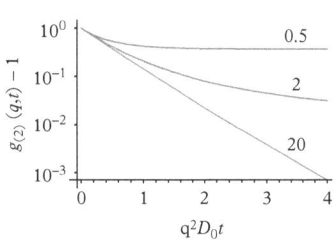

Fig. 5.34 *Semi-logarithmic plots of the calculated ensemble-average intensity time–correlation function for non-interacting particles that are harmonically bound with mean-squared displacements $0.5 < q^2\langle r^2 \rangle < 20$.*

In the limit the delay time goes to infinity, eqn 5.95 indicates that $\lim_{t \to \infty} g_{(2)}(q, t) = \exp(-q^2 \langle \Delta r^2 \rangle) \equiv w^2$.

The correlation functions calculated with eqn 5.95 are plotted semi-logarithmically in Fig. 5.34 for a range of stiffnesses, characterized by the displacement of the scattering particles relative to the wavevector $q^2 \langle \Delta r^2 \rangle$. For weakly-bound particles $(q\langle r^2 \rangle = 20)$, the *ensemble average* correlation function decays very nearly as free particles. As the stiffness increases, the lag times at which the correlation function takes on constant, non-zero values decreases.

5.7.2   Pusey and van Megen's method

Pusey and van Megen's method of determining the ensemble-average time–correlation function requires an experimenter to characterize the nonergodicity of a single time–correlation function measurement (Pusey and Van Megen, 1989). The *ensemble-averaged* field correlation function, $g_{(2)}^{E}(t)$ is related to the *time-averaged* intensity correlation function $g_{(2)}^{T}(t)$ by

$$g_{(2)}^{E}(t) = \frac{Y-1}{Y} + \frac{(g_{(2)}^{T}(t) - \sigma^2)^{1/2}}{Y} \tag{5.96}$$

where $Y \equiv \langle I \rangle_E / \langle I \rangle_T$ is the nonergodicity parameter, a ratio of the ensemble averaged intensity $\langle I \rangle_E$ and the time-averaged intensity $\langle I \rangle_I$, and

$$\sigma^2 = \frac{\langle I^2(0) \rangle_T}{\langle I(0) \rangle_T^2} - 1 \tag{5.97}$$

is the normalized variance of the time-averaged intensity. To use eqn 5.96, the Pusey and van Megen method requires the experimenter to make multiple measurements of the intensity average, then correct the time–correlation function based on the calculated nonergodicity parameter Y. This is accomplished, for instance, by making a time–correlation measurement in a single sample orientation, then rotating or translating the sample to different orientations in order to obtain the ensemble average intensity $\langle I \rangle_E$.

5.7.3   Ensemble of measurements

An alternate and practical method to that of Pusey and van Megen consists of making many statistically independent time-averaged intensity correlation functions (Xue *et al.*, 1992*a*). In this case, the ensemble average intensity time–correlation function is

$$\langle g_{(2)}^T(t)\rangle_E - 1 = \lim_{N\to\infty} \sum_{n=1}^{N} \langle g_{(2)}^T(t)\rangle_T. \qquad (5.98)$$

Since an ensemble is used, the average nonergodicity parameter is $\langle Y\rangle = \langle\langle I\rangle_E/\langle I\rangle_T\rangle = 1$.

While straightforward, the ensemble-averaged intensity requires a somewhat painstaking effort to make measurements at multiple positions in a sample, often by physically rotating or translating it relative to the incident beam and detector. In many DLS instruments equipped with a goniometer and cylindrical sample cells, rotation is the best option. DWS experiments are often performed with square cuvettes, which usually requires translating the sample. Each measurement of the ensemble is still a time-averaged correlation function, and is constrained by the same concerns of integration time and considerations of noise and statistical accuracy that determine the signal-to-noise we have discussed.

5.7.4 Optical mixing

Optical mixing is a DWS-based transmission experiment that uses a combination of two independent samples: The nonergodic sample of interest (n) and an ergodic reference sample (e) with slow relaxation dynamics (Scheffold *et al.*, 2001). The second sample introduces a decay time of the correlation function that serves to properly average the signal from the nonergodic sample. The samples are placed in series, as shown in Fig. 5.35, but can also be separated optically by introducing a lens between the sample and reference cells (Harden and Viasnoff, 2001).

Light incident on the first cell containing the ergodic medium is scattered and propagates into the second cell. The "composite" measured field correlation function $g_{(1)}^m(t)$ is calculated similar to

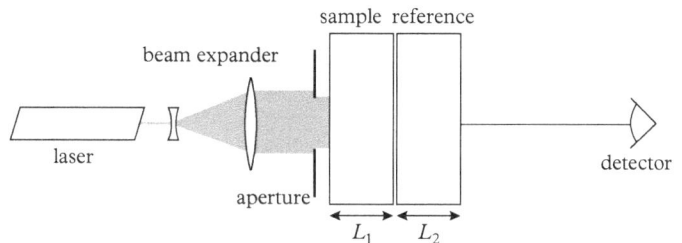

Fig. 5.35 *Optical mixing using a two-cell DWS transmission geometry. The reference cell is an ergodic medium with slow-relaxation dynamics.*

eqn 5.49, except that the diffusive paths propagate through both samples with path lengths s_n in the sample of interest and s_e in the slowly relaxing reference sample,

$$g_{(1)}^m(t) = \int_0^\infty ds_n \int_0^\infty ds_e \mathcal{P}(s_n; s_e)$$

$$\times \exp\left[-\frac{1}{3}k_0^2\left(\langle\Delta r_n^2(t)\rangle\frac{s_n}{l_n^*} + \langle\Delta r_e^2(t)\rangle\frac{s_e}{l_e^*}\right)\right]. \tag{5.99}$$

The cell thickness and photon mean-free path length do not necessarily have to be the same, but both should exhibit sufficiently strong multiple scattering to ensure that the diffusion approximation is met. Common reference samples include probe particles dispersed in highly-viscous glycerine solutions (Scheffold *et al.*, 2001; Wyss *et al.*, 2001) and foams (Durian *et al.*, 1991; Durian, 1997).[8]

The field-correlation function for two-cell scattering, eqn 5.99, can be evaluated by recognizing that there is a low probability that light will loop back through the first cell. Thus, the joint probability of the path lengths can be factored into their separate probability distributions, $\mathcal{P}(s_n; s_e) = \mathcal{P}(s_n)\mathcal{P}(s_e)$, resulting in a total intensity correlation function that is a product of the intensity-correlation functions of the two samples (Scheffold *et al.*, 2001),

$$g_{(2)}^m(t) = g_{(2)}^n(t) \cdot g_{(2)}^e(t). \tag{5.100}$$

Because the *composite* dynamics are ergodic, the Siegert relation can be used to convert between the field and intensity-correlation functions,

$$g_{(2)}^m(t) = 1 + \beta |g_{(1)}^m(t)|^2 \tag{5.101}$$

The Siegert relation implies that the multiplication rule, eqn 5.100, holds for the intensity correlation as well. One then simply calculates the intensity correlation function of the nonergodic sample of interest by dividing the composite-correlation function by the intensity-correlation function of the reference sample,

$$g_{(2)}^n(t) - 1 = \frac{g_{(2)}^m(t) - 1}{g_{(2)}^e(t) - 1}. \tag{5.102}$$

[8] Shaving cream is a good ergodic material with slow-relaxation dynamics.

The multiplicative property of the two-cell correlation functions that leads to eqn 5.102 is demonstrated by the data in Fig. 5.36.

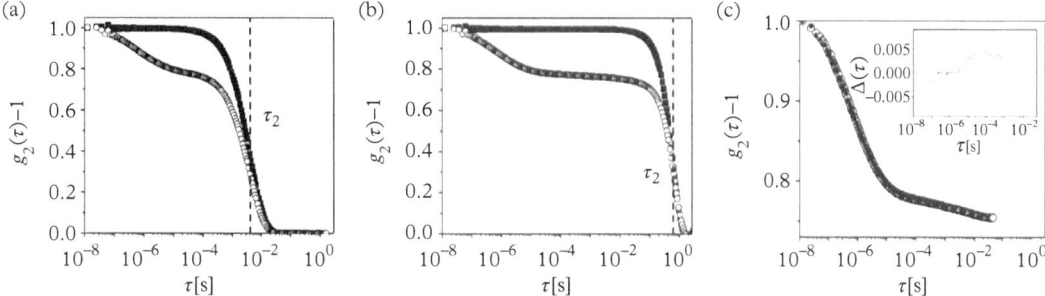

Fig. 5.36 *Optical mixing DWS. Open symbols are the composite correlation from a nonergodic sample and er-godic reference. Closed symbols show the reference in isolation. (a) and (b) show results using two characteristic relaxation time scales of the reference τ_2 which is adjusted by changing the concentration of glycerine. (c) Correla-tion functions of the nonergodic sample using the multiplication rule eqn 5.102. Reprinted figure with permission from Scheffold, F., Skipetrov, S. E., Romer, S., & Schurtenberger, P., Phys. Rev. E, 63, 61404 (2001). Copyright (2001) by the American Physical Society.*

Shown are the time–correlation functions of the two-cell experiment and the independently measured correlation function for the ergodic reference, particles dispersed in glycerine solutions. The character-istic decay time of the ergodic medium is given by $\tau_2 = \tau_0 (l_e^* / L_e)^2$ where $\tau_0 = \pi a \eta / k_B T k_0^2$ is the time for a particle in the reference fluid to diffuse k_0^{-1}. The multiplication rule is used to calculate the time–correlation function of the nonergodic sample. The results are identical for both ergodic reference samples. The optical mixing method gives accurate correlation functions of the nonergodic me-dium to lag times less than the characteristic relaxation time of the reference, $t < \tau_2$.

Optical mixing is easy to implement without having to modify the DWS hardware, however it may suffer from the low-intensity of transmitted light through the thick composite sample. Scheffold *et al.* (2001) provide useful suggestions for maximizing the transmit-tance and reducing the effects of reflections and light leakage between the samples. For instance, using scatterers in the reference sample with a particle size larger than the laser wavelength ensures that the photon mean-free path of the reference l_e^* is much larger than the scattering mean-free path, and hence, that the amount of light pass-ing through the reference without being scattered at all is significantly reduced for the same optical density L_e/l_e^*. The relaxation time of the reference must also be selected such that the dynamics of in-terest can be captured from the nonergodic sample. Probe particles dispersed in glycerol and foams provide relaxation times on the order of 10^{-3}–10^{-2} s (Durian *et al.*, 1991; Durian, 1997).

5.7.5 Multispeckle detection

A versatile method of producing an ergodic intensity-correlation function is to sample many speckles by mechanically rotating or translating a glass diffuser (Churnside, 1982; Viasnoff *et al.*, 2003). The analysis of mechanical dephasing is identical to the optical mixing experiment described in the previous section, but there is one key advantage of mechanical dephasing: The time of the correlation can be controlled precisely by changing the translation or rotation rate of the glass diffuser. An alternate method of mechanical dephasing is to physically rotate or translate the sample (Nisato *et al.*, 2000; Pham *et al.*, 2004).

A diagram of the rotating disk experiment is shown in Fig. 5.37. In this example, an expanded, collimated laser beam is incident on the disk, which is driven slowly on a shaft by a stepper motor. The motor is geared down to provide steady, slow movement. Here, disk rotation rates on the order of 10^{-6} rad/s (or 10 mrad/hr) give decorrelation times on the order of 1–10 s. A concave lens images the disk onto the sample. The aperture before the disk and the magnification of the lens (typically close to $M_T \approx 1$) are selected to maximize the illumination of the sample. As Fig. 5.37 illustrates, a second advantage of mechanical dephasing is the ability to measure dynamics in both transmission and backscattering, thus providing greater control over the path-length distribution and the ability to tailor the time- and length-scales probed by the microrheology experiment.

The dynamics of the nonergodic medium are calculated using eqn 5.102, substituting the correlation function of the ergodic sample $g_{(2)}^e(t)$ with that of the disk, $g_{(2)}^d(t)$, the measured correlation function is

$$(g_m^{(2)}(t) - 1) = (g_{(2)}^d(t) - 1)(g_{(2)}^n(t) - 1). \qquad (5.103)$$

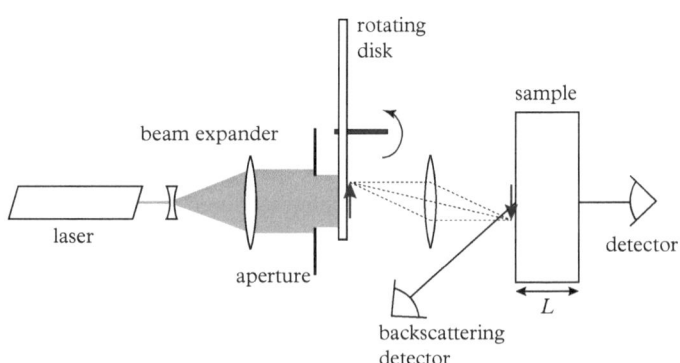

Fig. 5.37 *Mechanical dephasing using a rotating glass disk. A lens images the disk onto the face of the sample.*

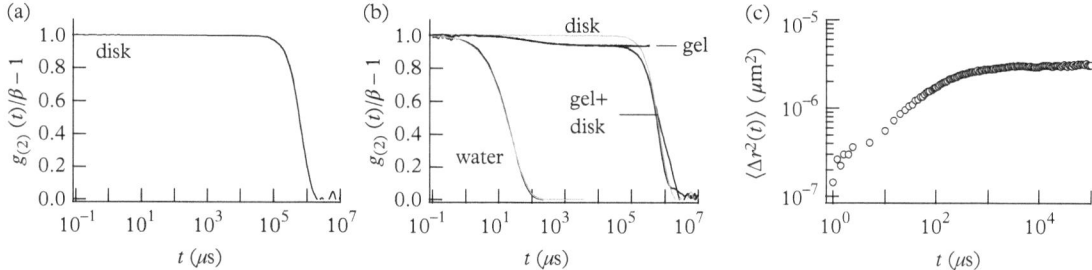

Fig. 5.38 *Correlation functions for DWS with mechanical dephasing. (a) The normalized disk correlation function. (b) Data collected from a gel sample, and for comparison, scattering from an ergodic sample of the same probes dispersed in water. (c) The MSD of the 1 μm diameter probe-particles in the gel. The elastic modulus is approximately 630 Pa.*

Like optical mixing, mechanical dephasing requires a separate measurement of the disk's correlation function. Fig. 5.38 shows an example rotating disk correlation and a measurement of a nonergodic sample. The sample is a peptide hydrogel (Schneider *et al.*, 2002; Veerman *et al.*, 2006). Polystyrene latex particles, 1 μm in diameter, are used as probes. Overlaid on the composite correlation function of the gel and disk are the separate disk correlation function and the gel correlation function calculated by eqn 5.102. The gel correlation function plateaus at a value $g_{(2)}(t) - 1 = 0.94$. The mean-squared displacement at the plateau is $\langle \Delta r^2 \rangle \approx 2 \times 10^{-18}$ m^2, which corresponds to an elastic modulus $G_0 \approx 630$ Pa. The data demonstrates the sensitivity of DWS—at the plateau, the probes have moved only about 1.4 nm, a small fraction—approximately one-thousandth—of their diameter.

Echo techniques

Closely related to mechanical dephasing are echo techniques in light scattering (Nisato *et al.*, 2000; Pham *et al.*, 2004). The DWS method of Zakharov *et al.* (2006) uses a rotating glass disk geometry similar to Fig. 5.37. The disk spins at a fast rate and the resulting correlation is a periodic function, shown in Fig. 5.39a. The initial correlation decays to zero as the disk rotates, and ultimately produces an "echo" peak after the disk completes one full rotation after a period of approximately 24 ms. Identical echoes are measured at integer periods for of the disk rotation. The peak widths are narrow, approximately 1 μs, but with a sample in place, slow relaxation of the material can be monitored by the decreasing amplitude of subsequent echoes, as illustrated in Fig. 5.39b. The advantage of echo DWS is similar to multispeckle detection that is discussed in the next section; namely, that correlation functions can be measured for slow-relaxation dynamics while

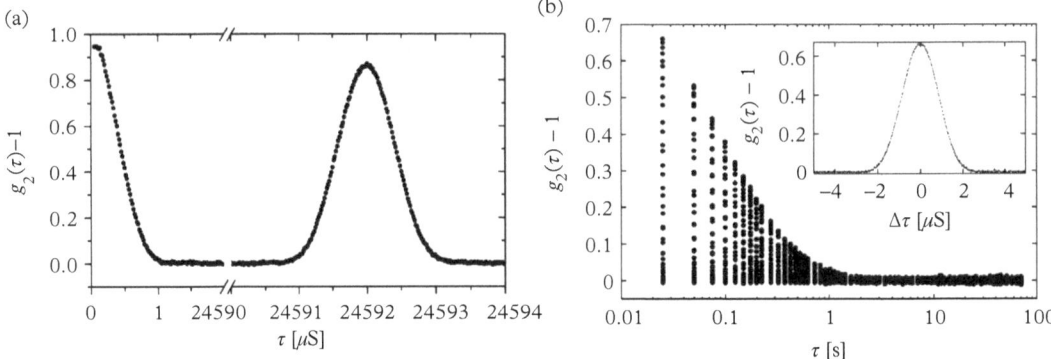

Fig. 5.39 *Echo DWS is performed by rapidly rotating a dephasing glass disk. (a) The rapid rotation produces "echoes" in the correlation function. (b) For a slowly-evolving sample these peaks decay and provide a measure of the intensity correlation function. Reprinted figure with permission from Zakharov, P., Cardinaux, F., & Scheffold, F., Phys. Rev. E, 73, 11413 (2006). Copyright (2006) by the American Physical Society.*

avoiding the long integration times necessary to calculate a time average correlation function. The data shown in Fig. 5.39b are obtained in just under two minutes. Such specialized detection methods are implemented in some commercial instruments.

5.7.6   Multispeckle imaging

Light scattering by multispeckle imaging replaces the PMT or APD detectors with a CCD camera (Wong and Wiltzius, 1993; Cipelletti and Weitz, 1999). Camera-based methods enable *ensemble*-average correlation functions to be calculated without the use of an ergodic reference or mechanical dephasing by simultaneously sampling a large number of speckles (Viasnoff *et al.*, 2002). The data acquisition in the form of image frames, each representing a time slice, is transferred to a computer. The time-intensity correlation function is calculated by the cross-correlation of speckles from different frames. The basic experimental setup is shown in Fig. 5.40.

An important advantage of multispeckle imaging is that it provides a true ensemble average and therefore does not require time averaging. Long time-scales of probe motion can be measured; for instance, the slow relaxation of a material that is normally beyond the range of time-averaging techniques with a correlator. Multispeckle imaging provides an instantaneous snapshot of the system dynamics, which moreover, do not have to be stationary.

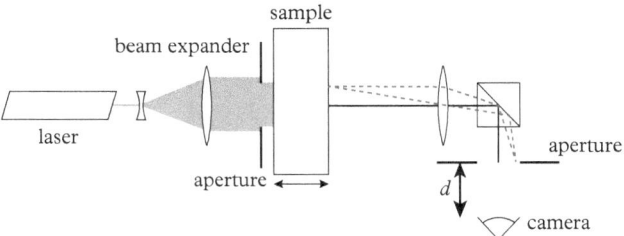

One of the only limitations of camera-based light scattering methods is the constraint that the dynamics that are measured are of course limited to time-scales greater than the time between image frames, and hence the shortest time-scales probed are set by the frame rate of the camera. Similar trade-offs between the signal-to-noise, camera integration times and frame rates that were discussed in Chapter 4 apply to multispeckle imaging. Because camera frame rates are slow compared to photon counting devices, the short-time scale motion that is a hallmark of the DWS measurements discussed until now are inaccessible to multispeckle imaging. However, as we will discuss in Section 5.8, multispeckle imaging can be used simultaneously with the other DWS techniques to extend the range of relaxation times probed in a material.

The basic scheme of multispeckle imaging is to capture many speckles scattered from a sample onto a CCD camera array. As shown in Fig. 5.40 a lens images the sample onto an aperture, which is placed a distance d in front of the CCD camera. The aperture diameter and distance d between it and the camera determine the number of speckles that are imaged and the speckle size in pixels. The best compromise between loss of statistics and loss of contrast is obtained for a speckle diameter of approximately 3 pixels, which was found to produce the best signal-to-noise in terms of the correlation intercept and spread (Fig 5.41).

The ensemble-averaged intensity-correlation function $g^e_{(2)}(t_0; t)$ is calculated by a pixel-to-pixel cross correlation between a reference image taken at t_0 and a subsequent image at time $t + t_0$ (Harden and Viasnoff, 2001),

$$g^e_{(2)}(t_0; t) = \frac{\langle I_i(t + t_0) I_i(t_0) \rangle_i}{\langle I_i(t_0) \rangle_i \langle I_i(t_0) \rangle_i}. \tag{5.104}$$

where $\langle \ldots \rangle_i$ indicates an average taken over all pixels. Thus, the correlation function is calculated by multiplying one frame by another, pixel-by-pixel, and dividing by the number of pixels. About $\sim 10^4$ speckles are obtained per image over $10^5 - 10^7$ frames. An example

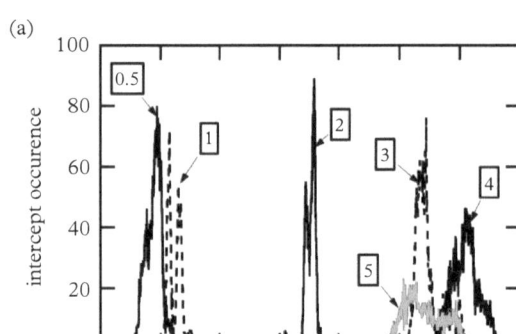
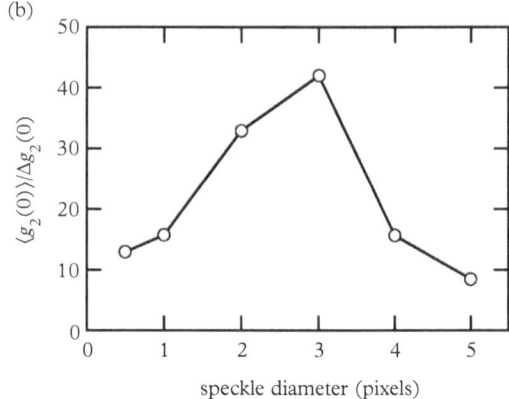

Fig. 5.41 *(a) Histogram of intensity correlation intercept values as the speckle size in pixels changes (boxed numbers). (b) Speckle diameters of about 3 pixels were found to produce the highest intercept with the narrowest spread. Reprinted from Viasnoff, V., Lequeux, F., & Pine, D. J., Rev. Sci. Inst. 73, 2336–44 (2002), with the permission of AIP Publishing.*

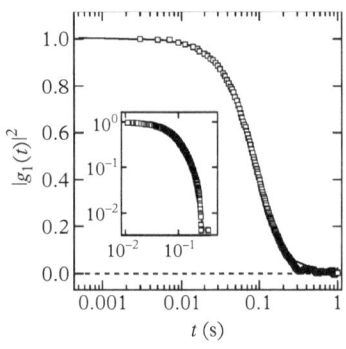

Fig. 5.42 *Probe particles in glycerol measured by a standard correlator (line) and multispeckle imaging (square symbols). Reprinted from Viasnoff, V., Lequeux, F., & Pine, D. J., Rev. Sci. Inst. 73, 2336–44 2002, with the permission of AIP Publishing.*

of multispeckle imaging data is shown in Fig. 5.42 for 0.5 μm diameter polystyrene particles dispersed at 1% in a glycerol solution. Viasnoff *et al.* (2002) report good agreement between between correlations from a standard correlator and multispeckle imaging data. Imaging requires only several seconds of sampling.

Multispeckle imaging is not limited to stationary dynamics, whereas the time averaging required by a correlator limits the sample to remain quasi-stationary and not change significantly on time scales shorter than the correlation acquisition time. In principle, each frame during imaging can be used as the reference time for subsequent correlation functions. Measuring $g_{(2)}(t_0; t)$ as a function of reference-time t_0 allows the dynamics of time-evolving materials to be followed.

A significant constraint for calculating simultaneous correlation functions from sequential frames is the computational time required by the pixel-to-pixel cross-correlation. To address this, Viasnoff *et al.* (2002) introduced an interlacing scheme in which successive frames are included in the calculation of separate correlation functions.

The method, illustrated schematically in Fig. 5.43, takes advantage of a quadratic spacing of subsequent correlation time points, or "channels," in the nomenclature of correlators, in order to interleave multiple-correlation functions. In Fig. 5.43, the vertical axis represents the jth of j_{max} correlations to be measured in parallel. Each jth correlation will begin at a different initial time t_0. The quadratic spacing actually begins after an initial linear spacing between frames (the first five correlation time points) in this implementation.

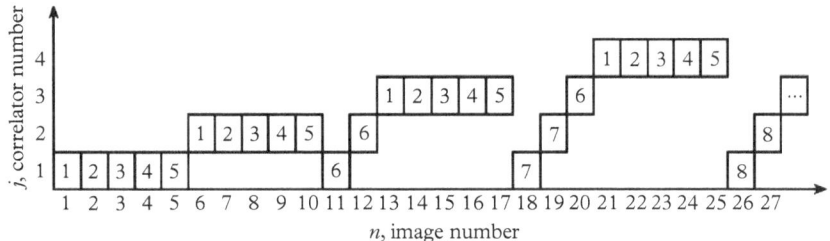

Fig. 5.43 *Calculation of the multispeckle correlation function (Viasnoff et al., 2002). In this example, N = 5. Reprinted from Viasnoff, V., Lequeux, F., & Pine, D. J., Rev. Sci. Inst. 73, 2336–44 (2002), with the permission of AIP Publishing.*

Let n_{ij} represent the ith frame of the jth correlation function. The reference frame ($i = 1$) of the jth correlation function is frame n of the consecutive image sequence, given by

$$n = n_{1j} = 1 + (N + 1)(j - 1),\qquad (5.105)$$

where N is typically in the order of 10 frames, and defines the initial period of linear sampling. The subsequent frames used to calculate the jth correlation function are written

$$n_{ij} = \begin{cases} n_{1j} + i - 1 = (N + 1)(j - 1) + i & \text{for} \quad 1 \leq i \leq N \\ (j - 1) + \frac{1}{2}i(i - 1) - \frac{1}{2}N(N - 1) & \text{for} \quad i > N \end{cases}.\qquad (5.106)$$

When multiple correlation functions are obtained with multispeckle imaging, they can be used to make microrheological measurements of time-dependent nonstationary processes, such as gelation,

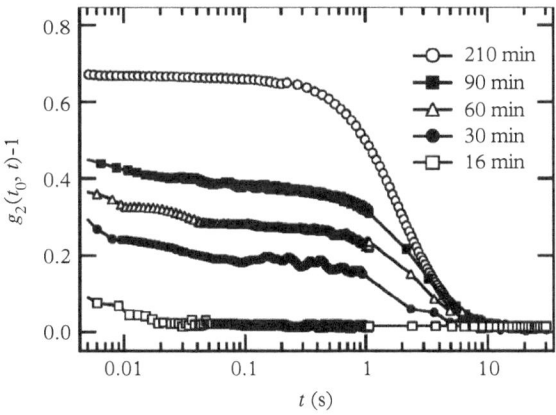

Fig. 5.44 *Multispeckle imaging of probe particles dispersed in gelatin as the fluid gels. Reprinted from Viasnoff, V., Lequeux, F., & Pine, D. J., Rev. Sci. Inst. 73, 2336–44 (2002), with the permission of AIP Publishing.*

degradation, or aging. The *time* interval between the j and $j + 1$ correlation functions in this method is approximately N/f for the frame rate f. This interval is tuned to accommodate the evolution time scale of the material. For example, Fig. 5.44 shows the gelation of gelatin as reported by the normalized intensity correlation function. At 16 minutes after initiating the gelation by cooling the gelatin sample, there is little correlation reported. As a fluid, the relatively fast material relaxation time allows sufficient probe motion to decorrelate the intensity. With increasing time, however, the probe motion slows.

5.8 Broadband microrheology

It is possible to extend the range of length and time-scales probed in a single microrheology experiment by combining DWS transmission and multispeckle or echo backscattering (Viasnoff *et al.*, 2002; Cardinaux *et al.*, Cardinaux *et al.*). Eqns 5.73 and 5.74 showed that the ratio of the characteristic length scales probed by these techniques is $l_B^2/l_T^2 \sim (9/25)(L/l^*)^2$. This separation makes what has been termed *broadband* microrheology possible.

An example experimental setup is shown in Fig. 5.45. Backscattered light incident on the sample is imaged using a CCD camera, while transmitted light is detected using a single-mode fiber optic. Here, the samples are ergodic in the transmission DWS geometry, but nonergodic materials can be used by incorporating a tandem reference cell for optical mixing (Viasnoff *et al.*, 2002) or a rotating disk

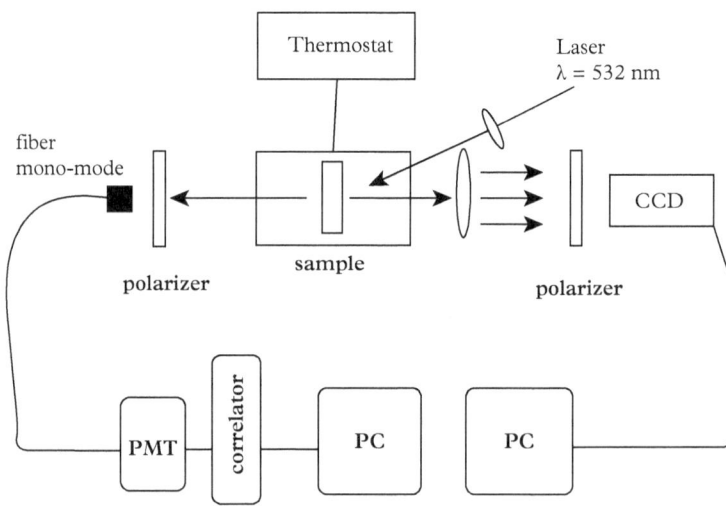

Fig. 5.45 *A combination of DWS transmission and multispeckle backscattering from Cardinaux* et al. *(2002). Reprinted with permission from* Europhys. Lett., *2002, Number 5, March, http://iopscience.iop. org/journal/0295-5075.*

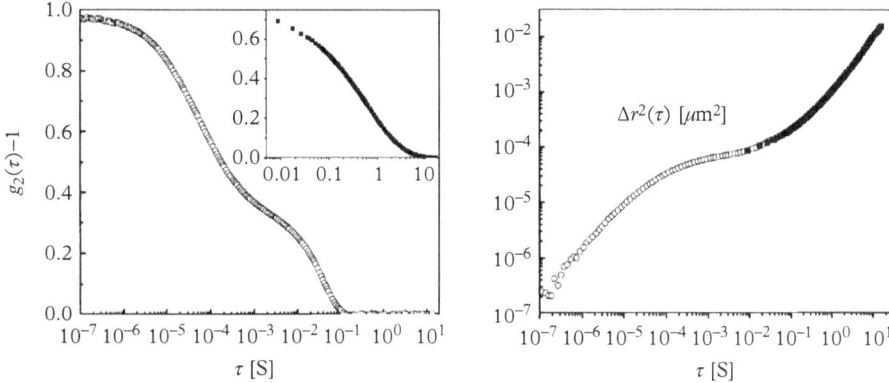

Fig. 5.46 *Worm-like micellar solution (100 mg/ml) measured using a combination of transmission DWS using an autocorrelator and camera-based multispeckle detection (Cardinaux et al., 2002). Reprinted with permission from* Europhys. Lett., *2002, 5, March, http://iopscience.iop.org/journal/0295-5075.*

for mechanical dephasing. These are placed between the sample and transmission detector to preserve the ability of the CCD camera to detect long-time correlations independent of the reference decay time. The correlation functions measured by the two techniques must be properly normalized by their respective dynamical contrast values, β.

The measured intensity correlation functions and corresponding mean-squared displacements using a broadband experiment are shown in Fig. 5.46 for probe particles dispersed in a surfactant worm-like micellar solution (Cardinaux *et al.*, 2002). Multispeckle detection in backscattering allows for the long-time relaxation behavior to be probed without extending the total sampling time, resulting in an impressive range of probe displacement—over five decades, from sub-nanometer to hundreds of nanometers. The transmission DWS data provides rheological data on time scales shorter than a microsecond, or frequencies on the order of several megahertz. Thus, the internal dynamics of the worm-like micelles captured at short time, as well as their long-time relaxation, can be simultaneously probed in a single sample.

5.9   Other DWS applications

Dynamic light scattering, in all of its forms, is frequently used as a tool to study and understand the microstructure dynamics that underlie the rheology of complex fluids. While such experiments are not strictly tracer probe microrheology, they are worth considering.

Of course, in the now long history of dynamic light scattering, measurements of colloid and polymer dynamics have provided numerous insights into their relaxation behavior and structure.

DWS has been especially useful for measuring the dynamics of sol-gel transitions, which are important both fundamentally and in numerous applications, from ceramics to foods (Scheffold and Schurtenberger, 2003; Mezzenga *et al.*, 2005). Examples include suspensions found in paints and coatings (Romer *et al.* 2000; Wyss *et al.* 2001) and the fluid-solid transitions of yoghurt- and cheese-making (Stradner *et al.*, 2001; Vasbinder *et al.*, 2001; Vasbinder and De Kruif, 2003; Alexander and Dalgleish, 2004; Alexander and Dalgleish, 2007).

The normalized intensity correlations from DWS experiments of alumina nanoparticles destabilized by a change in pH or ionic strength shown in Fig. 5.47 are a good illustration of the type of data collected and the nature of the sol-gel transition. Initially, the correlation function holds its shape, but shifts to a longer time. This initial change corresponds to growing aggregates, which diffuse more slowly. As the aggregates grow and join as a network, the dynamics change; particles are localized and the correlations, including the emergence of nonergodicity, are similar to the results one would expect for scattering probes dispersed in the material. A mean-squared displacement can even be calculated, but this should not be interpreted by the GSER. Instead, it provides a measure of the network

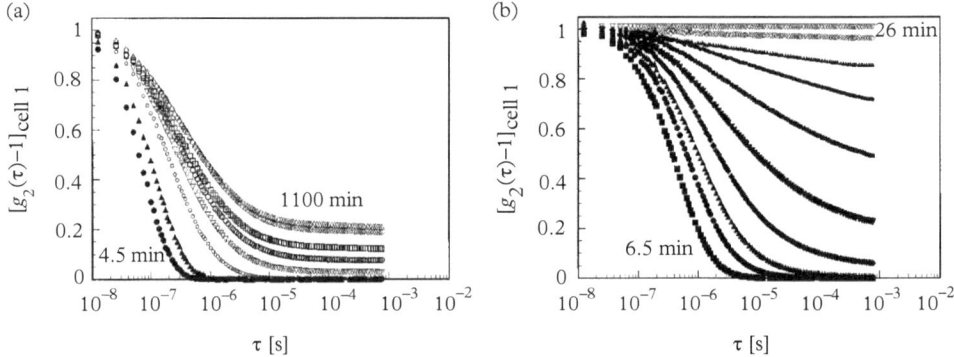

Fig. 5.47 *Intensity autocorrelation for alumina nanoparticle suspensions destablized by (a) pH and (b) ionic strength. A two-cell optical-mixing configuration is used on account of the transition to nonergodicity that accompanies gelation. Reprinted from J. Colloid Interface Sci., 240, Wyss, H. M., Romer, S., Scheffold, F., Schurtenberger, P., & Gauckler, L. J., Diffusing-wave spectroscopy of concentrated alumina suspensions during gelation, 89–97, Copyright (2001), with permission from Elsevier.*

elasticity after the sol-gel transition is traversed, as demonstrated in related work by Krall and Weitz (1998), who used DLS to characterize the dynamics of fractal colloidal gels from polystyrene nanoparticles in salt solutions. In the latter study, a model of the network dynamics provides a measure of its stiffness, reminiscent of the application of high-frequency microrheology to characterize semiflexible polymers that we discussed in Section 5.6.1.

Other DWS studies have examined the high-frequency diffusivity of colloidal suspensions (Qiu *et al.*, 1990; Xue *et al.*, 1992*b*), rearrangement dynamics in foams (Durian *et al.*, 1991; Durian, 1997), and the colloidal glass transition (Rojas-Ochoa *et al.*, 2002).

5.10 Summary

In this chapter we discussed passive tracer particle microrheology using dynamic light scattering. We saw that there are several key benefits and a few drawbacks of light scattering microrheology. Dynamic light scattering has good potential for microrheology experiments. DLS instruments are common in soft materials laboratories and well-supported by numerous companies. A chief drawback is the necessity to separate the scattering signal derived from the material and probes. The high multiple scattering limit of DWS has an advantage here, and enables rheology to be measured on short time scales (high frequencies) beyond the range of conventional mechanical rheometers and particle tracking microrheology. In general, however, sample volumes are larger than particle tracking and the data acquisition time is longer, in the order of minutes to hours, limiting the rate at which rheological changes can be measured. The long-integration time it takes to calculate the time-averaged time–correlation function limits the microrheology measurement to equilibrated or very slowly evolving systems that act as stationary processes. However, multi-speckle scattering can be used to characterize more rapidly evolving materials. Finally, it is straightforward to incorporate nonstandard sample environments, like high-pressure cells, into a scattering experiment.

Laser light scattering has been the predominant technique used for scattering-based tracer particle microrheology, but ongoing work is aimed at extending these methods to other radiation wavelengths. Recently, x-ray photocorrelation spectroscopy (XPCS) using high-flux sources, such as third-generation synchrotrons, has been adapted for microrheology (Leheny, 2012). The key advantage is the potential to access tracer particle dynamics over small length scales and long time scales that are particularly suited to studying highly viscous or stiff materials. Leheny (2012) estimates that XPCS using nanoparticle probes should be able to access moduli exceeding 10^6 Pa!

. . . .

EXERCISES

(5.1) **Field-correlation function.** The field-correlation function $g_{(1)}(t)$ is the Fourier Transform of the Van Hove self-space time–correlation function described in Chapter 4. Show that taking the Fourier Transform of $G_s(\mathbf{R}, t)$ yields eqn 5.29 as expected.

(5.2) Show that eqn 5.24 follows from the Fourier Transform of eqn 5.28. This function is called the *self-intermediate scattering function*.

(5.3) **Dynamic light scattering microrheology.** Use eqn 5.37 to calculate the dynamic light scattering intensity correlation function from 10 nm radius proteins in the presence of probe particles. Assume that the protein is roughly spherical and has a volume fraction of about $\phi = 0.05$. Vary the probe particle concentration and examine its effect on the calculated correlation. The laser is a doubled YAG.

(5.4) **Nonergodicity.** Why is nonergodicity an issue in light scattering microrheology, but rarely affects particle tracking microrheology? Under what conditions would particle tracking microrheology have to take into account nonergodicity?

(5.5) **Operating limits.** What is the highest viscosity of a Newtonian fluid that could be practically measured given a probe diameter of 1μm using DLS? What about for DWS in a backscattering or transmission geometry (assume a sample thickness of 4 mm and a probe volume fraction $\phi = 0.01$).

(5.6) **Finding the intercept and baseline values of a correlation function.** Calculate the field correlation function for 1.0 μm particles diffusing in water at $25°$C, and the corresponding intensity correlation function using the Siegert relation, assuming $\beta = 0.8$.

(5.7) **Photon mean-free path length.** Plot the field and intensity correlation functions for DWS transmission of 1 μm diameter polystyrene probe particles for a few volume fractions over the range $\phi = 0.001 - 0.05$. What effect does probe concentration have on the scattering mean-free path l^*? How does a larger or smaller value of l^* affect the length and time scales probed by the DWS experiment?

Interferometric tracking

6

The purpose of this chapter is to present a survey of passive microrheology techniques that are important complements to more widely used particle tracking and light scattering methods. Such methods include back-focal-plane interferometry and extensions of particle tracking to measure the *rotation* of colloidal particles.

6.1 Back-focal-plane interferometry

Back-focal-plane interferometry and related methods collect laser light scattered from a single particle in a microscope (Gittes *et al.*, 1997; Gittes and MacKintosh, 1998; Schnurr *et al.*, 1997). The origin of the method comes from its use as a non-imaging displacement measurement, which is often employed with optical traps (Denk and Webb, 1990). Here, we will focus on the non-trapping version of interferometry. Later, in Chapter 9, we discuss its use with optical tweezers and active microrheology.

The principle advantages of back-focal-plane interferometry are its fine spatial resolution and high bandwidth. Experiments are typically capable of measuring particle displacement on the order of nanometers at frequencies as high as 100 kHz. This puts back-focal-plane interferometry on par with the sensitivity and bandwidth of light scattering methods discussed in Chapter 5, while retaining some of the advantages that are typical for multiple particle tracking microrheology: Small-sample volumes and the ability to make two-point measurements.

6.1.1 Back-focal-plane experiment

A schematic of the back-focal-plane interferometry experiment is shown in Fig. 6.1. Laser light is focused on the specimen in a microscope. The scattered and unscattered light is collected and collimated by the microscope condenser or a second long working distance microscope objective. Light from the back-focal-plane of the condenser is then imaged onto a quadrant photodiode. By imaging the back-focal-plane, the resulting heterodyne scattering pattern, a mix

Microrheology. Eric M. Furst and Todd M. Squires, Oxford University Press (2017).
© Eric M. Furst and Todd M. Squires. DOI 10.1093/oso/9780199655205.001.0001

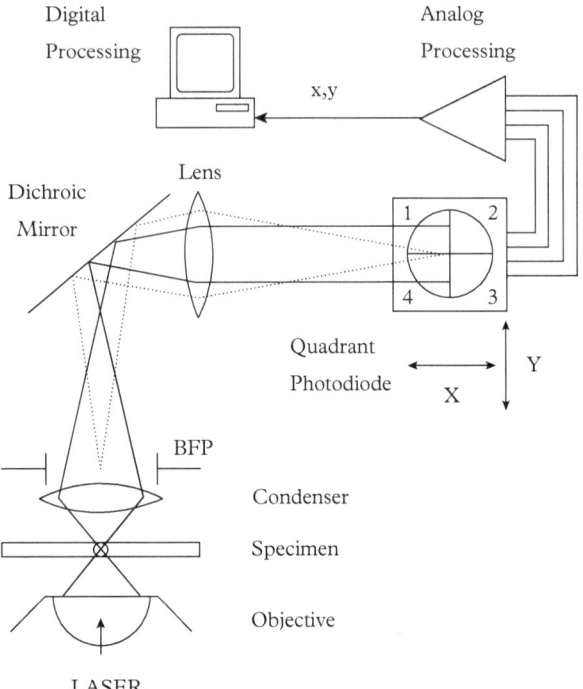

Digital
Processing

Analog
Processing

x,y

Lens

Dichroic
Mirror

1   2

4   3

Quadrant
Photodiode

Y

X

BFP

Condenser

Specimen

Objective

LASER

Fig. 6.1 *Schematic of the back-focal-plane detector. Scattered and unscattered light from the specimen plane at the back-focal-plane of the condenser is imaged onto a quadrant photodiode. Analog electronics convert the photocurrent to voltages and the signal differences between each half of the quadrant. Adapted from Gittes and Schmidt (1998).*

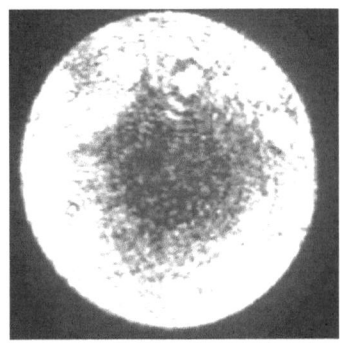

Fig. 6.2 *The back-focal-plane intensity pattern imaged by a ccd camera. Reprinted by permission from Macmillan Publishers Ltd:* Nature Protocols *Lee et al. (2007), copyright (2007).*

of scattered and unscattered light, is independent of the position of the particle in the sample image plane. The laser can be steered in the focal plane to scatter off any particle, rather than positioning the particle to align with a fixed beam (this is particularly useful for optical trapping to allow trapping over the entire imaging plane).

When the particle is centered in the beam, the resulting interference pattern at the back-focal-plane is symmetric. This pattern is shown in Fig. 6.2. Displacement of the particle transverse to the beam causes the scattering pattern to shift, as shown in Fig. 6.3. The asymmetric intensity of the resulting scattering pattern is detected by the quadrant photodiode by taking the difference between signals from each half of the quadrant. Displacement in the Y direction in Fig. 6.1 is detected by summing the voltage output from quadrants 1 and 2 and taking the difference with the sum of quadrants 3 and 4,

$$\Delta Y = \frac{(I_1 + I_2) - (I_3 + I_4)}{I_1 + I_2 + I_3 + I_4}. \tag{6.1}$$

Likewise, displacement in the X direction is given by

$$\Delta X = \frac{(I_2 + I_3) - (I_1 + I_4)}{I_1 + I_2 + I_3 + I_4}. \tag{6.2}$$

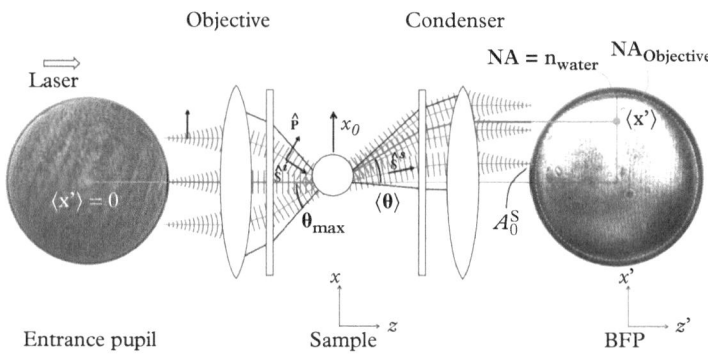

Fig. 6.3 *Adapted from Farré, Marsà and Montes-Usategui 2012.*

Both differences are normalized by the total intensity at the detector and are typically calculated by analog electronics of the quadrant detector circuit (Simmons *et al.*, 1996). Figure 6.5 shows an example of the quadrant detector response. Here, a 1.6 μm silica bead attached to a microscope slide is moved through the beam using a piezoelectric microscope stage. There is a range of displacements over which the response is linear, but then falls off. The detector response in each direction is in reasonable agreement with a simple interference model based on a Rayleigh approximation for the scattered field Gittes and Schmidt (1998),

$$\frac{I_+ - I_-}{I_+ + I_-} \approx \frac{32\sqrt{\pi}\alpha}{\lambda w_0^2} e^{-2(x/w_0)^2} \int_0^{x/w_0} e^{y^2} \, dy \qquad (6.3)$$

where w_0 is the $1/e$ radius of the beam's waist and x is the displacement of the bead transverse to the beam axis.

6.1.2 Detector sensitivity and limits

Here, we consider both the temporal and spatial limits of back-focal-plane detection.

Frequency bandwidth

Quadrant photodiodes achieve a high temporal bandwidth due to the strong incident intensity of scattered light. One limit on the detector bandwidth occurs due to the light absorption of silicon detectors above $\sim$850 nm. In a typical silicon *n*-type PN photodiode, photons are absorbed in the depletion region of the diode junction. The electron-hole pairs separate rapidly to the anode and cathode in the electric field of the depleted layer. Lasers that emit in the near-infrared wavelengths, often used for optical trapping and back-focal-plane detection (Neuman and Block, 2004), exhibit far lower absorption in

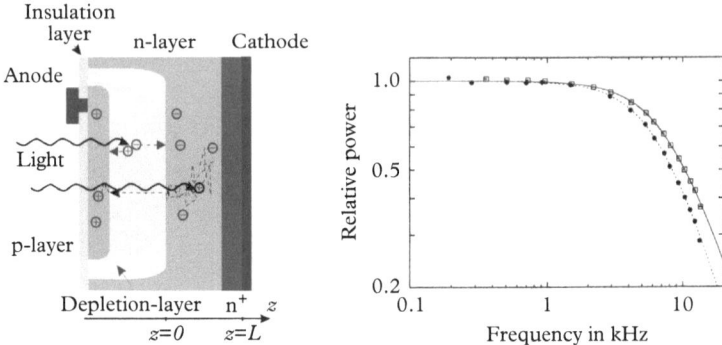

Fig. 6.4 *A typical silicon photodiode will absorb near-infrared light in the n-type region beyond the depletion-layer. The increased time for charge to reach the junction leads to a filtering effect that limits the bandwidth of a quadrant detector. Adapted from Berg-Sørensen et al. (2003).*

silicon detectors, which leads to a higher relative amount of absorption in the *n*-layer beyond the depletion region (Berg-Sørensen *et al.*, 2003; Peterman *et al.*, 2003).

The diffusion of the valence holes in the *n*-layer to the depletion region is slow and effectively becomes a low-pass filter at about 10 kHz at 1064 nm (Berg-Sørensen *et al.*, 2003). Figure 6.4 shows the geometry and spectral response of a typical silicon detector. Despite the limitations of silicon PN photodiodes, several remedies exist that can increase the bandwidth of a quadrant detection scheme to ∼100 kHz. For example, InGaAs PIN photodiodes or special purpose fully-depleted *p*-type silicon photodiodes are now available and do not suffer from the carrier diffusion effect (Peterman *et al.*, 2003). Detectors also frequently forward bias the photodiode to improve the response, at the cost of higher noise.

Spatial resolution

The useful spatial resolution of back-focal-plane detection is bounded in the lower limit by detector noise, and in the upper limit by the response function non-monotonicity (and non-linearity), as illustrated in Fig. 6.5. Shindel *et al.* (2013) find that their implementation of back-focal-plane tracking detects the motion with a noise floor of ∼2 mV. Given a typical QPD-sensitivity of 100 nm/V in the study, the noise floor means instantaneous displacements of the probe smaller than 2Å cannot be measured with statistical significance. The maximum extent of the linear response is reported to be ∼300 nm.

As an alternative to the quadrant photodiode, some investigators have found that high-speed CMOS cameras provide sufficient accuracy and bandwidth for particle measurements (Gibson *et al.*, 2008). The temporal resolution of camera-based tracking can be improved using high-intensity light sources, such as a dephased laser, to increase

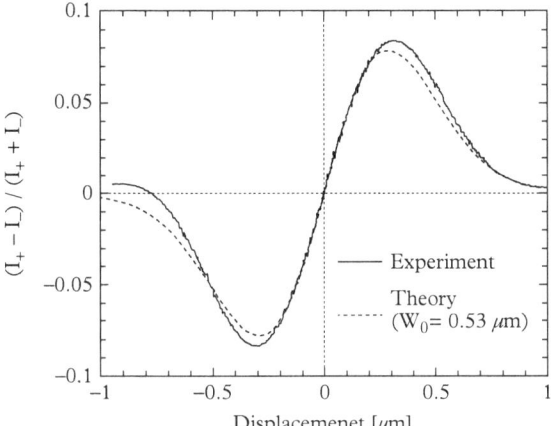

Fig. 6.5 *Quadrant photodiode-detector response for a 1.6 μm diameter silica particle moving through a focused 1064 nm beam (solid line). The dashed line is the theoretical signal from the interference pattern based on a Rayleigh approximation. Adapted from Gittes and Schmidt (1998).*

photon counts over short integration times of the device (Biancaniello and Crocker, 2006).

6.1.3 Linear response

Back-focal-plane interferometry is interpreted using a different form of the GSER that was discussed in Chapter 3. While our previous derivation connected the particle *displacement* to the complex modulus, interferometry's output is the *position* of the probe as a time-varying voltage signal from the quadrant photodiode electronics. Such measurements can be analyzed in the frequency domain, and back-focal-plane methods have generally been interpreted in terms of the position correlation, and in particular, the power-spectral density of position $\langle |\tilde{\mathbf{x}}(\omega)|^2 \rangle$.

The particle displacement in response to a time-dependent force $\mathbf{f}(t)$ in the absence of inertia is

$$\mathbf{x}(t) = \int_0^t \chi(t - t')\mathbf{f}(t')dt' \tag{6.4}$$

where χ is the particle displacement "susceptibility" to the force. A Fourier Transform gives the simple equation of motion in the frequency domain

$$\tilde{\mathbf{x}}(\omega) = \tilde{\chi}(\omega)\tilde{\mathbf{f}}(\omega). \tag{6.5}$$

and the susceptibility is a complex function, $\tilde{\chi}(\omega) = \tilde{\chi}'(\omega) + i\tilde{\chi}(\omega)$ (Chaikin and Lubensky, 2000). Similar to the resistance $\zeta(\omega)$, the susceptibility is related to the macroscopic rheology (in the linear limit) by

$$\tilde{\chi}(\omega) = \frac{1}{6\pi a G^*(\omega)}. \tag{6.6}$$

This expression of the GSER is valid when all of the relevant assumptions of microrheology are met, including continuum mechanics and incompressibility, discussed in Chapter 2. The susceptibility and memory function $\tilde{\zeta}(\omega)$ introduced in Section 3.1 are related by

$$\tilde{\chi}(\omega) = 1/i\omega\tilde{\zeta}(\omega), \tag{6.7}$$

which is shown by noting that $\tilde{v}(\omega) = i\omega\tilde{x}(\omega)$ and, by taking the Fourier Transform of eqn 3.1 neglecting inertia, $\tilde{f}(\omega) = \tilde{\zeta}(\omega)\tilde{v}(\omega)$. Eqn 6.7 is analogous to the relationship between the complex creep compliance and shear modulus introduced in eqn 1.38. Whereas the mean-squared displacement may be related to the compliance, the position fluctuations are a direct measure of the complex shear modulus.

The fluctuation-dissipation theorem ties the relevant experimental measurement of probe position to the GSER. It can be stated

$$\tilde{\chi}''(\omega) = \frac{\omega}{2k_B T}\langle|\tilde{x}(\omega)|^2\rangle \tag{6.8}$$

where $\langle|\tilde{x}(\omega)|^2\rangle$ is the power-spectral density (PSD) of the displacement. Like the storage and loss modulus, the real and imaginary parts of χ are not independent functions, but are connected by Kramers-Kronig relations (see Section 1.2.2)—knowledge of one determines the other. The real and imaginary response functions are given by

$$\tilde{\chi}'(\omega) = \frac{2}{\pi}\int_0^\infty \frac{\omega'\tilde{\chi}''(\omega')}{\omega'^2 - \omega^2}d\omega' \tag{6.9}$$

and

$$\tilde{\chi}''(\omega) = \frac{2}{\pi}\int_0^\infty \frac{\omega\tilde{\chi}'(\omega')}{\omega^2 - \omega'^2}d\omega', \tag{6.10}$$

respectively. Thus, $\chi'(\omega)$ can be calculated from the measured $\chi''(\omega)$.

Discrete transforms can be used to calculate $\tilde{\chi}'(\omega)$ from $\tilde{\chi}''(\omega)$ from the sampled data points of a long time-series (Schnurr *et al.*, 1997). In this case, eqn 6.9 is expressed in terms of the dispersion integral

$$\tilde{\chi}'(\omega) = \frac{2}{\pi}\mathcal{P}\int_0^\omega \frac{\omega'\tilde{\chi}''(\omega')}{\omega'^2 - \omega^2}d\omega' \tag{6.11}$$

where the $\mathscr{P}$ denotes the Cauchy principal value integral to account for the singularity at $\omega' = \omega$.[1] Conveniently, eqn 6.11 can be calculated by taking successive sine and cosine transforms of $\tilde{\chi}''(\omega)$ McQuarrie (2000),

$$\tilde{\chi}'(\omega) = \frac{2}{\pi} \int_0^\infty dt \cos(\omega t) \int_0^\infty d\omega' \sin(\omega' t) \tilde{\chi}''(\omega'). \qquad (6.12)$$

In an experiment, $\tilde{\chi}''(\omega)$ is found by computing the power spectrum of the position time-series $x(t)$ output from the back-focal-plane detector (eqn 6.8) and $\tilde{\chi}'(\omega)$ is subsequently calculated numerically by successive discrete transforms given by eqn 6.12. Once the complex susceptibility $\tilde{\chi}^*(\omega)$ is known, the storage and loss moduli follow from eqn 6.6,

$$G'(\omega) = \frac{1}{6\pi a} \left[\frac{\tilde{\chi}'(\omega)}{\tilde{\chi}'(\omega)^2 + \tilde{\chi}''(\omega)^2} \right] \qquad (6.13)$$

and

$$G''(\omega) = \frac{1}{6\pi a} \left[\frac{-\tilde{\chi}''(\omega)}{\tilde{\chi}'(\omega)^2 + \tilde{\chi}''(\omega)^2} \right]. \qquad (6.14)$$

6.1.4   Studies using interferometry

Despite requiring a more specialized instrument, back-focal-plane interferometry is powerful because it spans the operating regime of the two other microrheology methods we've discussed: Video-based particle tracking microrheology, and light scattering based tracer particle microrheology. Back-focal-plane interferometry measures particle displacements over a wider range of frequencies and with greater position sensitivity than conventional particle tracking. The low-frequency limit generally extends beyond those of DWS microrheology, but more importantly, requires much smaller sample volumes (Mason *et al.*, 1997*a*).

Results from an early passive microrheology study using back-focal-plane interferometry are shown in Fig. 6.6 for 0.9 μm diameter silica particles in cross-linked poly(acrylamide) gels and entangled F-actin networks. In the 2 wt% acrylamide gel (Fig. 6.6a), the power spectrum decreases with a logarithmic slope of approximately ~ -1.5 above about 100Hz, reflecting the Rouse-like dynamics of the polymer chains that make up the network, where $G'(\omega)$ and $G''(\omega)$ scale as $\sim \omega^{1/2}$. The slope of the power spectrum decreases below 100 Hz due to the onset of the elastic plateau. At higher polymer

[1] Again, see Section 1.2.2.

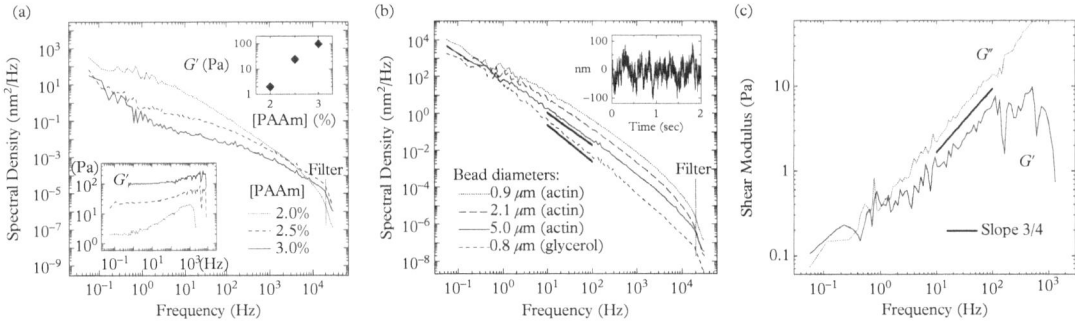

Fig. 6.6 *Back-focal-plane-interferometry microrheology results for cross-linked-polymer networks. (a) The power spectrum of probe position in cross-linked polyacrylamide gels. The lower inset shows the frequency-dependent storage modulus, and the upper inset shows the gel plateau modulus at 1 Hz. The dotted, dashed, and solid lines are results for 2.0, 2.5, and 3.0 wt% polymer. (b) Power spectrum for similar probe particles in glycerol and entangled actin networks (2 mg/ml) with (c) the corresponding frequency-dependent storage and loss moduli. Reprinted figure with permission from Gittes, F., Schnurr, B., Olmsted, P. D., MacKintosh, F. C., & Schmidt, C., F., Phys. Rev. Lett. 79, 3286–9 (1997). Copyright 1997 by the American Physical Society.*

concentrations, the relaxation time of the gel becomes too short to capture the cross-over to the terminal Rouse behavior. In contrast, Fig. 6.6b shows that the power spectrum decreases as $\sim \omega^{-2}$ for probes in glycerol and $\sim \omega^{-1.77}$ for actin. The latter value is the expected terminal frequency response for a semiflexible polymer, for which $G'(\omega) \sim G''(\omega) \sim \omega^{3/4}$ (Morse, 1998b; Gittes and MacKintosh, 1998) and are identical to high-frequency measurements using DWS discussed in Section 5.6. Laser tracking has been used to measure particle dynamics in epithelial cells by tracking the movement of endogenous lipid granules (Yamada *et al.*, 2000).

Passive microrheology with optical traps

Back-focal-plane detection works best when the particles cannot diffuse out of the tracking laser beam, and are limited to the linear regime of the detector response. This is fine if probe particles are suspended in cross-linked or tightly-entangled polymer networks, but measurements in Newtonian and viscoelastic fluids with relatively fast relaxation times suffer from the limited statistics of the short-time-series data. To overcome this limitation, experiments often use optical traps in conjunction with back-focal-plane interferometry to constrain probe displacement (Atakhorrami *et al.*, 2006; Koenderink *et al.*, 2006). We discuss optical trapping in Chapter 9 in the context of *active* microrheology, but it is instructive to review some results from studies using back-focal-plane interferometry with optical traps, especially since these incorporate high-bandwidth detection systems. With optical traps it is also possible to extend the back-focal-plane method to two-point microrheology (Starrs and Bartlett, 2003*a*; Atakhorrami and Schmidt, 2006).

Buchanan *et al.* (2005*a*) report microrheology measurements of the viscoelastic moduli for optically trapped 0.98 μm diameter silica particles in worm-like micellar solutions of the surfactant cetylpyridinium chloride (CPyCl) dissolved in 0.5 M NaCl and sodium salicylate (NaSal). As shown in Fig. 6.7, the measurements span 4–5 decades in frequency, from 0.1 Hz to 10^5 Hz. The optical trap stiffness is kept low by using a low laser power and its contribution is subtracted as an apparent, frequency-independent modulus of about G' ~0.1–1 Pa. Figure 6.7a shows the power spectrum of probe displacement for a 4 wt% CPyCl/NaSal solution, which is measured for different probe particles. The calculated viscoelastic moduli are plotted in Fig. 6.7b.

The power spectrum for a number of other surfactant concentrations ranging between 1–8 wt% are shown in Fig. 6.7c. Notice that the curve for water exhibits a low-frequency plateau. This deviation from the expected ω^{-2} power law behavior of a particle in water represents the plateau associated with the optical trap. The "corner frequency" represented by the cross-over is used to determine the optical trap stiffness, which is discussed in further detail in Section 9.5.3. The power spectrum for the water sample also clearly highlights the distinct high-frequency response of the surfactant network, which exhibits a power-law behavior similar the actin networks in Fig. 6.6.

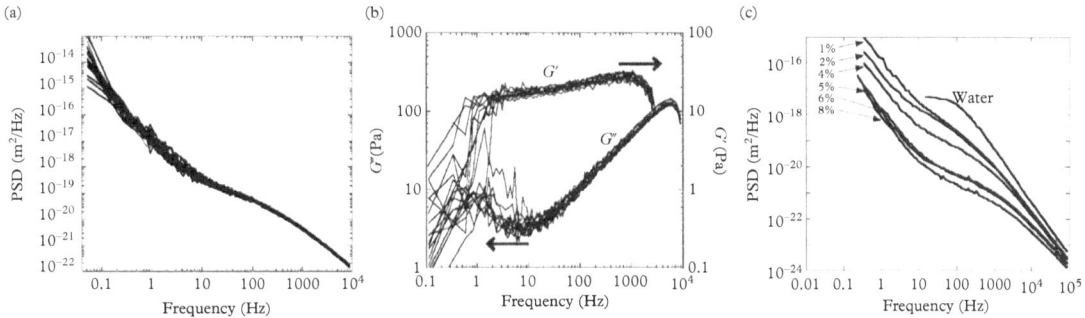

Fig. 6.7 *Back-focal-plane interfermoetry microrheology results for a worm-like micellar solution of the surfactant cetylpyridinium chloride (CPyCl) dissolved in 0.5 M NaCl and sodium salicylate (NaSal). (a) The power spectrum in 4 wt% CPyCl and (b) calculated viscoelastic moduli. (c) The power spectrum for water and CPyCl concentrations 1–8 wt%. Reprinted (abstract/excerpt/figure) with permission from Buchanan, M., Atakhorrami, M., Palierne, J. F., MacKintosh, F. C., & Schmidt, C. F., Phys. Rev. E 72, 11504 (2005). Copyright 2005 by the American Physical Society.*

6.2 Two-point interferometry

Tracking or trapping two particles simultaneously allows two-point microrheology to be performed with back-focal-plane interferometry by studying the *correlated* motion between the two-probe particles. The analysis of two-point microrheology and its applications in heterogeneous materials in which the continuum approximation of the Stokes equation fails, are discussed in Section 4.11. Again, the extended frequency range of back-focal-plane interferometry is one possible advantage over video-based particle tracking. We will return to this topic in Section 9.8.

6.3 Rotational diffusion microrheology

In the previous section, we saw that a laser can be used to track small displacements of a probe's translational Brownian motion. Since laser light is polarized, the *orientation* of birefringent particles can also be detected, and thus, the rotational Brownian motion of a probe can be measured. This phenomenon is the basis of *rotational* diffusion microrheology (Cheng and Mason, 2003).

Rotational microrheology requires birefringent particles. Anisotropic, micrometer-dimension wax disks were originally used by Cheng and Mason (2003). Light scattered from the particle depends on its orientation relative to the polarization of the beam, resulting in a fluctuating intensity as the particle rotates, like those shown in Fig. 6.8a–c. The correlation of the orientation with respect to time $\theta(t)$ is related to the local viscoelastic properties by a new, rotational form of the GSER,

$$\tilde{G}(s) = s\tilde{\eta}(s) = \frac{k_B T}{4\pi a^3 s \langle \tilde{\Delta\theta^2}(s) \rangle} \tag{6.15}$$

which is derived from the Langevin-torque equation

$$I\ddot{\theta} = L_b - \int_0^t dt' \zeta_R(t - t')\dot{\theta}(t') \tag{6.16}$$

given a random torque L_b in a manner analogous to the derivation of the GSER discussed in Chapter 3. The rotational resistance for a sphere is given by

$$\zeta_R = 8\pi \eta a^3 \tag{6.17}$$

in a Newtonian fluid (see Problem 2.1), which can be extended by the Correspondence Principle to viscoelastic materials. Like the

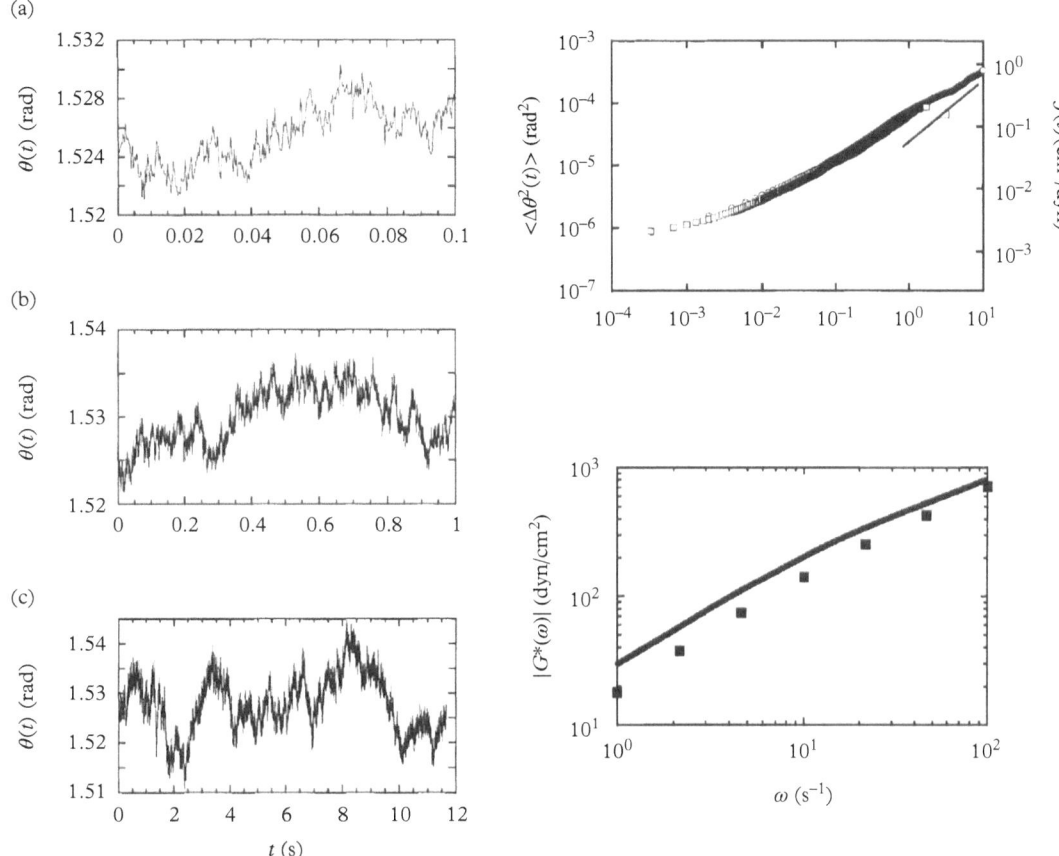

Fig. 6.8 *Rotational diffusion microrheology of wax microdisks (~ 1 μm) suspended in a 4 wt% 900kDa PEO solution. (a–c) The light intensity as the disk orientation changes with time. (d) The mean-squared angular rotation and modulus amplitude calculated by the rotational GSER. Reprinted figure with permission from Cheng, Z. & Mason, T. G., Phys. Rev. Lett., 90, 18304 (2003). Copyright 2003 by the American Physical Society.*

translational GSER, the inverse Laplace Transform shows that the mean-squared angular difference $\langle \Delta\theta^2(t) \rangle$ is proportional to the creep compliance,

$$\mathcal{J}(t) = \frac{4\pi a^3 \langle \Delta\theta^2(t) \rangle}{k_B T}. \tag{6.18}$$

Cheng and Mason (2003) measured the rotational diffusion of wax disks in PEO solutions (900kDa). The mean-squared angular rotation and derived modulus amplitude are shown in Fig. 6.8d. The microrheology results are in good agreement with bulk rheology measurements.

. . . .

EXERCISE

(6.1) **Rotational GSER.** Derive the rotational GSER (eqn 6.15) from the Langevin-torque equation (eqn 6.16) using the methods of Chapter 3. What samples are amenable to measurement by this technique? What is the operating regime of rotational diffusion microrheology?

Active microrheology

7.1 Introduction and overview

Thus far, we have centered our discussion on microrheology techniques that exploit spontaneous, thermally-fluctuating forces to drive colloidal probes. We now turn to other techniques that complement and extend this "passive" microrheology. In particular, external forces (most typically magnetic or optical) can be actively applied to force microrheological probes into motion. These techniques, collectively called "active" microrheology, short-circuit the Einstein component of passive microrheology.

Active microrheology provides an additional handle to probe material properties, and has been used both to extend the range of materials amenable to microrheological analysis, and to examine material properties that are inaccessible to passive microrheology. In particular, this chapter discusses three main topics:

- Active microrheology can be used to extend the range of passive microrheology, while maintaining many of the advantages (small sample size, wide frequency range, *etc.*). For example, active and linear microrheology can be used to probe materials that are so stiff that thermal fluctuations give too small a response, as discussed in Section 7.2.

- Active and linear microrheology can be used to complement passive microrheology in active systems, which convert chemical fuel to mechanical work, in order to elucidate the power provided by molecular motors, as discussed in Section 7.2.1.

- Finally, active and nonlinear microrheology has been used to investigate the nonlinear response properties of materials, as discussed in Section 7.3.

In framing the issues that arise in active microrheology, it is helpful to recall how and why the Generalized Stokes–Einstein Relation works. Under conditions laid out in Chapters 2 and 3, the Stokes and Einstein components enable $G^*(\omega)$ to be extracted quantitatively from the mean-squared displacement. In short, the Einstein

Microrheology. Eric M. Furst and Todd M. Squires, Oxford University Press (2017).
© Eric M. Furst and Todd M. Squires. DOI 10.1093/oso/9780199655205.001.0001

component captures the strength and statistics of the fluctuating thermal forces, and the Stokes component relates those forces into concrete and measurable probe translation and rotation.

The Einstein relation relates the fluctuations of a probe within an *equilibrium* material to its (deterministic) linear response behavior, irrespective of the specific rheology of the material, the validity of the continuum approximation, the specifics of the probe shape, or the type of response (*e.g.*, translation or rotation of individual probes, relative motion between multiple probes). The Einstein component holds provided the system is in equilibrium.

The Stokes component captures the (deterministic) response of a particle that is forced into small-amplitude oscillations within the material of interest. As discussed in Chapter 2, the Correspondence Principle enables the probe response to be determined for material of any rheology, so long as the material is isotropic, homogeneous, behaves as a continuum and remains in the linear response limit. The Stokes component holds for any type of forcing—*e.g.*, stochastic, thermal fluctuating forces, or externally-imposed forces—so long as the material remains in the linear response regime.

7.2   Active, linear microrheology

The most straightforward example of active microrheology involves the use of external forcing—*e.g.*, magnetic or optical tweezers—to drive a probe more strongly than would occur by thermal fluctuations alone. Consequently, active forcing expands the operating range of microrheology to include materials that are so stiff that thermal fluctuations give immeasurably small displacements.

Active, linear microrheology effectively amounts to a direct application of the Stokes component, without recourse to the Einstein component. Its success for microrheology is pinned to the applicability of the Correspondence Principle; so long as the oscillations are gentle enough—in a manner that must be checked self-consistently after the fact—the linear response properties of the material remain intact. The practice of identifying the linear response regime is common to bulk rheology experiments in the use of "strain sweeps." If such conditions are met, the Correspondence Principle still holds, and Stokes drag (2.75)

$$\zeta = 6\pi \eta a \tag{7.1}$$

can be generalized for linear viscoelastic materials to give the frequency-dependent Stokes resistance

$$\zeta(\omega) = 6\pi a \eta^*(\omega). \tag{7.2}$$

The complex viscosity $\eta^*(\omega)$, or equivalently the linear viscoelastic modulus $G^*(\omega)$, can thus be extracted using a Generalized Stokes Relation,

$$\eta^*(\omega) = \frac{G^*(\omega)}{i\omega} = \frac{F_0}{6i\pi\omega\Delta X_0}. \tag{7.3}$$

An example of active, linear microrheology is shown in Figure 7.1. In this experiment, Hough and Ou-Yang (2006) used laser tweezer microrheology to characterize an associating polymer solution consisting of a telechelic polyethylene oxide, terminated on both ends by hydrophobic hexadecyl groups. The storage and loss modulus for a 2.5 wt% aqueous solution, measured microrheologically using 1.6 μm diameter silica probe particles, agrees well with bulk rheology measurements where measurement frequencies overlap (0.1–100 rad/s). Some deviation is evident between the low-frequency storage moduli measured using the two techniques, which is to be expected—when the modulus and frequency are low, the phase angle retrieved from the lock-in amplifier is noisy. This example highlights several ways in which active, linear microrheology push the boundaries where rheology measurements are possible. The active, linear microrheology extends the frequency range available for linear response measurements, allowing measurements at frequencies as high as 4×10^4 rad/s (6 kHz), about two orders of magnitude higher than mechanical rheometry. Both bulk and micro-rheology show the Maxwell-like relaxation of the fluid at low frequencies due to the lifetime of the hydrophobic interaction between micelles, whereas the extended frequency range of active microrheology captures the Rouse-like dynamics of the telechelic supramolecular assemblies.

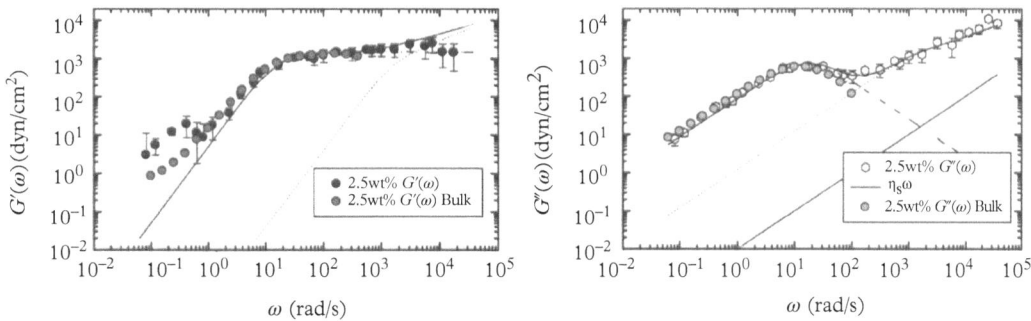

Fig. 7.1 *Telechelic PEO (number average molecular weight M_n = 67 kDa) dispersed in water at 2.5 wt%. Reprinted figure with permission from Hough, L. A. & Ou-Yang, H. D., Phys. Rev. E, 73, 31802 (2006). Copyright 2006 by the American Physical Society.*

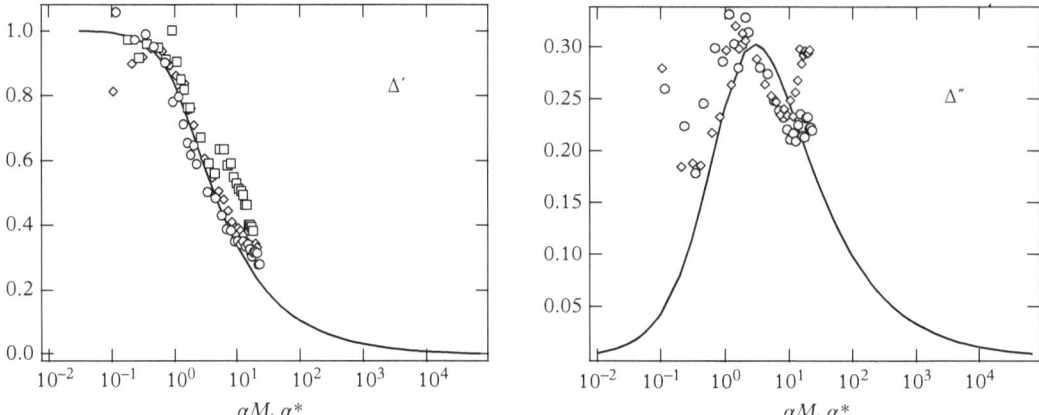

Fig. 7.2 *Comparison of the calculated viscoelastic increment Δ with experimental data for an FEP suspension. All measurements are made using 2 μm probe particles. Circles, diamonds, and squares represent volume fractions of 0.33, 0.29, and 0.24, respectively. Sriram, I., DePuit, R., Squires, T. M. & Furst, E. M., J. Rheol., 53, 357–81 (2009). Copyright 2009, The Society of Rheology.*

Sriram *et al.* (2009) provide another example (Fig. 7.2) where active methods extend the material range accessible to microrheology. Laser tweezer microrheology was used to probe the rheology of a suspension of small colloids, the results of which agreed well with behavior *computed* for such suspensions. This highlights not only the success of active microrheology for very dilute, weak materials, but also the success with which theory can be used to predict their properties.

The relative simplicity with which suspension rheology can be modeled, combined with the richness of the resulting material response, has motivated their use as model materials for studies into active, nonlinear microrheology. We will develop these ideas when moving into the active, nonlinear response limit, where a variety of issues arise that complicate the enterprise significantly.

7.2.1   Active microrheology of active (non-equilibrium) materials

Actively driving the probe—even while remaining within the linear response limit—enable fascinating measurements on active, non-equilibrium materials that use chemical reactions to do mechanical work. Active materials convert some external energy (*e.g.*, from chemical reactions) into mechanical work. Obvious and important examples include molecular motor proteins in biopolymer networks, which typically derive energy from the chemical fuel ATP to exert mechanical stresses and drive motion. An example is a network of actin and myosin filaments, shown in Fig. 7.3. Myosin binds to nearby actin filaments, and through a process of ATP hydrolysis, pulls its neighbor about 10 nm with a force of approximately 5 pN. Microrheology has been used to study the dynamics and rheology of such "active networks" (Le Goff *et al.*, 2002; Mizuno *et al.*, 2007).

Comparisons between the response of probes that are weakly-forced and those that are thermally fluctuating reveal the fluctuation-dissipation theorem to be violated in such systems (Mizuno *et al.*, 2007; Hoffman *et al.*, 2006; Martin *et al.*, 2001). In particular, the response of an actively-driven probe encodes the frequency-dependent, mechanical response of the material. The stochastic, fluctuating response of the probe encodes the energy dissipated within the material (*i.e.*, both thermal and chemical), in addition to the mechanical response of the material. In an equilibrium (non-active) material, the two responses would agree—as expected from the fluctuation-dissipation theorem and the Einstein component of the GSER. In an active material, by contrast, the difference between the active and passive-probe response can be related directly to the work performed by the motors, and ultimately to their power spectrum. It is difficult to imagine other ways of obtaining this information.

Both passive and active laser tweezer microrheology experiments of a model cytoskeleton network by Mizuno *et al.* (2007) are shown in Fig. 7.4. The power spectra of the probe was measured for under passive and active conditions in a control system depleted of the molecular motor myosin. This control system is a non-active, equilibrium system, and the two measurements agree, as expected. Moreover, little changes initially when myosin is added: The passive and active responses agree, indicating that the motors have not yet begun to impact the mechanics of the material (Fig. 7.4b). Figure 7.4c shows that clear differences between the active and passive responses appear after several hours, however, indicating the breakdown of the fluctuation-dissipation theorem (or, analogously, the Einstein component). At high-enough frequencies, the two methods agree quantitatively, indicating that the molecular motors have no impact on the mechanical properties on such short time scales. At lower frequencies, however, disagreements between the active and passive responses reveal the power input of the motors on these longer time scales. In other words, the motors actively do work on the material on long enough time scales, adding energy which is ultimately dissipated away as the material relaxes. On these time scales, the energy contributed by ATP hydrolysis-powered motors acts in addition to what would normally be contributed from thermal fluctuations. The

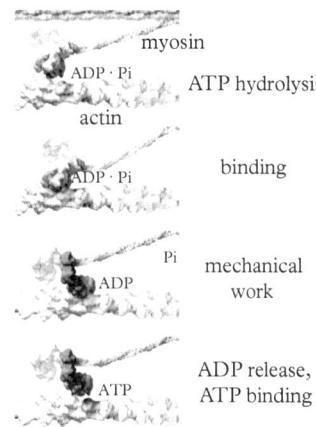

Fig. 7.3 *The proteins actin and myosin perform mechanical work by hydrolyzing ATP. The release of the phosphate P_i is associated with the "power stroke" of the myosin head group, which moves about 10 nm and can pull with a force on the order of 5 pN. The two filaments remain together in a "rigor" state after releasing ADP until a fresh ATP molecule binds to the myosin. From Vale, R. D. & Milligan, R. A. Science, 288, 88–95 2000. Reprinted with permission from AAAS.*

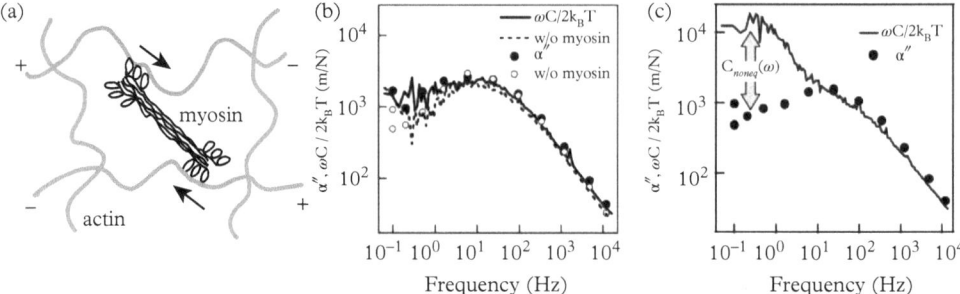

Fig. 7.4 *Comparisons between the responses of actively-forced and passively-fluctuating probes in a model cytoskeletal network (a) of actin filaments and myosin molecular motors. Signs indicate the polarity of the actin filaments. (b) In the absence of the motor protein (or before it has begun to act), the passive and active responses agree quantitatively, as expected from the GSER for equilibrium materials. (c) After several hours of myosin activity, the power spectrum for the passive response is much higher at low frequencies than the active response, directly revealing the additional power contributed by the motors. From Mizuno, D., et al.,* Science, *315, 370–3 2007. Reprinted with permission from AAAS.*

difference between the power spectra, measured using active and passive microrheology, is directly related to the power introduced by the motors at these frequencies.

7.3   Active and nonlinear microrheology

The remainder of this chapter focuses on active and nonlinear microrheology, in which probes are driven with enough force to elicit a nonlinear response from the material, with the goal of measuring the material's nonlinear rheology. As we discussed in Chapter 1, nonlinear rheology plays an essential role in many materials and products. For example, shampoo and honey have similar zero shear viscosity, but the viscosity of shampoo thins dramatically with increasing shear rate. Mayonnaise and jello have similar linear rheology, but mayonnaise yields and flows under sufficiently strong stress. A suite of macroscopic rheometry techniques have been developed over the past century to quantitatively characterize these properties.

There would be clear advantages and applications if analogous techniques could be developed for microrheology. During the design and formulation of new materials, nonlinear rheology like yield stress and shear thinning could be screened using only small sample volumes, before scaling up production.

7.3.1 Measuring nonlinear rheology

Before discussing nonlinear microrheology, it is worthwhile examining what nonlinear rheological properties can be measured using macroscopic, mechanical methods. Although difficulties certainly remain, and development continues, well-established techniques have been developed to make quantitative, reproducible measurements of rate-dependent shear viscosity (*e.g.*, shear thinning and thickening), as well as yield stresses. With more difficulty, normal stress differences can also be measured under steady-shear deformations, and rate-dependent uniaxial extensional viscosity measurements. Notably, these are *intrinsic* material properties—meaning that different measurement equipment, and even different measurement strategies that specifically probe these material properties would report the same results. On the basis of these measured properties, falsifiable predictions can be tested, and materials and products may be designed and engineered.

How does macroscopic rheometry succeed in doing so, and when does it fail? Successful experimental geometries, like those illustrated in Fig. 1.9 and 7.5, are carefully designed to excite rheologically "pure" flows, from which flow curves and constitutive relations may be extracted. Examples include those familiar to any rheologist: *e.g.*, a cone rotating above a plate exerts a uniform shear stresses across a sample placed in between. Likewise, materials placed between two concentric cylinders in relative rotation (also known as Couette cells), are sheared inhomogeneously, but with a shear stress profile that is not homogeneous, but which can be determined. Normal stress differences can be measured with pressure transducers integrated into the shearing plates—either to measure the total force driving the plates apart (for N_1) or by measuring the normal force profile along the plate (from which N_2 can be determined). Key to the success of both geometries is that symmetry and momentum conservation are sufficient to determine shear stress profiles exactly from the geometry and kinematics of the measurement, independent of the constitutive relation of the material. Of course, exceptions can arise: Elastic stresses can drive instabilities in Taylor-Couette flows even at vanishingly small Reynolds numbers (Shaqfeh, 1996), and shear banding can arise in systems where uniform profiles would otherwise be expected (Ovarlez et al., 2009). If, however, the stress and velocity fields behave as expected, then flow curves and constitutive relations for shear rheology may be extracted directly from these measurements. As discussed in Section 1.14, entirely different geometries are required for extensional rheometry.

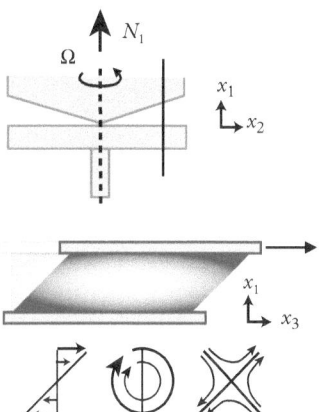

Fig. 7.5 *Rheometry tools like the cone and plate are designed to excite rheologically "pure" deformations. Each material element along the plane indicated by the line in the figure experiences a shear deformation consisting of rotation and extension. Normal forces can also be measured by the rheometer.*

These techniques can be contrasted with—and are complemented by—a suite of methods that are often cheaper, simpler, and easier to use, yet which do not give precision measurements of intrinsic material properties. Instead, these "index" techniques elicit a response that gives a measure of some rheological signature—for example, key time or stress scales relevant to a material. The results may not be sufficient for detailed calculations or predictions, but they are often enough for practitioners "in the field." Examples include the so-called fifty cent rheometer for yield stress measurements (Pashias *et al.*, 1996), where a material in a cylindrical container is placed on a surface, then the cyindrical shell removed, causing the material to yield and flow to an extent relatable to the yield stress. Falling ball viscometry does not excite pure shear flows, with well-defined or known stress profiles, yet nonetheless give some sense for low-shear viscosity, and onset shear rates for thinning or thickening.

To summarize, various macroscopic, mechanical methods have been developed to characterize the nonlinear flow properties of materials. Those that offer precise, quantitative measurement of intrinsic material properties employ geometries that have been specifically designed to ensure homogeneous, controllable deformations of a certain rheological type (*i.e.*, pure shear, or pure extension). Other methods are simpler to employ, and hold value as characterization tools, yet do not necessarily provide precise, quantitative measurements of intrinsic material properties. At present, it is not known which options are available for microrheology. Various issues have been identified in attempting to develop these tools, but significant questions remain, and research is underway.

7.3.2 Nonlinear microrheology: The issues

As with linear microrheology, the quantity that is actually measured in any nonlinear microrheology measurement is the *resistance* ζ of a probe to some motion within a material. For nonlinear microrheology, however, the resistance depends on probe velocity, rather than frequency. One might simply hope that a "Generalized Stokes Resistance" (GSR) would hold—*i.e.*, that

$$\eta(V) = \frac{F}{6\pi a V} \tag{7.4}$$

which would seem to provide a measurement of a shear-rate-dependent viscosity.

Unfortunately, the Correspondence Principle does not hold when materials are deformed beyond their linear response limit. In what follows, we will describe issues that introduce ambiguities, uncertainties, or even impossibilities, into the endeavor. Even in the best-case scenario where linear microrheology can be proven to quantitatively recover macroscopic rheology, severe complications arise; for example, what shear rate should be attributed to a probe pulling speed, given that a continuous spectrum of strain rates are excited by steady probe motion? Both extension thickening and shear thinning may occur in a nonlinear microrheology experiment, yet the contributions of each are wrapped into a single quantity (probe drag), measured as a function of a single quantity (velocity). Can the two be deconvolved?

Still, the results of active microrheology experiments do encode the material's nonlinear rheological properties; whether it is possible to extract such properties, however, remains a subject of active research in this nascent field. To provide context for future development, we now lay out issues and complications that are known to arise.

7.3.3 Nonlinear microrheology of continuum materials: Known sources of discrepancy

We start with the "best-case" scenarios, for which linear microrheology is known to work: Materials that obey the continuum approximation, are isotropic and homogeneous (meaning, additionally, that the probe does not create any structural heterogeneities within the material), and obey the no-slip condition. Under the weak stresses of linear microrheology, the Correspondence Principle can thus be expected to hold for these materials. Stresses in nonlinear microrheology, on the other hand, are strong enough (by design) to elicit the nonlinear response of a material, which renders the Correspondence Principle inappropriate.

This section largely follows Squires (2008), who identified mechanisms that cause discrepancies between linear and nonlinear microrheology, and DePuit and Squires (2012a,b), who detailed theoretical studies that quantified those discrepancies. In particular, heterogeneous strain rates, Lagrangian unsteadiness in the flow fields, and mixed rheological flows all introduce measurable differences between the (velocity-dependent) microviscosity $\eta_\mu(VU)$ measured microrheologically, and the shear-rate-dependent macroviscosity $\eta_M(\dot{\gamma})$ measured conventionally. Moreover, direct collisions between the probe and microstructure influence the probe resistance ζ in both probe-specific and velocity-dependent manners, further complicating the probe-material interactions, *e.g.*, discussed in Sections 2.2 and 3.12.

Mixed flow fields

Flow fields excited around moving probes are not rheologically "pure," meaning they are neither purely shear nor extension, but rather different mixtures in different regions (Fig. 7.6). While linear response properties do not depend on the type of deformation, nonlinear rheological responses do. In particular, polymer solutions often shear thin, but thicken under extensional flows. Consequently, regions of strong shear (*e.g.*, along the "equator" of the probe) may exert a relatively weak stress on that part of the probe, whereas regions of strong extension (*e.g.*, at the "poles") may exert extremely strong stresses. Worse still, such stresses may change even the qualitative character of the flow—strong extensional rheology can give rise to "negative wakes" in front of moving spheres, for example.

The dependence of nonlinear rheology upon flow type can be seen in Fig. 7.7, which shows the shear and uniaxial extensional viscosity increments computed by DePuit and Squires (2012*a*) for a dilute suspension of Brownian ellipsoids. The Peclet number describes the ratio of flow strength (which acts to orient the ellipsoids) against rotational diffusion (which acts to randomize their orientation),

$$Pe = \frac{|\nabla \mathbf{v}|}{D_R} \sim \frac{\text{rotational diffusion time}}{\text{flow orientation time}}, \qquad (7.5)$$

so that $Pe \ll 1$ gives the linear response limit, and $Pe \gg 1$ corresponds to flows strong enough to cause the orientational distribution of ellipsoids to depart significantly from equilibrium (giving a nonlinear rheological response). In the low-Pe limit, both shear and extensional flows give rise to the same viscosity increment, as expected from linear response rheology. At steady state, the shear viscosity of the suspension *decreases* as the shear rate (Pe) increases, reflecting shear thinning.

Fig. 7.6 *An illustration of the inhomogeneity of Stokes flow around a probe. The local relative velocity streamlines are shown with corresponding microstructural perturbation for low- and high-frequency oscillations, α. Reprinted with permission from Sriram, I., DePuit, R., Squires, T. M. & Furst, E. M., J. Rheol., 53, 357–81 (2009). Copyright 2009, The Society of Rheology.*

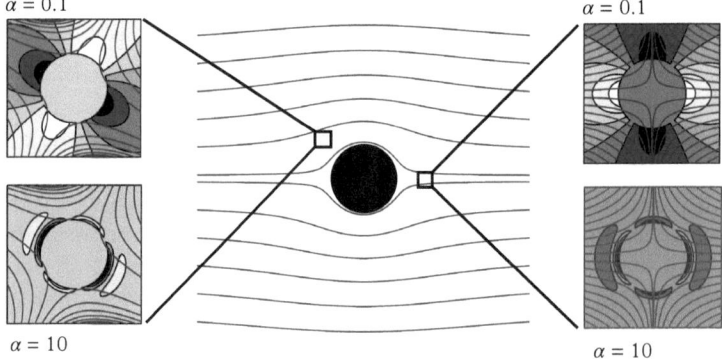

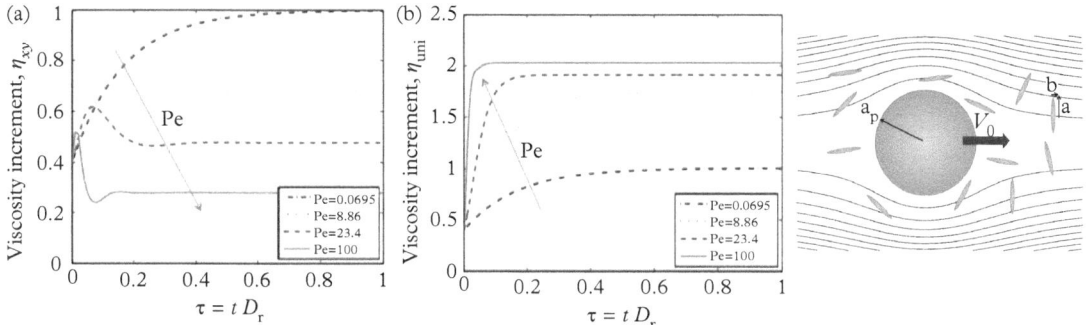

Fig. 7.7 *Transient rheology of Brownian ellipsoids, whose microstructure relaxes to equilibrium via rotational diffusion, with characteristic time scale $\tau \sim D_R^{-1}$. (a) For shear deformations, the viscosity increment $\Delta\eta_{xy}$ increases in time over a time scale D_R^{-1} for $Pe \ll 1$ (i.e., the linear response limit), but more quickly (like $D_R^{-1} Pe^{-1/2}$) for $Pe \gg 1$. The steady shear viscosity of these suspensions decreases with increasing shear rate. (b) For uniaxial extensional flows, the (extensional) viscosity increment reaches a steady state over a time scale D_R^{-1} at low Pe, but with a time scale $D_R^{-1} Pe^{-1}$ at high Pe—more rapidly than for shear flows. Moreover, the extensional viscosity increases with increasing extension rate, in contrast to the shear viscosity. Adapted from DePuit and Squires (2012a)* ©*IOP Publishing. Reproduced with permission. All rights reserved.*

By contrast, the extensional viscosity *increases* with extension rate (*Pe*), revealing extension thickening.

Lagrangian unsteadiness

From a reference frame moving with the probe, the flow field in the material is steady in time. Material elements, on the other hand, do not see things that way. In their "Lagrangian" frame, this flow field is *unsteady*. To understand this, imagine the flow from the standpoint of a small bit of fluid far in front of the particle, almost along the axis of motion (refer back to the streamlines plotted in Figs. 7.6 or 7.7c). For some time, this fluid element feels very little stress, and undergoes a vanishingly weak deformation. As the fluid element approaches the probe, however, the stress it feels grows stronger and stronger, and the fluid element deforms more and more significantly. During this time, a biaxial extensional stress is exerted on the fluid element, tending to compress it along the flow axis, and stretch it radially in the perpendicular directions. As the fluid element passes the probe, the extensional character of the deformation gives way to a predominantly shear stress that ramps up from nearly zero near the pole, to some maximum, then back to zero. Once the fluid element has passed the probe and is departing, shear gives way to bi-axial compression (stretching along the axis of probe motion, compressing in the perpendicular directions) which decays from some maximum value to vanishingly small as the probe moves farther away.

Lagrangian unsteadiness has no impact on the steady Stokes flow of Newtonian fluids, which deform instantaneously in response to stress. For complex fluids with nontrivial rheology, however, Lagrangian unsteadiness may impact the material response to probe motion significantly. This can be understood when one considers the fact that various deformation and relaxation processes in complex fluids occur on different characteristic time scales: Polymers in solution relax via some Rouse time, colloids diffuse apart on some diffusive time, emulsion droplets relax on a capillary time scale, and so on. Again, we refer to Fig. 7.7: The transient rheological response computed for a dilute suspension of Brownian ellipsoids depends on both the strength and type of the flow.

The microstructure of a complex fluid element—and thus the rheological stresses it exerts—responds over some finite time scale τ_R in response to any change it experiences in the flow it experiences. If the time scale τ_L of the Lagrangian unsteadiness (with a simple estimate $\tau_L \sim a/U$) significantly exceeds the characteristic relaxation time(s) τ_R of the material element, one might hope to neglect the impact of Lagrangian unsteadiness, and instead assume that the material elements everywhere evolve quasi-steadily with the local stress. This ratio of time scales is encapsulated by the Deborah number,

$$De = \frac{\tau_R}{\tau_L}. \tag{7.6}$$

When $De \ll 1$, the material relaxes so much more quickly than the flow conditions change that the material evolves quasi-steadily, and Lagrangian unsteadiness plays an insignificant role. When $De \sim 1$ or larger, on the other hand, the deformation state and stress exerted by each material element does not attain the value expected from a fully-developed, steady-state system. Instead, material elements are constantly deforming in response to to the stresses they experience at each moment, but those stresses change before the material manages to fully adapt.

Heterogeneous strain rates

The velocity field around the probe decays with distance from the probe—as does the strain rate. Far enough from the probe, stresses (or strain rates) are weak enough that the linear response limit is valid. Closer to the probe, however, the rheological response becomes non-linear, with the nonlinearity increasing as the probe is approached. A range of nonlinear rheological responses is thus excited, complicating the interpretation. Squires (2008) highlighted these effects for

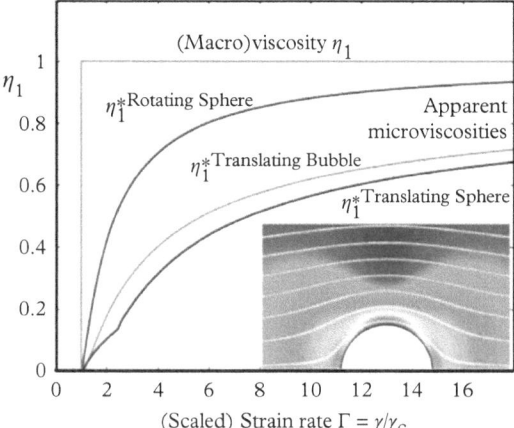

Fig. 7.8 *Microviscosities obtained from a fluid with abruptly shear thickening viscosity η_1 using a rotating sphere (RS), translating sphere (TS), and translating bubble (TB). Although shear thickening begins at the correct strain rate, the microviscosities approach the macroviscosity only slowly due to the inhomogeneous rate-of-strain field around the probe. Reprinted with permission from Squires, T. M., Langmuir 24, 1147–59 (2008). Copyright 2008 American Chemical Society.*

a particularly simple, generalized Newtonian fluid, described by a scalar velocity that depends on the overall strain rate: $\eta(\sqrt{\dot{\epsilon}:\dot{\epsilon}})$. A probe that is moving with some velocity V has highest stress near the probe, which gives the highest shear rate $\dot{\epsilon}$ (Fig. 7.8), and therefore the strongest non-Newtonian response. At some distance from the probe, however, the stresses are weak enough that the linear response constitutive relationship holds, and the GSR would work just fine.

If one hopes to extract nonlinear rheology from such a measurement, one must disentangle multiple, simultaneous unknowns: (1) The size of the "nonlinear" region is not known *a priori*, but instead depends on the material's (unknown) constitutive relation, and the details of the flow field. (2) The nonlinear flow field within the non-linear region requires a full non-Newtonian fluid-mechanics solution, which in turn requires $\eta(\sqrt{\dot{\epsilon}:\dot{\epsilon}})$ be known; and (3) the Generalized Newtonian framework must be known to hold. The drag on a probe moving with velocity U though a "weakly non-Newtonian" material,

$$\eta(\sqrt{\dot{\epsilon}:\dot{\epsilon}}) \approx \eta_0 + \epsilon\eta_1(\sqrt{\dot{\epsilon}:\dot{\epsilon}}),\tag{7.7}$$

can be computed (*e.g.*, using the Lorentz reciprocal relation, Leal 2007). If one assumed a GSR to hold, one would extract an effective viscosity

$$\eta^*(U) \approx 1 + \frac{\int \eta_1(\sqrt{\dot{\epsilon}_0:\dot{\epsilon}_0})\dot{\epsilon}_0:\dot{\epsilon}_0 dV}{\int \dot{\epsilon}_0:\dot{\epsilon}_0 dV},\tag{7.8}$$

where $\dot{\epsilon}_0$ is the rate of strain that would be driven around a probe moving in a *Newtonian* liquid. Even in this simplest of limits for the simplest of complex fluids, the "microviscosity" that emerges from the GSR does not recover the true $\eta(\sqrt{\dot{\epsilon}:\dot{\epsilon}})$; instead, it reflects a weighted average of all non-Newtonian responses excited by the probe, dominated by the rheology closest to the probe. Figure 7.8 shows the apparent microviscosities η^* that would be measured by probes moving at different velocities thorough a model material can be directly compared the the (specfied) rheology of that material. While the apparent microviscosity does broadly track the true viscosity, the agreement is far from quantitative, since eqn 7.8 "smooths" features of the nonlinear rheology. For some probes, eqn 7.8 can be inverted exactly to extract $\eta(\dot{\gamma})$ from $\eta^*(\dot{\gamma})$; whether it can be numerically inverted in general is not known.

Discrepancies in bulk stresses: Quantitative impact

In the previous section, we identified several processes that cause the apparent nonlinear microviscosity to differ from the macroscopically-measured nonlinear shear viscosity. "*In silico*" experiments provide a natural way to determine the quantitative impact of these phenomena. In particular, the flow of a model material may be computed in response to a translating probe, and an apparent microviscosity determined from the drag computed on the probe using an assumed GSR, eqn 7.4, then compared to the nonlinear shear viscosity of that same material, as computed under homogeneous, steady-shear flows.

Many studies of active and nonlinear microrheology have used colloidal suspensions as model materials, owing to the relatively well-developed theoretical machinery for their treatment. Puertas and Voigtmann (2014) and Wilson and Poon (2011), for example, review the nonlinear microrheology of colloidal suspensions, including experiments, dilute theories, Stokesian dynamics simulations, and mode-coupling theories.

DePuit and Squires (2012*a*) describe computational experiments for active and nonlinear microrheology of dilute suspensions of Brownian ellipsoids, and deconvolve the various sources of discrepancy already identified by explicitly turning them off, one by one, in a series of computations (Fig. 7.9). Brownian ellipsoids are particularly appealing in this regard because their rotational anisotropy gives rise to a non-Newtonian response even at $\mathcal{O}(\phi)$, simplifying the computational challenge considerably. The apparent "microviscosity" $\eta_{app}(V)$ computed for the full nonlinear microrheolog experiment ("x" symbols), agrees with the shear-dependent "macroviscosity" $\eta(\dot{\gamma})$ (black line) at at sufficiently small Pe, as expected in the linear response limit, but differences appear in the nonlinear limit as $Pe \sim 1$: η_{app} thins only at a higher Pe, and slightly more weakly with Pe.

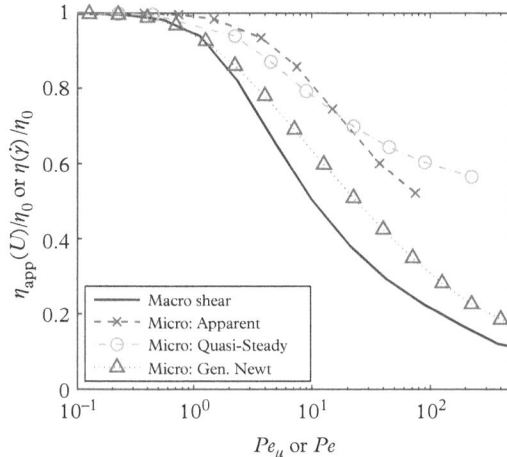

Fig. 7.9 *Computational nonlinear microrheology "experiments" for a dilute suspension of Brownian ellipsoids. The viscosity increment $\eta(\dot{\gamma})$ computed for a steady-shear geometry (as would be imposed by a rheometer) shows shear thinning around where $Pe = \dot{\gamma}/D_R \sim 1$. The apparent microviscosity η_{app} force-thins at slightly higher $Pe_\mu = U/(a_p D_R)$, and thins more gradually with Pe_μ than the macroviscosity. The onset of shear thinning agrees better when Lagrangian unsteadiness is turned off, but the thinning occurs much more gradually due to the extension thickening. Omitting extension-thickening (by treating as a quasi-steady, generalized Newtonian fluid) shows the correct onset for thinning and more accurate thinning rate. The thinning is nonetheless weaker than in macro-rheometry, owing to the spectrum of nonlinear rheological responses excited by the probe. Adapted from DePuit and Squires (2012a)*

Lagrangian unsteadiness can be omitted by computing the probe drag under the assumption that each material element attains its fully-developed microstructural response to the local deformation rate. In that case, η_{app} (circles) initially tracks $\eta(\dot{\gamma})$ somewhat more closely, but ultimately decays even less slowly than the Lagrangian-unsteady material. The increase in apparent microviscosity reflects the extension-thickening nature of the ellipsoids (Fig. 7.7), as the local (extensional) viscosity near the poles of the probe *increases* with pulling speed, whereas the macroscopic rheometry excites pure shear flows, wherein viscosities thin with shear rate. In this case, Lagrangian unsteadiness *reduces* the effect of extension-thickening, evidently because the microstructure does not have time to fully develop near the poles of the probe.

Mixed flow rheology effects may be neglected by assuming the material to respond identically to shear and extension—and could thus be treated as a generalized Newtonian liquid. In that case, the quasi-steady apparent microviscosity η_{app} agrees even more closely with the macroscopic shear viscosity, but still decays somewhat more slowly owing to the non-uniformity of the flow strength around the probe (Fig. 7.8).

Understanding the factors that give rise to micro-macro discrepancies naturally suggests strategies to lessen their impact. To more faithfully capture macroscopic shear rheometry in the microrheological context, for example, requires probes excite deformations that are steadier in the Lagrangian sense, and around which deformations are more predominantly shear, rather than extensional or mixed. For example, Fig. 7.10 shows analogous computational experiments for prolate ellipsoidal probes, revealing that the apparent microviscosity (computed in the fully Lagrangian-unsteady framework) approaches the macro-viscosity more strongly for higher-aspect ratio probes. This makes sense, in that rapid changes in flow type occur only near the probe tips, followed by long stretches of that are very nearly shear in character, and whose magnitude changes relatively slowly.

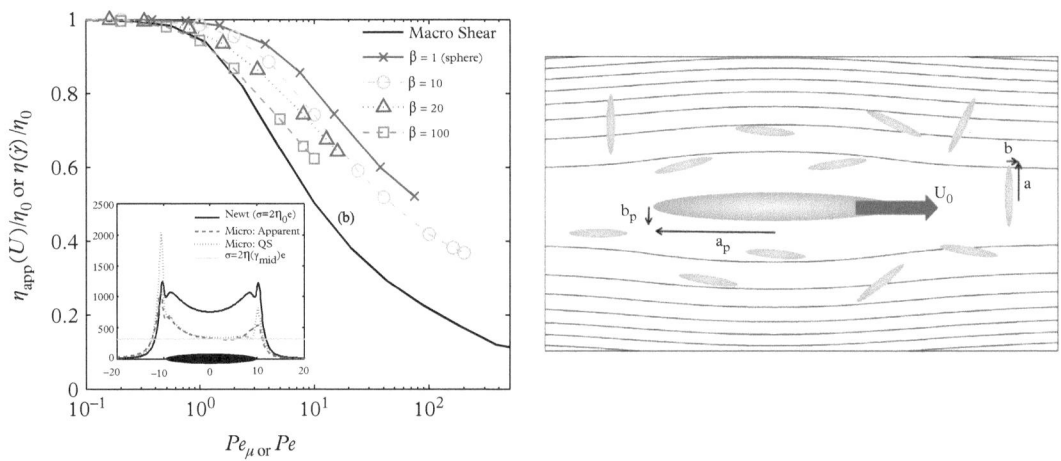

Fig. 7.10 *Ellipsoidal probes establish flow fields that are steadier in the Lagrangian sense, and more dominated by shear flows. Increasing the aspect ratio of an ellipsoidal probe from 1 (spherical) to 100 reveals increasingly better agreement with the macroscopic shear viscosity. Local stresses exerted by Brownian ellipsoids advecting past an ellipsoidal probe in various model materials: A Newtonian fluid, a suspension of Brownian ellipsoids treated as fully Lagrangian-unsteady and as quasi-steady. Adapted from DePuit and Squires (2012b)*

7.3.4   Direct probe-material interactions

The last source of drag on an actively-driven microrheological probe that we will discuss comes from direct interactions between the probe and the material it moves through. In the context of suspensions, direct collisions between the probe and suspended particles represent one example, as shown in Fig. 7.11. Other examples include adsorption and desorption of polymers or particles, or the compression/expansion of gels as probes translate through them (Uhde *et al.*, 2005). Notably, such effects do not impact macroscopic shear rheometry in the same way. Collisions between suspended colloids and rheometer plates contribute to normal stresses, but not shear stresses, whereas collisions with a microrheological probe and suspended colloids do increase the drag on the probe. While these direct interactions may play an important role in nonlinear microrheology experiments, it is important to recognize that they have no analog

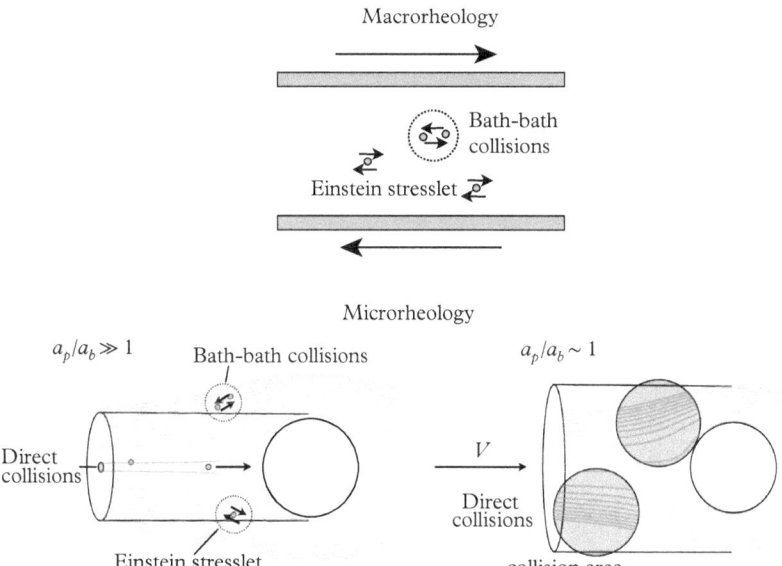

Fig. 7.11 *Sources of stress in rheology measurements of colloidal suspensions in a (macro)rheometer and in nonlinear microrheology. In both cases, an Einstein contribution arises because colloids cannot shear with the fluid and bulk contributions occur when bath colloids collide. The bulk contribution is non-Newtonian, typically reducing from one plateau at high Pe to another at low Pe. A third source of stress arises in nonlinear microrheology when the probe collides directly with a bath colloid, the strength of which will depend on the relative sizes of the bath and probe particles.*

in macroscopic rheometry, and therefore should not be expected to capture macroscopically-measured shear rheology in any quantitative sense, although in some special cases one might hope that appropriate connections may be drawn between direct interactions and processes that are intrinsic to the material itself.

A "direct" Peclet number,

$$Pe_D = \frac{V_p(a_p + a_b)}{D_{rel}} \qquad (7.9)$$

emerges in describing collisions between the probe and suspended "bath" colloids, where a_p and a_b are the radii of the probe and bath particles respectively, and D_{rel} is the relative diffusivity for the pair. The fact that Pe_D depends on the probe radius a_p serves as a warning that the measurement depends on probe size, so that care is required in determining how (or whether) these measurements reflect intrinsic material properties of the suspension. To probe the continuum response of a material, probe radii a_p are chosen to be much larger than material length scales a_b, in which case

$$Pe_D \sim \frac{V_p a_p}{D_b}. \qquad (7.10)$$

The Pe relevant for intrinsic suspension dynamics, on the other hand, would be

$$Pe_M = \frac{\dot{\gamma} a_b^2}{D_b}. \qquad (7.11)$$

As discussed in Section 7.3.3, a range of shear rates are excited around the translating probe; a characteristic (maximal) shear rate, $\dot{\gamma} \sim V_p/a_p$ gives a characteristic "intrinsic" Peclet number,

$$Pe_M \sim \frac{V_p a_b^2}{a_p D_b} \sim \frac{a_b^2}{a_p^2} Pe_D. \qquad (7.12)$$

As a probe is pulled through a suspension, it collides more frequently with colloids upstream than downstream, increasing the concentration in front of the probe, and decreasing it behind, as in Fig. 7.12. The character of these local concentration variations changes qualitatively as Pe_D increases. For $Pe_D \ll 1$ the variations are weak and show dipolar symmetry, whereas when $Pe_D > 1$, a strong but thin boundary layer forms along the front, and a particle-depleted wake trails behind (Fig. 7.12, from Khair and Brady (2006)). Meyer

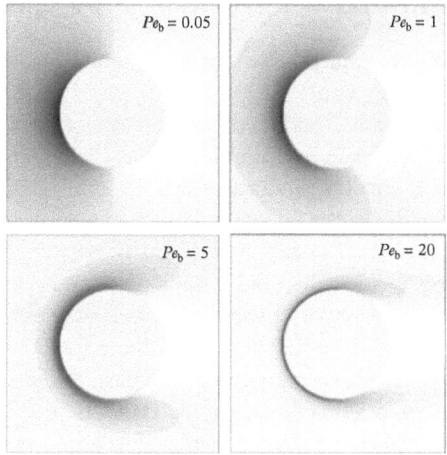

Fig. 7.12 *The calculated average microstructure of a suspension around a translating probe particle. Khair, A. S. & Brady, J. F., Single particle motion in colloidal dispersions: A simple model for active and nonlinear microrheology. J. Fluid Mech., 557, 73–117 2006, reproduced with permission.*

et al. (2006) and Sriram *et al.* (2010) measured these collisional boundary layers and wakes directly with laser tweezer microrheology simultaneous with confocal microscopy imaging (Fig. 7.13).

Associated with these qualitative changes in the direct collisional microstructure is a qualitative change in their contribution to the drag on the probe, generally decreasing with increasing Pe_D ("force-thinning"). Because probe-bath collisions are similar to bath-bath collisions, one might hope that the direct probe-bath interactions might act as analogs to the full suspension rheology. Any such comparisons must be made with real caution, as the direct probe-material interactions do not, by nature, reflect rheological processes intrinsic to the material itself. The direct interactions may thus be distinguished from the intrinsic material rheology in both magnitude and time scale: The direct microstructure "thins" at pulling speeds $(V_p/U_b)^2$ slower than the material rheology itself does; and the relative contributions of the two effects depend on ϕ in different ways. For example, in the dilute limit, direct probe-bath interactions scale linearly with suspension volume fraction ϕ, whereas the non-Newtonian behavior in a bulk suspension first arises at $\mathcal{O}(\phi^2)$. In the context of colloidal suspensions, explicit knowledge of suspension micro-mechanics (*e.g.*, how various effects scale with probe and/or particle radii, suspension volume fraction, the different Pe numbers, and so on) suggest strategies to determine intrinsic material rheological properties from probe-bath collisions. How successful—and how general—such strategies can be remains a topic of active research.

(a)

(b)

(c)

(d)

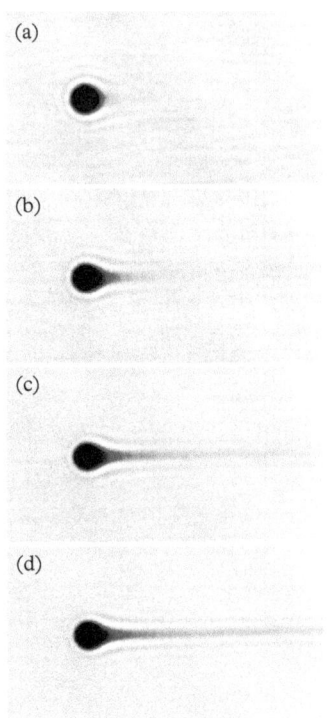

Fig. 7.13 *Averaged confocal micros-copy images of the suspension micro-structure developing around a probe particle as it is pulled through a qui-escent suspension of bath particles at a solids volume fraction* $\phi = 0.1$. *Reprinted from Sriram, I., Meyer, A., & Furst, E. M., Phys. Fluids, 22, 62003 2010 with the permission of AIP Publishing.*

7.3.5   Nonlinear microrheology: Experiments

The discrepancies detailed in Sections 7.3.3 and 7.3.4 reveal that non-linear microrheology experiments should not be expected to quanti-tatively reproduce macroscopic measurements of intrinsic, nonlinear rheological properties in any simple manner. Still, experiments have produced a few successes, suggesting that it is possible to at least characterize nonlinear rheological response in at least some materials.

For example, Meyer *et al.* (2006) used laser tweezers to pull pol-ystyrene probe particles of radii $a_p = 0.5$ and 1.5 μm through suspensions of fluorinated ethylene propylene (FEP) particles, with average particle radius 79 ± 10 nm, and volume fractions ranging from $\phi = 0.23$ to $\phi = 0.37$. Because the probe particles were 10–30 times larger than the suspended particles, the Peclet number Pe_D for direct collisions was 100–1000 × larger than the Peclet number Pe_M for the bath-bath interactions that give the complex rheological response in-trinsic to the material. Interpreting the apparent microviscosity using the simple GSR eqn 7.4, plotted against the intrinsic Peclet number $Pe_M \sim Ua_b^2/(a_pD_b)$, reveals a suprisingly strong micro/macro agree-ment (Fig. 7.14). By contrast, if Pe_D were used as the dependent variable, the apparent microviscosities would have thinned at values of Pe_D that were 100–1000 times smaller.

How well nonlinear microrheology captures macrorheometry de-pends on the relative impact of these phenomena, which, in turn, depends on the specific materials being probed. Theoretical studies of the shear and extensional non-Newtonian rheology of dilute sus-pensions of hard spheres reveal both shear and extensional viscosities to thin with increasing flow strength, lessening the impact of mixed flow types on the nonlinear microrheology of suspensions of sphere (Brady and Morris, 1997). Additionally, rheological stresses within the bulk of the material (which are responsible for the rheology in-trinsic to the material) must impact the drag on the probe far more strongly than direct probe-bath collisions.

The picture becomes more complicated for more complex ma-terials; other effects may arise that are not identified here, so care should be taken. With an appropriate model for the material under study, one might compute the expected microrheological response, then compare such predictions with experiments to check the pre-dictive capabilities of the model, and to extract parameters for such models.

Computational nonlinear microrheology studies in concentrated, soft-particle pastes by Mohan *et al.* (2014) reveal yield stresses that agree well with those computed for (macroscopic) shear rheome-try; moreover, apparent (velocity-depentent) microviscosities agree

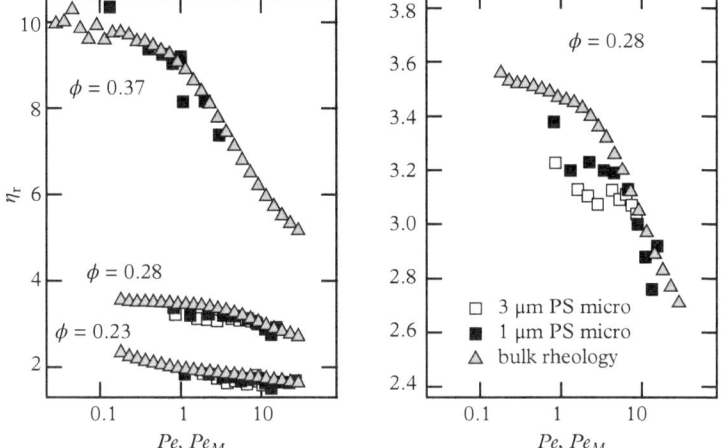

Fig. 7.14 *Bulk and laser tweezer microrheology of a fluorinated ethylene propylene (FEP) suspension with particle radii a_b = 78 ± 10 nm, at solids volume fractions ϕ =0.23, 0.28, and 0.37. Probe particles with radii a_p 0.5 and 1.5 µm are dragged at a steady velocity through the bath suspension. The bulk shear thinning is measured by steady shear in a Couette cell. The right-most plot is the ϕ = 0.28 data. Replotted based on the data by Meyer et al. (2006).*

reasonably well with (shear-rate-dependent) viscosities, although the η_{app} exceed η_M slightly and systematically (Fig. 7.15), as might be expected from the spectrum of rheological responses excited around the probe (*e.g.*, Fig. 7.8). Mohan *et al.* (2014) attribute this agreement to the observation that the number of particles in contact with the translating probe, and the degree of compression of these particles, is approximately the same as in simple shear flow, despite differences in the detailed structure. Rich *et al.* (2011*b*) used magnetic forces to pull magnetic probes through Laponite suspensions, and reported yield stresses and shear thinning that agreed reasonably well with macroscopic measurements.

In making such comparisons, however, there is the need to consider the complex (and unknown) flow field, for example as computed by Beris *et al.* (1985) around a sphere in a Bingham fluid. We see in Fig. 7.16 the stream lines generated by a particle translating in a viscous Newtonian fluid compared with a sphere translating steadily in a Bingham fluid. In the latter, a "fluid" region of plastic deformation is surrounded by a yield surface. Such descriptions incorporate shear thinning into calculations, heterogeneities within the material, and other issues as factors that complicate the interpretation (and impact the agreement).

Nonlinear microrheology studies of worm-like micellar solutions have more mixed success. In particular, Gomez-Solano and Bechinger (2014) used laser tweezers to pull probes through WLM solutions, finding good agreement in the linear response regime, but a shear thinning that was delayed by up to a decade, relative to the macroscopic measurement. The systematic and detailed studies

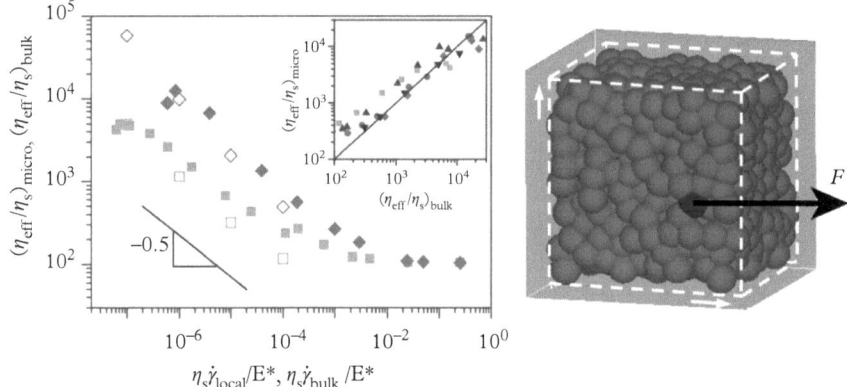

Fig. 7.15 *Computational experiments comparing the nonlinear microrheology and "macroscopic" simple shear rheology of concentrated, soft particle pastes, from Mohan* et al. *(2014).*

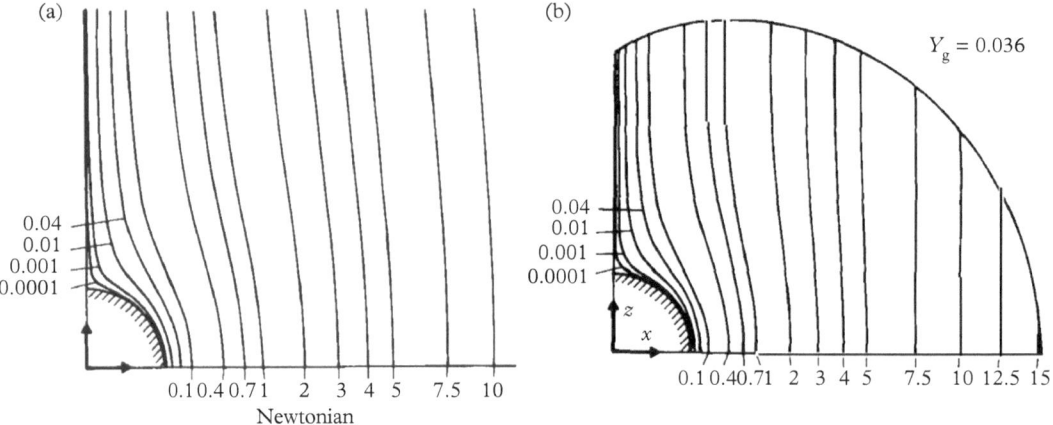

Fig. 7.16 *The calculated streamlines for a spherical particle translating in (a) a Newtonian fluid and (b) a yield stress fluid. Beris, A. N., Tsamopoulos, J. A., Armstrong, R. C., & Brown, R. A., Creeping motion of a sphere through a Bingham plastic,* J. Fluid Mech., *158, 219–44 1985, reproduced with permission.*

of Mohammadigoushki and Muller (2016), using gravity to drive spherical probe translation, even revealed cases where probe velocity was never steady, but oscillated periodically. They attributed these and other phenomena to the strong, qualitative differences between shear and extensional flows in WLM solutions, including the formation of "negative wakes" (Arigo and McKinley, 1998), WLM scission in the strongly extensional regions and the effect of strain history on that scission (Bhardwaj *et al.*, 2007), and even probe inertia. Contrasting these results are experiments studying probes translating through entangled biopolymer solutions, for which the shear thinning in the bulk flow curve is in good agreement with the GSR-derived apparent viscosity (Cribb *et al.*, 2013). We discuss these experiments further in Section 8.5.2.

The examples discussed in this chapter, however, suggest that nonlinear microrheology can be effective in measuring nonlinear rheological properties for some materials, under certain circumstances. In other cases, both quantitative and qualitative differences appeared. In all cases, however, some knowledge of the material itself is required to properly interpret the results. Given that nonlinear microrheology using translating-spherical probes is the small-scale analog of falling-ball viscometry, an index rheometry method, this should not be surprising.

7.4 Looking ahead

In this chapter, we've established several important principles for active microrheology. Of foremost importance are several clear differences from passive microrheology. Active microrheology methods are not limited by the relatively weak thermal motion and, in principle, expand the operating regime of microrheology to stiffer materials while retaining small sample volumes and other advantages. Second, as we've seen, active microrheology can be used to measure active non-equilibrium systems. Finally, like mechanical rheometry, active microrheology can drive a material out of equilibrium, potentially leading to active nonlinear microrheology measurements. However, the startling ability of passive microrheology to quantitatively capture linear response rheology, as enabled by the Correspondence Principle, does not generally translate to nonlinear microrheology. Clearly, more research is needed to establish what information can be determined with confidence using nonlinear microrheology.

In Chapters 8 and 9, we will discuss active microrheology experiments, including their implementation, limits, and some examples from the literature. In Chapter 8 we focus on magnetic bead microrheology and in Chapter 9 on experiments that use laser tweezers.

Magnetic bead microrheology

Magnetism produces a convenient force for actively pulling colloidal particles in a material. Many materials of interest in a microrheology experiment have a negligible magnetic susceptibility, and so embedded magnetic particles can be subject to strong forces by fields imposed from outside of the sample. These are usually generated by electromagnets, but can also include the use of permanent magnets, or a combination of both.

Like laser tweezer experiments discussed in the next chapter, these so-called magnetic tweezers are used as sensitive force probes. Capable of generating forces ranging from femtonewtons to nanonewtons, magnetic tweezers have been used to study mechanics of many soft biological materials and systems. Of course, these experiments date back to pioneering work in the early twentieth century, as was discussed at the beginning of Chapter 1. Measurements like those of Heilbronn (1922) and Freundlich and Seifriz (1923) in the 1920s, especially work focused on the mechanics of the cell, were repeated in the following decades as the physics, chemistry, and biology of cells became clearer, and newer methods, especially imaging technologies, developed (Yagi, 1961; Holliday, 1947; Hiramoto, 1969; White, 1980; Sato *et al.*, 1983). Crick and Hughes (1950), for instance, performed magnetic bead microrheology experiments on chick embryos, but employed high-speed film to track the particle displacements.

Contemporary experiments, such as the one shown in Fig. 8.1, include computer-controlled electromagnets and video microscopy or laser interferometry to track the displacement of probe particles. Like laser tweezers, discussed in the next chapter, magnetic tweezers are broadly used in biophysics and soft condensed matter as micromanipulators and dynamometers with piconewton force resolution. Materials of interest are natural and artificial biopolymers, protein assemblies, molecular motors, and individual cells (Gosse and Croquette, 2002). Such studies include measurements of the extension and twist of DNA (Haber and Wirtz, 2000; Kruithof *et al.*, 2008), the stiffness of cells and cellular constructs like the cortical cytoskeleton

Microrheology. Eric M. Furst and Todd M. Squires, Oxford University Press (2017).
© Eric M. Furst and Todd M. Squires. DOI 10.1093/oso/9780199655205.001.0001

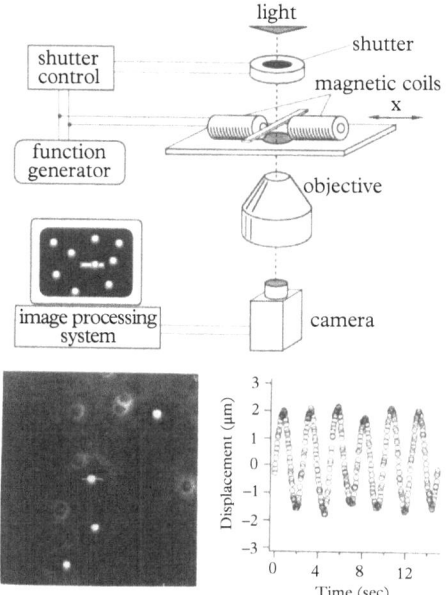

Fig. 8.1 *Magnetic tweezers manipulate microrheology probe particles using a magnetic field. Reprinted from* Biophys. J., *66, Ziemann, F., Radler, J., & Sackmann, E., Local measurements of viscoelastic moduli of entangled actin networks using an oscillating magnetic bead micro-rheometer, 2210–16, Copyright (1994), with permission from* The Biophysical Society.

(Zaner and Valberg, 1989; Bausch *et al.*, 1998; Fabry *et al.*, 2001), and molecular-scale interactions, including the bonding forces of ligands and receptors Neuman and Nagy (2008). Nonetheless, not all of the early active microrheology experiments focused on biological systems like cells. In one work, Freundlich and Roder (1938) pulled millimeter-diameter metal spheres through a slurry of quartz powder in attempt to measure the thixotropy and dilatancy of these particle suspensions. We will see that several contemporary studies attempt to study similar non-Newtonian rheology.

We begin this chapter with a brief review of magnetism and magnetic particles. Following this, we will discuss aspects of the experimental design of magnetic bead microrheology experiments. We will then present a few of its applications.

8.1 Magnetism

We begin by reviewing a few essential concepts of magnetism and magnetic materials. For more in-depth discussions, the reader is referred to comprehensive texts on the subject, as well as more general treatments. We've found those by Bozorth (1978), Spaldin (2011) and of course Feynman, Leighton, and Sands (1964) to be particularly helpful.

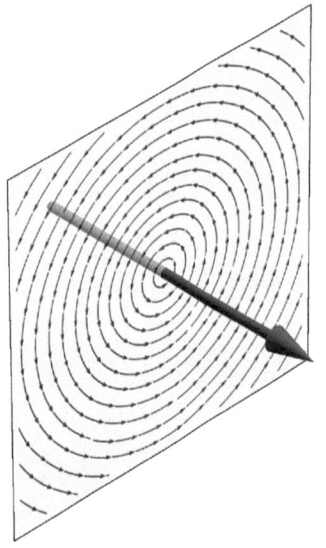

Fig. 8.2 *A representation of the magnetic field **B** around a wire carrying a current.*

<hr/>

[1] We will use SI units. See Table 8.1 for common unit conversions. Recall the relation between the speed of light, vacuum permeability, and vacuum permittivity, $c^2 = \frac{1}{\varepsilon_0 \mu_0}$.

8.1.1 Fields generated by electrical currents

Moving charges induce magnetic fields. This phenomenon was first reported by Hans Christian Oersted in 1820 and summarized in the Law of Biot-Savart for electrical currents conducted by wires in cases where the current does not vary with time—the domain of *magnetostatics*. We will summarize a few convenient results.

The magnetic field **B** generated by an electrical charge q moving with velocity **v** is[1]

$$\mathbf{B}(\mathbf{r}) = \frac{\mu_0}{4\pi} \frac{q\mathbf{v} \times \hat{\mathbf{r}}}{r^2} \qquad (8.1)$$

where the vacuum permeability is $\mu_0 = 4\pi \times 10^{-7} \text{N/A}^2$. The vector sum of charges moving in a conductor with current I over a path **s** is

$$\mathbf{B}(\mathbf{r}) = \frac{\mu_0}{4\pi} \int \frac{I d\mathbf{s} \times \hat{\mathbf{r}}}{r^2}, \qquad (8.2)$$

the law of Biot–Savart. From this relation, one can show that current flowing through a long, straight wire generates a magnetic field

$$B(r) = \frac{\mu_0 I}{2\pi r}. \qquad (8.3)$$

The field lines of this vector field are plotted in Fig. 8.2 in a plane perpendicular to the conductor. The field strength decays as $\sim 1/r$ from the wire.

Table 8.1 *Magnetic units.*

Quantity	SI unit	cgs unit	conversion
magnetic field, **B**	tesla (T) *or* $\text{Nm}^{-1}\text{A}^{-1}$	gauss (G)	$1\,\text{G} = 10^{-4}\,\text{T}$
field intensity, **H**	A/m	oersted (Oe)	$1\,\text{Oe} = (10^3/4\pi)\,\text{A/m}$
magnetization, **M**	A/m	gauss (G) *or* emu/cm^3	$1\,\text{G} = 10^3\,\text{A/m}$
magnetic moment, **m**	Am^2		
susceptibility, χ	dimensionless	$\text{emu/(cm}^3\,\text{Oe)}$	1 (cgs) = 4π (SI)

vacuum permeability (SI) $\mu_0 = 4\pi \times 10^{-7} \text{N/A}^2$

In microrheology, electromagnets are normally wound in circular coils. The field around a current loop is shown in Fig. 8.3. The magnetic field is torroidal. The field strength along the central axis of a current loop with radius b is

$$B_z = \frac{\mu_0 I b^2}{2(z^2 + b^2)^{3/2}} \tag{8.4}$$

and a coil composed of N loops is

$$B_z = \frac{\mu_0 N I b^2}{2(z^2 + b^2)^{3/2}}. \tag{8.5}$$

The field gradient is

$$\frac{\partial B_z}{\partial z} = -\frac{3\mu_0 N I b^2 z}{2(z^2 + b^2)^{5/2}}. \tag{8.6}$$

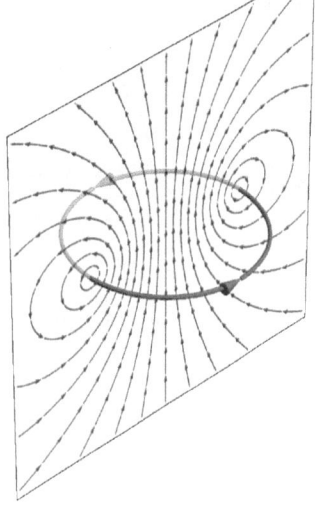

Fig. 8.3 *The magnetic field of a current loop.*

Two coils arranged along their axes with a separation of one radius is called a Helmholtz coil. The superposition of the field from both coils (provided the current flows in the same direction) leads to a uniform field at their center with a vanishing field gradient. Pairs of Helmholtz coils arranged orthogonally can be used to generate rotating magnetic fields when the current to each pair is oscillated sinusoidally and 90 degrees out of phase.

Most coils incorporate a metal such as "soft iron" or specialty alloys in their core like mu-metal, permalloy, and supermalloy. The metal core provides a much higher inductance compared to air, strengthening and focusing the magnetic field. The mechanisms of this induction are discussed next, as well as some implications, both in the design and operation of magnetic bead microrheology experiments.

A typical electromagnet used in a microrheology experiment generates a maximum magnetic field on the order of 0.1 T. To put this value in some perspective, the Earth's magnetic field ranges from 25 to 65 μT (0.25 to 0.65 G) on its surface, and a strong refrigerator magnet has a maximum field of about 0.01 T (100 G). Some permanent magnets and electromagnets with metal cores and sharpened poles can produce fields in the order of 1 T. Superconducting magnets used for nuclear magnetic resonance (NMR) and magnetic resonance imaging (MRI) are much stronger, on the order of several teslas.

8.1.2 Magnetic materials

Magnetism in a material is ultimately a quantum mechanical phenomenon. In addition to carrying charge, electrons have a magnetic spin. In magnetic materials, the spins of electrons in unfilled atomic or molecular orbitals respond to a magnetic field, like tiny magnetic dipoles. The spins tend to align in a field (or, for *diamagnetic* materials, antiparallel to the field). The dipole ordering in paramagnetic, ferromagnetic, anti-ferromagnetic, and ferrimagnetic materials are illustrated in Fig. 8.4.

A *paramagnetic* material is one in which thermal fluctuations of the dipoles maintain a random overall alignment, so there is no net magnetization in the absence of a field. In other materials, like *ferromagnets*, the spins spontaneously align, and the material can have a magnetization in the absence of a field, at least for temperatures below the Curie temperature—the transition temperature above which these materials lose their permanent magnetic properties. The spins tend to correlate over length scales called Weiss domains. *Anti-ferromagnetic* materials contain alternating dipoles, while in materials alternating dipoles differ in strength. Because of the presence of antiparallel spins, ferrimagnetic materials may exhibit magnetization reversal on heating.

When a magnetic material is placed in a magnetic field with intensity **H**, the total magnetic field **B** is the vector sum of the magnetic field strength $\mu_0\mathbf{H}$ and the material's intrinsic *magnetization* **M**,

$$\mathbf{B} = \mu_0(\mathbf{H} + \mathbf{M}). \tag{8.7}$$

At low field intensities, the magnitude of the magnetization is proportional to H,

$$M = \chi_0 H \text{ as } H \to 0. \tag{8.8}$$

where χ_0 is the magnetic susceptibility of the material. The magnetization, though, tends to grow nonlinearly, and more importantly, it is bounded. At high field intensities, the magnetization reaches a maximum,

$$M = M_{sat}, \tag{8.9}$$

known as the *saturation magnetization*. In this case, increasing the magnetic field does not yield further induction. All of the spins have aligned in the material. Alternatively, it is common to define the magnetic susceptibility over the entire range of field strengths as the ratio

$$\chi(H) = M(H)/H. \tag{8.10}$$

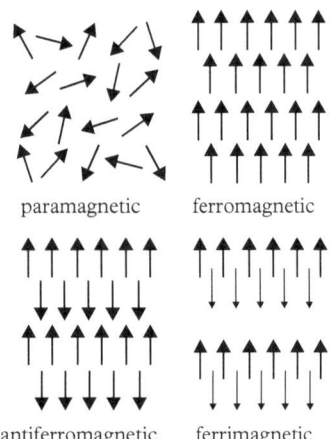

paramagnetic   ferromagnetic

antiferromagnetic   ferrimagnetic

Fig. 8.4 *The arrangement of magnetic dipoles in paramagnetic, ferromagnetic, anti-ferromagnetic, and ferrimagnetic materials.*

The low- and high-field intensity limits of the magnetization curve are captured well by the empirical Fröhlich–Kennelly equation (Bozorth, 1978),

$$M = H \left(\frac{1}{\chi_0} + \frac{|H|}{M_{sat}} \right)^{-1}, \tag{8.11}$$

which is plotted in Fig. 8.5.

A material's magnetization is often characterized by the permeability μ, a quantity that relates the *total* magnetic field to the field intensity by

$$B = \mu H, \tag{8.12}$$

or by the dimensionless *relative permeability*, K_m,

$$B = \mu_0 K_m H, \tag{8.13}$$

where

$$K_m = 1 + M/H. \tag{8.14}$$

In the limit of a weak field intensity,

$$K_m = 1 + \chi_0. \tag{8.15}$$

Tables of material magnetic properties will often list the permeability or susceptibility.

Magnetization like that described by the Fröhlich-Kennelly equation 8.11 applies to *paramagnetic* materials. *Ferromagnetic* materials, by contrast, exhibit *coercivity*. The magnetization curve has hysteresis and the material can remain magnetized in the absence of a field—it is a "permanent" magnet. An example of magnetic hysteresis is

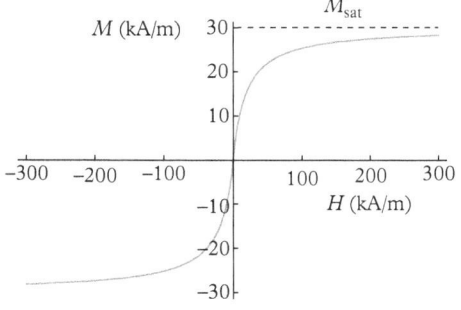

Fig. 8.5 *A magnetization curve calculated by the Frölich–Kennelly equation (8.11) for a material with magnetic susceptibility $\chi_0 = 1.6$ and saturation magnetization $M_{sat} = 30$ kA/m.*

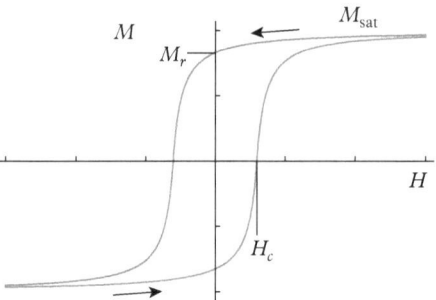

Fig. 8.6 *Magnetization of a ferro-magnetic material with saturation magnetization M_{sat}, and coersion H_c. The arrows indicate the direction of the magnetization hysteresis loop. When there is no applied field, the material maintains a remanent or residual magnetization M_r.*

plotted in Fig. 8.6. In a positive field H, the material magnetizes and eventually the magnetization saturates. Lowering the field from this point, the magnetization doesn't return to zero as $H \rightarrow 0$. Instead, it retains a *remanent* or *residual* magnetization M_r. The magnetization only returns to zero if the field is reversed to the coercive field intensity H_c.

Ferromagnetic materials are classified as hard or soft depending on their coercivity and remanence. Soft magnets are more easily saturated, but are also more easily demagnetized. These materials are useful in electromagnets and transformer coils, which require the direction of magnetization to be easily reversed. Materials with smaller coercivity are also used in magnetic storage devices, like hard drives and, in the past, magnetic tape and magnetic-core memory. Hard magnetic materials have high remanence and require a large field to reduce their inductance to zero—a useful property for permanent magnets.

8.2 Magnetic tweezers

"Magnetic tweezer" experiments use an external field to drive the translation (and possibly rotation) of a magnetic probe particle. The displacement of the probe is imaged by a microscope and measured most commonly by brightness-weighted centroid tracking methods that are the basis of multiple particle tracking (see Section 4.4), but laser tracking (discussed in Chapter 6) is also possible.

The force acting on a magnetic bead depends on whether the particle has a permanent magnetic moment or if the moment is induced by the external field. The particle magnetic moment **m** has SI units $A \cdot m^2$. Its magnitude for spheres is $m = V_p M$, where M may be the residual magnetization or induced magnetization.

The energy of a polarizable particle in an external magnetic field is $-\mu_0 \mathbf{m} \cdot \mathbf{H}$. The work it takes to bring the particle to this field strength is

$$w = -\mu_0 \int_0^H \mathbf{m} \cdot \hat{\mathbf{H}} dH, \tag{8.16}$$

which is equal to the magnetic potential energy of the particle, U_{mag}. The total magnetic energy of the particle depends on whether its moment is induced by the field, which would be the case for a paramagnetic probe, or constant like a ferromagnetic particle.

In the case that the particle moment is permanent, then its magnetic energy is simply

$$U_{mag} = -\mu_0 \mathbf{m} \cdot \mathbf{H}, \tag{8.17}$$

and the force acting on it is

$$\mathbf{F} = -\nabla U_{mag} = \mu_0 \nabla (\mathbf{m} \cdot \mathbf{H}). \tag{8.18}$$

Assuming that the moment aligns with the field (since it is torque free), the force acting on the particle pulls it along the field gradient,

$$\mathbf{F} = -\nabla U_{mag} = \mu_0 m \nabla H. \tag{8.19}$$

If the magnetic moment of the particle is *induced* by the field, then the force acting on it depends on the gradient of the *square* of the field. We can understand this dependence by considering the limiting case of a weak field, such that the magnetic moment of the particle is linear with the applied field strength,

$$\mathbf{m} = V_p \chi_0 \mathbf{H} \tag{8.20}$$

where V_p is the volume of the polarizable particle and χ_0 is the magnetic susceptibility (we will discuss the magnetization of particles in greater detail shortly). Equation 8.16 gives the magnetic energy

$$U_{mag} = -\mu_0 V_p \chi_0 \int_0^H H dH = -\frac{1}{2} \mu_0 V_p \chi_0 H^2 \tag{8.21}$$

and force

$$\mathbf{F} = \frac{1}{2} \mu_0 V_p \chi_0 \nabla (H^2). \tag{8.22}$$

In general, when the magnetic moment $\mathbf{m}$ depends on field strength (and may saturate) and the field is a vector field of position $\mathbf{H}(\mathbf{x})$,

$$\mathbf{F} = \mu_0 \mathbf{m} \cdot \nabla \mathbf{H}. \tag{8.23}$$

This follows when we make the substitution $d\mathbf{H} = \nabla H \cdot d\mathbf{x}$ in eqn 8.16, rewriting the integral in terms of the spatial coordinate instead of the field strength.

8.2.1   Magnetic probes

While early microrheology experiments used ferromagnetic particles, typically iron oxide or nickel particles separated from a fine powder, the most common magnetic probe particles used today are paramagnetic. As we saw in Section 8.2, unlike ferromagnetic particles, which hold a magnetic moment, paramagnetic particles have no remanent magnetization.

In this section, the magnetization of colloidal probes is discussed in terms of their mechanisms and model equations. Since microrheology experiments typically calibrate the force on a magnetic bead using a material of known rheology, the probe magnetization is not critical, unless the bead motion is used to map out the field strength, as Rich *et al.* (2011*a*) demonstrate, or during the initial design of an instrument and its operating range. It is good practice to measure the magnetic characteristics of the probes when possible using instruments such as a superconducting quantum interference device (SQUID) or vibrating sample magnetometer (VSM).

Superparamagnetic colloids

Paramagnetic probe particles, or *super*paramagnetic as they are often called,[2] are typically monodisperse latex spheres with iron oxide nanoparticles dispersed throughout their matrix. The nanoparticles are typically maghemite, γ-Fe_2O_3 or possibly magnetite, iron(II,III) oxide, Fe_3O_4 (Fonnum *et al.*, 2005). The small dimensions of the nanoparticles causes these normally ferrimagnetic iron oxides to exhibit strong paramagnetism, as first described by Néel (1949) and Brown (1963). The nanoparticles are embedded in the glassy polymer matrix, but their magnetic moments thermally fluctuate. The average lifetime of the fluctuation depends on the volume of the nanoparticle, and is given by the Néel-Arrhenius time,

$$\tau_N = \tau_0 e^{KV/k_B T} \tag{8.24}$$

where τ_0 is typically on the order of 10^{-10}–10^{-9}s (Zhang *et al.*, 1996), K is the magnetic anisotropy energy density,[3] and V is the volume of the particle. The small volume and dispersion of the nanoparticles ensure that their moments fluctuate on time scales $\sim 1 - 10$ ns. Because the nanoparticles and their moments have a random orientation at any instant (see Fig. 8.7), the larger composite colloid lacks a net magnetic moment in the absence of a field. The application of a field causes the paramagnetic nanoparticles magnetize preferentially in the field direction. The paramagnetic behavior modeled in Fig. 8.6 is based on values reported for 4.5 μm diameter magnetic-latex particles with an

[2] Paramagnetism is normally a weak effect compared to ferromagnetism. Paramagnetic materials have susceptibilities on the order of 10^{-5}, while ferromagnetic (or ferrimagnetic) materials have susceptibilities that are many orders of magnitude larger. Ferromagnetism and ferrimagnetism both hold a spontaneous magnetization at sufficiently low temperatures (below their Curie temperatures). These properties differ mainly in their temperature dependence.

[3] Johansson *et al.* (1997) report $K = 4.5 \times 10^4 \text{J/m}^3$ for 8 nm iron oxide nanoparticles.

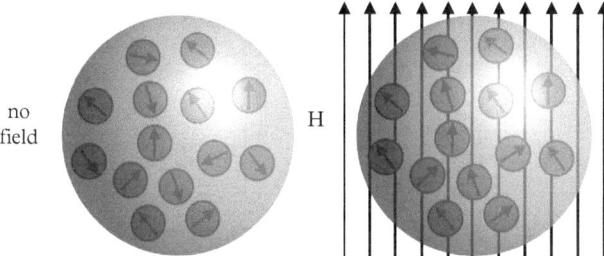

Fig. 8.7 *The magnetization of paramagnetic nanoparticles embedded in the polymer matrix of a latex particle are the basis for superparamagnetic collodial probes used in many magnetic tweezer microrheology experiments. In the absence of a field, the probe has no moment due to the random orientation of the nanoparticle moments. In an applied field, the nanoparticles magnetize preferentially in the field direction.*

iron nanoparticle content of about 20 wt% (Rich *et al.*, 2011*a*). The susceptibility of the particles in this case is reported to be $\chi_0 = 1.6$ and the saturation magnetization is $M_{\text{sat}} = 30$ kA/m. Due to the short Néel-Arrhenius time of the nanoparticles, the susceptibility of superparamagnetic latex particles has a negligible dependence on the field frequency in the range of interest to microrheology experiments (Kuipers *et al.*, 2008).

Next, we will calculate the magnetic force acting on a paramagnetic probe. In the case of paramagnetic particles, the magnetic moment is induced solely by the external field,

$$\mathbf{m}(H) = V_p M(H)\hat{\mathbf{H}}, \tag{8.25}$$

where $V_p = \frac{4}{3}\pi a^3$ is the particle volume. The magnetization implicitly accounts for the demagnetizing field of a sphere.[4] The induced moment points in the field direction, indicated by the unit vector $\hat{\mathbf{H}}$. Substituting the Fröhlich-Kennelly equation 8.11 for M and using eqn 8.16, the force generated on the particle is

$$\mathbf{F} = \mu_0 V_p \left(\frac{1}{\chi_0} + \frac{H}{M_{\text{sat}}} \right)^{-1} \mathbf{H} \cdot \nabla \mathbf{H} \tag{8.26}$$

which can also be written in terms of the gradient of the squared-field magnitude,

$$\mathbf{F} = \frac{1}{2}\mu_0 V_p \left(\frac{1}{\chi_0} + \frac{H}{M_{\text{sat}}} \right)^{-1} \nabla (H^2). \tag{8.27}$$

[4] If the bulk magnetic latex (as a dispersion of iron nanoparticles in the low-permeability polystyrene matrix) has an intrinsic susceptibility χ_i, then a spherical particle will have a susceptibility $\chi = \chi_i/(1 + \chi_i/3)$ (Landau *et al.*, 1984).

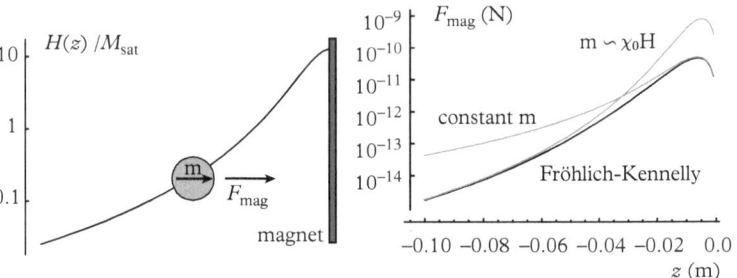

Fig. 8.8 *The force exerted on a probe particle with diameter 4.5 μm, M_{sat} = 31 kA/m and χ_0 = 1.6 for a magnetic field generated along the axis of a 1 inch diameter coil with 100 turns and a core composed of soft iron.*

We see that both limits represented by eqns 8.19 and 8.22 are recovered when the field saturates $\chi_0 H \sim M_{sat}$ and in the limit of a weak field, $H \ll M_{sat}$.

An example force calculation is plotted in Fig. 8.8 for a particle in a magnetic field generated by an electromagnet with a core of soft iron. The black curve represents the force using the full Frölich–Kennelly equation, 8.26, for the magnetization. The dashed lines represent forces calculated for a constant magnetization using the saturation magnetization, $m = V_p M_{sat}$ and another when saturation is ignored, $m = V_p \chi_0 H$. In this case, the force exhibits a maximum and then decreases as the field becomes more constant near the magnet.

The magnetization of superparamagnetic probes can also be modeled by the Langevin model of paramagnetism (Langevin, 1905), which provides further physical insight into the mechanism. Each nanoparticle in the probe colloid has a magnetic moment $\mathbf{m}_{np}$. The energy of the nanoparticle magnetic moment in an applied field is

$$U = -\mu_0 \mathbf{m}_{np} \cdot \mathbf{H}. \tag{8.28}$$

The magnetization of the probe is related to the number of nanoparticles it contains, N, and the Boltzmann-weighted average $\langle \hat{\mathbf{m}}_{np} \cdot \hat{\mathbf{H}} \rangle$,

$$M = N m_{np} \langle \hat{\mathbf{m}}_{np} \cdot \hat{\mathbf{H}} \rangle \tag{8.29}$$

where

$$\langle \hat{\mathbf{m}}_{np} \cdot \hat{\mathbf{H}} \rangle = \frac{\int_0^\pi e^{\mu_0 m_{np} H \cos\theta / k_B T} \cos\theta \sin\theta \, d\theta}{\int_0^\pi e^{\mu_0 m_{np} H \cos\theta / k_B T} \sin\theta \, d\theta}. \tag{8.30}$$

Integrating, this equation becomes

$$\langle \hat{\mathbf{m}}_{np} \cdot \hat{\mathbf{H}} \rangle = \coth\left(\frac{\mu_0 m_{np} H}{k_B T}\right) - \frac{k_B T}{\mu_0 m_{np} H}, \tag{8.31}$$

which is a form called the Langevin function,

$$f_L(x) = \coth x - \frac{1}{x}. \tag{8.32}$$

Thus, the magnetization is

$$M = Nm_{np} \left[\coth \left(\frac{\mu_0 m_{np} H}{k_B T} \right) - \frac{k_B T}{\mu_0 m_{np} H} \right]. \tag{8.33}$$

Equation 8.33 is functionally similar to the empirical Fröhlich-Kennelly equation (8.11) in that it exhibits an initial linear increase, than plateaus at a saturation value $M_{sat} = Nm_{np}$ that is a product of the number of nanoparticles and their magnetic moment. A Taylor series expansion of the Langevin function, about $x = 0$ gives $f_L(x) = x/3 + \mathcal{O}(x^3)$, and enables us to calculate the initial susceptibility

$$\chi_0 = \frac{\mu_0 N m_{np}^2}{3 k_B T}. \tag{8.34}$$

Thus, we see that the Langevin model also captures the general temperature trend of the magnetization—the susceptibility decreases as the temperature increases, a relation known as Curie's Law. At higher thermal energies, the magnetic moments of the nanoparticles are able to populate more energetically costly orientations, which decreases the net moment of the parent magnetic colloid. With the initial susceptibility and saturation magnetization one can estimate the number of nanoparticles and the nanoparticle moment (Amblard *et al.*, 1996; Fonnum *et al.*, 2005). Finally, we can rewrite equation 8.33 as

$$M = M_{sat} \left[\coth \left(\frac{3\chi_0 H}{M_{sat}} \right) - \frac{M_{sat}}{3\chi_0 H} \right]. \tag{8.35}$$

The magnetization of two commercial superparamagnetic bead chemistries are shown in Fig. 8.9 and reported parameters, including the initial susceptibility and saturation magnetization, are summarized in Table 8.2 for these and several other particles. The data in Fig. 8.9a confirm the paramagnetic properties of the particles—there is no coersion or remanence as the field is ramped up and down.[5] Comparing the SQUID measurements of Amblard *et al.* (1996) and the VSM measurements of Fonnum *et al.* (2005) in Fig. 8.9b, we see that the magnetic properties of the M-280 Dynabeads have been remarkably consistent over time.

Despite the theoretical basis of the Langevin model (eqns 8.33 and 8.35), the magnetization of superparamagnetic probes tends to

[5] Shevkoplyas *et al.* (2007) report measurements of at one commercial "superparamagnetic" probe chemistry that did exhibit a weak remanence. Verifying the magnetic properties of probes is always prudent.

Table 8.2 *The magnetic properties of commercial superparamagnetic microrheology probes.*

Bead	diameter (μm)	CV (%)	density (g/cm^3)	iron content (mg/cm^3)	χ_0	M_{sat} (kA/m)	reference
Dynabead M-280	2.83	1.4	1.4	118	0.76	15	Fonnum *et al.* (2005)
					1.0	14	Amblard *et al.* (1996)
Dynabead M-450	4.40	1.2	1.6	202	1.6	31	Fonnum *et al.* (2005)
MagSense	~ 1		2.5		1.8	132.4	Lipfert *et al.* (2009)
MyOne	1.05	1.9	1.7	255	1.3	40	Fonnum *et al.* (2005)
					1.5	43.3	Lipfert *et al.* (2009)

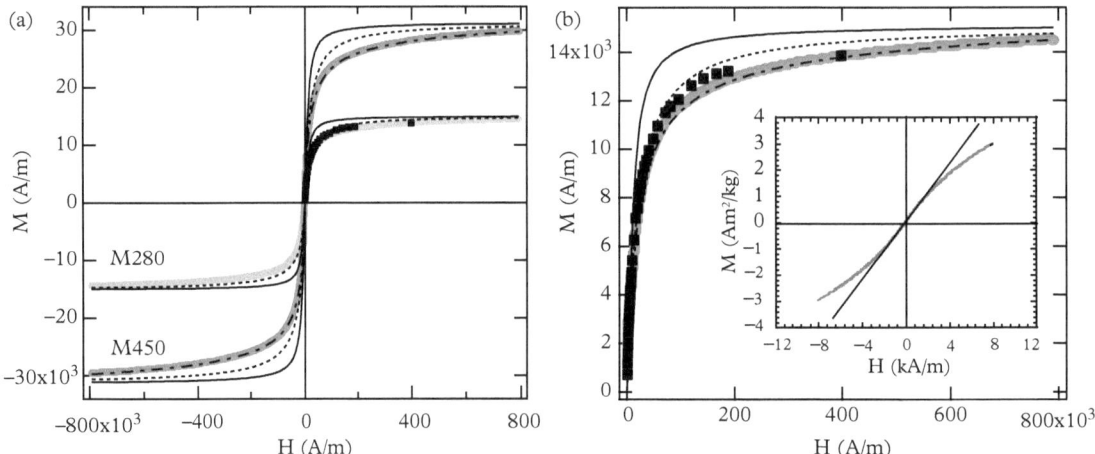

Fig. 8.9 *(a) Magnetization curves reported by Fonnum* et al. *(2005) for commercial superparamagnetic-probe particles (M-280 and M-250 Dynabeads, Invitrogen). The experimental measurements (gray symbols) are compared to the Langevin equation (8.35, solid line) and the the Frölich–Kennelly equation (8.11, dashed line) using the initial susceptibilities and saturization magnetizations given in Table 8.2. The solid symbols are data from Amblard* et al. *(1996). The range of the applied-field strength corresponds to ±1T. (b) The magnetization of M-280 beads, focusing on the top-right quadrant of the magnetization curve. The inset highlights the linear region, indicated by the straight black line. Note the lack of coercivity and remanence. In both plots, the dashed-dot lines are an empirical relation for the magnetization, equation 8.36.*

increase with the applied-field intensity more weakly than expected. The possible reasons for this deviation include the distribution of nanoparticle sizes, magnetic anisotropy of the nanoparticles, and nanoparticle clustering in the polymer matrix, which introduces correlations between the magnetic moments (Johansson *et al.*, 1997).

The empirical Frölich-Kennelly relation, eqn 8.11, is in better agreement, but it still tends to produce higher values of the magnetization, at most on the order of 5–10% mid-way between the linear magnetization regime and the saturation magnetization. An empirical function that fits the magnetization curve well is a modified form of the Fröhlich–Kennelly equation,

$$M(H) = H \left(\frac{1}{\chi_0} + \left[\frac{H}{M_0} \right]^\alpha \right)^{-1}. \tag{8.36}$$

This equation is plotted in 8.9a and b as the dash-dot line for M-450 ($\chi_0 = 1.6$, $M_0 = 25300 \pm 200$ A/m and $\alpha = 0.95 \pm 0.002$) and M-280 ($\chi_0 = 0.76$, $M_0 = 12500 \pm 100$ A/m and $\alpha = 0.96 \pm 0.002$) beads.

The magnetization of individual beads is suspected to vary, because of both the size dispersity of the particles and possible variations in the amount of magnetic material. Reddy *et al.* (1996) report that the standard deviation of the magnetization for Dynabead M-450 is 5%, while the standard deviation for M-280 particles was found later to be much higher, about 70% (Baselt *et al.*, 1998).

Ferromagnetic particles

Like paramagnetic particles, the translational force on a ferromagnetic particle is

$$\mathbf{F} = \mu_0 \mathbf{m} \cdot \nabla \mathbf{H}, \tag{8.37}$$

but here, **m** is generally a "permanent" moment, with a magnitude related, for instance, to the remanent magnetization,

$$m = V_p M_r \tag{8.38}$$

pointing in the direction of the unit vector $\hat{\mathbf{m}}$. By eqn 8.37, the force exerted on the particle depends on the angle between the moment and the applied field. But because ferromagnetic particles have a permanent magnetic moment, a torque is also generated,

$$\mathbf{L} = \mu_0 \mathbf{m} \times \mathbf{H} \tag{8.39}$$

and the particle moment will therefore tend to align in the field direction. This second property leads to the ability to use rotating fields to perform rotational microrheology, which will be discussed later in the chapter, but for anisotropic probes, such as rod-shaped particles.

8.2.2   Probe interactions

The probe concentration in a microrheology experiment must minimize the magnetic interactions between the particles. The induced field of a probe is dipolar, leading to the interaction potential

$$U_{\alpha\beta} = \frac{\mu_0}{4\pi\, r_{\alpha\beta}^3} \left(\mathbf{I} - 3\hat{\mathbf{r}}_{\alpha\beta}\hat{\mathbf{r}}_{\alpha\beta}\right) : \mathbf{m}_\alpha\mathbf{m}_\beta. \tag{8.40}$$

The maximum interaction occurs when particles are aligned in the field direction. In this orientation, the interaction potential is

$$U_{\alpha\beta} = -\frac{\mu_0 m^2}{2\pi\, r_{\alpha\beta}^3}. \tag{8.41}$$

8.3   Instrument designs

8.3.1   Electromagnet tweezers

In most magnetic tweezers, the field is generated by an electromagnet, although some recent designs use strong permanent magnets (Lin and Valentine, 2012a). The design can incorporate one or more magnets in typical configurations of one, two, or four poles, as shown in Fig. 8.10. The design of magnetic tweezers needs to balance the ability to generate the desired field (gradient) to generate sufficient forces with the sometimes competing need to incorporate an imaging system, including short working distance, high-numerical aperture microscope objectives.

Using one magnet, the magnetic tweezer experiment is similar to a bulk creep test, since one relates the displacement of the probe to a known *force* given by eqn 8.26. The single-magnet experiment was used by early investigators. The instrument of Freundlich and Seifriz (1923) is shown in Fig. 8.10a and serves as a general illustration of the design. An electromagnet generates the field. A high-permeability metal is included in the coil core to increase the field strength, which is calculated by the Law of Biot-Savart, eqn 8.2, replacing the vacuum permeability (in fact, the permeability of air) with the permeability of the metal. Thus, the axial field of a coil with N loops of radius b becomes

$$B_z = \frac{\mu N I b^2}{2(z^2 + b^2)^{3/2}} = \frac{K_m \mu_0 N I b^2}{2(z^2 + b^2)^{3/2}}. \tag{8.42}$$

The metal core increases the magnetic field strength by several orders of magnitude (see eqn 8.7). Core metals include several high-performance alloys, but soft iron is commonly used in magnetic

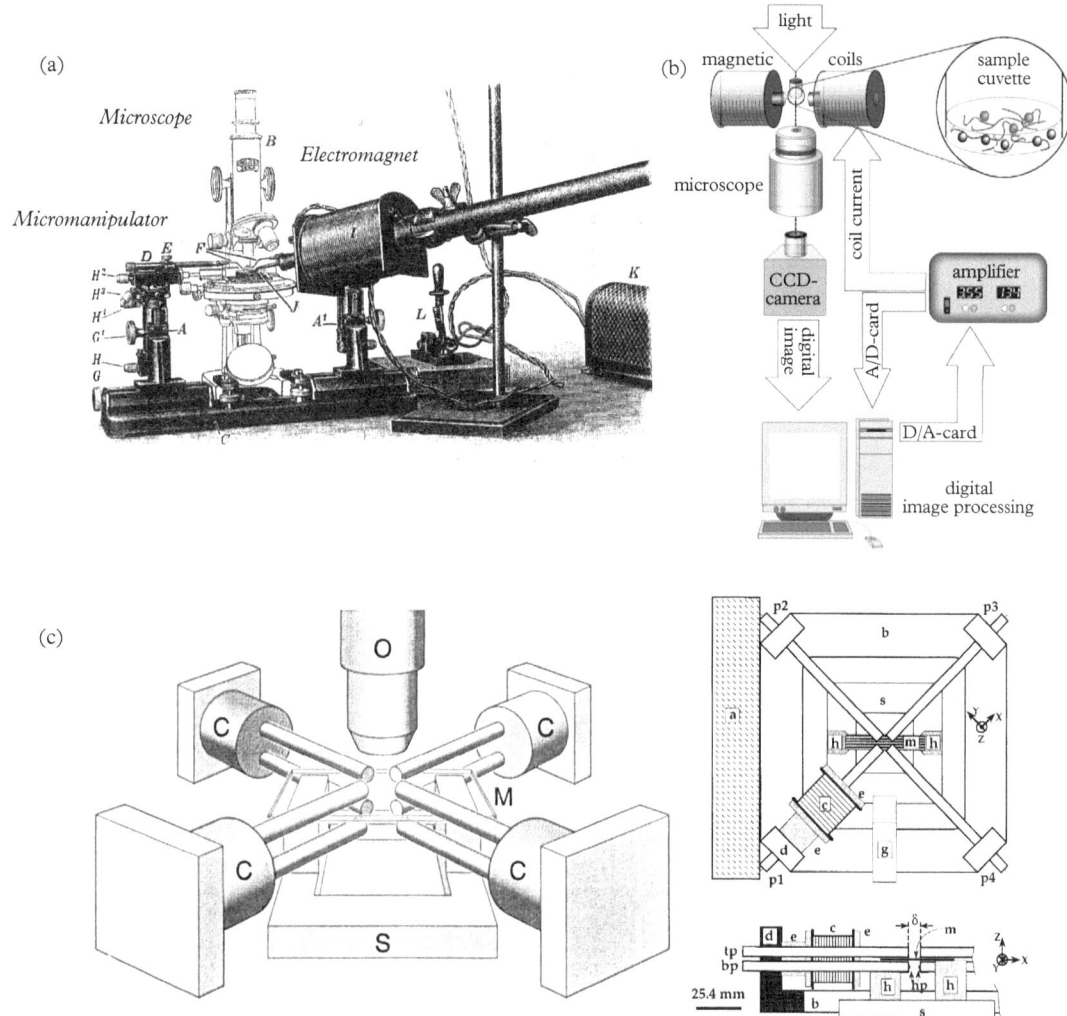

Fig. 8.10 *Magnetic tweezer experiments that incorporate one, two, and four electromagnets. (a) The instrument used by Freundlich and Seifriz (1923) was a microscope equipped with an electromagnet to pull nickel particles. A micromanipulator was used to position the particle. (b) A modern two-magnet tweezer is shown schematically on the right (Keller et al., 2001). (c) The design of Amblard et al. (1996) incorporates four magnets in a configuration of eight poles and can generate a rotating field. Reprinted from Amblard, F., Yurke, B., Pargellis, A., & Leibler, S. A., Rev. Sci. Instrum., 67, 818–27 (1996) with the permission of AIP Publishing.*

tweezers due to its lack of remanence and coercivity and relatively-high saturation magnetization.

Large field gradients are generated by sharpening the core to a tip and and inserting it directly into the material under test, sometimes very close to the probe particles. A microscope is used to view the probe displacement. A contemporary example of one such device, from Rich *et al.* (2011*a*), is shown in Fig. 8.11. Here, the soft iron core (CMI-C metal, CMI Specialty Products, Bristol, CT) forms a sharp, flat edge and is placed in the sample about 100 μm from the beads that will be tracked. The angle of the edge and magnet are such that the field gradient is strongest in the focal plane of the microscope. The core is wrapped with 300 turns of AWG 19 copper-magnet wire (bare diameter 0.912 mm). Magnet wire is insulated by a thin-enamel coating to allow for higher cross-sectional densities in the wound coil.

In the one-pole experiment, the probe can translate steadily (although not necessarily at a constant velocity), providing a measure of the medium's viscosity. Or the probe may exhibit a finite displacement in a viscoelastic solid. In the latter case, the recoverable compliance can be determined when the field is removed.

Incorporating a second electromagnet provides the means to generate oscillatory forces. In Fig. 8.10b, the apparatus of Keller *et al.* (2001) is shown in comparison to that of Freundlich and Seifriz (1923), and is representative of a modern design. A computer (or alternatively, a function generator) is used to control the output of a

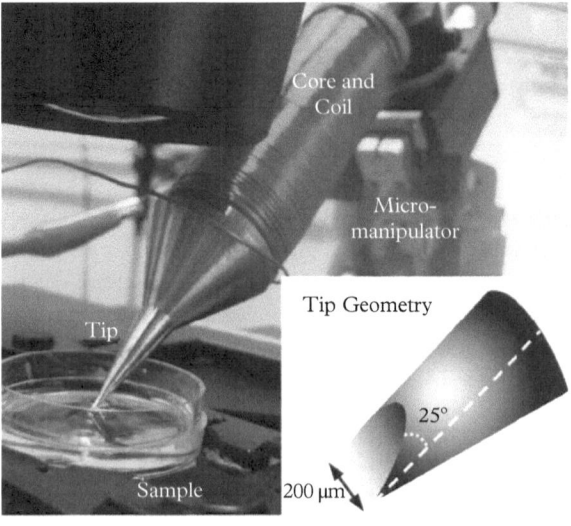

Fig. 8.11 *An electromagnet designed for high field strengths and gradients. A soft iron core is machined to a sharp, flat tip, which is inserted directly into the sample. Reproduced from Rich* et al. *(2011a) with permission of The Royal Society of Chemistry.*

power amplifier operating in current mode. Digital images are captured concurrently by a CCD camera and computer. The probe motion is tracked with respect to the corresponding time-dependent force, allowing its phase and amplitude to be measured.

Four electromagnets arranged at right angles provide control of the magnetic-field orientation in the focal plane. One example is the design of Amblard *et al.* (1996), shown in Fig. 8.10c. The instrument uses two soft-iron poles for each electromagnet in a configuration that aligns the field and field gradient with the microscope imaging plane. This arrangement enables the field to be rotated. Particles with permanent moments or anisotropic particles that align with the field (including self-assembled aggregates from multiple particles due to magnetic interactions—see eqn 8.41) can then be used for rotational microrheology. The separation of the coils from the sample also reduces heating due to resistive losses in the coils.

The choice of magnet geometry largely depends on the stress requirements of the experiments. A weak gel with a modulus on the order of only 0.1 Pa may require only a 0.1 T magnetic field and 10 T/m field gradient, leading to the requirement to generate a field of 0.1 T that varies over a distance of 1cm (Amblard *et al.*, 1996). External coils with multiple poles are sufficient. Materials in which probes need to generate stresses on the order of 100Pa require much higher field strengths and gradients, and favor sharpened poles inserted close to the beads (Rich *et al.*, 2011*a*).

An elegant solution for generating high field gradients in a compact geometry ideal for microscopy was developed by Fisher *et al.* (2006). The design, shown in Fig. 8.12, separates the pole tips, composed of either a lithographically deposited metal or micromachined thin-film alloy (*e.g.*, Permalloy) from the coils. The design is based on the insight that the sample gap makes the greatest contribution to the magnetic circuit reluctance, and hence, that the magnetic performance does not suffer with small gaps between the coils and poles. A backiron ring assures that the magnetic-flux circuit is completed (Huang *et al.*, 2002). The other advantage of this design is its flexibility for fabricating different pole geometries. Shown in 8.12c is a tip-flat geometry that maximizes the field gradient and applied forces, but other geometries can be used to induce torques or produce forces in all directions in the sample plane. The reported sample thickness ranges from 100–500 μm, which is sufficient to minimize interactions with the sample boundaries.

8.3.2 Tweezers with permanent magnets

Permanent magnets have been used recently as an alternative to electromagnets in magnetic-tweezer instruments (Lin and Valentine,

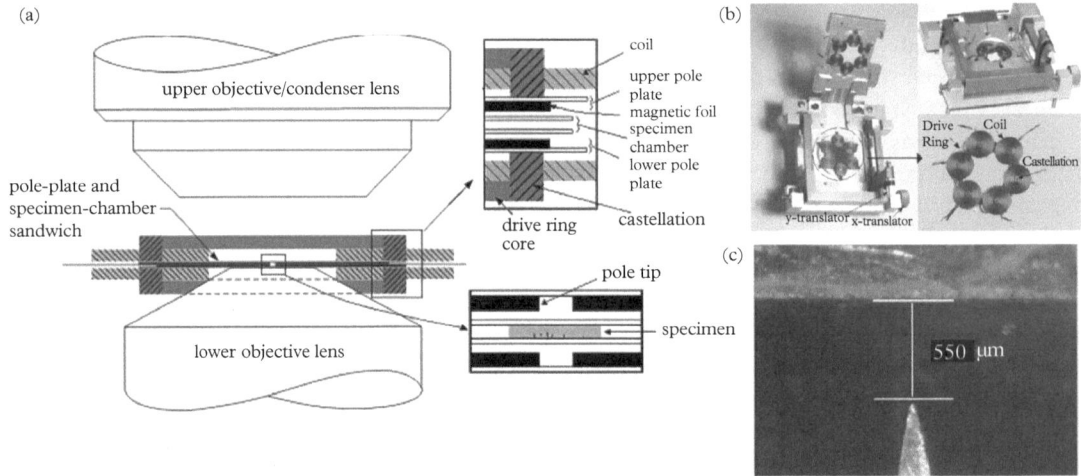

Fig. 8.12 *The design of thin-foil magnetic tweezers. (a) A side view of the design showing the thin magnetic film poles, drive rings, and coils. (b) Images of the assembly, which can be opened to exchange samples. (c) Two poles in a tip-flat geometry. Reprinted from Fisher, J. K. et al.,* Rev. Sci. Instrum., *77 (2006), with the permission of AIP Publishing.*

2012*a*,*b*; Zacchia and Valentine, 2015), a practice that extends from their numerous applications in biophysical measurements of cellular and single molecule mechanics (Lipfert *et al.*, 2009). Instead of controlling the magnetic field strength and induced force on probes by changing the flow of an electrical current, permanent magnets are physically moved relative to the particles—bringing the magnet closer to or further from the probes to control the field strength in the region of the particles. Rare earth magnets like neodymium-iron-boron (NdFeB) are usually used. These magnets have residual fields on the order of $B_r = 1$ T.

An example of a permanent magnet tweezer assembly is shown as a schematic in Fig. 8.13a. The device is mounted to a normal inverted microscope and consists of a translation stage that positions the magnet relative to the imaging plane. An LED-based illumination source is incorporated into the stage. The magnet assembly consists of several permanent magnets arranged to focus the field through a set of low carbon steel tips or "yokes" and low carbon steel backing. The field gradient is generated perpendicular to the image plane, and so particles are drawn up. This arrangement differs from most translation microrheology experiments, which attempt to drive probe motion in the focal plane. An image of the bead is used to track the displacement. A number of yoke and magnet geometries have been examined to optimize the design to provide the greatest field gradient (Zacchia and Valentine, 2015).

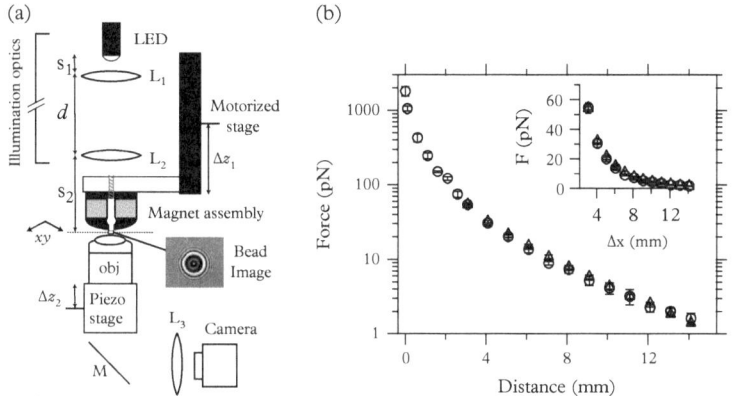

Fig. 8.13 *A permanent magnet assembly for magnetic tweezers. (a) A schematic of the assembly and (b) the calibrated force exerted on 4.5 μm diameter Dynabead M-450 probes versus the distance of the magnet to the sample plane. Reprinted from Lin, J., & Valentine, M. T., Rev. Sci. Instrum., 83, 53905 (2012a), with the permission of AIP Publishing.*

The yoke design is a key element to achieving high field gradients. As shown in the calibrated force plot in Fig. 8.13b, such tweezers are capable of generating on the order of 1 nN forces using larger 4.5 μm paramagnetic probes. The maximum force is limited by the field strength, which is governed by the magnetic saturation of the yoke material and the magnetic saturation of the probe particles. Because the force is generated by changing the magnet assembly's position, the performance of the mechanical translation system and its positioning accuracy is one of the key limits to generating reproducible forces.

Permanent magnets enable rather compact devices. Simple magnetic tweezers can be designed to be portable for mounting in a variety of imaging systems. So, for instance, tweezers capable of generating calibrated forces on probes can be more easily implemented in core imaging facilities or a collaborator's laboratory (Yang *et al.*, 2011). As a result, the rheology of soft materials can be studied while simultaneously measuring their microscale deformation or imaging other environmental cues, like the gradients of cross-linking or degradation agents, or the local mechanical activity of cells (Schultz, et al., 2013; Schultz et al., 2015).

8.3.3 Force calibration

The force generated by a magnetic particle is given by the probe magnetization, magnetic field strength, and field gradient. Field strengths and gradients can be calculated, for instance using eqns 8.5 and 8.6. But the presence of high-permeability cores and poles, often with complex geometries like sharpened tips, significantly increases the complexity of the calculation. Two approaches then to calibrate the force are to measure the field directly or use the Stokes drag of a particle in a rheological standard as a known force.

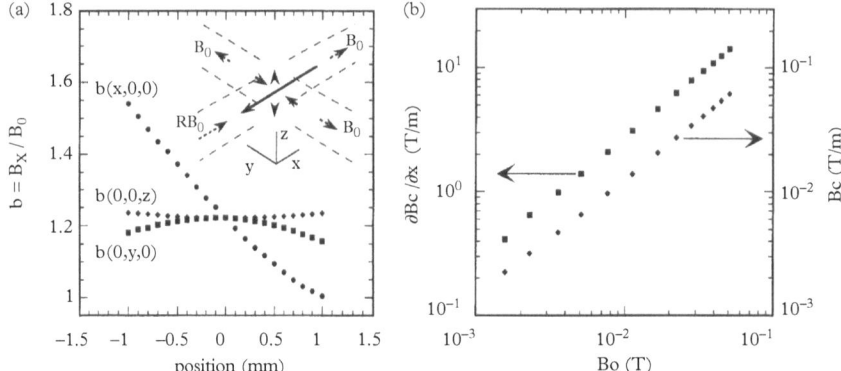

Fig. 8.14 *Measurements of the (a) field intensity and (b) field gradient of the instrument shown in Fig. 8.10c. B_0 = 51 mT is the pole face boundary field, which can be varied between 2–60 mT. Reprinted from Amblard, F., Yurke, B., Pargellis, A., & Leibler, S. A., Rev. Sci. Instrum., 67, 818–27 (1996) with the permission of AIP Publishing.*

Field measurements

An open electromagnet geometry like that of Amblard *et al.* (1996) (see Fig. 8.10c), which uses widely separated coils and multiple poles to focus the field close to the imaging plane, enable direct measurements of the field strength with a Hall probe. In Fig. 8.14, the field strength is plotted for all three axes when a gradient is generated along one pair of coils (x-axis) in the four-coil assembly. The field is linear along the x-axis and exhibits negligible gradients in the y- and z-directions over the imaging area of the microscope. The gradient increases with the field strength at the pole boundary face B_0, which can be adjusted from 2–60 mT by changing the current supplied to the coils.

Based on direct measurements of the field, the force exerted on the probes can be calculated by eqn 8.26 using their measured magnetization (shown in Fig. 8.9) combined with the field gradient. The forces exerted on Dynabead M-280 particles using the values shown in Fig. 8.14 are modest, ranging up to the order of 1–2 pN. The forces are confirmed by measuring the velocity of the beads as they move in a viscous, density matched solution of 4.81 molar solution of $CaCl_2$ with a specific gravity of 1.38 and viscosity 7.9 mPa·s at 20°C.

Stokes drag

When it is not feasible to accurately measure the field and field gradient or when it's desirable to verify the magnetic forces acting on a probe particle, the force can be calibrated by its movement through

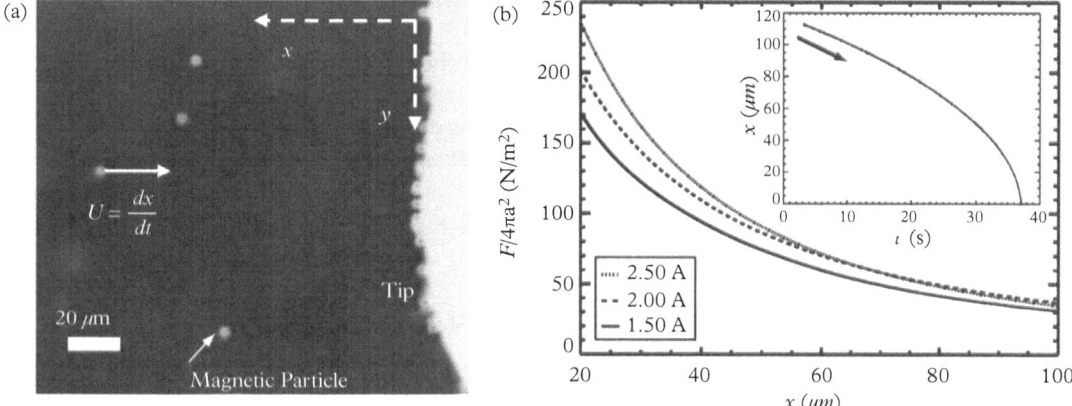

Fig. 8.15 *(a) Dynabead M-450 probe particles near the edge-shaped core tip of the device shown in Fig. 8.11. (b) The calibrated force generated by the probes scaled by the particle area $4\pi a^2$ as a function of distance from the core tip. The inset shows an example particle trajectory used to calculate the probe velocity. Reproduced from Rich et al. (2011a) with permission of The Royal Society of Chemistry.*

a material of known rheology, such as a Newtonian fluid. In this case, one relates the displacement of the particle as a function of time and from the (instantaneous) probe velocity V, simply calculates the (local) magnetic force from the Stokes drag,

$$F_m = 6\pi a\eta V. \qquad (8.43)$$

Example calibration data are shown in Fig. 8.15 for M-450 Dynabeads near the tip of the soft iron core in the electromagnet tweezer device shown in Fig. 8.11. The beads are dispersed in trimethylsiloxy-terminated polydimethylsiloxane (DMS-T43, Gelest, Morrisville, PA), a Newtonian liquid with a viscosity $\eta = 29.5$ Pa·s. The inset of Fig. 8.15 plots a sample bead trajectory as a function of time as the distance from the core tip (another example of calibration trajectories is shown in Fig. 8.16). By fitting the trajectory, the velocity relative to the distance from the tip $V(x)$ is calculated, and from this, the magnetic force. The force in Fig. 8.15 has been divided by the particle surface area $4\pi a^2$ to give an average stress exerted by the probe. For the highest coil current, 2.50 A, the force ranges from 0.6–4 nN. The force is three orders of magnitude greater than those in Fig. 8.14, owing to the proximity of the sharpened tip as well as the larger probe size. Decreasing the current coil by 40% reduces the force by about 30%. This dependence suggests that the particles are close to their saturation magnetization in the field generated by the tip.

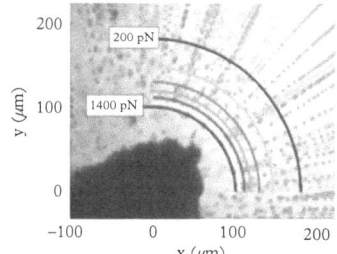

Fig. 8.16 *A time-lapse image of 4.5 μm paramagnetic beads moving near a magnetic pole. Calibration contours are superimposed. Adapted from Spero et al. (2008).*

With sharpened cores close to the probe particles, there is the likelihood that the field, and therefore force, experienced by probe particles is spatially inhomogeneous. Rich *et al.* (2011*a*) demonstrate this variation by mapping out probe motion with respect to the lateral edge of their sharpened core. (The tip geometry is illustrated in the inset of Fig. 8.11.) Significant variations in the magnetic force are measured for probes that are below about 20 μm separation from the magnet tip. This uncertainty limits the range of probe distances and forces that one can use to obtain accurate microrheology results.

Calibration uncertainty

As we saw earlier, the probe magnetization is important to characterize, but individual probe particles may vary in the content of magnetizable material and total magnetization. Indeed, Amblard *et al.* (1996) report that the force generated by their coils at the maximum-field strength for 30 beads is 1.9 pN with a standard deviation of 0.8 pN.

8.4   Linear experiments

8.4.1   Creep response

A typical experiment using magnetic tweezers measures the response of probe particles as the magnetic field is applied and removed by either turning on or off the electromagnet or quickly positioning a permanent magnet (Crick and Hughes, 1950; Ziemann *et al.*, 1994). The experiment is straightforward to implement because it only requires one magnet or pole.

An analysis of the experiment is analogous to a creep and recovery experiment. Starting with the equation of motion in the absence of inertia (see eqns 2.60 and 2.61),

$$\mathbf{F}_m + \int_{-\infty}^{t} \zeta(t-t')\mathbf{V}(t')dt' = 0, \tag{8.44}$$

the Laplace or Fourier Transform gives

$$\tilde{\mathbf{F}}_m + \tilde{\zeta}\tilde{\mathbf{V}} = 0, \tag{8.45}$$

in terms of the particle velocity, or

$$\tilde{\mathbf{F}}_m + s\tilde{\zeta}\tilde{\mathbf{X}} = 0 \tag{8.46}$$

in terms of its position. Solving for $\tilde{\mathbf{x}}$,

$$\tilde{\mathbf{X}} = \frac{\tilde{\mathbf{F}}_m}{s\tilde{\zeta}} \tag{8.47}$$

or

$$\tilde{\mathbf{X}} = \frac{\tilde{\mathbf{F}}_m}{6\pi a \tilde{G}(s)}. \tag{8.48}$$

With a step change in the applied magnetic force $F_m(t) = F_0 H(t)$, for which

$$\tilde{\mathbf{F}}_m = \frac{F_0}{s} \tag{8.49}$$

the displacement becomes

$$\tilde{\mathbf{X}} = \frac{F_0}{6\pi \, as \tilde{G}(s)} \tag{8.50}$$

or

$$\tilde{\mathbf{X}} = \frac{F_0}{6\pi a} \tilde{\mathcal{J}}(s). \tag{8.51}$$

Simply taking the inverse transform gives

$$\mathbf{X} = \frac{F_0}{6\pi a} \mathcal{J}(t). \tag{8.52}$$

The displacement is proportional to the creep compliance of the material. When the applied magnetic force is time-dependent, the particle displacement becomes a convolution of the *rate* of the applied force and the compliance,

$$\mathbf{X}(t) = \frac{1}{6\pi a} \int_0^t \mathcal{J}(t - t') \dot{\mathbf{F}}_m(t') dt' \tag{8.53}$$

in the linear response limit, analogous to the strain measured for a macroscopic shear stress (Ferry, 1980),

$$\gamma(t) = \int_0^t \mathcal{J}(t - t') \dot{\sigma}(t') dt'. \tag{8.54}$$

Plotted in Fig. 8.17 are several responses of probe particles in an entangled solution of F-actin proteins filaments. Probes are subjected to pulses of three durations. For the shortest, 0.5 and 1 s, the 0.5 pN

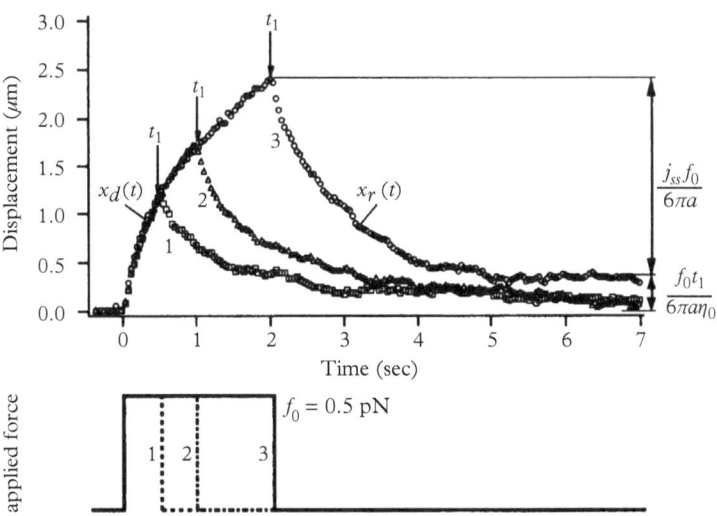

Fig. 8.17 *Bead motion in an en-tangled F-actin solution (0.3 mg/ml) in response to an applied field. The beads are 4.5 μm diameter superparamagnetic latex particles. Reprinted from* Biophys. J., 66, *Ziemann, F., Radler, J., & Sackmann, E., Local measurements of viscoelastic moduli of entangled actin networks using an oscillating magnetic bead micro-rheometer, 2210–6, Copyright (1994), with permission from The Biophysical Society.*

force causes the particles to displace, but they exhibit an elastic recovery to their initial positions. A longer pulse, 3 seconds in duration, pulls the particles into a terminal, linear trajectory, reflecting a viscous response, presumably beyond the longest-relaxation time of the material. When the field is removed at time $t = t_1$, only a portion of the particles' displacements are recovered,

$$X_r(t \to \infty) = \frac{J_e F_m}{6\pi a}. \tag{8.55}$$

The final displacement is non-zero,

$$X(t \to \infty) = \frac{F_m t_1}{6\pi a \eta_0} \tag{8.56}$$

due to viscous dissipation in the material. So, we can interpret the curves in terms of the recoverable and non-recoverable creep, associated with the elastic modulus and viscosity of the sample.

There is one important caveat for the creep experiment. Equation 8.52 is valid only in the linear response limit—the limit of small deformations or rates of deformation. If the deformation (or deformation rate) is not small, such that the viscoelastic behavior becomes nonlinear, the relation may not hold. But there is no way to tell *a priori* whether an applied force and subsequent response fall within the linear limit. The experiment should be performed with several applied magnetic forces to verify that the measured creep response remains independent of the force. For example, O'Brien *et al.* (2008) present

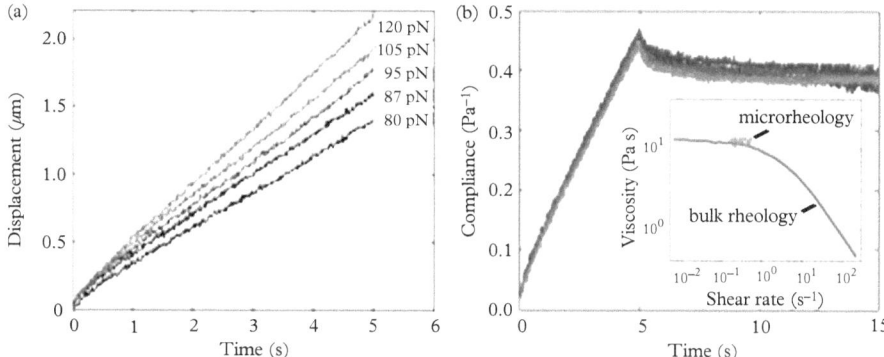

Fig. 8.18 *Magnetic bead microrheology of 10 mg/ml hyaluronic acid solutions. (a) Bead displacements measured at different applied forces. (b) The calculated compliance and a comparison of the viscosity to bulk measurements. Reprinted from Methods in Cell Biology, 89, OBrien, E. T., Cribb, J., Marshburn, D., Taylor, R. M., & Superfine, R., Magnetic Manipulation for Force Measurements in Cell Biology, 433–50, Copyright (2008), with permission from Elsevier.*

data for 1 μm diameter magnetic beads embedded in a 10 mg/mL solution of hyaluronic acid, which is reproduced in Fig. 8.18a. The solutions exhibit significantly less elasticity than the F-actin samples shown in Fig. 8.17. After a short transient period, the beads translate at a steady velocity which depends on the applied force. Plotting the data as the creep compliance (rearranging eqn 8.52), shows that the curves follow the same linear response (Fig. 8.18b). The inset confirms that the viscosity derived from the microrheology experiment is in good agreement with the low-shear viscosity reported using bulk rheology (Krause *et al.*, 2001). Also note the small degree of elastic recovery the compliance exhibits when the field is turned off.

8.4.2   Oscillating magnetic tweezers

An oscillating magnetic tweezer microrheology experiment requires two poles or magnets capable of pulling the probe particle with a force

$$F(t) = F_0 e^{i\omega t}. \tag{8.57}$$

We substitute this force into the rearranged frequency-domain equation of motion

$$\tilde{\mathbf{X}}(\omega) = \frac{1}{i\omega\tilde{\zeta}(\omega)}\tilde{\mathbf{F}}(\omega), \tag{8.58}$$

or recalling that $G^*(\omega) = 6\pi a i \omega \tilde{\zeta}(\omega)$,

$$\tilde{X}(\omega) = \frac{1}{6\pi a G^*(\omega)} \tilde{F}(\omega). \tag{8.59}$$

In the time-domain, this gives a general solution for the particle position

$$X(t) = D(\omega) e^{i[\omega t - \delta(\omega)]} \tag{8.60}$$

in terms of the frequency-dependent amplitude of the probe motion, $D(\omega)$ and phase lag $\delta(\omega)$ relative to the applied magnetic force. The storage and loss modulus for each frequency is then calculated by

$$G'(\omega) = \frac{F_0}{6\pi a D(\omega)} \cos \delta(\omega) \tag{8.61}$$

$$G''(\omega) = \frac{F_0}{6\pi a D(\omega)} \sin \delta(\omega). \tag{8.62}$$

These relations are found simply by rearranging eqn 8.59,

$$G' + iG'' = \frac{\tilde{F}}{\tilde{X}} = \frac{F_0}{6\pi a D(\omega)} e^{i\delta(\omega)} \tag{8.63}$$

and applying Euler's identity. Oscillatory magnetic bead microrheology has been used to study solutions of tightly-entangled F-actin (Ziemann *et al.*, 1994; Keller *et al.*, 2001) in addition to the creep experiments shown in Fig. 8.17.

Before performing an experiment, we can turn the analysis around and ask what the *expected* amplitude and phase lag are for a given viscoelastic modulus. This exercise helps us to understand the operating limits of the oscillatory magnetic tweezer microrheology method. The amplitude is calculated from the real and imaginary components of the response function $\chi(\omega) = 1/i\omega\tilde{\zeta}(\omega)$. We find that it decreases with an increase in the modulus amplitude $|G^*(\omega)| = \sqrt{G'^2 + G''^2}$ as

$$D(\omega) = F_0 \left[\chi'^2 + \chi''^2 \right]^{1/2} = \frac{F_0}{6\pi a |G^*(\omega)|}. \tag{8.64}$$

The phase angle of the bead position relative to the oscillating magnetic force is the loss tangent of the viscoelastic response,

$$\tan \delta(\omega) = \chi''/\chi' = G''(\omega)/G'(\omega). \tag{8.65}$$

Induction in coils

The induction of a coil or electromagnet stores energy in the magnetic field it produces. By Lenz's law, the self-induced electro-motive force (emf) of a coil opposes the change in current that is its source. Self-induction makes it more difficult for variations in current to occur and limits the frequencies and resolution of step changes that can be made by electromagnets.

The self-induction of an electric coil with N turns is

$$L = N\Phi_B/I \tag{8.66}$$

where the coil magnetic flux is approximately $\Phi_B \sim \pi b^2 B_z$ and B_z is given by eqn 8.5.[6] Thus, $L \sim \pi b \mu_0 N^2$. A coil behaves as an inductor-resistor (L-R) circuit with a time constant L/R which governs the rate at which current changes at an applied emf. Electromagnets with high-permeability cores have especially high inductances, and magnetic tweezers designed to generate sinusoidal or time-varying forces must use high-performance amplifiers that are capable of controlling the current precisely against this reaction (Keller *et al.*, 2001).

8.4.3   Operating diagram

The lowest compliance (highest modulus) that can be measured by magnetic-tweezer microrheology is determined by the largest force that can be imposed on the probe particle F_{max} and the smallest displacement of its motion that can be resolved $\Delta\varepsilon$. By eqn 8.52,

$$\mathcal{J} > \frac{6\pi a \Delta\varepsilon}{F_{max}}. \tag{8.67}$$

We've seen that the magnetic forces can be as large as 0.5–4 nN using sharpened single-pole electromagnets or permanent magnets. For particles with radius 2.25 μm and magnetic characteristics typical of those in Table 8.2, the smallest measurable compliance is on the order of $\mathcal{J}_{min} \sim 10^{-2}$–$10^{-3}$ Pa^{-1} with the minimum displacement set to a fraction of the probe radius, $\Delta\varepsilon \sim 100$ nm. Similarly, the shortest time scale is usually determined by the frame rate of a digital camera, similar to particle tracking $t_{min} \sim 10^{-2}$ s. But this limit is optimistic. Rich *et al.* (2011*a*) estimate this time scale based on the maximum velocity they could track translating beads at 20× magnification, and reached a value $t_{min} \sim 10^{-1}$s. Both the time and spatial resolution could be improved by incorporating laser tracking, which Fisher *et al.* (2006) do, at the expense of limiting tracking at larger displacements.

[6] The SI unit of inductance is the henry. An inductor of 1 henry (1 H) produces an electromotive force (emf) of $\mathcal{E} = 1$ V when the current through the inductor changes at the rate of 1 A/s. Thus, 1 H = V·s/A = 1 J/A^2.

The upper-compliance limit is determined by the Brownian motion of the probe, or when the magnetic force is on the order of magnitude as the Brownian force, $F_0 \sim k_B T/a$ (essentially reverting to a passive microrheology experiment). In this case, $\mathcal{J} < 6\pi a^3/k_B T$ gives the upper compliance boundary $\mathcal{J} < 10^4$ Pa^{-1}. Finally, the longest time scale is governed by the ability to track a particle for up to tens of minutes.

The magnetic-microrheology operating limits are illustrated in Fig. 8.19 by the solid black lines, plotted together with the operating ranges of the passive-microrheology experiments, multiple particle tracking (MPT), diffusing wave spectroscopy (DWS), and laser tracking (LT). Magnetic tweezer microrheology generally enables microrheology measurements at lower compliances or higher shear moduli than are accessible by thermal forces alone. A magnetic tweezer experiment also doesn't require time averaging like multiple particle tracking and light scattering, and so it can extend the time window considerably, as long as the conditions of an incompressible, homogeneous material are met (see Section 2.7).

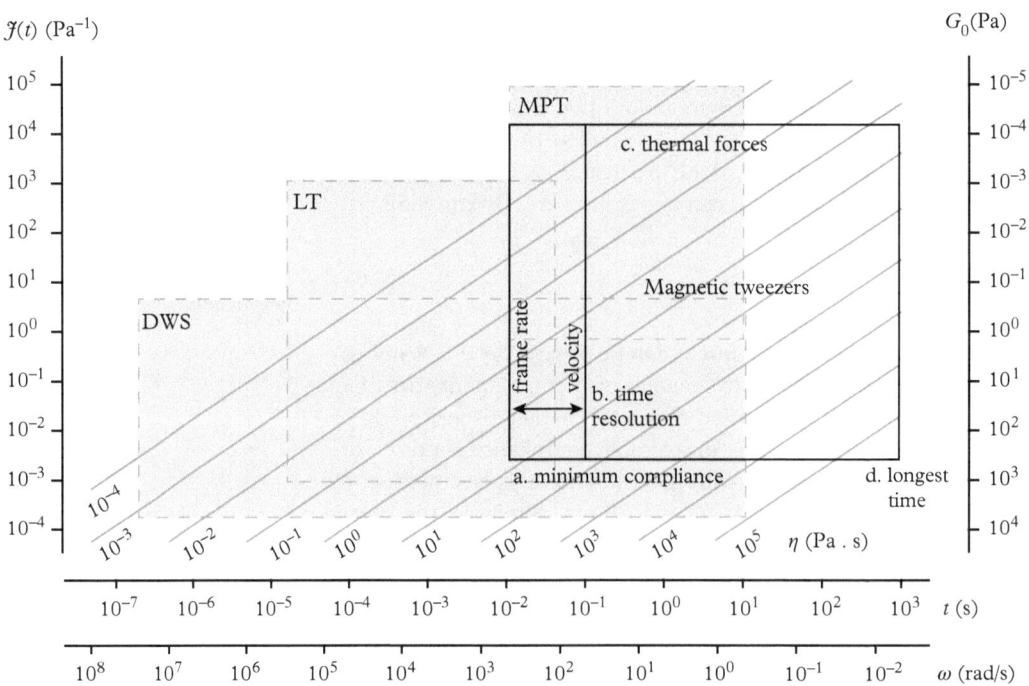

Fig. 8.19 *The operating regime of magnetic tweezer microrheology compared to passive microrheology techniques based on probe characteristics discussed in the text.*

Experiments using single-pole electromagnets or permanent magnets can generate relatively high magnetic forces. Oscillatory experiments, which usually use multiple coils that are placed further from the sample, generate forces on the order of piconewtons. This lower force increases the minimum compliance by several orders of magnitude. An exception is represented by recent experiments that mechanically move a permanent magnet orthogonal to the focal plane (Lin and Valentine, 2012*b*). These appear to be promising advances for generating higher oscillatory forces in magnetic bead microrheology. Overall, the dynamic response of oscillatory magnetic tweezer experiments and their acquisition times are not as good, but the small sample volumes required by magnetic bead microrheology still make it an important technique.

8.5 Nonlinear measurements

8.5.1 Yield stress and jamming

Yielding is one of the nonlinear rheological properties studied with magnetic bead microrheology. Yielding represents a good target application of magnetic tweezers because multiple beads can be manipulated at once and their location in the magnetic field simultaneously induces a range of forces on the probes. An example is shown in Fig. 8.20a, in which the positions of five magnetic particles in a Laponite gel are plotted as a function of time. All five particles exhibit some degree of slow motion—creep—but three eventually reach

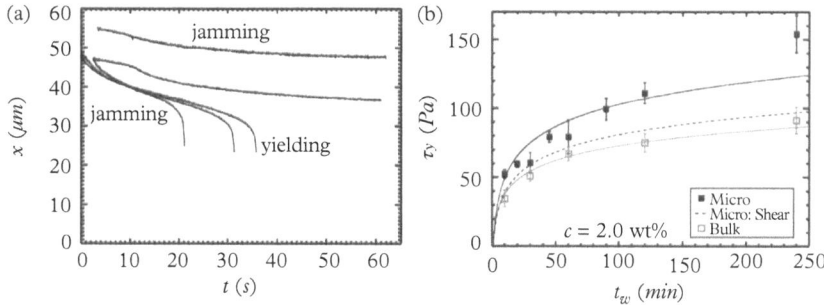

Fig. 8.20 *(a) Bead trajectories in a yield stress fluid, 2 wt % Laponite. (b) Yield stresses calculated from the applied magnetic force on probe particles using eqn 8.70 and bulk rheology measurements as a function of waiting time t_w. The dashed line is the calculated yield stress from microrheology using eqn 8.73. Reproduced from Rich et al. (2011a) with permission of The Royal Society of Chemistry.*

positions in which the magnetic force exceeds the yield stress of the material, causing them to move rapidly towards the magnetic pole.

We can understand the motion of the probes in the yielding material by considering the stresses generated on (and by) the particle. For the sake of simplicity, let us first consider the stresses exerted by a Newtonian fluid. At least in the linear limit, this should be generalized to viscoelastic materials by the Correspondence Principle. By eqn 2.75, the traction on a bead surface of a translating sphere is

$$\mathbf{t} = \mathbf{T} \cdot \hat{\mathbf{r}} = \frac{3\eta}{2a} \mathbf{V}_0 \qquad (8.68)$$

which is constant over its surface. Integrating the traction is one way to show that the drag force is given by the familiar Stokes result,

$$\mathbf{F}_0 = \int_S \mathbf{t} dS = 4\pi a^2 \mathbf{t} = 6\pi a\eta \mathbf{V}_0. \qquad (8.69)$$

Because the traction (and stress) is constant over the surface of the sphere, one could consider the magnitude of the average stress acting on the particle (and the particle acting on the fluid) as the drag force is divided by its area,

$$T = \frac{F_0}{4\pi a^2} = \frac{3\eta V_0}{2a}, \qquad (8.70)$$

but this expression includes both the shear and normal components. If one focuses solely on the *shear* stress, given by the r-θ component of $\mathbf{T}$, we see that it has a magnitude

$$T_{r\theta} = -\frac{3\eta V_0}{2a} \sin\theta, \qquad (8.71)$$

over the surface. When $T_{r\theta}$ is integrated over the sphere, it gives an average

$$\bar{T}_{r\theta} = \frac{3\pi\eta V_0}{8a}. \qquad (8.72)$$

This average shear stress is lower than that calculated naively with eqn 8.70 by a factor $\pi/4$, or about 20%, meaning that the yield stress derived from the force F_y that the probe particle first moves is

$$T_y = \frac{\pi}{4} \cdot \frac{F_y}{4\pi a^2}. \qquad (8.73)$$

Results for 2 wt% Laponite suggest that eqn 8.73 is in better agreement with bulk rheological measurements. The calculated yield stress from the displacement of magnetic beads is shown in Fig. 8.20b. Values of the microrheological yield stress using eqn 8.70, indicated by the solid symbols, are higher than bulk yield stress measurements, shown by the open symbols. The microrheology results interpreted by eqn 8.73, plotted as a dashed line, is in better agreement with the bulk rheology. Both micro- and bulk-derived values of the yield stress depend on the waiting time t_w after the sample is prepared.

A rigorous calculation of the translation of spheres in a Bingham fluid (cf. eqn 1.45) was performed by Beris *et al.* (1985), who calculated the minimum force to be overcome before the sphere would move. Their expression, rearranged, gives a macroscopic yield stress

$$T_y = 0.286 \frac{F_y}{4\pi a^2} \tag{8.74}$$

which is a factor of approximately 2.7 lower than eqn 8.73.

The few studies of yield stress fluids using microrheological methods yield promising results. Laponite, like many complex yield stress fluids, introduces a number of potential artifacts (Oppong *et al.*, 2008; Rich *et al.*, 2011a). It has a heterogeneous microstructure, to a point that the continuum approximation breaks down, and it is thixotropic—the modulus and yield stress are time-dependent. That said, in many applications differences between microrheological and macrorheological measurements of yield stress may also be important. If the aim of the yield stress fluid is to suspend particles and prevent them from sedimenting or creaming, then probe-based measurements are among the most direct methods of quantitatively testing this design objective.

8.5.2   Shear thinning

A number of magnetic bead microrheology experiments have been reported for which the material "shear thins"—that is, the *apparent* viscosity η_{app}, calculated by assuming the Stokes equation (eqn 7.4) is valid, decreases with increasing velocity of the probe. In Section 7.3.2, we already discussed the issues that accompany the nonlinear measurement problem. Nonetheless, sometimes exceptional agreement between bulk and micro measurements are reported. More importantly, like yielding, micro-scale experiments can give insight into the transport of micrometer-scale matter (including particles, bacteria, and cells) in complex fluids, and provide experimental tests for models based on constitutive relations derived from bulk rheology measurements.

In one study, Cribb *et al.* (2013) measured the translation of magnetic probes in three entangled polymer solutions: DNA, hyaluronic

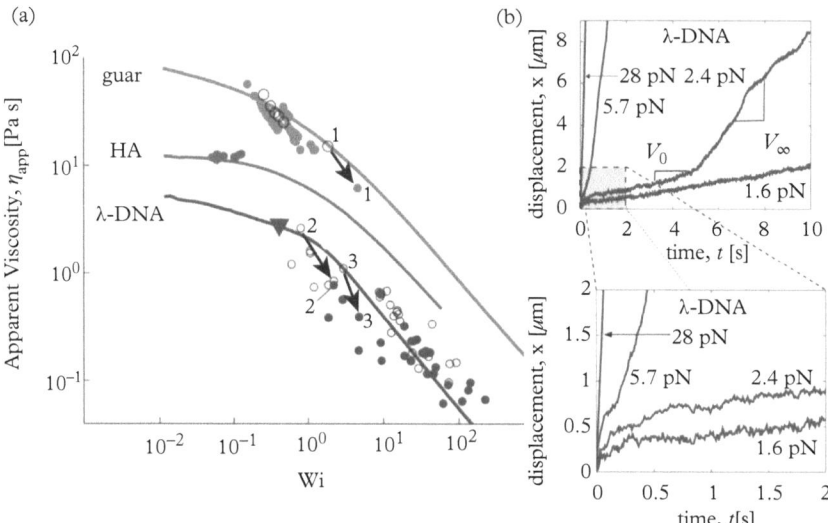

Fig. 8.21 *Magnetic bead microrheology for entangled polymer solutions of DNA, hyaluronic acid (HA), and guar. (a) The apparent viscosity from microrheology experiments (symbols) and bulk rheology flow curves (lines). Arrows indicate values for initial V_0 and final V_∞ probe velocities. (b) Particle trajectories showing the two velocity regimes. Reprinted from Cribb, J. A. et al. J. Rheol., 57, 1247–64 (2013) with the permission of The Society of Rheology.*

acid, and guar. For each material, the apparent viscosity, shown in Fig. 8.21a, generally tracks the flow curve of bulk measurements as a function of Weissenberg number

$$\mathrm{Wi} = \tau_D \dot{\gamma}, \tag{8.75}$$

where τ_D is the longest relaxation time of the polymer solution and the characteristic shear rate of the translating probe particle is taken as

$$\dot{\gamma} = 3V/\sqrt{2}a, \tag{8.76}$$

from the maximum of the strain rate tensor for a spherical particle in a Newtonian fluid. However, the probe motion in guar and DNA solutions exhibits two velocity regimes: An initial velocity V_0 when the force is first applied, and a final, lower, velocity V_∞ (Fig. 8.21b). Surprisingly, both velocities give viscosities that lie on the bulk rheology flow curve.

8.6 Nanorods in steady and rotating fields

Given the potential limitations of active nonlinear microrheology experiments conducted with spheres, including the effects of mixed
flows and Lagrangian unsteadiness (see Section 7.3.3), experiments
using magnetic rod nanoparticles as probes are especially promising.
Such methods take advantage of advances in the synthesis of uniform
nickel (Ni) nanowires.

Ferromagnetic nanowires with radius of about 180 nm and lengths
ranging from 5 to 50 μm are synthesized by electrochemical dep-
osition into a nanoporous alumina filter template, after which the
wires are extracted and magnetized along their long axes (Tanase
et al., 2001). The magnetic properties of such nanowires often exhibit higher remanance and coercivity compared to micrometer-size
or bulk metal Fert and Piraux (1999). Others have synthesized iron
oxide nanowires by controlling the assembly of $\gamma - Fe_2O_3$ nanoparticles, which retain their paramagnetic properties (Fresnais *et al.*, 2008;
Yan *et al.*, 2011).

Experiments have used nanowires suspended in worm-like micellar solutions to characterize the linear and nonlinear rheology of these
complex fluids (Fig. 8.22). In a strongly nonlinear regime, Cappallo
et al. (2007) noted a "rotational thinning" when nanowires are rotated continuously, but found poor agreement between the apparent
microrheological shear viscosity compared to the bulk shear viscosity.
They also noted unusual features such as the generation of out-of-
plane torques, which cause the rods to rotate out of the field (and
focal) plane. Using paramagnetic iron oxide nanorods, Chevry *et al.*
(2013) probed the linear response regime of similar worm-like micellar solutions, with results that agree well with bulk measurements.
Cribb *et al.* (2010) also used nanorods to study shear thinning in
entangled polymer solutions.

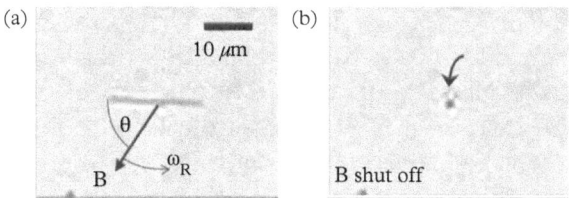

Fig. 8.22 *(a) Ferromagnetic nickel nanorods rotated in a worm-like micellar solution. (b) Torques cause the rods to rotate out of plane. Reprinted figure with permission from Cappallo, N., Lapointe, C., Reich, D. H., & Leheny, R. L., Phys. Rev. E, 76, (2007). Copyright 2007 by the American Physical Society.*

8.7 Summary

Magnetic tweezers are the oldest form of microrheology. It is versatile and innovative developments in instruments and probes are still being introduced. The significant strength of magnetic tweezer microrheology lies in the expanded single-particle operating regime. Higher forces can be imposed on magnetic probes, enabling the characterization of materials with lower compliances (higher moduli). It is also the first of two active microrheology methods we discuss. Driven probes can access regimes where nonlinear rheological phenomena begin to affect probe mobility, like yielding and shear thinning. Magnetic tweezers can also be used in cases where the Generalized Einstein Relation breaks down to measure the mechanics of *active systems* like living cells. Despite advances, no two-point magnetic tweezer microrheology experiment has been developed.

. . . .

EXERCISES

(8.1) **Magnet coils.** Using the Law of Biot–Savart, show that the axial field of a current loop is given by equation 8.4.

(8.2) **Magnetic force on a paramagnetic probe.** The magnetic force exerted on a paramagnetic particle is sometimes written as

$$\mathbf{F}_{mag} = \frac{1}{2}\nabla(\mathbf{m} \cdot \mathbf{B}). \qquad (8.77)$$

(a) Show that this expression gives a force consistent with eqn 8.23 in the low-field limit, $\chi_0 H \ll M_{sat}$.

(b) Does this expression hold as the field strength approaches the saturation magnetization of the particle? Why or why not?

(8.3) **Force and magnetization.** Calculate the force exerted on paramagnetic particles as a function of position along the axis of a magnetic coil with radius $b = 1.27$ cm, $N = 100$ turns, current $I = 1$ A, and a core composed of a metal with relative permeability $\mu = 100$.

(a) Calculate the force using the Fröhlich-Kenneley equation and for particles with diameter 4.5 μm, $M_{sat} = 31$ kA/m, $\chi_0 = 1.6$.

(b) Calculate the force using the empirical equation 8.36 for the same particles with $M_0 = 25.3$ kA/m, $\chi_0 = 1.6$ and

$\alpha = 0.95$. How does this force differ from that calculated in part (a)?

(8.4) **Equation of motion.** Show that eqn 8.59 leads to an equation of motion in terms of the elastic F_e and viscous F_v contributions to the force acting on a probe particle $F_m = F_0 \cos \omega t$,

$$F_e + F_v + F_m = 0 \qquad (8.78)$$

where

$$F_e = -6\pi \, a G' x(t) \qquad (8.79)$$
$$F_v = -6\pi \, a (G''/\omega) \dot{x}(t). \qquad (8.80)$$

When do the maximum elastic and viscous restoring forces occur on the probe? How are the forces related to G' and G''?

Laser tweezer microrheology

To many, the idea that light can be used to hold and manipulate matter is probably quite foreign. The photon is a seemingly evanescent particle; its interactions with matter are weak. But while it has no rest mass, a photon carries momentum. Photons exert forces on material objects, but the forces are weak compared to those we experience in the every day world: Gravity, atmospheric pressure, and the stresses imparted by flowing liquids and gases. Nevertheless, on small length scales, optical forces can be significant relative to thermal and viscous forces.

In 1970 Arthur Ashkin reported observations of the acceleration of freely suspended particles by the radiation pressure of a visible light laser (Ashkin, 1970*a*). An illustration of this effect, taken from his seminal paper, is shown in Fig. 9.1. Based on these principles, optical traps or laser tweezers were later developed. Optical traps are created by focusing a laser beam using a microscope objective, and enable particles to be held stably in three-dimensions (Ashkin et al., 1986; Ashkin et al., 1987).

Optical traps have become important, if still somewhat specialized tools to measure forces on nanometer to micrometer length scales. They have been used in micromechanical studies of proteins, molecular motors (Finer *et al.*, 1994; Visscher *et al.*, 1999), polymers (Quake *et al.*, 1997; Smith and Chu, 1998), and colloids (Grier, 1997; Furst, 2005; Sainis *et al.*, 2007). In this chapter, we discuss linear and nonlinear microrheology with optical traps. While there are significant limitations on the samples that can be studied with laser tweezer microrheology, a topic we will discuss in more detail later, it has several important strengths. First, the method requires far fewer probe particles, since only one is trapped at a time; the sample volumes are (in theory) vanishingly small. More importantly, the probe can be oscillated or dragged through a sample with high precision in ways that mimic the oscillatory or steady deformation of bulk rheology. Like magnetic bead microrheology (Chapter 8), this gives one the ability to measure microrheology in the *nonlinear* regime, in contrast to passive microrheology, which is inherently limited to measuring

Microrheology. Eric M. Furst and Todd M. Squires, Oxford University Press (2017).
© Eric M. Furst and Todd M. Squires. DOI 10.1093/oso/9780199655205.001.0001

the linear rheological response. For these reasons, and others, laser tweezer microrheology has seen a growing interest (Yao *et al.*, 2009).

We will begin this chapter by presenting the physical principles of optical trapping, followed by a discussion of the important parts of an optical trapping instrument. Our next topic, calibration of the optical trap force, is critical to their application as a *dynamometer*—an instrument that measures force. From there, we discuss active oscillatory microrheology and "steady-drag" experiments, as well as the operating regimes of the experiments.

9.1 Radiation forces and Gaussian beams

The momentum carried by a photon of wavelength λ is $\mathbf{p} = \hbar\mathbf{k}$, where $\mathbf{k}$ is the wavevector of the light with magnitude $|k| = 2\pi/\lambda$ and $\hbar$ is the reduced Planck constant.[1] Photons scattered by a macroscopic object will impart a radiation pressure

$$P_{rad} = \frac{2QI_0}{c} \tag{9.1}$$

where I_0 is the incident irradiance in units Wm^{-2} and c is the speed of light. Q is a scattering efficiency that typically has a value on the order of 0.1. For incident radiation with an intensity similar to that of sunlight at sea level ($I_0 \sim 1000\ Wm^{-2}$), this radiation pressure is $P_{rad} \approx 10^{-8}\ Pa$. Since the atmospheric pressure at sea level is about $10^5\ Pa$, it is little wonder that we are not accustomed to thinking much about radiation pressure. However, as an object becomes smaller, the radiation force becomes more significant relative to other forces. If we consider a colloidal particle with diameter approximately $2a = 1\ \mu m$ in a 10 mW laser beam, then the radiation force is approximately $F_{rad} \approx 10^{-11}\ N$, or 10 piconewtons. Although the force is still quite small, it is significant relative to other forces that act on colloids. For comparison, the characteristic Brownian force on a one micrometer particle is $F_B \sim k_B T/a \approx 10^{-14}\ N$. Radiation forces can be many times greater than the random thermal force exerted on the particle.

9.2 A focused Gaussian beam in the diffraction limit

Before discussing the mechanisms of optical trapping, we briefly review the characteristics of focused Gaussian beams, including

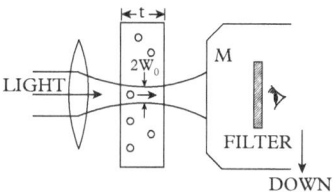

Fig. 9.1 *An illustration showing the principle of manipulating colloidal particles using the radiation pressure of a laser. Ashkin's work serves as the foundation for optical trapping and laser tweezer microrheology. Adapted from Ashkin (1970a).*

---

[1] Recall $\hbar = h/2\pi$ and Planck's constant is $h = 6.626068 \times 10^{-34}$ kg m²/s.

the radiant field and irradiance. Important quantities for our later discussion include the beam width w_0 and Rayleigh length z_R.

9.2.1   Radiant field

Consider a Gaussian beam with vacuum wavelength λ focused by an optical system with numerical aperture NA. The radiant field of the beam in the diffraction limit is

$$E(r, z) = E_0 \frac{w_0}{w(z)} e^{i2\pi\kappa z} e^{-r^2/w(z)^2} e^{i\pi\kappa r^2/R(z)} e^{-i\eta(z)}, \qquad (9.2)$$

where r and z are the radial and axial coordinates, $\kappa = n/\lambda$, and

$$w(z) = w_0\sqrt{1 + z^2/z_R^2} \qquad (9.3)$$

$$R(z) = z\left(1 + z_R^2/z^2\right) \qquad (9.4)$$

$$\eta(z) = \tan^{-1}(z/z_R). \qquad (9.5)$$

The radiant field has a width w_0 in the diffraction limit given by

$$w_0 = \frac{\lambda}{2NA}. \qquad (9.6)$$

The Rayleigh length

$$z_R = \pi w_0^2 \kappa \qquad (9.7)$$

is the axial distance $z > z_R$ after which the beam diverges with the geometric angle given by the numerical aperture

$$NA = n\sin\theta. \qquad (9.8)$$

Note that the radiant field as defined has units $E(r, z) [=] \sqrt{W}\,m^{-1}$. We multiply the radiant field by $\sqrt{c\epsilon_0/2}$ to obtain the electric field, which has units $E [=] V\,m^{-1} [=] J\,C^{-1}\,m^{-1} [=] N\,C^{-1}$. Also, the wavefront is nearly planar for $z < z_R$, indicated by the almost-uniform phase over the radial coordinate r.

9.2.2　Irradiance and laser power

The radiant field is related to the intensity, or more precisely, the irradiance or radiant-flux density, by

$$I(r, z) = \langle E(r, z)E^*(r, z)\rangle, \tag{9.9}$$

where the asterisk denotes the complex conjugate of $E(r, z)$ and the angle brackets indicate an average taken over a time greater than the period of the electric field, $T = n\lambda/c$. The intensity is the power incident on a surface, and has units $I(r, z)\ [=]\ \mathrm{W\,m^{-2}}$. The irradiance is therefore

$$I(r, z) = \frac{I_0 e^{-2r^2/w(z)^2}}{1 + z^2/z_R^2}, \tag{9.10}$$

which can be related to the beam (laser) power

$$P = \pi \int_0^\infty I(r, z) r\, dr \tag{9.11}$$

$$= \frac{\pi w_0^2 I_0}{4} \tag{9.12}$$

$$= \frac{\pi \lambda^2 I_0}{16 NA^2}. \tag{9.13}$$

Thus, the incident irradiance is related to the beam power by

$$I_0 = \frac{16 NA^2 P}{\pi \lambda^2}. \tag{9.14}$$

The irradiance given by eqn 9.10 is plotted in Fig. 9.2. The dashed lines represent the ray cone set by the numerical aperture, eqn 9.8. The circle is 2 μm in diameter.

9.3　Optical trapping

9.3.1　Rayleigh regime

In scattering theory, the Rayleigh regime corresponds to cases in which a particle is much smaller than the wavelength of light—so small that it experiences a uniform electric field. A particle of in the Rayleigh regime would clearly be much smaller than the 2 μm diameter circle shown in Fig. 9.2. The electric field causes the particle to polarize, inducing a simple dipole. The dipole will move in response

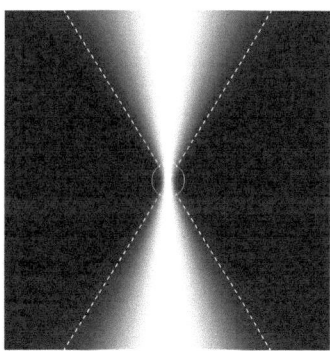

Fig. 9.2 *Irradiance of a Gaussian beam with vacuum wavelength* λ = 1.064 μm *in a medium with refractive index* n = 1.33 *for an optical system with a numerical aperture* NA = 1.1. *The dashed lines depict the geometric light cone defined by the numerical aperture. The circle is* 2 μm *diameter.*

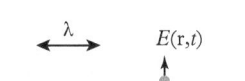

Fig. 9.3 *In the Rayleigh regime, a particle that is small relative to the wavelength λ experiences a uniform, time varying electric field* $E(\mathbf{r}, t)$.

to the strong field gradient of the Gaussian beam. This is the principle of optical trapping in the Rayleigh regime, as in Fig. 9.3. Although such small particles are rarely used for microrheology, an analysis of the Rayleigh regime provides several important and general insights into optical trapping—in particular its dependence on the particle size and optical properties of the trapped particle and surrounding medium (Harada and Asakura, 1996).

An electric field will induce a dipole moment

$$\mathbf{p} = \alpha^* \mathbf{E} \tag{9.15}$$

in a particle, where $\alpha^* = \alpha' + i\alpha''$ is the particle's frequency-dependent polarizability. Here, we use the conventional electric field units instead of the units of the radiant field. The units of the dipole moment and polarizabilty are $p \,[=]\, \mathrm{C\ m}$ and $\alpha \,[=]\, \mathrm{C\,m^2\,J^{-1}} \,[=]\, \mathrm{C\,m\,N^{-1}}$, respectively. Another way of expressing eqn 9.15 is

$$\mathbf{p} = |\alpha| e^{i\phi} \mathbf{E}. \tag{9.16}$$

In the electric field generated by the laser illumination with wavelength λ,

$$\mathbf{E} = \mathbf{E}_0 e^{i\omega t}, \tag{9.17}$$

where $\omega = 2\pi\nu = 2\pi c/\lambda$, the dipole moment is

$$\mathbf{p} = |\alpha| \mathbf{E}_0 e^{i(\omega t + \phi)}. \tag{9.18}$$

The energy of this induced dipole in the electric field is the scalar product

$$U = -\mathbf{p} \cdot \mathbf{E}, \tag{9.19}$$

which can be written

$$U = -|\alpha| E^2 \cos(\omega t + \phi) \cos(\omega t), \tag{9.20}$$

dropping the subscript for the electric field amplitude. Averaging the interaction over one period,

$$\bar{U} = -|\alpha| E^2 \frac{1}{2\pi} \int_0^{2\pi} \cos(\omega t + \phi) \cos \omega t \, d(\omega t) \tag{9.21}$$

yields

$$\bar{U} = -\frac{1}{2} E^2 |\alpha| \cos\phi = -\frac{1}{2} \mathrm{Re}\{\alpha\} E^2 \tag{9.22}$$

recognizing that $\mathrm{Re}\{\alpha\} = |\alpha| \cos\phi$. The *gradient force* is then

$$\mathbf{F}_g = -\nabla \bar{U}, \tag{9.23}$$

or

$$\mathbf{F}_g = \frac{1}{2}\mathrm{Re}\{\alpha\}\nabla(E^2). \tag{9.24}$$

From eqn 9.24, we see that in the Rayleigh regime, the trapping force arises due to a gradient in the electric field of the laser radiation.

The polarization of a dielectric sphere with radius a in an electric field is Jackson (1998)

$$\alpha^* = 4\pi\bar{\epsilon}_s\epsilon_0 a^3 \left(\frac{\epsilon_p - \epsilon_s}{\epsilon_p + 2\epsilon_s}\right), \tag{9.25}$$

where ϵ_p and ϵ_s are the complex particle and medium permittivities, respectively, and the overbar indicates the real component of the medium permittivity is used, $\bar{\epsilon}_s$. This yields the gradient force

$$\mathbf{F}_g = 2\pi\bar{\epsilon}_s\epsilon_0 a^3 \, \mathrm{Re}\left[\frac{\epsilon_p - \epsilon_s}{\epsilon_p + 2\epsilon_s}\right]\nabla(E^2). \tag{9.26}$$

Since we are interested in optical wavelengths, for which the polarizability of the particle is due to electronic transitions in the material, we can re-write this expression in terms of the refractive indices of the particle n_p and medium n_s, noting that $n \approx \sqrt{\epsilon}$ in the visible spectrum,

$$\mathbf{F}_g = 2\pi\epsilon_0 a^3 \left(\frac{n_s^2(n_p^2 - n_s^2)}{n_p^2 + 2n_s^2}\right)\nabla(E^2). \tag{9.27}$$

Next, the laser irradiance $I = cn_s\epsilon_0 E^2/2$ is substituted, where n_s is the refractive index of the medium. This expression assumes that the magnetic susceptibility of the medium is negligible. The units of the irradiance are $I\,[=]\,\mathrm{W\,m^{-2}}$. The gradient force is then

$$\mathbf{F}_g = \frac{4\pi a^3}{c}\left(\frac{n_s(n_p^2 - n_s^2)}{n_p^2 + 2n_s^2}\right)\nabla I(r, z). \tag{9.28}$$

The trapping force can then be calculated by taking the gradient of the Gaussian beam irradiance, given by eqn 9.10. For a particle displaced in the radial direction, transverse to the beam axis at the trap center $(z = 0)$ the gradient force is

$$F_g(r) = -\frac{512NA^2a^3P}{c\lambda^3}\left(\frac{n_s(n_p^2 - n_s^2)}{n_p^2 + 2n_s^2}\right)(r/w_0)e^{-2(r/w_0)^2}. \qquad (9.29)$$

Along the beam axis $(r = 0)$, the gradient force is

$$F_g(z) = -\frac{512NA^4a^3P}{c\pi\lambda^3}\left(\frac{n_p^2 - n_s^2}{n_s(n_p^2 + 2n_s^2)}\right)\frac{z/z_R}{1 + z^2/z_R^2} \qquad (9.30)$$

Here are the important facts that we learn from eqns 9.29 and 9.30: First, the trapping strength increases with the volume of the particle in the Rayleigh regime. Larger particles are subject to substantially larger trapping forces. The trapping force eventually plateaus as the particle size approaches the laser wavelength and the optical forces are more accurately described in the ray optic limit. Second, the equations show that the trapping force increases as the refractive index contrast between the particle and medium increases. Obviously, this increase with contrast cannot apply without limit, since the *scattering force* offsetting the gradient force will also increase; however, the use of highly polarizable particles (*i.e.*, metals such as gold) can compensate for the smaller gradient force when particles approach tens of nanometers in diameter, which leads to stable trapping (Svoboda and Block, 1994*b*). For dielectric particles, the trapping force is substantially lower. Third, the trapping force increases with the laser power P.

The dependence of the optical trapping force on the medium refractive index n_s is also important. A dispersed material such as polymer, protein, or surfactant contributes to the refractive index of the medium. Optical microrheology experiments must take special care to account for changes in the trapping force with the dispersed material concentration. An increase in the medium refractive index leads to less contrast with the trapped particle, and a corresponding *decrease* in trapping force. The lower trapping force could be mistaken for higher viscoelastic moduli of the medium. The calibration of an optical trapping force will be discussed in Section 9.5.

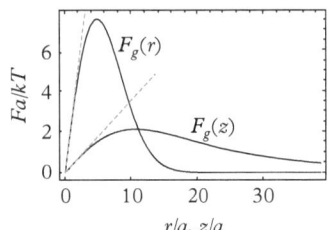

Plotting eqns 9.29 and 9.30 shows a few additional features, as illustrated in Fig. 9.4. First, there is a small region in which the particle in the optical trap experiences a force proportional to its displacement from the trap center. This Hookean regime is limited to displacements that are just a fraction of the beam width. In the radial direction, the Hookean trap stiffness is

$$\kappa_t = \frac{1024NA^4a^3P}{c\lambda^4}\left(\frac{n_s(n_p^2 - n_s^2)}{n_p^2 + 2n_s^2}\right). \qquad (9.31)$$

Fig. 9.4 *Trapping force calculated in the Rayleigh regime and scaled by the characteristic Brownian force k_BT/a for a particle radius $a = 50$ nm, refractive index $n_p = 1.46$, medium refractive index $n_s = 1.33$, laser wavelength $\lambda = 1064$ nm, laser power $P = 100$ mW, and numerical aperture, $NA = 1.1$.*

The trap stiffness is substantially higher in the radial direction than along the z-axis. In the scaled dimensions of the Fig. 9.4, the axial trapping force is on the order of the characteristic thermal force $k_B T/a$, and so particles of this size are unlikely to be held for long by the trap. The other feature to note is the maximum trapping force, or *escape force*, along both the radial and axial dimensions. The trapping force ultimately determines the largest resistance imparted on the probe by the surrounding medium that can be measured by laser tweezer microrheology.

9.3.2 Ray optic regime

The ray optic regime represents the other extreme of trapping in which the particle diameter is on the order of the wavelength of light. The irradiance of a focused Gaussian beam in Fig. 9.2 shows that the light impinging on a particle at the center of the focus basically follows paths described by geometric rays. Reflections and refractions of these rays induce a momentum change in the light that is, in turn, imparted to the particle.

Fig 9.5 illustrates trapping in the limit of the geometrical optics approximation. Incident rays simply pass through a spherical particle that sits directly at the center of the focused beam, since the light arrives and leaves normal to the particle surface. A particle that

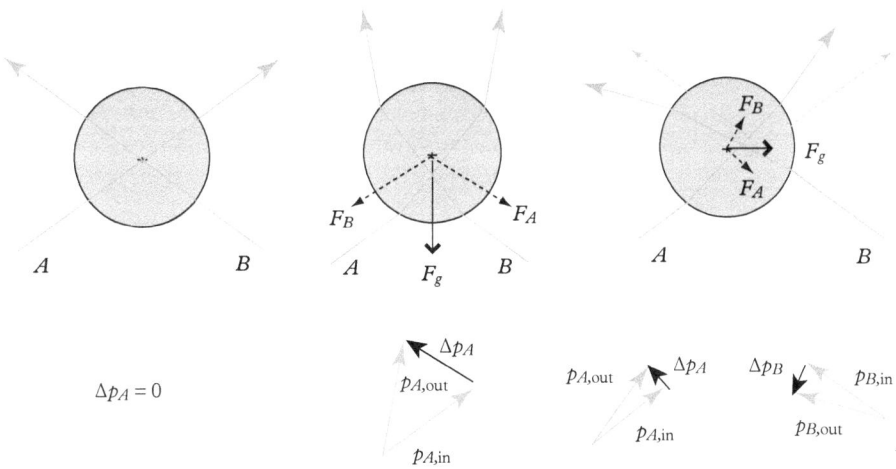

Fig. 9.5 *The geometrical optics approximation of a laser tweezer. The momentum difference between incident and refracted rays imparts a force in the opposite direction. Summing over all incident rays results in a net force that on a particle that pushes it back towards the center of the trap.*

is offset from the trap center, however, causes rays to refract by Snell's law,

$$\sin \theta_1 = (n_2/n_1) \sin \theta_2 \qquad (9.32)$$

where θ_1 and θ_2 are the incident and refracted rays at the first surface, respectively. The rays experience additional refraction as they leave the particle, as well as reflections at each of the interfaces (Ashkin, 1992). Each ray has an incident momentum per unit time $n_1 P/c$. The force, $F = Q(n_1 P/c)$, is given in terms of the dimensionless efficiency Q. The maximum radiation force of a ray reflected perpendicularly by a perfectly reflecting mirror corresponds to $Q = 2$ (Ashkin, 1992).

Although the total force exerted by a trap is a sum of the contributions from all reflected and refracted rays, consider only the first refracted ray for a particle offset from the trap center along the beam axis. The incident ray A in the figure leaves the particle at a new angle. The incident and refracted rays define a momentum change $\Delta p_A = p_{A,\text{out}} - p_{A,\text{in}}$, which by conservation of momentum, imparts a force on the particle in the opposite direction F_A. Summing over all rays gives a gradient force that pulls the particle towards the center of the trap. Similar ray diagrams can be constructed for particles displaced laterally from the trap; again, the resulting gradient forces act to restore the particle to the trap center.

The dimensionless trapping force was calculated for circularly polarized light by Ashkin (1992) in the geometrical optics limit for displacements transverse to the trap axis. Some results of Ashkin's calculations are reproduced in Fig 9.6. The total dimensionless trapping force Q_t is a sum of the gradient Q_g force (considered above) and additional scattering force Q_s. The latter force arises from reflected light at each interface between the particle and medium, which are given by the Fresnel reflection coefficients for parallel polarized light

$$R_s = \left[\frac{\sin(\theta_2 - \theta_1)}{\sin(\theta_2 + \theta_1)} \right]^2 \qquad (9.33)$$

and perpendicularly polarized light (Hecht, 2001)

$$R_p = \left[\frac{\tan(\theta_2 - \theta_1)}{\tan(\theta_2 + \theta_1)} \right]^2. \qquad (9.34)$$

Figure 9.7 illustrates the refracted and reflected rays from a single incident ray. Note that the refracted ray that we have considered has a power PT^2, where $T = 1 - R$ is the transmittance. The total gradient force is actually a sum of refracted rays of successively decreasing power, $PT^2, PT^2 R, PT^2 R^2, \ldots$ and so on.

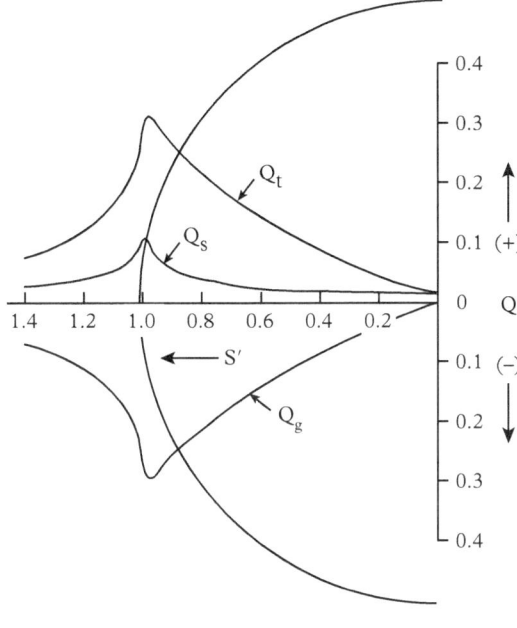

Fig. 9.6 *The dimensionless total trapping force* Q_t *and the contributions from gradient* Q_g *and scattering* Q_s *forces as a function of the transverse displacement of a trapped particle. The numerical aperture is NA = 1.25, and refractive index of the particle* n_p = *1.2. Reprinted from Biophys. J., 61, Ashkin, A., Forces of a single-beam gradient laser trap on a dielectric sphere in the ray optics regime, 569582, Copyright (1992), with permission from The Biophysical Society.*

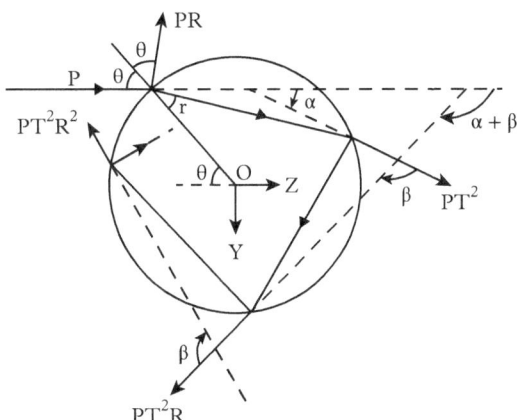

Fig. 9.7 *The refraction and reflection of a single incident ray incident on a scattering particle. Reprinted from Biophys. J., 61, Ashkin, A., Forces of a single-beam gradient laser trap on a dielectric sphere in the ray optics regime, 569582, Copyright (1992), with permission from The Biophysical Society.*

Similar to the Rayleigh trapping regime, the calculated trapping force plotted in Fig. 9.6 has a maximum. An external force acting on the particle that is greater than the maximum trapping force will pull the particle from the trap. An important characteristic of the trapping force in the ray optic trapping regime that distinguishes it from the Rayleigh regimes is the relatively large range of displacements over which the total trapping force increases linearly. The trap behaves as a Hookean spring with a trap stiffness κ_t over a reasonably large range of displacements. A second important feature is that the trap becomes *stiffer* towards the edge of the particle. In fact, large particles with a diameter exceeding about $2a > 10\lambda$ will tend to be held at their edges, rather than at the center of the trap.

9.3.3   Laser tweezer microrheology samples

We've discussed some basic calculations of the trapping forces exerted by a focused Gaussian beam on a dielectric particle. The calculations bring to light some of the key parameters that affect optical trapping, and thus its applicability as a microrheology tool. Optical trapping depends on the optical system (especially the numerical aperture) and trapping is sensitive to aberrations. These conditions will be explored further in the next section on optical trap instruments. Trapping is also affected by the laser wavelength, the particle optical properties, particle shape, and the medium optical properties. As we will see shortly, the trap stiffness is a key attribute for performing and interpreting laser tweezer microrheology experiments, so the established practice is to *empirically* calibrate the trap stiffness. Such calibration methods are discussed in Section 9.5.

It should also become clear from our discussion that there are limitations on the optical properties of samples used in laser tweezer microrheology. Primarily, samples must be optically transparent in the laser wavelength and have a reasonably homogeneous index of refraction that contrasts the probe's index. Transparent biological samples, biofluids, polymer solutions, hydrogels, and the like have been the primary focus of laser tweezer microrheology experiments.

The laser wavelengths used in trapping experiments are often selected based on the absorption of water and chromophores in biological samples (Svoboda and Block, 1994*a*). Otherwise, samples must be *designed* for tweezer experiments. Colloidal suspensions, for instance, must be index-matched to the surrounding solvent. Fluorinated polymer particles are nearly index-matched in water, so these have seen some use (Koenderink *et al.*, 2001; Meyer *et al.*, 2006) as well as PMMA particles dispersed in index matching solvents like decalin or cyclohexylbromide (Sriram *et al.*, 2010).

9.4   An optical-trapping instrument

Optical traps are not terribly difficult to build. Several commercial systems are on the market as of this writing, but a simple optical trap can be made with little more than an existing microscope and laser, or even using a handmade microscope core (Appleyard *et al.*, 2007).

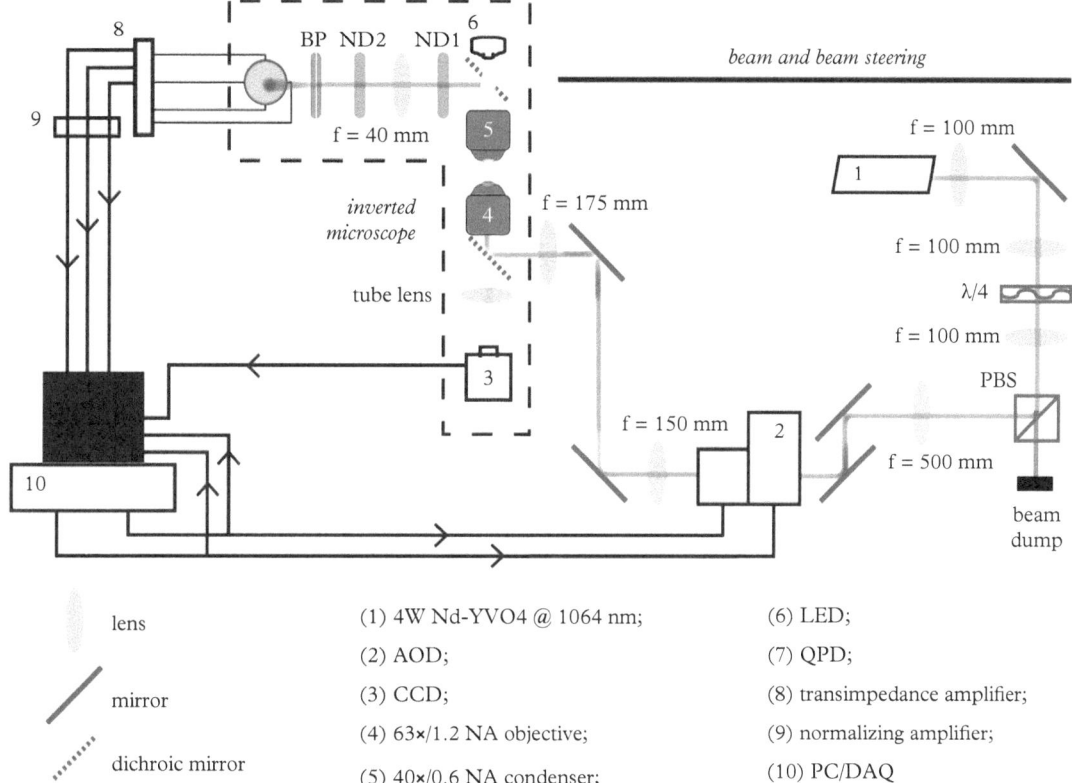

lens	(1) 4W Nd-YVO4 @ 1064 nm;
mirror	(2) AOD;
dichroic mirror	(3) CCD;
	(4) 63×/1.2 NA objective;
	(5) 40×/0.6 NA condenser;

(6) LED;
(7) QPD;
(8) transimpedance amplifier;
(9) normalizing amplifier;
(10) PC/DAQ

Fig. 9.8 *An optical tweezer instrument. The laser, beam expander and beam steering are on the right side. On the left is an inverted microscope with optics installed for back-focal-plane detection. The instrument is controlled by a personal computer, which generates the signal that steers the trap, as well as recording the particle position from the back-focal-plane detector.*

Such a setup would suffice for interferometric-passive microrheology. To actively drive the optical trap requires a bit more effort in the way of beam-steering optics that control the incident angle at the entrance pupil (or back aperture) of the objective.

Figure 9.8 is a schematic of an optical-trap experiment. There are several key components: (1) The laser, (2) the beam-steering device, a microscope with (3) a ccd camera, (4) objective, (5) condenser, and (6) LED light source, and a back-focal-plane detector with (7) a quadrant photodiode, and (8, 9) amplifiers that send signals to (10) a PC. The PC also controls the optical trap position. The laser, in this case a near-infrared diode-pumped type, generates the trapping beam. The first optical subsystem after the laser collimates, modulates the intensity, and expands the beam. The second subsystem after the acousto-optic deflector (AOD) directs the beam into the back aperture of the microscope objective. This is accomplished by a periscope and dichroic mirror positioned just below the objective. A third optical subsystem images the back-focal-plane of the condenser onto the quadrant photodiode, while attenuating the high-beam intensity with neutral density (ND) filters and using a band pass (BP) filter to select the beam wavelength. More details of the major instrument components are discussed in Section 9.4.1.

The instrument design in Fig. 9.8 is representative of the majority of trapping systems reported in the literature. Here, a computer controls the trap position and records the particle response, but these functions are easily replaced by a function generator to generate the oscillating trap and lock-in amplifier, which records the amplitude and phase lag of the quadrant photodiode signal.[2]

9.4.1   Major components

Laser

Since Ashkin's pioneering studies of radiation pressure in the 1970's (Ashkin, 1970*b*; Ashkin, 1970*a*), optical traps have been built using continuous wave (cw) lasers[3] with nearly every available visible and near-infrared wavelength, including gas lasers such as argon ion (Ar$^+$, λ = 488 nm, 514.5 nm) and helium-neon (HeNe, λ = 632.8 nm), and solid-state lasers like frequency doubled Nd:YAG (neodynium-yttrium-aluminum-garnet, λ = 532 nm), and Nd:YVO$_4$ (neodynium-orthovanadate) lasers that emit at λ = 1064 nm. More recently available near-infrared solid-state diode and fiber lasers in the 750–1400 nm spectral band that are used in the telecommunications industry and manufacturing provide excellent power and stability at low cost.

The choice of laser wavelength will be dictated by the optical properties of the sample. The foremost concern is to avoid sample damage due to laser absorption. Lasers in the near-infrared spectrum have generally prevailed because of the window of transparency between many biological cytochromes and the absorption of infrared energy by water (Svoboda and Block, 1994*a*). Typical laser powers are on the order of several hundred milliwatts to 1 watt measured at the entrance pupil of the objective. Higher laser powers may damage the optics of the microscope objective.

Along with an appropriate choice of wavelength, the trapping laser should have a single transverse mode, typically Gaussian (TEM$_{00}$), with good pointing and power stability. Stability is a significant source

---

[2] The root mean-squared amplitude reported by a lock-in amplifier is related to the measured amplitude by $D = \sqrt{2}D_{\mathrm{rms}}$.

[3] Lasers that emit with a continuous power and not in discrete pulses, such as q-switched, gain-switched, or mode-locked lasers.

of fluctuations, as illustrated by the excess power in the low-frequency spectrum of the power-spectral density plotted in Fig. 9.18.

Microscope objective

Trapping theory prescribes the use of a high-numerical aperture microscope objective to enable stable, three-dimensional trapping, Fig. 9.9. Light enters the back aperture pupil and is focused into the sample. The laser light should overfill the back aperture to provide the stiffest possible optical trap by matching the $1/e^2$ intensity to the objective entrance pupil (Neuman and Block, 2004). This ensures that rays with the highest angle of convergence have sufficient power, and hence, momentum transfer to the trapped particle, to create good trapping stability along the beam axis.

Typical high-numerical objectives used for optical trapping are oil or water immersion objectives with $NA = 1.2$–1.4. If the samples are aqueous, then water-immersion objectives are preferred. These increase the trapping strength by decreasing spherical aberrations (fig. 9.10) caused by the index mismatch between the immersion fluid and medium (Neuman and Block, 2004). Many immersion objectives incorporate correction collars to account for a variable thickness of the cover glass, another source of spherical aberration. The potential variability in the trapping force from sample-to-sample due to even minute variations in the cover-glass thickness and correction collar settings makes the *in situ* calibration methods of Section 9.5.4 that much more useful.

High-magnification objectives have small working distances, which limits trapping close to the microscope cover glass. A typical $100\times$ NA 1.4 oil immersion objective will trap particles 10–20 μm from the cover glass, resulting in strong hydrodynamic interactions with the no-slip surface, a poor situation for microrheology. Lower power oil and water-immersion objectives are capable of trapping ~ 100 μm from the cover glass.

With optical trapping becoming more common, many high-performance objectives are now designed with coatings that pass near-infrared wavelengths. In early trapping studies, as little as half the laser intensity at the back-aperture would pass through an objective. Most modern objectives are also infinity-corrected. A collimated beam entering the back aperture will be focused by the objective.

Trap steering

The optical trap position in the focal plane is controlled by the beam angle at the entrance pupil of the objective (see Fig. 9.11). A beam

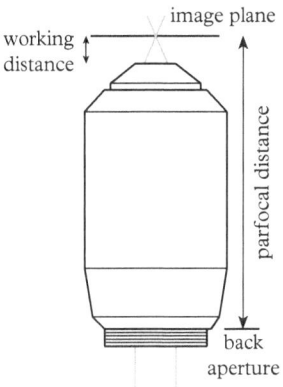

Fig. 9.9 *A microscope objective. Working distance is measured from the front-lens element of an objective to the specimen-imaging plane.*

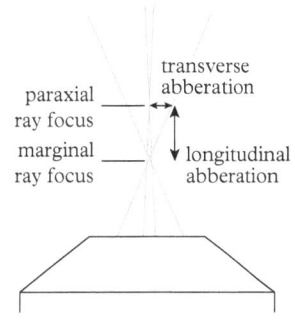

Fig. 9.10 *Spherical aberration results when marginal rays focus at a different point than paraxial rays.*

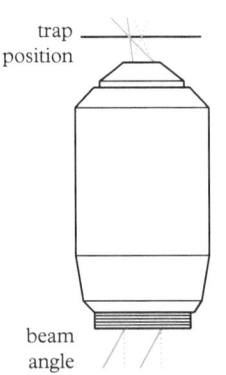

trap position

beam angle

Fig. 9.11 *The optical trap position* x_t *in the focal plane is controlled here by changing the angle of the collimated beam incident at the entrance pupil of the objective. The beam is expanded to overfill the aperture.*

steering device controls this angle. A number of methods can be used steer the beam. Slow methods include using a parfocal gimbal mirror or translating a motorized lens (Svoboda and Block, 1994a). These methods are easy to implement and very stable over time, but are limited to low frequencies.

Typical microrheology instruments employ faster means of controlling the beam angle. A decade ago, fast beam steerers were primarily galvonometers and mirrors driven by piezoelectric actuators (Mio *et al.*, 2000). Acousto-optic deflectors (AODs) and modulators are now more common due to their fast switching frequency and improvements in light transmission through the device. An AOD works by scattering light from a refractive index gradient generated by a standing acoustic wave in a solid crystalline material such as tellurium dioxide (TeO_2). The crystal response to the standing wave acts like a diffraction grating. This standing wave is the origin of the beam steering mechanism. The first-order diffraction peak has an angle that depends on the standing wave period, which can be quickly changed on the order of the time it takes an acoustic wave to propagate across the crystal—the resulting frequency response on the order of 100 kHz (or higher) is possible, depending on the crystal size.

A spatial light modulator is another method of generating and controlling an optical trap (Dufresne and Grier, 1998). Instead of steering a Gaussian beam, the spatial light modulator modulates the phase of an expanded beam wavefront, essentially generating a phase-space hologram, or kinoform, to create specific patterns of optical traps in or near the focal plane (Grier, 2003). By programming an evolution of patterns in the spatial light modulator, which can be a liquid crystal device or array of micromechanical mirrors, the trap pattern is changed to move trapped particles. Preece *et al.* (2011) switch kinoforms to generate step displacements of an optical trap, analogous to the *in situ* calibration method discussed in Section 9.5.4.

Microscope

A commercial microscope is not an absolute requirement for optical trapping. Home-built microscope cores, consisting of a simple illumination device and objective, can provide excellent performance for microrheology experiments at relatively low cost (Appleyard *et al.*, 2007). The advantage of using a microscope comes from its imaging capabilities—confocal, epifluorescence, and bright-field imaging can be combined with microrheological experiments to simultaneously image the material structural response.

Inverted microscopes are generally preferred because the weak gravitational acceleration exerted on a probe particle towards the objective somewhat offsets the radiation pressure from scattering that

pushes the particle in the beam direction. Since particles accumulate at the bottom of the sample, it is easier to locate them and move
them into the center of the sample cell in less-viscous materials. A
drawback, however, is that particles below the focal plane are often
pushed by the radiation pressure of the beam, and may eventually
be swept into the trap causing a disruption to the experiment. Light
scattered from these out-of-plane particles can also interfere with
back-focal-plane detection.

Back-focal-plane-interferometry

Position detection of the trapped particle using back-focal-plane
interferometry follows from the methods described earlier for interferometric passive microrheology in Chapter 6. The laser trap itself
can serve as the detected beam at the photodetector. In this case, the
particle position relative to the actively moving optical trap leads to
a *moving* or *translating* reference frame. A second, co-aligned laser is
sometimes used to detect the particle displacement. In such cases, the
particle motion in response to the moving trap is relative to a *fixed* reference frame. We discuss the analysis of experimental data collected
using these two references frames in Section 9.5.

9.5 Trapping force calibration

Quantitative measurement of rheology using a laser tweezer depends
on the accurate calibration of the optical trap force exerted on the
probe. One of the important challenges is to determine the trap
stiffness κ_t in the material of interest. This is not necessarily straightforward. The trap stiffness is typically characterized by measuring the
displacement of the probe due to a known applied force. A convenient
and immediately accessible force is the drag due to either translating
the sample (uniform flow) or movement of the trap itself. Because the
rheology of the sample is in question, the drag force is not immediately
available.

   One solution to the calibration problem is to trap probes in a Newtonian fluid of known rheology and use this calibration in subsequent
samples, not unlike using calibration fluids in a mechanical rheometer. However, as we have seen, the material itself will change the
trapping force through changes in the index of refraction. Accounting for such variation may be non-trivial and leads to uncertainty in
the measurement. Trapping is also sensitive to the thickness of the
coverslip and position of aberration-correcting optics in the trapping
microscope objective, which makes a comparison between a calibration sample and a material sample less certain. What is needed is a

calibration method that will work in the material of interest regardless of its rheological characteristics. Here, we will first describe simple methods of calibrating the optical trap stiffness. We then show that combinations of these methods can be used to calibrate traps in complex media. Since the displacement of the trapped particle in response to a known force is the principal quantity of interest, the calibration is limited to the position detection scheme. In the following, we will generally assume that the trapped particle displacement is measured by a back-focal-plane detector, such as a quadrant photodiode, as described in Chapter 6. Some calibration methods, such as dragging the particle in a viscous fluid, are amenable to analysis using video microscopy and particle tracking, similar to the techniques of Chapter 4.

9.5.1   Drag in a viscous fluid

One straightforward method of generating a known force on a colloidal particle is to pull it with a velocity V through a fluid with (Newtonian) viscosity η. The drag force is of course $F_d = 6\pi a \eta V$, provided the particle moves far from stationary boundaries and the movement has reached a steady state.[4] One measures the displacement of the trappedparticle relative to its equilibrium position ΔX with increasing relative velocity, and finds the force-displacement relation by the force balance

$$\kappa_t \Delta X = 6\pi a \eta V. \tag{9.35}$$

This force balance is illustrated in Fig. 9.12. A uniform flow is easily created by translating the microscope stage while holding the optical trap stationary.

A representative force-displacement plot is shown for 3.2 μm diameter polystyrene particles in water. The trap force over this range of

---

[4] The need to trap far from a sample boundary is limited by the working distance of the objective.

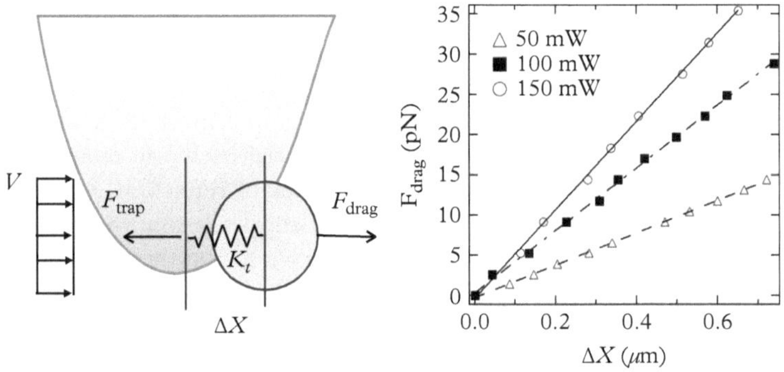

Fig. 9.12 *Trap calibration by means of translating the particle through a fluid of known viscosity η. The displacement from the trap is measured with increasing velocity. A force-displacement plot is shown for 3.2 μm diameter polystyrene particles trapped in water by a 1064 nm wavelength laser. The laser power is measured at the back aperture of the microscope objective, a 63× NA 1.2 water immersion objective.*

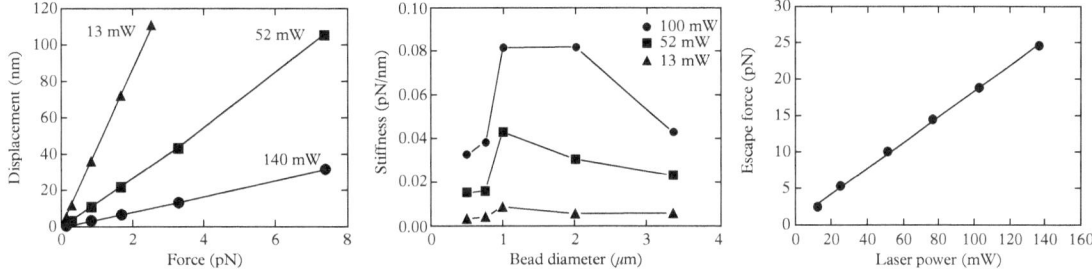

Fig. 9.13 *Trap stiffness for polystyrene particles in a 63× magnification oil immersion objective with a numerical aperture of 1.25 NA and laser wavelength λ = 1064 nm. The left panel shows the displacement-force curve dependence on laser power for 1 μm diameter particles. The center panel plots the trap stiffness dependence on the particle size. The right panel shows the escape force with laser power, again for 1 μm particles. Reprinted from* Biophys. J., 70, *Simmons, R. M., Finer, J. T., Chu, S., & Spudich, J. A., Quantitative measurements of force and displacement using an optical trap, 1813–22, Copyright (1996), with permission from The Biophysical Society.*

displacements is linear, and the trap stiffness κ_t increases with the laser power. This magnitude of trapping force is typical for a near-IR laser (1064 nm), which can usually generate forces of tens of piconewtons on such particles. The trap stiffness is similar to results reported by Simmons *et al.* (1996), shown in Fig. 9.13. The trap forces are similar for the 1 μm diameter-polystyrene particles held in a 63× magnification oil immersion objective with a numerical aperture of 1.25 NA. The trap stiffness is reported for particle sizes between 0.5 and 3.5 μm. The stiffness increases rapidly with particle size, consistent with the volume dependence of the trapping force we found earlier in the Rayleigh regime. As diameter approaches the laser wavelength, λ =1064 nm, the trap stiffness plateaus. For the largest particle diameter, the trap stiffness decreases somewhat as particles tend to trap at their edges.

Simmons *et al.* (1996) also report the maximum trapping force for 1 μm diameter particles. The average force that particles escape from the optical trap increases with laser power. The highest-trapping force is about 25 pN at a power of 140 mW. The maximum trapping forces reported in the literature rarely exceed 100 pN, typically for near-IR lasers (like the 1064 nm wavelength used here), polystyrene particles (which have a reasonably large index of refraction), and particle diameters in the 1–2 μm range.

Escape forces can be difficult to measure accurately. Particles are held in the trap's potential well and subject to an external force. The escape force represents an inflection point in the trap potential energy well, and normally the drag would have to exceed this force. However, escape at lower forces is made possible by the particle's random thermal motion, much like Kramer's theory of reaction rates (Kramers, 1940). A *distribution* of escape forces will be measured (Swan *et al.*, 2012).

9.5.2   Oscillating trap in a viscous fluid

A second method for calibrating the optical trap stiffness is to measure the particle response to an oscillating optical trap as a function of oscillation frequency. This method requires an ability to move the trap in or near the focal plane, usually with a sufficiently high frequency that requires fast beam steering optics. The mechanics of the instrument will be discussed in Section 9.4. Here, we will focus on the basis of the oscillating trap calibration.

We will first consider the equation of motion for a particle in an oscillating optical trap. The trap is driven sinusoidally with amplitude A and angular frequency ω relative to a fixed reference frame $x = 0$ such that its position with time is

$$X_t = A\cos(\omega t). \tag{9.36}$$

The forces acting on the trapped particle are shown in Fig. 9.14. The trap imposes a force on the particle $F_t = \kappa_t[X_t - X(t)]$, where $X(t)$ is

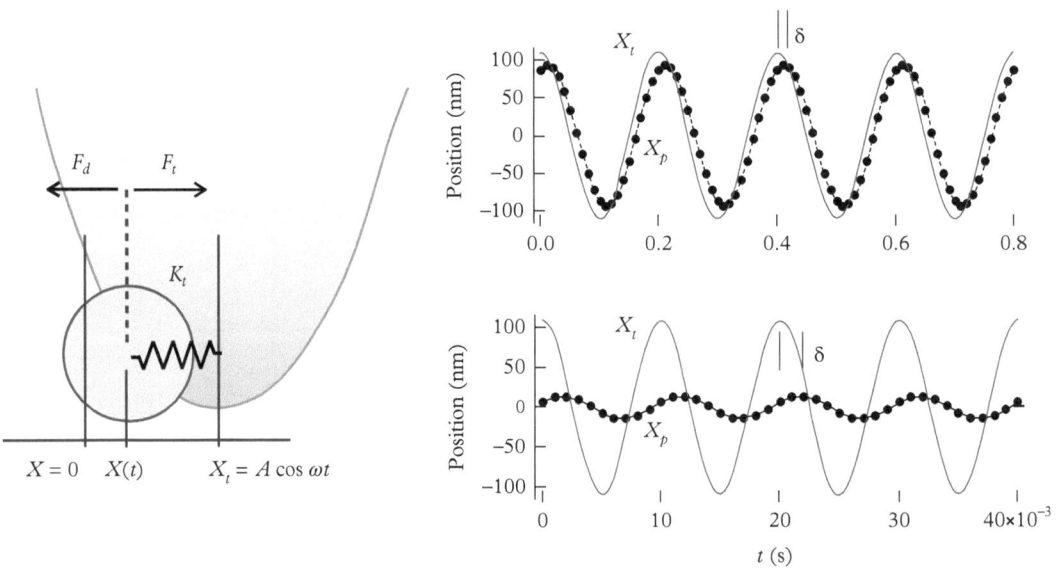

Fig. 9.14 *The forces acting on a trapped particle in an oscillating trap. The positions of the trap X_t and particle $X(t)$ are measured relative to a fixed laboratory reference frame $X = 0$. The right-hand plot shows the position with time of an oscillating optical trap (solid line) and 2 μm diameter probe particle (symbols). The trap is driven at a frequency $f = 5$ Hz (top) and 1 kHz (bottom).*

the position of the particle, again relative to the fixed reference frame. The drag resistance to this force is $F_d = -\zeta \dot{X}$. Neglecting particle inertia, the equation of motion for the particle is

$$F_t + F_d = 0 \tag{9.37}$$

in which case the governing equation becomes

$$\zeta \dot{X} + \kappa_t X = \kappa_t X_t. \tag{9.38}$$

The general solution to this equation of motion,[5]

$$X(t) = D^*(\omega)e^{i[\omega t - \delta(\omega)]} \tag{9.39}$$

can be substituted into eqn 9.38 to find the (real) particle amplitude

$$D(\omega) = \frac{A}{(1 + \omega^2 \tau^2)^{1/2}} \tag{9.40}$$

and phase lag

$$\delta(\omega) = \tan^{-1} \omega \tau \tag{9.41}$$

where $\tau = \zeta / \kappa_t$. In a Newtonian fluid, $\tau = 6\pi a\eta / \kappa_t$.

An example of a particle response to an oscillating trap is shown in Fig. 9.14. The particle position lags the optical trap by δ. In one case, for an oscillation frequency $f = 5$ Hz, the phase lag is small. The particle amplitude is also nearly the amplitude of the trap, A, as $\omega \tau \sim 1$. At $f = 1$ kHz, however, the trapped particle barely responds. The amplitude now is much smaller and the phase lag larger.

Equations 9.40 and 9.41 can also be expressed in terms of the complex susceptibility (Hough and Ou-Yang, 2002),

$$\chi^*(\omega) = \frac{D(\omega)e^{i\delta(\omega)}}{A\kappa_t} = \frac{1}{\kappa_t} \left[\frac{1}{1 + \tau^2\omega^2} + \frac{i\tau\omega}{1 + \tau^2\omega^2} \right]. \tag{9.42}$$

One can then fit eqn 9.42 to a measured particle response to find the trap stiffness κ_t. Fig. 9.15 shows such a fit to the real $\chi'(\omega)$ and imaginary $\chi''(\omega)$ components of the susceptibility for a 1.6 μm diameter silica particle trapped with a near-IR laser. The trap stiffness is $\kappa_t = 12.5$ pN/μm.

There are two things to note about the oscillatory calibration (which are also relevant to oscillatory experiments discussed later). First, the analysis presented assumes that the particle displacement relative to the trap does not exceed the linear regime of the trap's restoring force. Second, the maximum trapping force of the laser can have a subtle effect on oscillatory measurements. At a fixed amplitude,

[5] See Appendix A.5.

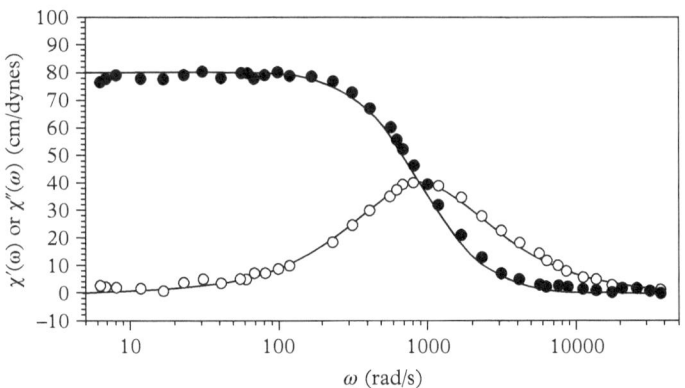

Fig. 9.15 *The real (solid symbols) and imaginary (open symbols) components of the complex susceptibility for an oscillating optical trap. Reprinted figure with permission from Hough, L. A. & Ou-Yang, H. D., Phys. Rev. E, 65, 21906 (2002). Copyright (2002) by the American Physical Society.*

the maximum velocity of the trap increases with its oscillation frequency as $v_{max} = A\omega = 2\pi A f$. At a critical frequency, the particle is no longer able to follow the trap—it exhibits aperiodic kicks from the passing optical potential. Then, at still higher frequencies, the particle experiences an average optical potential. This latter regime is the basis of line scanning optical traps, which are useful for measuring the interactions between colloidal particles (Crocker *et al.*, 1999).

The transition from the "phase lock" trapping regime used for microrheology to the "phase slip" regime of aperiodic kicks was demonstrated in experiments by Faucheux *et al.* (1995). Using a circular pattern for the optical trap (see Fig. 9.16) the point at which

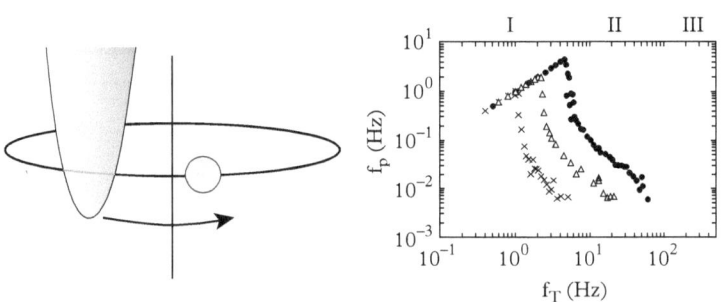

Fig. 9.16 *A 2 μm diameter polystyrene particle is held by an optical trap rotating in a 12.4 μm diameter circle. At a critical angular frequency of the optical trap, here denoted f_T, the particle angular frequency f_P abruptly decreases, marking the "phase slip" regime (II). In the third regime (III), the particle experiences an average optical potential. The three symbols represent laser powers of 150 mW (crosses), 300 mW (triangles) and 700 mW (circles). Reprinted figure with permission from Faucheux, L. P. et al., Phys. Rev. E, 51, 5239–50 (1995). Copyright (1995) by the American Physical Society.*

the particle-angular frequency begins to decrease relative to the trap angular frequency can be identified.[6] The critical frequency f_c occurs when the drag force (for the circular pattern) $-\zeta \pi D f_T$ exceeds the maximum-trapping force. The average angular velocity in the phase slip regime above the critical frequency $f_P > f_C$ can be predicted by accounting for the particle inertia in the equation of motion (Faucheux *et al.*, 1995).

9.5.3 Thermal motion in a stationary trap

Optical tweezers may be calibrated *passively* by measuring the displacement distribution of the trapped particle or its fluctuations. This method will work for any *fluid* where the equilibrium position of the trapped particle is solely a function of the trap stiffness. A viscoelastic fluid might suffice provided that the position sampling time was longer than the material relaxation time. Even so, the calibration methods described for complex fluids are better for non-Newtonian samples.

Equipartition

The simplest passive method for calibrating an optical trap comes by invoking the equipartition of energy, $\frac{1}{2}\kappa_t \langle X^2 \rangle = \frac{1}{2}k_B T$. The measured mean-squared position of the particle position is related to κ_t by

$$\kappa_t = \frac{k_B T}{\langle X^2 \rangle}. \tag{9.43}$$

Similarly, the probability of observing a position (in one dimension) of a particle in a stationary trap is given by the Boltzmann equation,

$$P(X) \sim \exp(-U/k_B T) \tag{9.44}$$

where the trap potential in the linear regime is $U(X) = \frac{1}{2}\kappa_t X^2$. From sampled displacements, the probability distribution function gives

$$U = -k_B T \ln P(X) \tag{9.45}$$

which of course is parabolic in X and can be fit to yield κ_t.

Fluctuation dissipation

The particle fluctuations in the trap can be used to calibrate its stiffness. The power spectrum of the position fluctuations is given by fluctuation-dissipation (eqn 6.8),

$$\langle |X(\omega)|^2 \rangle = \frac{2k_B T}{\omega} \chi''(\omega). \tag{9.46}$$

[6] Creating a circular trapping pattern is straightforward—one only needs to move the trap in the orthogonal direction 90 degrees out of phase.

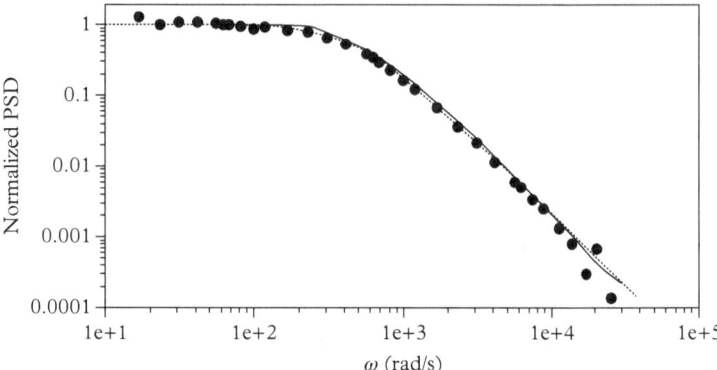

Fig. 9.17 *The normalized power-spectral density of a particle position in an optical trap. The dashed line is calculated by eqn 9.47 with $\kappa_t =$ 12.5 pN/μm. Reprinted figure with permission from Hough, L. A., & Ou-Yang, H. D., Phys. Rev. E, 65, 21906 (2002). Copyright (2002) by the American Physical Society.*

For a particle trapped in a viscous fluid, the response function is eqn 9.42. This gives a power spectrum with a Lorentzian form

$$\langle |X(\omega)|^2 \rangle = \frac{2k_B T}{\kappa_t} \left(\frac{\tau}{1 + \tau^2 \omega^2} \right). \tag{9.47}$$

An example of the power spectrum is shown in Fig. 9.17. For frequencies $(\tau\omega)^2 \ll 1$, the power-spectral density is constant. At high frequencies, $(\tau\omega)^2 \gg 1$, the power spectrum decreases as $\sim \omega^{-2}$. These two scaling regimes define a "corner frequency" or "rolloff frequency" of the power spectrum,

$$\omega_0 = 1/\tau = \kappa_t / 6\pi a \eta. \tag{9.48}$$

One advantage of estimating the trap stiffness from the corner frequency is that the position detector does not need to be absolutely calibrated (Svoboda and Block, 1994a), whereas eqn 9.45 will exhibit error proportional to the detector position calibration constant, and by eqn 9.43, to the calibration constant squared. Another advantage over calibration by equipartition is the strong effect low-frequency noise has on the latter. Figure 9.18 shows one such example, in which the low-frequency power spectrum exhibits a ballistic regime, where $\langle |X(\omega)|^2 \rangle \sim \omega^{-1}$, due to the the laser pointing and output power stability and stability of the beam-steering devices. The higher values of the power spectrum are perceived as greater particle displacements, and result in an apparent weaker trapping stiffness (Shindel *et al.*, 2013). As the laser power increases, the low-frequency noise tends to become stronger at increasing frequency relative to the fluctuations caused by the probe motion, which may finally obscure the corner frequency.

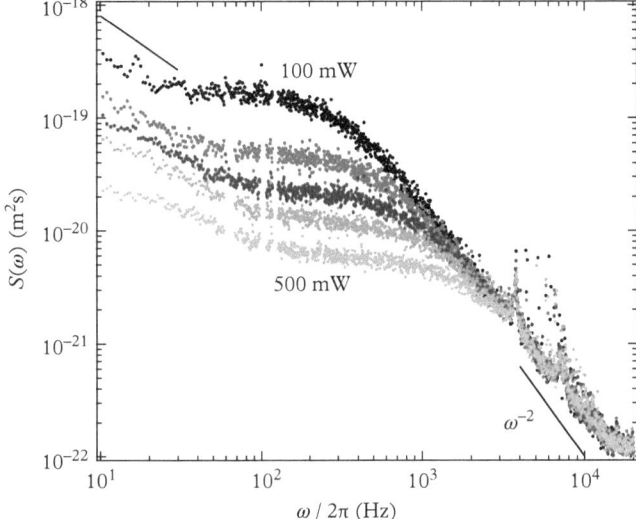

Fig. 9.18 *The power-spectral re-*
sponse of trapped 1.57 μm diameter
PMMA particles in water. The laser
wavelength is 1064 nm and the re-
sponse is measured at five laser pow-
ers, 100, 200, 300, 400, and 500
mW. Rheol. Acta, Calibration of
an optical tweezer microrheome-
ter by sequential impulse response,
52, 2013, 455–65, Shindel, M.
M., Swan, J. W., & Furst, E.
M., © Springer-Verlag Berlin Hei-
delberg 2013 "With permission of
Springer".

9.5.4 *In situ* calibration in a complex fluid

The drag, oscillatory, and passive calibration methods described in Section 9.5.3 are good methods for characterizing the optical trap stiffness, maximum trapping strength, and trap nonlinearity. However, they all require that the surrounding medium has a known rheology (usually a Newtonian fluid). For fluids with an increasing concentration of a component, such as a polymer solution, the methods could be used because the solvent (presumably of known viscosity) can be measured in the "infinitely-dilute" limit. Nonetheless, it is a problem that the material refractive index may change substantially with composition. There are still differences in probe size, composition, and sample-to-sample variation that limit the precision of such calibration schemes.

What is desired is a means to calibrate an optical trap in a material of unknown rheology. Fortunately, a combination of passive and active methods provides a solution (Fischer *et al.*, 2010; Shindel *et al.*, 2013). Once a particle is trapped, the calibration proceeds in two general stages: First, the trap is held stationary and the equilibrium thermal fluctuations are measured, identical to the methods described in Section 9.5.3. During the second stage, the particle is forced by an active displacement of the trap or stage, variations on the ideas presented in Sections 9.5.1 and 9.5.2. In the example that follows, the optical trap position is repeatedly "jumped" along one axis, essentially executing a rectangle function. The transient relaxation of the

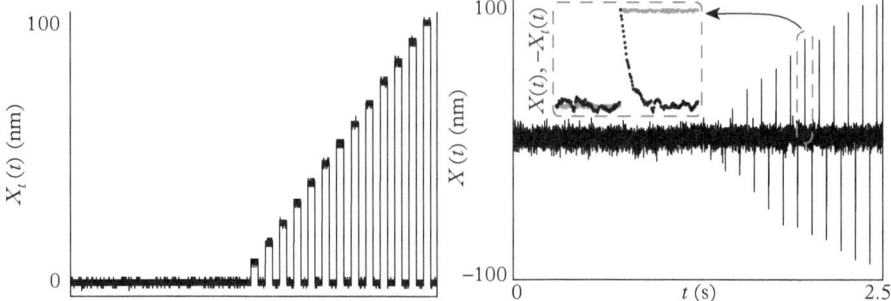

Fig. 9.19 *Trap position x_t and particle displacement w during sequential steps of an optical trap. The inset shows an expanded view of one step. Rheol. Acta, Calibration of an optical tweezer microrheometer by sequential impulse response, 52, 2013, 455–65, Shindel, M. M., Swan, J. W., & Furst, E. M., © Springer-Verlag Berlin Heidelberg 2013 "With permission of Springer".*

particle provides a second, independent characterization of the linear viscoelasticity and trap stiffness.

The two-step process calibration is illustrated in Fig. 9.19. The top plot shows the trap position X_t. Initially the trap is stationary. After a second passes, the trap is displaced in a series of steps back and forth with increasing amplitude. The displacement of the particle relative to the trap $w = X - X_t$ is is illustrated in the lower-half of Fig. 9.19. The particle randomly samples positions during the stationary part, then follows the steps that the trap takes. On the resolution of the larger plot, the displacement has a delta-function quality as the particle is initially far from the trap center, then recovers back to its equilibrium position. The inset of the plot magnifies one such step and shows that the transient response as the probe moves back to the trap is a nearly exponential decay.[7] Provided the step distance of X_t is known and the position sampling rate sufficiently high, the initial value of w provides a means to calibrate the detector response—the voltage of a quadrant detector for instance can be correlated with the step distance. This is the reason that steps of increasing magnitude are used.

The two-stage calibration is analyzed as follows. Like the analysis of interferometric passive microrheology in Chapter 6, we proceed by calculating the power-spectral density. However, instead of invoking the fluctuation-dissipation theorem, we take the Fourier Transform of the Langevin equation (eqn 3.4). This yields the power-spectral density

[7] In the absence of Brownian motion, or averaged many times, the curve is exponential in a Newtonian fluid.

$$S(\omega) = \frac{2k_B T \mathrm{Re}[\tilde{\zeta}(\omega)]}{\left|i\omega\tilde{\zeta}(\omega) + \kappa_t\right|^2}. \tag{9.49}$$

The measured power-spectral density is calculated from the particle positions recorded over a period T

$$S(\omega) = \lim_{T \to \infty} \frac{2}{T} |X(\omega)|^2 = \lim_{T \to \infty} \frac{2}{T} \left| \int_0^T X(t) e^{i\omega t} dt \right|^2 \qquad (9.50)$$

in units $S(\omega)$ [=] $L^2 t$ (such as m²Hz) (Brau *et al.*, 2007). The resistivity is unknown, but by assuming that the material exhibits some viscous dissipation when steadily deformed, the power spectrum is finite and $\omega \tilde{\zeta}(\omega) = 0$ as $\omega \to 0$. The power spectrum at low frequencies can be approximated by a second-order Taylor expansion

$$S(\omega) \approx S(0) + \frac{1}{2} S''(0) \omega^2, \qquad (9.51)$$

where

$$S(0) = \frac{2 k_B T}{\kappa_t} \left[\frac{\tilde{\zeta}(0)}{\kappa_t} \right], \qquad (9.52)$$

and

$$S''(0) = -\frac{2 k_B T}{\kappa_t} \left[2 \left(\frac{\tilde{\zeta}(0)}{\kappa_t} \right)^3 + 4i \frac{\tilde{\zeta}(0)}{\kappa_t} \frac{\tilde{\zeta}'(0)}{\kappa_t} - \frac{\tilde{\zeta}''(0)}{\kappa_t} \right]. \qquad (9.53)$$

$S(\omega)$ is an even function of frequency, therefore the first-order derivative in the expansion of eqn 9.51 must be zero. The bracketed terms in eqns 9.52 and 9.53 can be determined from the transient response of the particle following a laser jump, leaving a single unknown, the trap stiffness κ_t.

In the second stage, the transient response of each jump of the laser trap, modeled as a Heaviside step function $X_t(t) = hH(t)$, is governed by the equation of motion

$$-\tilde{\zeta}(\omega) \left(i\omega \tilde{w}(\omega) + h \right) - \kappa_t \tilde{w}(\omega) = 0 \qquad (9.54)$$

where $w = X - X_t$ is the displacement of the particle from the optical trap. In eqn 9.54, the Brownian force is neglected; the experiment is averaged over many realizations. Solving for the resistance,

$$\tilde{\zeta}(\omega) = -\frac{\kappa_t \tilde{g}(\omega)}{i\omega \tilde{g}(\omega) + 1}. \qquad (9.55)$$

Here,

$$\tilde{g}(\omega) = \langle \tilde{w}(\omega) \rangle / h \qquad (9.56)$$

is the ensemble-averaged transient response normalized by the impulse magnitude h. In the limit that $\omega \to 0$, it follows from eqn 9.55 and the definition of the Fourier Transformation that

$$\left[\frac{\tilde{\xi}(0)}{\kappa_t}\right] = -\tilde{g}(0) = -\int_0^\infty g(t)\, dt. \tag{9.57}$$

The lower limit of integration is set by the fact that $g(t) = 0$ for $t < 0$ by definition; then

$$-\left[2\left(\frac{\tilde{\xi}(0)}{\kappa_t}\right)^3 + 4i\frac{\tilde{\xi}(0)}{\kappa_t}\frac{\tilde{\xi}'(0)}{\kappa_t} - \frac{\tilde{\xi}''(0)}{\kappa_t}\right]$$
$$= -\tilde{g}''(0) = \int_0^\infty t^2 g(t)\, dt. \tag{9.58}$$

Substitution for these bracketed terms in eqn 9.51 relates the transient response and the power-spectral density,

$$S(\omega) \approx \frac{k_B T}{\kappa_t}\left[-2\left(\int_0^\infty g(t)\, dt\right) + f^2\left(\int_0^\infty t^2 g(t)\, dt\right)\right]. \tag{9.59}$$

After the integrals of the transient response function $g(t)$ are calculated, the trap stiffness is determined from a one parameter fit of the power-spectral density in the limit of low frequencies (*i.e.*, below the corner frequency ω_0).

Figure 9.20 shows three ensemble averaged transient responses for 200 kDa PEO polymer solutions. At the lowest concentration (1.25 wt%), $g(t)$ is indistinguishable from a Newtonian fluid—the function decays exponentially. As the polymer concentration increases, the non-Newtonian rheology of the sample begins to emerge, as is evident from the more complex relaxation process.

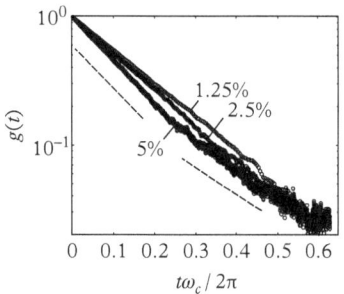

Limitations of the calibration method

A shortcoming of the current two-step *in situ* calibration is the requirement that $S(0) \to 0$; that is, that the medium is a viscoelastic fluid. Calibration methods for viscoelastic solids and fluids with long relaxation times, such as strongly entangled semiflexible polymers, remain to be developed.

One step of the calibration method depends on determining the power-spectral density of the probe's thermal motion. In Fig. 9.18, a low-frequency noise signal is obvious. While it is difficult or impossible to eliminate entirely, the extraneous noise decays faster than the signal generated by the particle. At a critical frequency, ω^*, the low-frequency noise diminishes. If the contribution to the power spectral density from the low-frequency noise is A/ω, then its intersection with the plateau of the Brownian contribution to the power spectrum scales as

$$\omega^* = \frac{A\kappa^2}{2k_B T \tilde{\zeta}(0)}.$$ (9.60)

An accurate calibration requires that ω^* is much smaller than the corner frequency given by eqn 9.48.[8] This ensures that there is a large portion of the plateau in the power-spectral density to fit for the trap stiffness, or

$$A \ll k_B T/\kappa.$$ (9.61)

The magnitude A characterizes the mean-squared fluctuation due to extraneous noise and that this must be smaller than the Brownian signal, which is identified from equipartition as the mean-squared fluctuation in particle position due solely to thermal motion. Increasing the trap stiffness and further localizing the particle will eventually violate this criterion. The power spectrum then becomes saturated with extraneous noise. Interestingly, the viscoelastic moduli play no direct role in defining the measurement limits. It is a dynamical quantity governing only how quickly relaxation happens.

As discussed in Section 6.1.2, the lowest detectable displacement using back-focal-plane interferometry is typically on the order of one nanometer, and the maximum useful displacement measurable is set by the linear range of the back-plane detector. If the limit of linearity is 300 nm, then the product of displacements with the trap stiffness set upper and lower bounds on the forces that can be measured—a range from 2 fN to 100 pN is possible.

A number of variations of two-stage calibration methods for optical tweezers have been developed. Grimm *et al.* (2012) present a novel *in situ* method for calibrating optical traps using the *velocity* correlation function and position mean-squared displacement. The method can resolve the trap stiffness and either the particle size or the medium viscosity, similar to the *in situ* method described, but it remains to extend the approach to complex fluids.

9.5.5   Trap stiffness and index of refraction

In our discussion of optical trapping forces, we found that the trap force (and stiffness) should scale with the difference of the squares of the refractive index of the trapped particle and medium (cf. eqns 9.29 and 9.31). This dependence on refractive index contrast must be accounted for in order to obtain accurate rheological measurements by

[8] In a non-Newtonian medium, $\omega_0 = k_B T/\tilde{\zeta}(0)$.

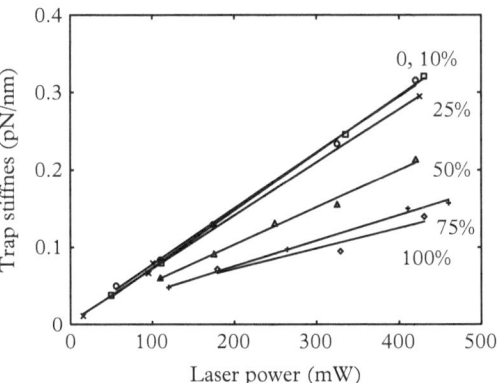

Fig. 9.21 *The effect of medium index of refraction on trap stiffness. Trap stiffness is measured as a function of laser power and increasing glycerol concentration in glycerol-water mixtures. The particles are 490 μm diameter polystyrene. Replotted data from Brau* et al. *(2007).*

laser tweezers. The *in situ* calibration methods described in the previous section are one way of accounting for the unknown rheology and optical properties of a material under test. Several studies have also reported the stiffness and maximum trapping force as a function of index contrast. Resnick (2003) details trap strength measurements for polystyrene and poly(methyl methacrylate) particles in hydrocarbon mixtures of specific refractive of index, ranging from 1.45 to 1.57.

Similar studies examining the trapping strength of polystyrene particles in water-glycerol mixtures are reported by Brau *et al.* (2007). Their data on the stiffness of a near-infrared trap with polystyrene particles at several laser powers and concentrations of water-glycerol mixtures is shown in Fig. 9.21. The refractive index of the solutions increases from n_s = 1.33 for pure water to 1.47 for pure glycerol. With increasing glycerol concentration, the trap stiffness decreases due to the lower contrast between the medium and the particle refractive index. Using the *in situ* calibration method described in the last section, Shindel *et al.* (2013) find that the trap stiffness (laser wavelength λ = 1064 nm,) decreases from κ_t = 48 pN/μm in 1.25 wt% PEO solutions to κ_t = 23 pN/μm at 5.0 wt%.

9.6 Active oscillatory microrheology

9.6.1 Fixed reference frame

The fixed reference frame applies when the probe motion is measured with microscopy or by back-focal-plane interferometry with a co-aligned, non-trapping beam. The position of the particle is directly measured with respect to its displacement from a fixed, equilibrium position. Figure 9.14 illustrates the fixed reference frame geometry.

In a fixed reference frame, active microrheology of a viscoelastic material is similar to the oscillatory trap calibration, with the Fourier Transform of the time-dependent equation of motion, eqn 9.37, now including the elastic term F_e,

$$F_t + F_d + F_e = 0. \tag{9.62}$$

In terms of a complex viscosity, the equation of motion becomes

$$-6\pi a\eta^*(\omega)\dot{X} - \kappa_t(X - X_t) = 0 \tag{9.63}$$

in the absence of inertia.

Similar to the analysis of oscillatory calibration in a Newtonian fluid (Section 9.5.2), the solution to this equation of motion remains eqn 9.39, but with a particle amplitude $D(\omega)$ and phase lag $\delta(\omega)$ in response to the driving trap oscillation $x_t = A\cos(\omega t)$ that now depends on the complex, frequency-dependent viscoelastic modulus. The particle amplitude and phase lag are measured, using a lock-in amplifier, for instance (Hough and Ou-Yang, 2002; Sriram et al., 2009), and these quantities are then related back to the storage and loss moduli by the equation of motion. The reference signal for the lock-in amplifier is simply taken from the frequency synthesizer or function generator driving the optical trap.

For the elastic modulus, consider the equation of motion when the particle has reached its maximum displacement at $t = \delta/\omega$. At this point, its instantaneous velocity is zero, but it experiences an elastic restoring force $F_e = -6\pi aG'(\omega)X$. The force balance on the particle becomes

$$-6\pi aG'D(\omega) - \kappa_t[D(\omega) - A\cos\delta(\omega)] = 0. \tag{9.64}$$

which yields

$$G'(\omega) = \frac{\kappa_t}{6\pi a}\left[\frac{A}{D(\omega)}\cos\delta(\omega) - 1\right]. \tag{9.65}$$

Similarly, $X = 0$ occurs at $t = (\delta + \pi/2)/\omega$. Here, the particle experiences a maximum dissipation, but no elastic restoring force as it passes through its equilibrium point. The force balance becomes

$$-6\pi a\omega\eta'D(\omega) - \kappa_t A\sin\delta(\omega) = 0 \tag{9.66}$$

which gives

$$G''(\omega) = \frac{\kappa_t A}{6\pi aD(\omega)}\sin\delta(\omega) \tag{9.67}$$

with the relation $G''(\omega) = \omega\eta'(\omega)$.

It may be tempting to interpret the phase lag of the particle position relative to the trap position $\delta(\omega)$ in terms of the loss tangent, the ratio of the storage and loss moduli. But eqns 9.65 and 9.67 show clearly that this notion is incorrect *for fixed-frame laser tweezer microrheology*,

$$\tan \delta(\omega) \neq G''(\omega)/G'(\omega). \tag{9.68}$$

Instead, we find that

$$\tan \delta(\omega) = \frac{G''(\omega)}{\frac{\kappa_t}{6\pi a} + G'(\omega)}. \tag{9.69}$$

As Brau *et al.* (2007) show, the ratio $G''(\omega)/G'(\omega)$ is related to the phase lag between the particle position and the *force* exerted on it, and that is exactly what we derived for magnetic bead microrheology (cf. eqn 8.65). We will return to this point in section 9.6.2.

We finally consider the expected probe position amplitude and phase angle given a viscoelastic response. We've already calculated the phase lag in eqn 9.69. The corresponding probe amplitude $D(\omega)$ is also found from eqns 9.65 and 9.67. From the trigonometric relation given by the phase angle (cf. Fig. 9.22), we see that $\sin \delta = G''/[(\kappa_t/6\pi a + G')^2 + G''^2]^{1/2}$ and

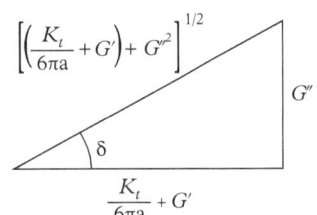

Fig. 9.22 *Trigonometric relation of the phase angle for a fixed reference frame.*

$$D(\omega) = \frac{\frac{\kappa_t}{6\pi a}A}{\left[\left(\frac{\kappa_t}{6\pi a} + G'\right)^2 + G''^2\right]^{\frac{1}{2}}}. \tag{9.70}$$

9.6.2 Moving reference frame

In the translating reference frame, the probe particle position is measured as a displacement from the oscillating optical trap, $w(t) = X - X_t$, rather than a fixed position, like a co-aligned beam or stationary plane of a microscope image. The solution to the equation of motion is identical to eqn 9.39 with an amplitude $D'(\omega)$ and phase lag $\delta'(\omega)$,

$$w(t) = D'(\omega)e^{i[\omega t - \delta'(\omega)]}. \tag{9.71}$$

The signal reported from a quadrant detector will be (after a calibration) to the moving frame amplitude and phase lag. These quantities must be related back to the storage and loss moduli.

Since we know the particle position in the laboratory reference frame $X(t)$, the displacement from the moving trap $w(t)$ can be written

$$w(t) = D(\omega)\cos(\omega t - \delta) - A\cos(\omega t). \tag{9.72}$$

Using Euler's identity, the real part of the function

$$w(t) = \text{Re}\left(De^{i(\omega t - \delta)} - Ae^{i\omega t}\right) = \text{Re}\left(D'e^{i(\omega t - \delta')}\right). \tag{9.73}$$

Multiplying the equation by $e^{-i\omega t}$, we have

$$De^{-i\delta} - A = D'e^{-i\delta'}, \tag{9.74}$$

then taking the magnitude of both sides yields

$$D'^2 = D^2 - ADe^{-i\delta} - ADe^{i\delta} + A^2 = D^2 - 2AD\cos\delta + A^2. \tag{9.75}$$

That gives us the magnitude of $w(t)$, but we also need its phase. To find it, take the real (or imaginary) terms of eqn 9.74,

$$D(\cos\delta - i\sin\delta) - A = D'(\cos\delta' - i\sin\delta'). \tag{9.76}$$

With the real terms of the equation, we find

$$\cos\delta'(\omega) = \frac{D(\omega)\cos\delta(\omega) - A}{D'(\omega)}, \tag{9.77}$$

and the imaginary terms yield

$$\sin\delta'(\omega) = \frac{D\sin\delta(\omega)}{D'(\omega)}. \tag{9.78}$$

The tangent of the phase lag can also be calculated

$$\tan\delta'(\omega) = \frac{D(\omega)\sin\delta(\omega)}{D(\omega)\cos\delta(\omega) - A}. \tag{9.79}$$

Next, we calculate G' and G'' in terms of the measured quantities of the translating reference frame experiment, the amplitude $D'(\omega)$ and phase lag $\delta'(\omega)$. Remember, these functions are defined relative to the input signal of the trap motion. The amplitude $D'(\omega)$ is a real and positive number and $\delta'(\omega)$ is the phase lag of $w(t)$ with respect to the trap position $X_t(t)$. Starting with the loss modulus given by eqn 9.67 and substituting for D^2 and and $\cos\delta$, we find

$$G''(\omega) = \frac{\kappa_t}{6\pi a}\left[\frac{AD'(\omega)\sin\delta'(\omega)}{D'(\omega)^2 + 2AD'(\omega)\cos\delta'(\omega) + A^2}\right]. \tag{9.80}$$

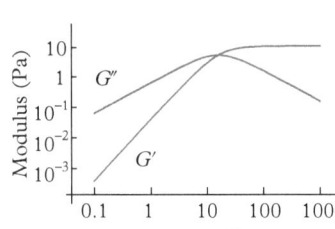

Fig. 9.23 *The frequency dependent moduli of a Maxwell fluid with G =* 10 Pa, τ = 10⁻²s *used for the calculations shown in Fig. 9.24.*

The storage modulus follows by a similar method. Starting with eqn 9.65, substitute for $\cos \delta(\omega)$, which gives

$$G'(\omega) = -\frac{\kappa_t}{6\pi a}\left[\frac{D'(\omega)^2 + AD'(\omega)\cos\delta'(\omega)}{D'(\omega)^2 + 2AD'(\omega)\cos\delta'(\omega) + A^2}\right]. \qquad (9.81)$$

The signal from $w(t)$ is π out of phase with X_t, and thus $\cos\delta'(\omega) < 0$. In the limit of a very stiff gel, where $D' \sim A$, the function becomes undefined, $G' \sim 0/0$. For a solid with a low modulus where $D' \ll A$, $G' \sim (\kappa_t/6\pi a)(D'/A)$.

Consider the calculated response for a probe particle oscillating in a Maxwell fluid for both the stationary and translating reference frames. This exercise will give us a better understanding of the limits and frequency dependence of the amplitude and phase in both regimes. The frequency-dependent shear modulus is given by eqn 3.113 and represented in Fig. 9.23.

The probe position $X(t)$ and displacement from the trap $w(t)$ are plotted in Fig. 9.24 for three frequencies: $f = 0.1, 4,$ and 10 Hz. For both reference frames, the probe trajectory at the lowest frequency is indistinguishable. In the fixed frame, $X(t)$ follows the optical trap position X_t closely. Consequently, in the moving reference frame, we barely observe any displacement $w(t)$. As the frequency increases, the particle position lags the trap position—$X(t)$ decreases and becomes

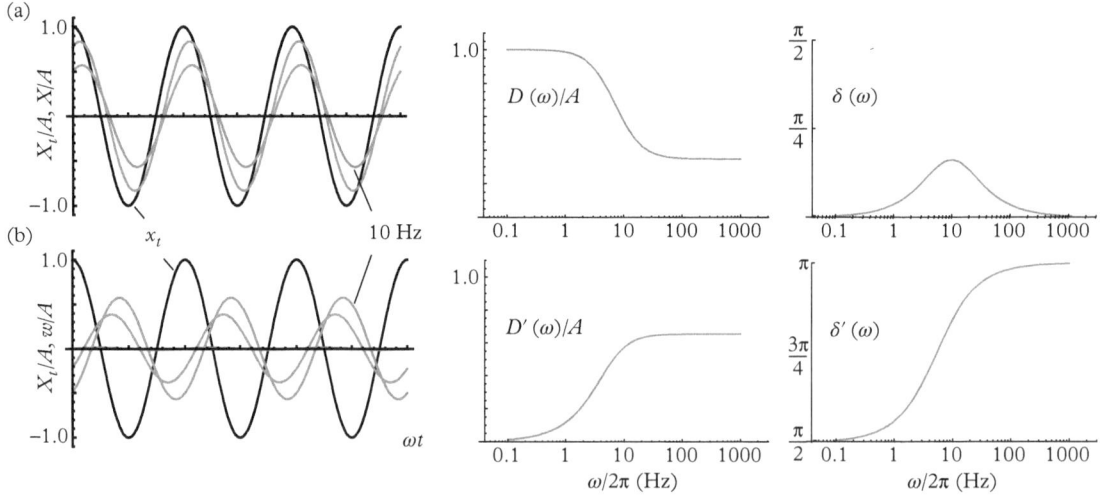

Fig. 9.24 *The calculated response for a probe particle oscillating in a Maxwell fluid represented in Fig. 9.23 for (a) fixed reference frame detection and (b) a moving reference frame. The laser amplitude is A = 0.1a with a trap stiffness of κ_t = 50 pN/μm. The particle diameter is 1 μm.*

more out of phase, and $w(t)$ increases in amplitude at 4 and 10 Hz. In the moving frame, at 10 Hz, $w(t)$ is approaching its limit of 180 degrees out of phase with X_t and there is a corresponding increase in amplitude.

The trajectories provide a visual sense of the amplitudes and phase angles plotted in the figure over a wide range of frequencies. Such plots are useful for understanding the response as well as the limits of the active oscillatory measurement.

We mentioned in the previous section that the difference in phase angles between the particle position and the particle displacement from the trap, $\Delta\delta = \delta - \delta'$, can be used to calculate the loss tangent

$$G''/G' = \tan^{-1} \Delta\delta. \tag{9.82}$$

The difference in phase angles $\Delta\delta$ represents the phase lag between the probe position and the force exerted on it, $f = -\kappa_t w(t)$ (Brau et al., 2007). The relation follows from the equation of motion, eqn 9.39, but substituting $w(t) = X(t) - A\cos\omega t = D'(\omega)\cos[\omega t - \delta'(\omega)]$. At $t = \delta/\omega$,

$$\cos(\delta - \delta') = G' \frac{6\pi a}{\kappa_t} \frac{D}{D'} \tag{9.83}$$

and at $t = (\delta + \pi/2)/\omega$

$$\sin(\delta - \delta') = G'' \frac{6\pi a}{\kappa_t} \frac{D}{D'}. \tag{9.84}$$

In the translating reference frame experiment $\delta'(\omega)$ is measured and δ is calculated by eqns 9.77 and 9.78,

$$\delta = \tan^{-1}\left(\frac{D'\sin\delta'}{D'\cos\delta' + A}\right). \tag{9.85}$$

9.6.3 Active oscillatory examples and limits

Several examples of active oscillatory microrheology using laser tweezers demonstrate its effectiveness, especially as a sensitive, small-volume experiment that can measure rheology over a wider range of frequencies than conventional rheometry. Earlier, we saw measurements of complex fluids like polymer solutions (Fig. 7.1) and colloidal suspensions (Fig. 7.2) as examples of linear, active microrheology. We can take a closer look at the latter study to see some of the operational limits of the technique.

In Fig. 9.25, active oscillatory microrheology measurements of a colloidal suspension of fluorinated ethylene propylene (FEP) particles are shown. Amplitude sweeps were conducted and real part of the

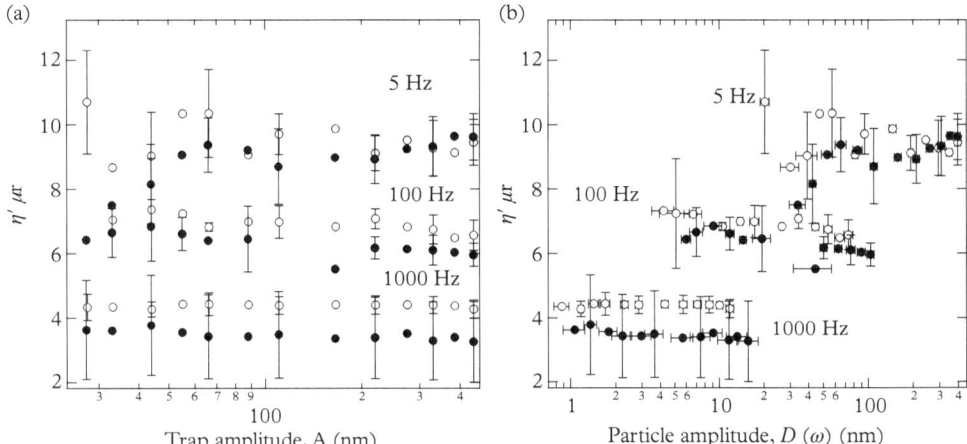

Fig. 9.25 *Oscillatory microrheology amplitude sweep measurements of a colloidal suspension. Open symbols are measurements made using 2 μm diameter silica probes while closed symbols are made with 3 μm diameter polystyrene particles. Reprinted with permission from Sriram, I., DePuit, R., Squires, T. M., & Furst, E. M., J. Rheol., 53, 357–81 (2009). Copyright 2009, The Society of Rheology.*

reduced complex viscosity $\eta'_r = \eta'/\eta_s$ is plotted for three frequencies, 5, 100, and 1000 Hz. The amplitude of the trap ranges from 10–400 nm, which for the 2 and 3 μm diameter probes, goes from less than 1% of the diameter to just under 25%. One expects the measurement to be within the linear response regime, and since there is no dependence on the amplitude at each frequency, nor on the probe size, this appears to be true.

Notice that the reduced viscosity decreases as the frequency increases. The suspension *frequency thins*, as expected, whereas all of the measurements would overlap if the material were a Newtonian fluid. The second thing to notice in the data is the increase in the measurement error that occurs at lower amplitudes, where the displacement signal becomes noisier. Increasing the frequency improves the averaging of the lock-in amplifier, so these measurements tend to be more accurate. The error also depends on the bead size. The larger probes tend to scatter less and result in a lower signal to noise.

Lastly, the reduced viscosities are plotted as a function of the measured probe amplitude $D(\omega)$, which is striking, since we clearly see that the probe amplitude decreases with increasing frequency. The range of the amplitude is governed by eqn 9.70, with the complication in this case that the viscosity is also decreasing with increasing frequency.

Stochastic, multiwave, and wideband microrheology

Among the limits of active oscillatory microrheology is the data acquisition time of the experiment. The response at discrete frequencies must be measured, with sufficient time averages performed to produce an adequate determination of the amplitude and phase angle. The measurements are more time consuming than passive microrheology and analogous to oscillatory bulk rheometry. The longer acquisition times significantly limit the ability to measure samples that are changing with time, such as gelators.

One technique that has been introduced by Lee *et al.* (2012) to reduce the acquisition time of active laser tweezer microrheology is "stochastic" microrheology. Instead of driving the probe at discrete frequencies and amplitudes, the trap is displaced randomly, but with a distribution of forces that are much higher than the Brownian force. This approach retains the benefit of characterizing the mechanical properties of samples in which Brownian motion is too small to detect, while efficiently sampling the frequency domain, similar to particle tracking or back-focal-plane interferometry. One could think of it as an active version of interferometry, in fact. Figure 9.26 is a plot of the random force distribution of the optical trap and an example of a measurement in a poly(acrylamide) gel. The forces, which can be

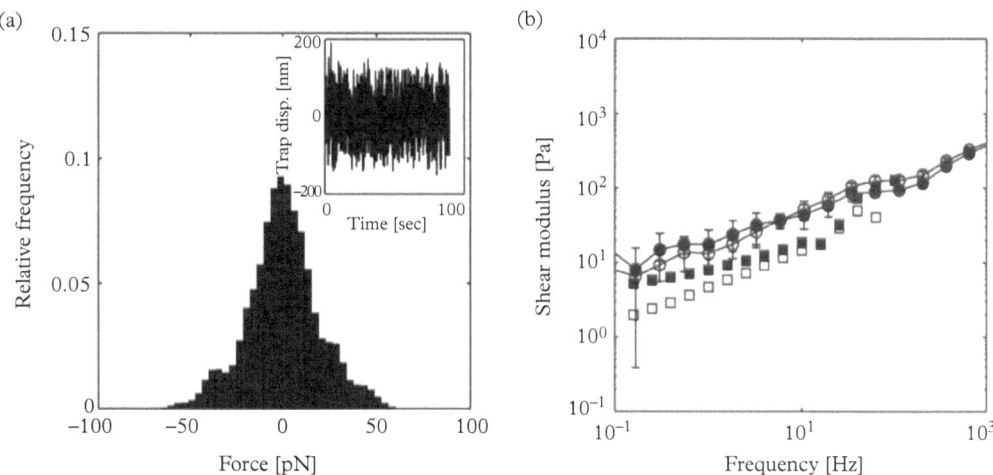

Fig. 9.26 *Stochastic laser tweezer microrheology uses (a) a Gaussian random distribution of active forces generated by driving the optical trap. (b) The storage (closed circle symbols) and loss (open circle symbols) moduli of a 3% polyacrylamide network with 0.01% bis-acrylamide near the sol-gel transition. Adapted from Lee* et al. *(2012).*

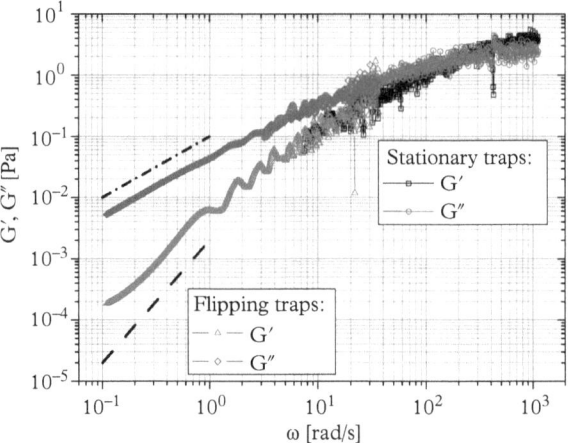

Fig. 9.27 *Wideband microrheology combines active and passive techniques to probe low and high frequencies. Here 5 μm diameter probe particles in a 1% polyacrylamide solution are subjected to active steps or "flips." High-frequency passive data is acquired when the traps are stationary. Reprinted with permission from Preece* et al. *(2011), J. Opt., 2011, 13, http://iopscience. iop.org/journal/2040-8986*

as high as tens of piconewtons, are several orders of magnitude larger than the corresponding Brownian force.

Another laser tweezer microrheology method implements the idea of "multiwave" or "Fourier Transform" rheometry (Holly *et al.,* 1988; In and Prud'homme, 1993). The optical trap is driven by a frequency modulated signal (Shindel and Furst, 2015). Like its macrorheology analog, the method measures the probe response simultaneously at several frequencies instead of consecutively at discrete frequencies, improving the throughput of the experiment.

One final but nice example of efficiently combining both passive and active microrheology is the "wideband" methods introduced by Preece *et al.* (2011). In this example, the material compliance is measured using step changes in the trap position, which probes long relaxation times, and is augmented by passive interferometry tracking in a stationary optical trap to measure the response at short times. An example of data collected with wideband microrheology is shown in Fig. 9.27.

9.7 Steady drag microrheology

Perhaps the simplest experiment to perform with optical traps is to pull the particle through a material at a constant velocity. We've already considered this experiment in Section 9.5.1 as a method of calibrating the optical trap stiffness and maximum trapping force. The motion of the probe is achieved by holding the it stationary and

translating the sample. The experiment resembles falling ball rheometry or magnetic bead microrheology, with the exception that the
velocity and not the *force* is imposed on the probe. Steady drag is
only useful for fluids, since the constant velocity would soon cause the
particle to leave the trap as the maximum trapping force is exceeded.

Steady drag experiments have been used to perform nonlinear microrheology experiments presented in Chapter 7, and in particular, to
test whether shear thinning can be measured in suspensions (Meyer
et al., 2006; Wilson *et al.*, 2009).

The limits of steady drag microrheology are bounded by the maximum trapping force F_{max} (or escape force), the stiffness of the optical
trap κ_t (which sets the lower bound of force measurement), and the
maximum and minimum velocities of the instrument (such as the motorized translation stage). The optical trap is held stationary, while the
stage is translated. In the reference frame of the probe, it translates
with a steady velocity V. The displacement from the optical trap provides the measure of the drag force. Experiments typically report the
"apparent" viscosity derived from the Stokes drag equation.

Here we consider the limits of drag microrheology in the context
of passive microrheology. The upper limit of viscosity η and probe
velocity V of drag microrheology are determined by the maximum
trapping force. This sets the lowest compliance, $\mathcal{J}(t) = t/\eta$ as

$$\mathcal{J} > \frac{6\pi a^2}{F_{max}} \tag{9.86}$$

with a strain rate approximated by $\dot{\gamma} = V/a$. The upper compliance
limit is calculated similarly by considering the minimum resolvable
force F_{min} such that

$$\mathcal{J} < \frac{6\pi a^2}{F_{min}}, \tag{9.87}$$

with Brownian forces $F \sim k_B T/a$ setting the extreme limit.

In Fig. 9.28 we plot the operating range of laser tweezer drag microrheology in the space defined by passive microrheology methods,
although we are more concerned here with steady-shear properties.
Recall that the boundaries for the compliance were computed using a $1\mu m$ diameter probes. Similarly, for 1 μm diameter particles
and an escape force $F_{max} \sim 30$ pN, the minimum compliance us-
ing laser tweezers is $\mathcal{J} > 0.1\,\text{Pa}^{-1}$. The minimum force is estimated
by $F_{min} = \varepsilon\kappa_t$, with $\varepsilon \sim 1$ nm a lower estimate of the trap displacement that can be measured by back-focal-plane interferometry and
$\kappa_t \sim 60\,\text{pN}/\mu m$ as the trap stiffness. Typical speeds of a motorized
microscope translation stage are estimated as $V_{max} = 900\,\mu\text{m/s}$ and

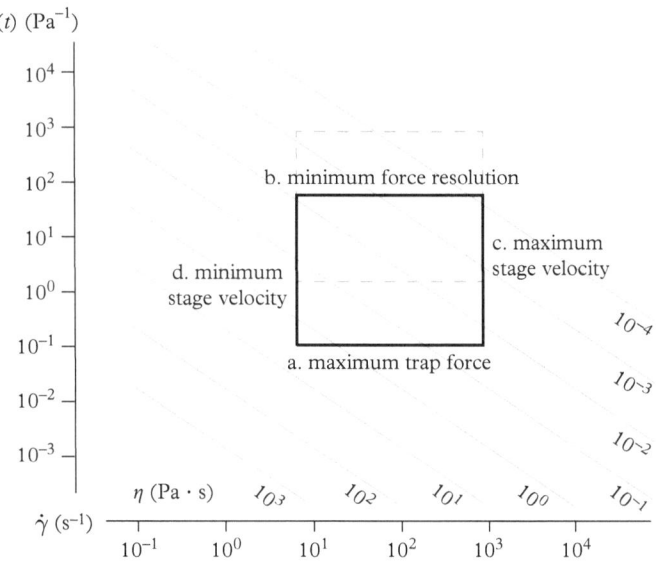

Fig. 9.28 *The operating regime of laser tweezer drag microrheology (black lines) are set by bounds given by the (a) maximum trap force, eqn 9.86, (b) minimum force resolution, eqn 9.87, and (c) upper and (d) lower-velocity limits of the transla-tion stage. The limits are calculated for a 1 μm diameter probe with trap stiffness $\kappa_t = 60$ pN/μm, es-cape force $F_{max} \sim 30$ pN, minimum force $F_{min} = 0.06$ pN, and mini-mum and maximum velocity $V_{max} = 900$ $\mu m/s$ and $V_{min} = 10$ $\mu m/s$, respectively.*

$V_{min} = 10\,\mu m/s$. Decreasing the optical trap laser power decreases the trap escape force and the trap stiffness, pushing the operating regime to higher compliances, illustrated by the dashed gray box in Fig. 9.28.

Polarized light also *exerts* a torque on birefringent particles. With higher intensity beams, it is possible to implement an *active* ver-sion of the rotational tracking microrheology experiment discussed in Chapter 6, analogous to steady drag microrheology by the transla-tion of probes. In a promising proof-of-concept, Wilking and Mason (2008) used optically trapped wax microdisks to measure the linear and nonlinear microrheology of gelatin solutions.

9.8 Two-point microrheology with tweezers

In Chapter 4, we discussed the potential of the two-point microrheol-ogy response, and highlighted its independence from probe-material interactions or even the shape and size of the probe particles. Laser tweezers have been used in both active and passive configurations to measure two-point microrheology. The fundamental principles of the measurement remain the same—the local disturbance generated by one particle is measured in the displacement of a neighboring particle. There are several advantages for using optical tweezers: The separa-tion between particles can be fixed, relatively long averaging times are possible while the probes are held by traps, and the probes (usually just one) can be actively driven to provide better signal and enable lock-in amplification for signal detection.

It is straightforward to construct an optical trapping instrument capable of two or more particles simultaneously. The polarization of a single laser beam can be rotated 90 degrees and split into parallel and perpendicular polarizations, each of which is controlled separately. A second method is to *time-share* a single optical trap. In a viscous fluid in which the Van Hove correlation function of a particle is given by eqn 4.25, the trap must return with a frequency given by (Mio *et al.*, 2000)

$$f = \tau^{-1} = \frac{2k_B T [\text{erf}^{-1}(p_n)]^2}{3\pi \eta a^3 n^2}. \tag{9.88}$$

Here, p_n is the probability that n particles remain trapped after τ seconds. The scanning rates of acousto-optic deflectors, piezo-controlled mirrors, and galvonometers are well within the typical frequencies needed for time-sharing. Methods that use holographic optical traps controlled by a spatial-light modulator are also possible.

Using laser tweezers, Hough and Ou-Yang (2002) measured the correlated motion of two trapped probes in a Newtonian fluid. One particle was driven while the response of a neighboring particle was measured. Similar studies, but using the *passive* coupling between two trapped particles, as in Fig. 9.29, have been reported for Newtonian fluids (Henderson *et al.*, 2001, 2002; Dufresne *et al.*, 2000; Lele *et al.*, 2011), F-actin (Koenderink *et al.*, 2006), semi-dilute solutions of polystyrene in decalin (Starrs and Bartlett, 2003*a,b*), and worm-like micellar solutions (Buchanan *et al.*, 2005*a,b*; Atakhorrami *et al.*, 2006; Atakhorrami and Schmidt, 2006)—the latter were also discussed in Section 6.1.4.

Studies of surfactant solutions in particular have demonstrated good agreement between bulk rheology, one-point microrheology, and two-point microrheology. This is notable because trapping with back-focal-plane interferometry provides the only current means to measure both one- and two-point responses at high frequencies. Interestingly, Atakhorrami and Schmidt (2006) showed the effect of fluid inertia on the two-point correlation by resolving the short-time vortex flow around particles and its corresponding evolution. The vortex propagation, which is normally diffusive in a Newtonian fluid, is faster in a viscoelastic fluid of worm-like micelles, reflecting the frequency dependence of the shear modulus.

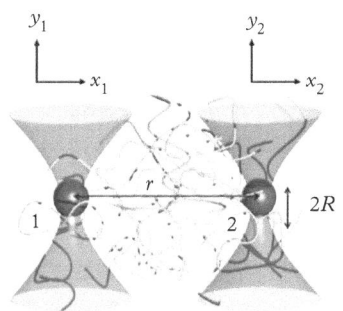

Fig. 9.29 *Two-point microrheology using optical traps.* Rheol. Acta. *High-bandwidth one- and two-particle microrheology in solutions of wormlike micelles, 45, 2006, 449–56, Atakhorrami, M. & Schmidt, C. F., ©Springer-Verlag 2006 "With permission of Springer."*

. . . .

EXERCISES

(9.1) **Piconewtons.** Optical traps generate forces on the order of piconewtons (10^{-12} N). What is a piconewton in the macroscopic world? Consider that the force of gravity acting between two bodies of mass m_1 and m_2 is

$$F_g = G\frac{m_1 m_2}{r^2} \qquad (9.89)$$

where $G = 6.67384 \times 10^{-11}$ m^{3}kg^{-1}s^{-2}.

Estimating that the diameter of a human fist is approximately 8 cm ($a = 0.04$ m), and assuming the density of each fist is on average $\rho \approx 1000$ kg/m^3 (the density of lean pork is quoted as 1030 kg/m^3; the average human body density is 1062 kg/m^3) show that two fists held at a separation $r = 1.7$ m apart gives

$$F_g \approx 2\,\text{pN} \qquad (9.90)$$

and that a piconewton is roughly (half) the gravitational attraction acting between two fists outstretched at arms length.

(9.2) **Viscosity estimate.** Estimate the viscosity from the data presented in Fig. 9.14 for a 2 μm diameter probe particle and a laser trap stiffness $\kappa_t = 12.5$ pN/μm.

(9.3) **Trap calibration—elastic solid.** A colleague proposes calibrating an optical trap by active oscillation in an elastic solid of known modulus G_0 instead of a Newtonian fluid as was described in Section 9.5.2. What would the response function $\chi^*(\omega)$ be in this case?

(9.4) **Drag experiment.** Using the Cross model (eqn 1.43) for shear thinning of a suspension, design a laser tweezer microrheology experiment to test whether you can measure its shear thinning.

(a) Estimate the low-shear η_0 and high-shear η_∞ suspension viscosities with the empirical Krieger-Dougherty equation for the reduced viscosity,

$$\eta_r = \left(1 - \frac{\phi}{\phi_m}\right)^{-(5/2)\phi_m} \qquad (9.91)$$

with the maximum packing volume fraction $\phi_m = 0.64$ for low-shear data and $\phi = 0.71$ for high-shear data. The solvent viscosity is $\eta_s = 1$ mPa· s. Plot several viscosity curves through the shear thinning.

(b) Superimpose the operating range of the laser tweezer experiments onto the calculated shear rheology. What trap stiffness and maximum trapping forces are necessary to measure the viscosity thinning (assuming it could be measured) using 1 μm and 3 μm diameter probe particles? Assume the probe velocities are set by the movement of a microscope stage with a speed range of 10–1000 μm/s.

10

Microrheology applications

Throughout this book, we've examined the general operating conditions of both passive and active microrheology. It has been clear from the outset that microrheology can have a distinct place in the rheologist's toolbox—one that can has some unique advantages, such as small material volumes, rapid acquisition methods, and high-throughput sample processing, to name a few.

We discussed several microrheology applications with their corresponding experimental methods. These were highlighted as *application notes* and included:

- Characterizing rheological heterogeneity, Section 4.10
- Rheological microscopy, Section 4.11.4
- Viscosity of protein solutions, Section 5.3.2
- Relaxation of polymer solutions, Section 5.4.8
- High-pressure microrheology, Section 5.5.1
- Molecular stiffness of semiflexible polymers, Section 5.6.1

In this chapter, we will discuss several more applications in greater depth. We start with a general approaches to planning microrheology experiments by considering the operating regimes of the different methods and their complementarity to mechanical rheometry.

The sensitivity of microrheology to *incipient* rheology—small changes that accompany an initial increase with concentration or the state of aggregation—make it ideally suited to characterize materials such as hydrogelators and protein solutions. Experiments to study gelation, degradation, and viscosity are discussed in greater detail in this chapter. Other applications make use of the small sample size requirements. Cell rheology is perhaps the oldest application of microrheological techniques, dating to pioneering experimental work in the early twentieth century. This broad field is beyond our scope, but we provide some starting points, keeping in mind what we have learned about passive and active microrheology. We conclude with promising directions for future work, including interfacial microrheology.

Microrheology. Eric M. Furst and Todd M. Squires, Oxford University Press (2017).
© Eric M. Furst and Todd M. Squires. DOI 10.1093/oso/9780199655205.001.0001

10.1 Planning a microrheology experiment

Before embarking on a microrheology experiment, it's worthwhile to consider its advantages (and disadvantages) and whether these justify the measurement. We've presented several operating regimes of microrheology techniques throughout the book: Fig. 3.16 for general passive microrheology, Fig. 5.19 for DWS microrheology, Fig. 8.19 for magnetic tweezer microrheology, and Fig. 9.28 for drag laser tweezer microrheology. Here, we compare these to the limits of rotational shear or drag rheometry. This isn't the only rheometry method available—others include pressure-driven and inhomogeneous rheometers such as capillary viscometers—but a typical rheology laboratory will be equipped with a sensitive and versatile rotational device—so it's a useful and familiar benchmark. We first review some basic limits of a shear rheology experiment. For a comprehensive reference of a number of rheometry methods, including their strengths and drawbacks, the reader is referred to Macosko (1994). The basic concepts that we survey in the next section for mechanical rheometry are presented in greater depth by Ewoldt *et al.* (2014), especially the limitations imposed by small samples and soft materials.

10.1.1 Mechanical rheometry

Earlier, we wrote the material rheological functions in terms of the relationships between the stress and deformation (strain) or rate of deformation (strain rate). Stress and strain, however, are not measured *directly*, but are instead derived from measurements that depend on the rheometry instrument being used. A rotational shear rheometer, for instance, measures torque τ, angular displacement Θ, or angular velocity $\Omega = d\Theta/dt$. At the very least, the sample and tool geometry must provide sufficient resistance to exceed a minimum measurable torque and displacement, or rate of displacement.

Minimum torque

Consider that the viscosity η, given by ratio of the shear-stress component T_{21} to the rate of strain $\dot{\gamma}_{21} = d\gamma_{21}/dt$, is related to the rheometer's measurement quantities by

$$\eta = \frac{T_{21}}{d\gamma_{21}/dt} = \frac{F_T \tau}{F_\gamma (d\Theta/dt)}, \tag{10.1}$$

where F_T and F_γ are functions that depend on the particular tool geometry. For the cone and plate and concentric cylinder geometries shown in Fig. 1.9, these functions are

$$F_T = \frac{3}{2\pi R^3} \quad \text{cone and plate} \tag{10.2}$$

$$F_T = \frac{1}{2\pi R^2 L} \quad \text{concentric cylinders.} \tag{10.3}$$

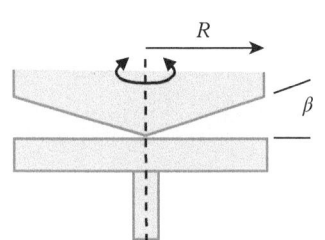

Fig. 10.1 *A cone and plate rotational shear rheometry tool.*

For a cone-and-plate geometry (Fig. 10.1),[1] $F_\gamma \approx \beta^{-1}$. We can use eqn 10.1 to understand the sensitivity limits that define part of the operating regime of bulk rheology as well as its potential failure modes.

In a rotational shear rheometer, the viscosity must be higher than a value determined by the minimum torque τ_{min} the rheometer can measure,

$$\eta > \frac{F_T \tau_{min}}{\dot{\gamma}} \tag{10.4}$$

Manufacturers often report τ_{min}, which for many instruments is between 0.001–0.1 μN · m. With a 50mm diameter cone and plate ($\beta = 0.1$ rad) and minimum instrument torque on the order of $\tau_{min} \sim 0.1$ μN · m, viscosities need to exceed $\eta > 30$ mPa · s at a shear rate $\dot{\gamma} = 0.1$, but this limiting viscosity decreases with increasing shear rate, which can reach as high as 10^4 s^{-1} in our example given a maximum rotational velocity on the order of $\Omega_{max} \sim 10^3$ rad/s and $F_\gamma = 10$. Similarly, in an oscillatory rotational shear rheology experiment, the modulus must exceed

$$G > \frac{F_T \tau_{min}}{\gamma_0} \tag{10.5}$$

where γ_0 is the strain amplitude. With the same geometry and $\gamma_0 \sim 1$ (not an unusually high value for soft materials) the minimum modulus is on the order of $G \sim 10^{-3}$ Pa. Of course, this lower value increases as the strain amplitude decreases, which may be necessary to lower in order to stay within the linear response limits of the material.

Inertial limits

Both steady and oscillatory shear experiments are also limited by inertial effects, which include contributions from the material and the instrument when the torque is measured at the moving boundary. Inertial contributions include secondary flows generated for steady

[1] For a cone and plate tool, the gap is $h = R \tan \beta$ as a function of distance R from the center. Since $\beta \ll 1$ and $\tan \beta \approx \beta$, $\dot{\gamma} = R\dot{\Omega}/R\beta = \dot{\Omega}/\beta$ at all radial points in the geometry; therefore, $F_\gamma = 1/\beta$.

shear in a rotational geometry and shear wave propagation during oscillation. However, in the latter case, the inertia of the instrument I will typically be more limiting.

In one common rheometer design, the stress is measured at the moving boundary, and a time-dependent measurement like oscillation requires the instrument and rheometer tool to accelerate. Hence, the load imposed on the measurement device is a combination of that generated by the sample and contributions from the instrument inertia. A minimum modulus,

$$G > I\omega^2 F_T/F_\gamma \qquad (10.6)$$

can be calculated based on the point at which the torque generated by the material is greater than the torque due to inertia (Ewoldt *et al.*, 2014), which is the combined inertia of the instrument and tool. These values are supplied by the rheometer manufacturer.[2]

In steady shear, secondary flows occur in rotational geometries. In a cone-and-plate geometry, acceleration of the fluid creates a radial velocity component at the rotating boundary. This leads to the condition,

$$\eta > \frac{R^2\beta^3}{\text{Re}_{\text{crit}}}\rho\dot{\gamma} \qquad (10.7)$$

where $\text{Re}_{\text{crit}} \approx 4$ (Ewoldt *et al.*, 2014).

Comparison of operating regimes

Using the operating limits previously described, we plot examples of a rheometer limits on the operating regime of several microrheology techniques in Fig. 10.2. We can see that rotational rheometers are quite sensitive—*given a sufficient volume of material*, one can easily measure the rheology of materials that are of within the operating range of microrheology, with the exception of the high frequencies that are accessible to DWS and, to a lesser extent, laser tracking. So, deciding whether a microrheology experiment should be pursued instead of a bulk rheology experiment will often depend on other factors: The amount of material available, the number of measurements to be made, and the time required for each measurement.

Volume limits and other considerations

Shear rheology measurements become more challenging when the amount of a material available becomes a limiting factor. Mechanical rheometers can be designed for small volumes, including those that use small-gap sliding plates (Clasen and McKinley, 2004; Clasen

[2] The 50 mm cone-and-plate measurement cited here has a reported instrument inertia $I_{\text{drive}} = 9.3 \times 10^{-5}$ N m s^2 and tool inertia $I_{\text{tool}} = 7.6 \times 10^{-6}$ N m s^2. These values are typically added together for a total inertia $\sim 10^{-4}$ N m s^2 and $IF_T/F_\gamma \approx 0.3$ Pa · s^2. Another example is the concentric cylinder geometry and instrument used by Ewoldt *et al.* (2014), who report $IF_T/F_\gamma \approx 2.9 \times 10^{-2}$ Pa · s^2.

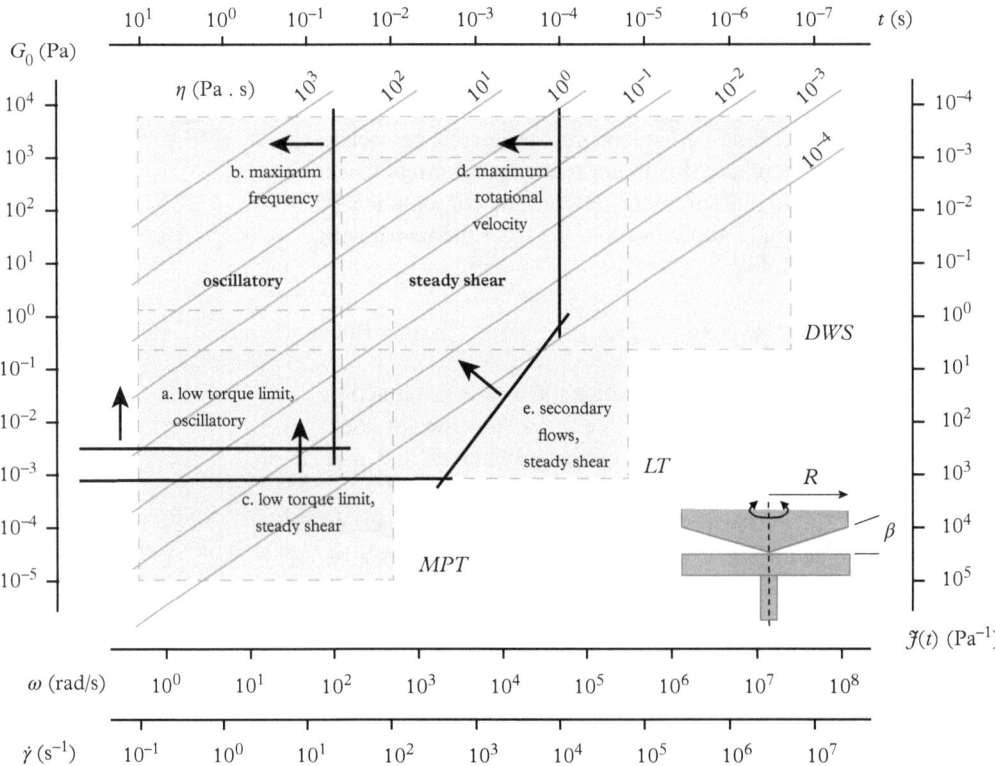

Fig. 10.2 *Operating limits calculated for both oscillatory and steady-shear experiments using a 50 mm cone and plate rheometer with cone angle $\beta = 0.1$ rad and minimum torque $\tau_{min} \sim 0.1$ $\mu N \cdot m$. Solid black lines show the limits on oscillatory rheology calculated by (a) eqn 10.5 for the torque limit and (b) the maximum drive frequency, here taken to be ~ 100 rad/s. In steady shear, the limits are (c) eqn 10.4 for the torque, (d) the maximum rotational velocity $\dot{\Omega} \approx 10^3 s^{-1}$, and (e) the emergence of secondary flows (eqn 10.7). The solid gray areas and dashed lines delineate the operating regimes of three microrheology techniques: Multiple particle tracking (MPT), laser tracking (LT), and diffusing-wave spectroscopy (DWS) assuming a probe diameter 1 μm.*

et al., 2006*b*) or by adapting instruments like atomic force microscopes and the surface forces apparatus (Gavara and Chadwick, 2010; Granick *et al.*, 2003). More conventional measurements with rotational rheometers use small tools and may require as little as 10 μl. How does this change the operating regime in Fig. 10.2? The main effect of using a smaller geometry is that the sample will be incapable of generating as much torque, shifting the limits of the viscosity or modulus that were calculated by eqns 10.4 and 10.5, respectively.

Other factors play a role as the sample volume decreases. Instruments are sensitive, but additional torques can arise due to uneven filling of the geometry (Ewoldt *et al.*, 2014). With small samples, the effect of solvent loss or moisture uptake can be exacerbated, whereas sealed samples in microrheology can limit these losses. Finally, interfacially active materials, like protein solutions, can cause the development of viscoelastic layers at the free interface of a rheology sample, resulting in additional sources of stress independent of the bulk behavior of the material (Sharma *et al.*, 2011).

Despite the overlap in operating regimes of rotational shear rheometry and microrheology, the precision of measurements at higher compliances (lower viscosities) is another area where microrheology should be considered. Precision and accuracy of viscosity measurements by particle tracking are presented in Section 10.4. Finally, the time to prepare and exchange samples and acquire rheology data can be significantly shorter for some microrheology techniques, like particle tracking. Samples can be prepared and equilibrated in parallel, increasing the experimental throughput.

10.2 High-throughput microrheology

The small sample volumes microrheology requires, especially multiple particle tracking and active microrheology methods, makes it a superb technique for screening the rheological properties of rare or scarce materials or many samples that span a large composition space.

High-throughput screening using microrheology was first proposed by Breedveld and Pine (2003). Their work used passive techniques, including both diffusing wave spectroscopy (light scattering) and multiple particle tracking, to create a rheological water/salt/surfactant phase diagram of block copolypeptide libraries. The experiment employed a multi-well plate in a computer controlled stage, which was automated to move from sample to sample as data was acquired (Fig. 10.3). The data analysis was also automated, resulting in experiments that required little human interaction. More recent high-throughput microrheology experiments have focused on integrating microrheology and microfluidic devices, several examples of which were discussed in Section 4.3.

Small sample volumes required by particle tracking (both multiple particle and single particle interferometric), magnetic bead, and laser tweezer microrheology give rise to rapid heat and mass transfer. These short times allow samples to equilibrate quickly to new conditions, and can be exploited to measure the rheology of samples as the pH, temperature, ionic strength, or other condition is changed. We saw examples in Section 4.3, including sample chambers that enabled

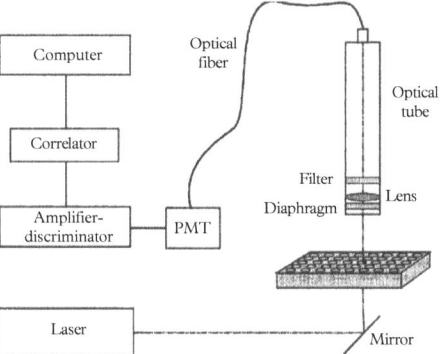

Fig. 10.3 *Diffusing wave spectroscopy for high-throughput rheology screening. J. Mat. Sci., Microrheology as a tool for high-throughput screening, 38, 2003, 4461–70, Breedveld, V. & Pine, D. J., © 2003 Kluwer Academic Publishers "With permission of Springer."*

rapid buffer exchange with rigid membranes and microfluidic devices for producing and processing microliter-scale droplets.

In microliter-volume samples, the characteristic length scales of the sample are on the order a millimeter, leading to characteristic mass and heat transfer times are dominated by diffusion and scale as

$$t \sim L_{sample}^2/\alpha \text{ or } \sim L_{sample}^2/D \qquad (10.8)$$

where α and D are the thermal and mass diffusivity, respectively. The thermal diffusivity of water is on the order of $\alpha \sim 10^{-7}$ m^2/s at 25 °C, producing equilibration times on the order of fractions of a second. Mass transfer is also rapid. For changes in pH and ionic strength, the binary diffusivities of ions in water, for instance, scale as $D \sim 10^{-9}$ m^2/s, leading to equilibration times that are typically shorter than a minute or so. One complication is mixing, since the fluid momentum also exchanges by diffusion-dominated processes in these small-scale laminar flows. Many creative solutions exist and continue to be developed for this problem (Squires and Quake, 2005).

10.3 Gelation

The gelation of a wide variety of materials has been studied using microrheology. These include physically and chemically cross-linked polymers such as poly(vinyl alcohol) (Narita *et al.*, 2013), acrylamides (Dasgupta and Weitz, 2005), and multifunctional acrylates (Boddapati *et al.*, 2011). Many scarce and emerging hydrogelators and several organogelators have been characterized using microrheology. These materials include proteins (Palmer *et al.*, 1998; Corrigan and Donald, 2009*b,a*, 2010; Mulyasasmita *et al.*, 2011), peptides

(Nowak *et al.*, 2002; Veerman *et al.*, 2006; Savin and Doyle, 2007*a*; Larsen *et al.*, 2009), polysaccharides (Heinemann *et al.*, 2004; Schultz *et al.*, 2009*b*,*a*), thermoreversible gels (Cingil *et al.*, 2015) and networks formed from carbon nanotubes (Chen *et al.*, 2010). Other materials that exhibit gelation macroscopically, such as colloidal suspensions, can be characterized using similar experiments, such as DWS (Wyss *et al.*, 2001; Rojas-Ochoa *et al.*, 2002). Since the focus of this book is probe-based microrheology methods, we emphasize such studies here.

A representative microrheology measurement of a gelling sample is shown in Fig. 10.4, in which the protein beta-lactoglobulin is induced form a gel by the addition of trifluoroethanol (Corrigan and Donald, 2009*b*). During gelation, the sample evolves from a viscous fluid to an elastic solid. The corresponding probe motion, captured by the MSD curves obtained from particle tracking microrheology, evolves continuously from free, Brownian motion in a viscous medium (the MSD has a logarithmic slope equal to one) to particles trapped in an elastic network (a constant MSD).

While the MSD curves appear to transition continuously in Figure 10.4, consider that the *incipient* gel that forms at the percolation transition marks the point at which the sample changes from a viscoelastic *fluid* to a viscoelastic *solid*. At the gel point, the longest relaxation time and viscosity of the material diverge, although it may not be readily apparent from the MSD curves in the figure. The discussion in this section will focus on the interpretation of microrheology experiments for gelling samples.

One of the principal concerns when implementing microrheology to study gelation is, of course, whether the Stokes criteria remains valid throughout the transition. If the formation of microstructures that are larger than the probe particles accompanies gelation, then

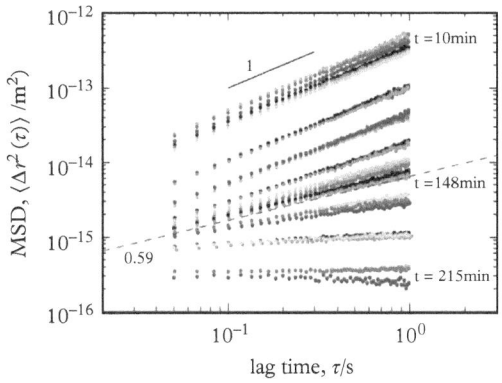

Fig. 10.4 *Microrheology measurements of a gelling beta-lactoglobulin sample. Reprinted with permission from Corrigan, A. M. & Donald, A. M. Langmuir 25, 8599–605 (2009). Copyright 2009 American Chemical Society.*

the discussion in section 10.3.1 may not apply, although similar basic features, such as decreasing probe mobility, will almost certainly be observed. Hydrogelators formed by small molecular components seem to meet the Stokes criteria the best. As with any rheology study, a thorough knowledge of the microstructures present in the sample is necessary for a proper interpretation of the experiment.

10.3.1   Critical gels

Similar to a continuous thermodynamic phase transition, the liquid-solid transition that accompanies gelation exhibits critical behavior, including a critical point (the gel point), a divergence of physical properties, and scaling behavior near the gel point. The critical gel is defined by the extent of reaction p at which the first percolating cluster spans the sample, which is denoted as p_c. Rheological properties, such as the zero shear viscosity η_0, equilibrium compliance $\mathcal{J}_e^0$ and longest relaxation time τ_L, diverge at the critical gel point (Stauffer *et al.*, 1982; Joanny, 1982). Defining the critical extent of reaction, or distance from the gel point, as

$$\varepsilon = \frac{|p - p_c|}{p_c}, \tag{10.9}$$

the scaling relationships are

$$\eta_0 \sim \varepsilon^{-k}$$
$$\mathcal{J}_e^0 \sim \varepsilon^{-z} \tag{10.10}$$
$$\tau_L \sim \varepsilon^{-y}.$$

where k, z and y are dynamic critical scaling exponents (Stauffer *et al.*, 1982; Martin *et al.*, 1988, 1987; Adolf and Martin, 1990). The scaling exponents are not independent, but are related by

$$y = z + k. \tag{10.11}$$

An equation such as 10.11 that relates critical scaling exponents is called a hyperscaling relation.

The critical scaling behavior near the gel point results in an unusual feature of gels, first identified by Winter and Chambon (1986): At the gel point, the viscoelastic moduli of the incipient network exhibit an identical power-law scaling with frequency, $G' \sim G'' \sim \omega^n$. This is a consequence of a power-law distribution of coupled relaxation modes, which leads to a "critical" relaxation modulus

$$G_c(t) = St^n. \tag{10.12}$$

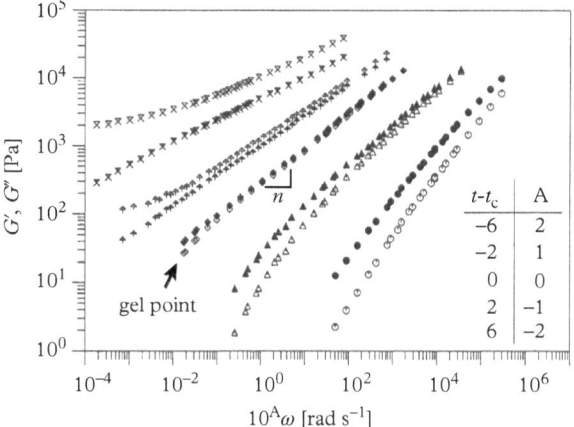

Fig. 10.5 *Storage (open symbols) and loss (closed symbols) moduli of cross-linked PDMS samples at different extents of reaction.* Adv. Polym. Sci. Rheology of Polymers Near Liquid-Solid Transitions, *134,* 1997, *165–234, Winter, H. H. & Mours, M.,* © *Springer-Verlag Berlin Heidelberg 1997 with permission of Springer.*

The rheology data of Winter and Chambon (1986) nicely illustrates the power-law scaling of the critical gel, as well as the pre- and post-gel rheology. Shown in Fig. 10.5 are their bulk rheology measurements for stoichiometrically balanced polydimethylsiloxane (PDMS) samples in which the cross-linking reaction has been stopped at different extents of reaction. In this case, the critical scaling exponent is $n = 0.5$, and $G' = G''$ over all frequencies at the gel point.

The critical relaxation exponent n is related to the critical exponents of the compliance and longest relaxation time by the scaling relationship $n = z/y$, and S is the *gel strength* with the unusual fractional units Pa · s^n, similar to the "consistency" prefactor of a power-law fluid. The power law behavior of the critical gel reflects mechanical self-similarity over a wide range of time scales due to the structural self-similarity, or fractal-like structure, of the incipient gel.

In bulk rheology, monitoring the loss tangent $\tan \delta = G''/G'$ of a material as a function of frequency and the extent of gelation is a robust method for identifying the gel point. Rather than measuring diverging properties, such as the viscosity, relaxation time or compliance (Winter and Chambon, 1986; Winter and Mours, 1997; Scanlan and Winter, 1991); instead, $\tan \delta$ passes through a single, frequency-independent value at the gel point,

$$\tan \delta = \tan \frac{n\pi}{2} \qquad (10.13)$$

as demonstrated by the data shown in Fig. 10.6. The loss tangent value at the gel point is given by the critical relaxation exponent.

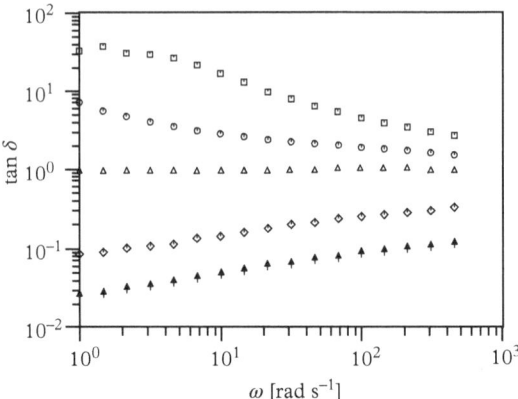

Fig. 10.6 *Loss tangent of quenched samples of a vulcanizing polybutadiene. Adv. Polym. Sci. Rheology of Polymers Near Liquid-Solid Transitions, 134, 1997, 165–234, Winter, H. H. & Mours, M., © Springer-Verlag Berlin Heidelberg 1997 with permission of Springer.*

The relation between compliance and modulus (1.38), is used to determine the microrheological response at the gel point from eqn 10.12. The critical creep compliance

$$\mathcal{J}_c(t) = \frac{\sin n\pi}{n\pi S} t^n \tag{10.14}$$

and corresponding MSD

$$\langle \Delta r^2(t) \rangle_c(t) = \frac{k_B T \sin n\pi}{a n \pi^2 S} t^n. \tag{10.15}$$

exhibit power-law behavior over all times. Recalling from eqn 3.148 that $\tan \delta = \pi \alpha(\omega)/2$, the gel point exhibits an MSD with logarithmic slope

$$\alpha = d \ln \langle \Delta r^2(t) \rangle / d \ln t = n \tag{10.16}$$

over all lag times.

10.3.2   Time-cure superposition

Close to the gel point, curves taken at different extents of reaction are shifted by dividing the MSD (compliance) and lag time by the scaling functions for the equilibrium compliance and longest relaxation time, Equations 10.10. This creates two master curves, one for the pre-gel and the other for the post-gel. Such rescaling is called time-cure superposition (Adolf and Martin, 1990).

Examples of time-cure superposition are shown in Fig. 10.7 for two hydrogel materials: A peptide hydrogelator that gels with time,

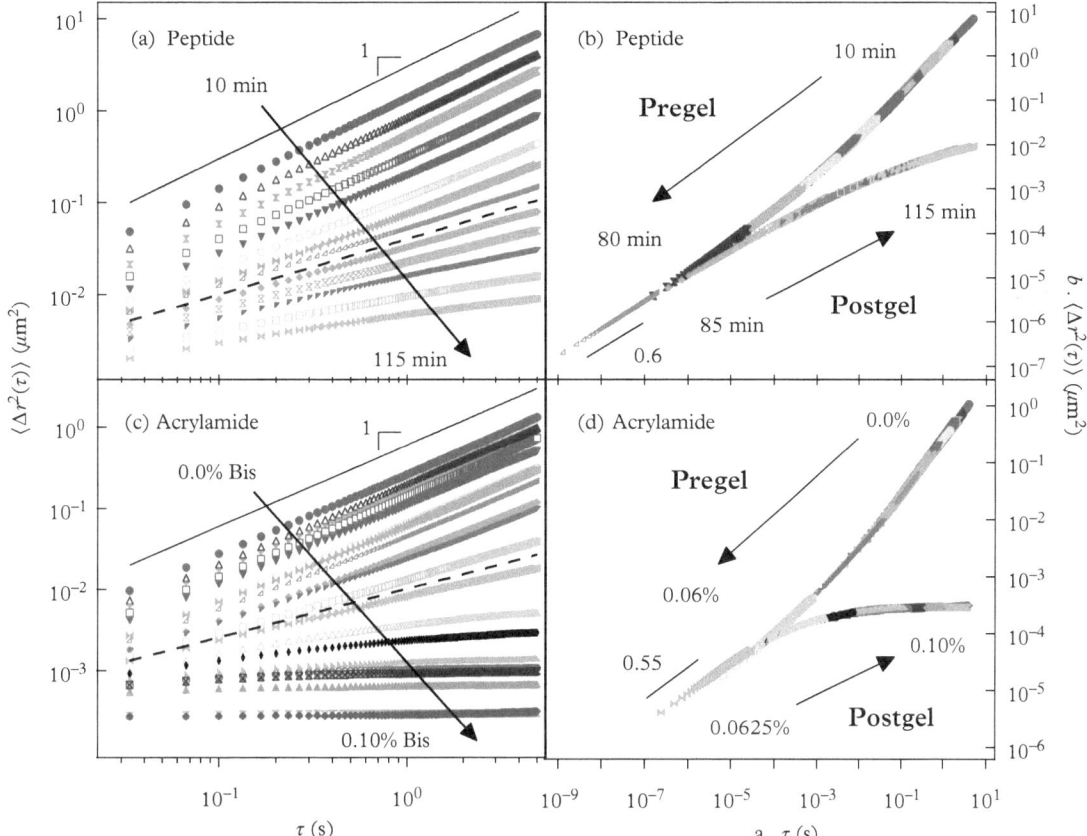

Fig. 10.7 *Gelation of a peptide hydrogelator as a function of cure time and acrylamide as a function of bis-acrylamide cross-linker concentration. Replotted from Larsen and Furst (2008) and Larsen et al. (2008). © 2008 The Korean Society of Rheology.*

and polyacrylamide cross-linked with *bis*-acrylamide. The gel point is reached at a critical time t_c for the peptide and with a critical amount of cross-linker (or a critical extent of reaction) p_c for the acrylamide. As expected, when the range of lag times in a microrheology experiment is sufficient to capture the longest relaxation time, it is possible to shift the curves to form two distinct branches, one corresponding to the pre-gel and the other to the post-gel. These converge at a single curve, representing the critical compliance, $\mathcal{J}_c$. The shifting is done empirically here by multiplying the MSD and lag times by the factors a and b, respectively. The logarithmic slope of the MSD closest to $\mathcal{J}_c$ identifies the critical exponent n, which is 0.6 for the peptide and 0.55 for acrylamide. These values are slightly higher than if the crossing of

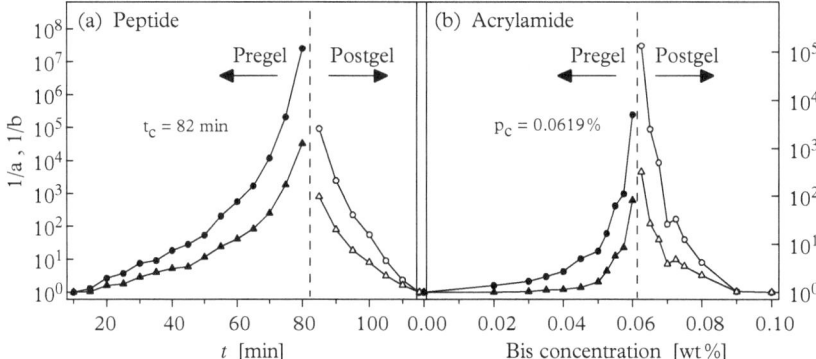

Fig. 10.8 *Reciprocal values of the longest relaxation time a and compliance b shift factors for plotted for two hydrogels, (a) a peptide hydrogelator and (b) cross-linked acrylamide. The divergence of the shift factors identifies the gel point. Replotted from data by Larsen and Furst (2008) and Larsen* et al. *(2008). © 2008 The Korean Society of Rheology.*

the storage and loss moduli, $G' = G''$, were assumed to indicate the gel point.

Plotting the reciprocal of the empirical shift factors a and b in Fig. 10.8 clearly shows the divergence of the longest relaxation time and compliance, respectively, and identifies the gel point in time or cross-linker concentration. Once values for t_c or p_c are known, the distance from the gel point ε (eqn 10.9) can be calculated. Then, plotting $\log a$ and $\log b$ versus $\log \varepsilon$ should produce lines that identify the scaling exponents y and z, as shown in Fig. 10.9. The ratio of y and z provide further confirmation of the critical exponent n over a wider range of the gel transition.

Time-cure superposition analysis of microrheology experiments relies on features in the MSD curves in order to shift them—an upturn or downturn that signifies the longest relaxation time τ_L of the material. Four microrheology measurements in polyacrylamide taken before and after the gel point are plotted in Fig. 10.10. Close to the gel point, the MSDs exhibit very nearly power-law behavior over the lag times. Further from the gel point, however, the curves turn markedly up or down to their terminal slope ($\alpha = 1$ and 0, respectively), but always start with a slope similar to the critical relaxation exponent.

Under some conditions, the shifting process may be ambiguous. For instance, in the early stages of gelation, τ_L is often significantly shorter than the minimum lag time, especially when using multiple particle tracking. The MSD would only capture the terminal regime with a logarithmic slope equal to 1. This ambiguity is apparent for the shift factors shown in Fig. 10.9b for polyacrylamide as ε goes to 0.

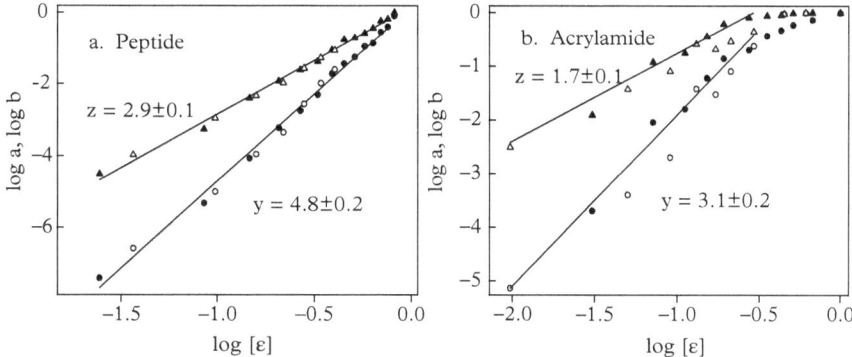

Fig. 10.9 *Time-cure superposition shift factors exhibit a power-law critical scaling before and after the gel point. Adapted from Larsen et al. (2008). © 2008 The Korean Society of Rheology.*

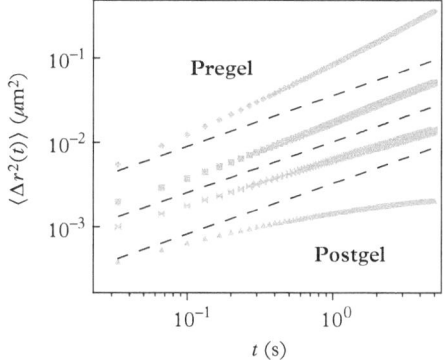

Fig. 10.10 *Two MSD curves measured before the gel point and two measured after the gel point. The curves have been shifted up and down for clarity. The expected critical scaling at the gel point is represented by the dashed lines. Adapted from Larsen et al. (2008).*

Here, the MSDs are linear over the lag times measured, but decreasing in magnitude. Without shifting the time axes, the curves are shifted with respect to the compliance. So the time shift factor $a = 1$ while the compliance shift factor b changes. The lower shift required (and hence, slope of b in Fig. 10.9b) for $\log \varepsilon > -0.5$ corresponds to $b \sim \varepsilon^k$ as the compliance decreases due to the increasing viscosity of the solution. Indeed, the apparent power-law exponent from $-0.5 < \log \varepsilon < 0$ is $b \sim \varepsilon^{1.4}$, which is close to $b \sim \varepsilon^{y-z}$. Once n, y and z are known, the data can be reshifted using these values for further confirmation over the entire range of ε.

10.3.3 Gelation critical scaling exponents

The dynamic scaling exponents y, z, k and n have been calculated for several universality classes based on different models of the molecular structure of gels and gel dynamics, including Flory-Stockmayer or mean-field class and the percolation class (Stauffer *et al.*, 1982; Daoud, 2000; Rubinstein and Colby, 2003).

It is possibly unwise to relate the measured values of dynamic scaling exponents to specific universality classes. Experimentally, the critical relaxation exponent n is found to vary over a wide range, from $n = 0.11$ for a bacterial polyester, poly(β-hydroxyoctanoate) (Richtering *et al.*, 1992), to $n = 0.92$ for end-linked poly(dimethylsiloxane) polymers (Scanlan and Winter, 1991). For a monodisperse solution of polymers exhibiting Rouse dynamics (freely-draining), the critical gel relaxation exponent is predicted by Muthukumar and Winter (1986) to be

$$n = d_f / (d_f + 2) \tag{10.17}$$

where d_f is the fractal dimension describing the cluster radius at the gel point for a cluster mass M, $R_M \sim M^{d_f}$. For $1 \leq d_f \leq 3$, the relaxation exponent has values $1/3 \leq n \leq 3/5$. Correcting for polydispersity of the critical gel results in

$$n = d_f (\tau - 1) / (d_f + 2) \tag{10.18}$$

where τ is the scaling exponent of the cluster number distribution with respect to cluster mass, $N_M \sim M^{-\tau}$. This correction leads to $n = 2/3$, noting that percolation theory predicts $\tau = 2.20$ and $d_f = 2.5$. The influence of excluded volume is to swell clusters from the percolation prediction to $d_f = 2$, resulting in $n = 1/2$. The full range of values for n appears to be explained by a combination of the precursor structure and its degree of hydrodynamic screening in the incipient gel (Muthukumar, 1989).

10.3.4   Logarithmic slope

A method for characterizing gelation kinetics with microrheology that appears often in the literature is to plot the logarithmic slope $\alpha = d \ln \langle \Delta r^2(t) \rangle / d \ln t$, typically at a single or small range of lag times, versus the cure time. The time at which $\alpha = n$, the critical relaxation exponent, is identified as the gel point. The method is approximate, but it may be the only option available. If the MSD spans a limited range of lag times, then the cross-over behavior around the longest relaxation time may not be captured adequately, leaving the MSDs without the features necessary for shifting using time-cure superposition.

From our discussion of critical gels (Section 10.3.1), we know that the logarithmic slope (as a measure of tan δ) only has a single value

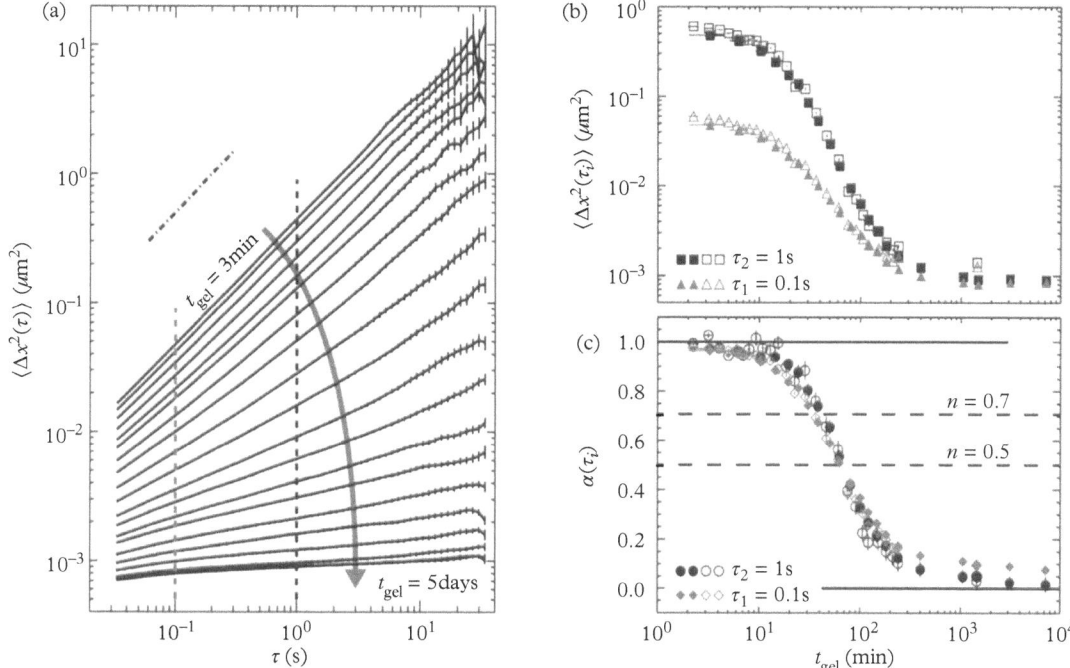

Fig. 10.11 *Gelation of a fibril-forming peptide showing the (a) mean-squared displacement and (b) the range of lag times used to calculate the (c) logarithmic slope to characterize the gelation kinetics. Adapted from Savin and Doyle (2007a) with permission of The Royal Society of Chemistry.*

for all times *precisely* at the gel point. Thus, the changing value of α is capturing some average of the cross-over region between the short- and long-time terminal mean-squared displacements of a pre- or post-gel state. An example of the logarithmic slope method is shown in Fig. 10.11 for a peptide hydrogelator (Savin and Doyle, 2007a).

Another shortcoming of the logarithmic slope method is the un-certainty of the value of n. In the previous section, we saw that the experimental and theoretical values of the critical relaxation exponent can range over nearly all of the possible values, from zero to one. Therefore, the gelation criteria $\alpha = n$ must assume a value, or range of values. For many hydrogels, values of n between 0.4–0.7 are rea-sonable as a first approximation (Schultz *et al.*, 2009b). The rate at which α changes during gelation will also depend on the lag time se-lected. In general, longer lag times would favor a faster decrease in α and smaller error for the gel point, but such choices also need to balance the poorer statistics and MSD accuracy at longer lag times.

In Fig. 10.11 we indicate two possible values of n with dashed lines, 0.5 and 0.7, to illustrate how the uncertainty in the critical relaxation exponent can change the perceived gel time. Nonetheless, the method is often sufficient for screening gelation kinetics under different conditions. Studies have used the method to screen peptide hydrogelators under varying conditions of pH, ionic strength, and changes to sequence (Savin and Doyle, 2007*a*; Larsen *et al.*, 2009).

10.3.5 Gelation screening

Consider a recent example of microrheology measurements to screen the rheology of covalently cross-linked high-molecular weight heparin-poly(ethylene glycol) (HMWH-PEG) hydrogels. With measurements of many samples, it is possible to identify gel compositions in a four-dimensional composition space consisting of the PEG cross-linker molecular weight, the number of cross-linkable sites on each backbone HMWH molecule, the total polymer weight percent of the hydrogel and ratio of HMWH and PEG.

Each hydrogel sample is prepared and equilibrated in parallel. Fig. 10.12 is the resulting gelation state diagram for hydrogels made with PEG M_n = 5000. Each subplot shows a different heparin backbone functionality, ranging from 3.9 to 11.8 cross-linkable maleimide sites per heparin. Each square represents one experimental condition. The color of each data point corresponds to the logarithmic slope of the mean-squared displacement, $\alpha = d\log\langle\Delta r^2(\tau)\rangle/d\log\tau$. For equilibrated hydrogels, knowledge of the critical relaxation exponent enables samples to be differentiated into gels ($\alpha < n$) and sols ($\alpha > n$), thus identifying the material compositions that form gels.

The black lines in Fig. 10.12 represent the lower and upper gelation limits (Flory, 1941, 1942; Flory and Rehner, 1943; Stockmayer, 1943). The lower-gelation limit describes the situation when one PEG cross-linker is attached to each HMWH backbone, $n_{PEG} = n_{hep}f_{hep}/(f_{hep}-1)$ where n_{PEG} and n_{hep} are the moles of PEG and HMWH, respectively, and f_{hep} is the functionality of the heparin backbone. The upper gelation limit describes the condition when there is one cross-linkable site is available for cross-linking on each backbone, $n_{PEG} = (f_{hep}-1)n_{hep}$. The empirical data identifies some discrepancies with theory, but also provides guidance for further processing. For intsance, the gel boundaries in Fig. 10.12 were used to create near-critical gels that could be electrospun (Schultz et al., 2012*b*).

10.3.6 Gel degradation

The reverse process of gelation, gel degradation, can also be characterized by microrheology. The ability to keep isolated samples

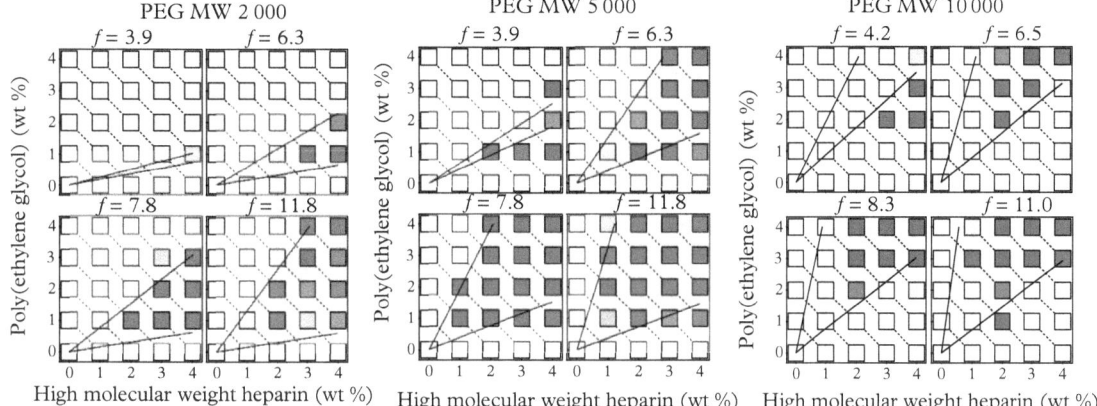

Fig. 10.12 *The logarithmic slope of the MSD is* $\alpha = d\log\langle\Delta r^2(\tau)\rangle/d\log\tau$. *Reprinted with permission from Schultz, K. M., Baldwin, A. D., Kiick, K. L., & Furst, E. M.* Macromolelcules, *42, 5310–16 (2009). Copyright 2009 American Chemical Society.*

for long times (weeks in some cases) and to monitor many samples simultaneously are unique features of these experiments (Schultz *et al.*, 2012*a*; Schultz and Anseth, 2013).

During degradation, the self-similar shape of the MSD curves, sol and gel master curves are created by empirically shifting the mean-squared displacements in a procedure analogous to time-cure superposition for a gelation reaction. An example for a hydrolytically degrading gel is shown in Fig. 10.13. The shift factors have identical meaning to those used in the analysis of gelation. The critical scaling with respect to the extent of reaction p of the longest relaxation time $a \sim \varepsilon^y$ and equilibrium compliance $b \sim \varepsilon^z$, where the extent of reaction p is represented by $\varepsilon = |p - p_c|/p_c$, the distance from the critical extent of reaction at the degradation point, p_c, and y and z are critical scaling exponents. The empirical shifting procedure is possible only if the range of MSD lag times captures the longest relaxation time of the pre- or postgel state and hence exhibits curvature on the logarithmic scale. Then, the intersection of the pre- and postgel master curves identifies the reverse percolation transition.

Methods like particle tracking microrheology will be sensitive closest to the depercolation point, when the material transitions from a solid to a fluid. A consideration of the operating regimes of microrheology and mechanical rheometry make it apparent that degradation studies can benefit from combining the two—macrorheology to track the initial degradation, and microrheology to capture the final stages. In Fig. 10.14, the elastic modulus from bulk rheology measurements,

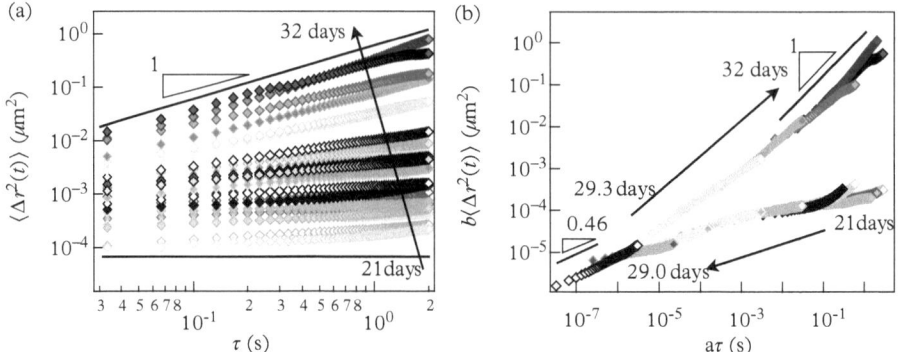

Fig. 10.13 *Mean-squared displacement of particles in a covalently cross-linked PEG hydrogel degrading slowly by the hydrolysis of ester linkages. (a) Over the first 21 days, the probe displacement remains outside the limits of particle tracking. During days 21–32, the probes exhibit a reverse percolation transition from a weak elastic solid to a viscous fluid. (b) The displacement curves can be shifted to a master curve, which identifies the reverse percolation transition at 29 days. Reprinted with permission from Schultz, K. M., Baldwin, A. D., Kiick, K. L., & Furst, E. M. ACS Macro Lett., 1, 706–8 (2012). Copyright 2012 American Chemical Society.*

Fig. 10.14 *Rheokinetic model of a degrading gel that interpolates data collected by macrorheology at early times and microrheology experiments near the degradation point. Reprinted with permission from Schultz, K. M., Baldwin, A. D., Kiick, K. L., & Furst, E. M. ACS Macro Lett., 1, 706–8 (2012). Copyright 2012 American Chemical Society.*

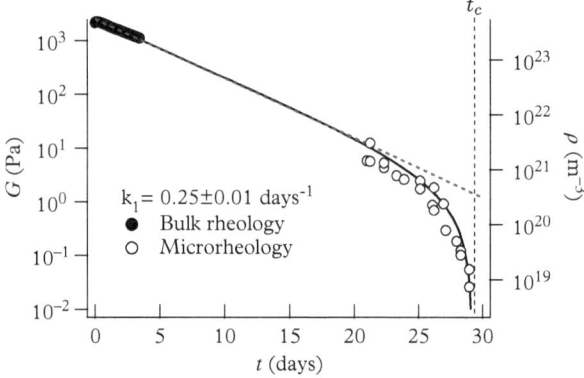

starting at about 2 kPa and degrading over a period of three days, marks one end of the data. The other, starting close to 21 days, picks up the elastic modulus close to the reverse-percolation transition. Using this combination of measurements, a single rheokinetic model is used to interpolate the data sets of the gel modulus with time of a hydrogel of covalently cross-linked poly(ethylene glycol). The gel modulus is estimated by

$$G' \sim \rho k_B T \left(|\rho - \rho_c| / \rho \right)^z \qquad (10.19)$$

where ρ is the density of hydrolyzing ester linkages in the gel. With first-order degradation kinetics of the hydrolysis reaction, $d\rho/dt = -k\rho$, where $k = 0.25$ days^{-1}, the modulus can be captured over 30 days and values that span five orders of magnitude.

More recent studies of degradation using microrheology have taken advantage of the spatial resolution of particle tracking to characterize the *local* degradation of protease-cleavable hydrogels as migrating cells move through the matrix (Schultz *et al.*, 2015).

10.4 Viscosity measurements

Viscosity measurements, especially for low viscosity fluids, are suitable for passive microrheology experiments like particle tracking and dynamic light scattering.[3] Bulk rheometry of such samples using rotational devices can be complicated by the low torque limit, inertial corrections, secondary flows, loss of sample at the edges, and uneven sample filling (Ewoldt *et al.*, 2014). Capillary viscometers and index flow methods like falling ball viscometry are often used instead. Two important applications for microrheology are measurements of the intrinsic viscosity of polymer solutions and the measurement of protein solution viscosities, particularly in the emerging therapeutics area. In this section, we will review a few details of the measurements for these applications, focusing on multiple particle tracking.

There are a few other issues to keep in mind: At lower viscosities, probe sedimentation becomes an important issue (see Section 1.3.4). Probe stability can also be compromised because particles diffuse relatively quickly and undergo more "collisions" that can lead to crossing a stability barrier (eqn 1.61) or by the formation of enough bridges under coagulating conditions of adsorbed polymer or protein (Section 1.3.3).

10.4.1 Measurement precision and accuracy

The measurement of viscosity η from particle tracking data requires a simple application of the Stokes–Einstein equation,

$$\langle \Delta r^2(t) \rangle = \frac{\mathbb{D}k_B T}{3\pi a\eta}t. \tag{10.20}$$

The mean-squared displacement is a straight line (provided there are no tracking errors or artifacts) the slope of which gives η (Fig. 10.15a). Simply fitting the mean-squared displacement by linear regression is usually adequate, but does not give the most precise value of η. By minimizing the residuals, the ordinary least squares

[3] See, for instance, the discussion in Section 5.3.2 on the use of DLS and the application note.

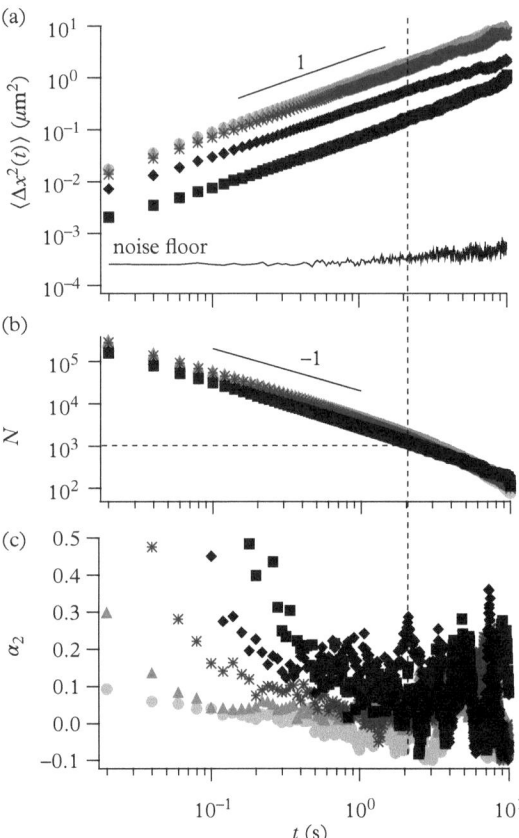

Fig. 10.15 *(a) Mean-squared displacement of 1 µm diameter probe particles in monoclonal antibody protein solutions with increasing concentration. The MSD here is reported for one-dimension ($\mathbb{D} = 1$) only. (b) The number of displacement observations with lag time. (c) The non-Gaussian parameter. Reprinted with permission from Josephson, L. L., Furst, E. M., & Galush, W. J., J. Rheol., 60, 531540 (2016). Copyright 2016, The Society of Rheology.*

method is weighted disproportionately to the longest lag times, where a lower number of displacement observations (see Fig. 10.15b) leads to greater uncertainty. Instead, we obtain a precise estimate of the viscosity by a least-squares fit of the Van Hove correlation function (see Section 4.5.1) to particle displacements at an optimal lag time.

What is the "optimal" lag time? We should consider two competing effects: At short lag times, where we benefit from a large number of particle displacement measurements, the motion of the probes may be close to the tracking resolution, ε. Even with an average mean-squared displacement significantly higher than ε, a sizable number of displacements will still fall within this range. The number of small displacements is the source of non-Gaussian statistics in the Van Hove correlation function described in Section 4.8.3. The non-Gaussian parameter α_2 (or excess kurtosis) is plotted in Fig. 10.15c. Choosing longer lag times in this case avoids the non-Gaussian statistics, but

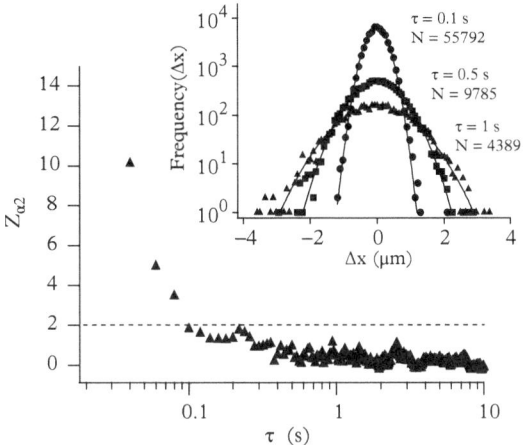

Fig. 10.16 *The excess kurtosis test statistic calculated for 1 μm diameter particles in a 1.05±0.10 mPa · s protein solution. Reprinted with permission from Josephson, L. L., Furst, E. M., & Galush, W. J., J. Rheol., 60, 531540 2016a. Copyright 2016, The Society of Rheology.*

at the cost of an increase in the error due to fewer displacement observations N. The most precise measurement of the viscosity can be made by selecting an appropriate lag time that minimizes these two contributions to the uncertainty (Josephson *et al.*, 2016a).

A robust means of picking an optimal lag time is needed. One idea is to use the test statistic $Z_{\alpha_2} = \alpha_2/\sigma_{\alpha_2}$, where σ_{α_2} is the standard error of kurtosis (SEK), given by eqn 4.54. When $|Z_{\alpha_2}| > 1.96$ at the 95% confidence level, the excess kurtosis is considered to be significantly different from zero. As an example, the Z_{α_2} test statistic is plotted in Fig. 10.16, along with Van Hove functions at three lag times, for the highest viscosity of the protein solutions previously discussed. The optimal lag time identified by the test statistic will of course depend on the viscosity, since higher values lead to a slower decrease in α_2. The standard deviation of a viscosity measurement is typically on the order of 0.1 mPa · s using this method.

In addition to being *precise*, viscosity measurements with particle tracking microrheology can be very *accurate*. Consider the viscosity of sucrose solutions shown in Fig. 10.17 that have been measured using the methods we have described. The measurement precision is similar to what we saw earlier; on the scale of the graph the uncertainty is not much greater than the size of the symbols. The viscosities derived from particle tracking are within 2% of tabulated reference data (Swindells *et al.*, 1958). In this case, the error from the reference data is random, but common sources of error are uncertainty in the probe size and systematic concentration errors, especially if samples are diluted from a single stock solution.

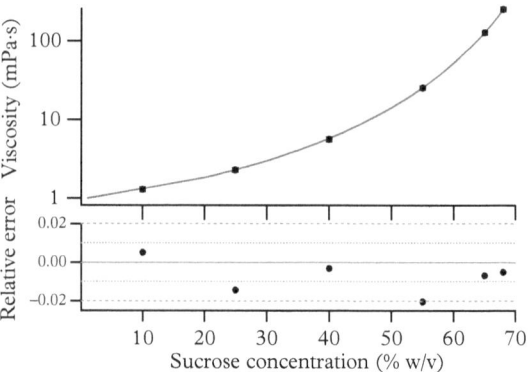

Fig. 10.17 *Viscosity of aqueous sucrose solutions measured by particle tracking microrheology. The lower plot shows the measurement accuracy relative to tabulated data. Reprinted with permission from Josephson, L. L., Furst, E. M., & Galush, W. J., J. Rheol., 60, 531540 2016a. Copyright 2016, The Society of Rheology.*

10.4.2   Measurement limits

In Section 3.11 we examined the limits of passive microrheology in general terms. Our key constraint was to unambiguously differentiate viscoelastic properties independent of the tracking precision. For Newtonian fluids, we can relax this heuristic to characterize samples at higher viscosities. With particle tracking, it is necessary to independently measure the static error ε and subtract it from the mean-squared displacements in order to apply eqn 10.20.

An example of microviscosity measurements at the limits of particle tracking are shown in Fig. 10.18. The samples are poly(ethylene oxide) (PEO) with a molecular weight of 2×10^6 daltons. While these are non-Newtonian fluids, the time scales captured using multiple particle tracking are beyond the longest relaxation time of

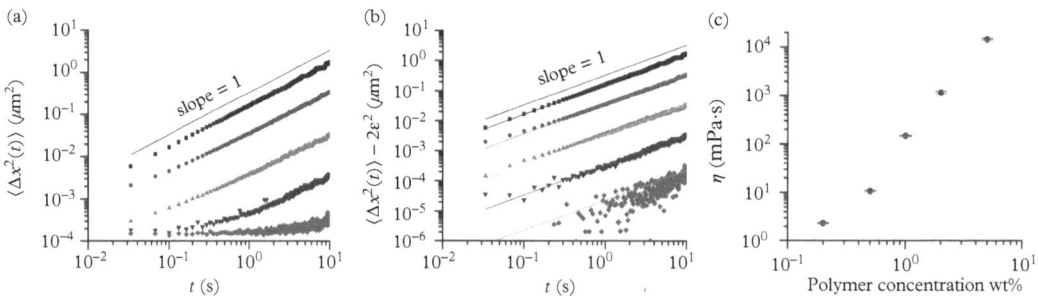

Fig. 10.18 *Multiple particle tracking experiments using 1 μm diameter polystyrene probes in solutions of 2 MDa PEO. Note that the particle trajectories are used to calculate one-dimensional mean-squared displacements in this case. The tracking error for these measurements is $\varepsilon = 12$ nm. (a) With increasing concentration, the mean-squared displacement decreases to the limit of the static tracking error. (b) Subtracting the static error recovers a mean-squared displacment with slope 1, indicating normal diffusive dynamics. (c) The Stokes–Einstein relation is used to calculate the polymer solution viscosity.*

the solutions,[4] so the measurements should report the "steady-shear" viscosity. But as the concentration increases, so does the viscosity, to a point that the probe motion is indistinguishable from the static tracking error ε (Fig. 10.18a). Subtracting the static error, the "true" mean-squared displacement is linear, and the Stokes–Einstein relation may be used to calculate the viscosity as a function of concentration (Fig. 10.18c). The maximum viscosity measured here is about 10 Pa · s, and given the noise of the data in Fig. 10.18b for this highest concentration, this is probably the outermost limit for multiple particle tracking, at least for the lag times used here. Nonetheless, the approach easily extends viscosity measurements an order of magnitude higher than the more conservative operating limits set earlier in Chapter 3.

10.5   Cell rheology

At the beginning of this book, we considered some of the earliest microrheology measurements. These studies were motivated by the interest in understanding the nature and mechanics of cells. Seifriz (1928) writes about the "matter of cells," identified at the time as the protoplasm: "Superficially, protoplasm everywhere looks very much the same, a soft, translucent, jelly-like substance, closely resembling the white of an egg, with not very striking differences in appearance whether seen in the petal of a flower or the brain of a human being." Yet, in this extraordinary material, "take place those peculiar reactions which distinguish the living from the non-living." The "protoplasm" is truly living soft matter.

Throughout the intervening century, and more recently as passive microrheology has grown in practice, many of the methods we've discussed—particle tracking, single-particle interferometry, magnetic and laser tweezer microrheology—have been applied to living and reconstituted biological systems. These measurements range from the mechanical deformation of individual cells to characterizing the rheology of the reconstituted cellular cytoskeleton. Given the role cell rheology has had in microrheology, we would be remiss not to mention a few important developments along with some of the challenges. Thorough reviews of recent advances and methods are provided by Crocker and Hoffman (2007) and Gal *et al.* (2013).

Passive microrheology largely developed together with studies of the rheology of cytoskeletal filaments: Actin, microtubules, and intermediate filaments. Proteins reconstituted in their pure forms were studied to gain insight into the mechanics of these protein assemblies and the networks they form as a basis to understand the mechanics of cell motility, division, and deformability. Many initial experiments

[4] See the application note in Section 5.4.8.

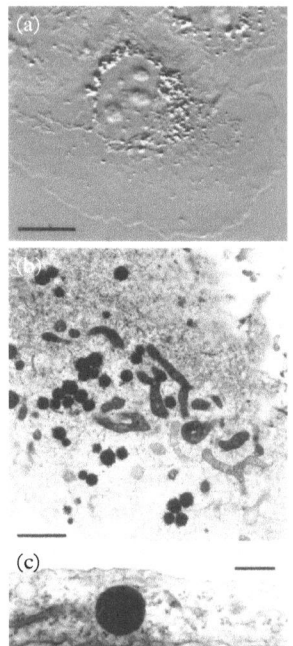

Fig. 10.19 *Endogeneous lipid granules act as laser tracking microrheology "probes" in endothelial cells. Reprinted from* Biophys. J., *78, Yamada, S., Wirtz, D., & Kuo, S. C., Mechanics of living cells measured by laser tracking microrhology, 1736– 47, Copyright 2000, with permission from The Biophysical Society.*

used DWS microrheology to investigate filament nanoscale mechanics, evident in the high-frequency rheology in highly entangled solutions (see Sections 5.6.1 and 6.1.4, for instance). Experiments demonstrated that F-actin is a semiflexible polymer and provided measurements of its persistence length (including the effects of binding proteins) and the concentration dependence of its elastic modulus (Palmer *et al.*, 1998, 1999; Gisler and Weitz, 1999; Mason *et al.*, 2000). Other experiments studied the the role of cross-linking proteins, like α-actinin, for stiffening networks and promoting bundling (Xu *et al.*, 1998b; Tseng *et al.*, 2002). At the same time, emerging multiple particle tracking and laser-tracking experiments were measuring the one- and two-point microrheology of these materials (Apgar *et al.*, 2000; Gardel *et al.*, 2003).

Moving towards living cells, the topics we've presented throughout the text should give us a sense of the strengths and limitations of microrheology methods when applied to cell rheology. Living cells are active and not at all at equilibrium. Each ATP-hydrolyzing event in the cell releases on the order of 20 $k_B T$ of energy. The GSER is violated! Cellular volumes are small, and in many cases the continuum approximation is not valid, a fact conveyed by images of endogenous lipid granules in endothelial cells (see Fig. 10.19) that are used as native "probes" in laser tracking experiments (Yamada *et al.*, 2000). Thus, "interpreting cell rheology measurements is often difficult due to uncertainties related to tracer boundary conditions and tracer/network association" (Crocker and Hoffman, 2007). Nevertheless, magnetic bead microrheology, laser tweezers, and even passive tracers play an important role towards understanding the deformation, motility mechanisms, internal transport, of cells as well as cell-material interactions that control their function and fate.

10.6   Interfacial microrheology

The focus of this book—and of the field in general—has remained almost exclusively on bulk, three-dimensional materials. Situations do arise, however, where complex fluid *interfaces* give rise to a nontrivial rheological response of their own. Examples range from the mundane—*e.g.*, the skin that forms as a tomato soup cools—to the molecular *e.g.*, phospholipid bilayers that comprise cell membranes, whose "fluidity" is regulated in order to control the diffusion of membrane proteins. In these and other systems, species organize themselves within a two-dimensional layer differently than they do in the bulk fluid below. When they are deformed, these complex fluid interfaces may exert some *surface excess* stress—e.g., a surface shear viscosity η_s.

Much like with bulk rheology, a variety of commercial instruments have been developed for interfacial rheometry measurements. As with macro-rheometry, these tend to require rather large quantities of sample, and are generally suitable for interfaces with sufficiently stiff moduli. The same factors that motivated the development of microrheology for bulk samples thus also apply to complex fluid interfaces. Fuller and Vermant (2012) provide a broad and insightful overview of recent developments in this field.

Whether the surface shear viscosity of an interface contributes appreciably to the drag or resistance on a probe is captured by the Boussinesq number

$$\text{Bo} \sim \frac{\text{resistance from interface}}{\text{interface from subphase}} = \frac{\eta_s P/L'}{\eta_b A/L''}, \tag{10.21}$$

Here, η_s is the interfacial shear viscosity, η_b is the subphase viscosity, P is the probe perimeter in contact with the interface, A is the area of the probe in contact with the bulk, and L' and L'' are the characteristic lengths over which the fluid velocity decays within the interface and the subphase, respectively. For a probe of radius a, the Boussinesq number takes the simpler form

$$\text{Bo} = \frac{\eta_s}{\eta_b a}. \tag{10.22}$$

Measurements are sensitive to surface rheology when $\text{Bo} \geq 1$. Notably, this implies that the minimum surface shear viscosity that can be reliably detected is of order

$$\eta_s^{\text{min}} \sim \eta_b a. \tag{10.23}$$

A further advantage of microrheology thus appears: The smaller the probe radius a, the more sensitively the probe is affected by the rheology of the interface. This cuts both ways, however: Smaller probes become increasingly difficult to force through stiff interfaces.

Ortega *et al.* (2010) review the two-dimensional analog of particle-tracking microrheology, wherein the MSD of probe particles diffusing within the interface are measured, and a GSER invoked to extract surface shear rheology. Just like in bulk microrheology, both the Stokes and the Einstein components must hold for the endeavor to work. A separate (and more difficult) problem must be solved for the Stokes component. In some cases, particle-tracking interfacial microrheology agrees quantitatively with macroscopic interfacial rheometry, but in others, the two techniques differ by orders of magnitude (Samaniuk and Vermant, 2014; Maestro *et al.*, 2011). These and many other questions remain open.

Lastly, we note that techniques have also been developed for active, interfacial microrheology. For example, magnetic fields can be use to torque magnetic needle probes (Lee *et al.*, 2010; Dhar *et al.*, 2010), or circular ferromagnetic "button" probes (Choi *et al.*, 2011). As discussed in Chapter 7, active microrheology introduces new complications. Nonlinear rheological responses often depend strongly on the nature of the flow; in this regard, rotating disks excite purely shear flows whereas rotating needles do not. Mixed flows complicate interfacial microrheology even in the linear-response limit: Even in the absence of any measurable surface shear viscosity, the tendency of surfactants to resist compression introduces Marangoni stresses, and impact the boundary conditions obeyed by the subphase. Surfactant incompressibility may thus be mistaken for surface shear viscosity based on the mixed-flow resistance to probe motion, even if the surfactant is surface shear inviscid (Fischer, 2004); this effect becomes even more pronounced for extended, interfacially-adsorbed filaments (Levine *et al.*, 2004).

Compared with bulk microrheology, interfacial rheology remains in its infancy, and many open questions remain.

10.7   Perspectives on future work

After nearly a century of use, microrheology continues to develop in methods and applications. Throughout this text, we've had the opportunity to study its fundamental underpinnings, established experimental practices, and many of its applications. Microrheology is a tool that has special potential for enabling rheological characterization in cases that have been difficult with conventional mechanical rheometry—small-volume samples, short time scales, or challenging sample environments.

Where does the field go next? It is one hope in writing this book that the methods and understanding the scientific community has generated, especially over the past two decades, are more readily accessible and available to catalyze new microrheology experiments.

In a way, however, microrheology stands at a point similar to mechanical rheometry in the mid twentieth century. Then, instruments were largely home-built and many standard rheometry practices—as well as the understanding of artifacts—had yet to be established.[5] Similarly, microrheology today is largely a "do-it-yourself" endeavor of assembling the right tools and materials. This presents a barrier to

[5] Macosko (2010) points out that Van Wazer *et al.* (1963) concluded their book with this summary of rheometry instruments: "Unfortunately, except for the Rheogoniometer, such equipment is not yet commercially available, although it is expected that some enterprising company will manufacture this kind of apparatus shortly."

entry for investigators who could use it most, especially in emerging materials development areas and industrial applications.

Several advances in instrumentation, however, are shifting this balance by making microrheology experiments more accessible to the non-specialist. Commercial light scattering and diffusing-wave spectroscopy instruments are available and have software capabilities that support microrheology measurements. Several can accommodate non-ergodic samples, like gels. In a more extended scope of microrheology than we have presented here, atomic force microscopy instruments are being adapted to rheology, and new microfluidic-based devices are entering the marketplace. Instruments using torsional resonators are also available (see Fig. 5.17, for instance, for a comparison of data taken using several of these devices). There are commercial laser tweezer and particle tracking instruments, as well.

The most mature microrheology methods are best suited to the characterization of linear shear rheology. The powerful Correspondence Principle (Chapter 2) assures us of the validity of measurements provided that equilibrium and continuum conditions are met. Extending microrheology to nonlinear properties is still a work in progress (Chapter 7) and requires careful attention to the fluid mechanics of moving probes in viscoelastic materials. Extensional rheology remains almost entirely unexplored, as have measurements of normal stresses, despite some promising theoretical results. Yet, nonlinear rheological properties—especially shear thinning, yielding, and strain hardening—are common characteristics of biological materials and often important design criteria for industrial applications. There is plenty more work to do, for everything flows—πᾶντα ῥεῖ.

Useful mathematics

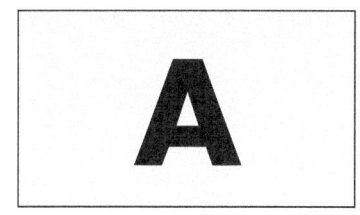

A.1 Fourier Transforms

The Fourier Transform is commonly used to analyze the dynamics of a system. The Fourier Transform of a function $f(t)$ is

$$\tilde{f}(\omega) = \mathcal{F}\{f(t)\} = \int_{-\infty}^{\infty} f(t)e^{-i\omega t}dt, \quad (A.1)$$

while the inverse Fourier Tranform is defined as

$$f(t) = \mathcal{F}^{-1}\{\tilde{f}(\omega)\} = \frac{1}{2\pi}\int_{-\infty}^{\infty} \tilde{f}(\omega)e^{i\omega t}d\omega. \quad (A.2)$$

Using integration by parts, it is straightforward to show that the Fourier Transform of $f'(t) = df/dt$ is

$$\int_{-\infty}^{\infty} f'(t)e^{i\omega t}dt = i\omega\tilde{f}(\omega). \quad (A.3)$$

Likewise, the Fourier Transform of $f''(t) = d^2f/dt^2$ is

$$\int_{-\infty}^{\infty} f''(t)e^{i\omega t}dt = -\omega^2\tilde{f}(\omega). \quad (A.4)$$

The Fourier Transforms and inverse transforms of many functions can be found in tables and classic texts like Bracewell (1986). The Fourier Transform of a convolution is particularly useful. The theorem states that Fourier Transform of the convolution of functions $f(t)$ and $g(t)$

$$f * g = \int_{-\infty}^{t} f(t')g(t-t')dt', \quad (A.5)$$

where $*$ denotes the convolution operation, is the product of the Fourier Transforms of those functions,

$$\int_{-\infty}^{\infty} (f * g)e^{i\omega t}dt = \tilde{f}(\omega)\tilde{g}(\omega). \quad (A.6)$$

Table A.1 summarizes several other useful Fourier-Transform pairs. However, many functions of interest to rheology do not have a Fourier Transform because eqn A.1 fails to converge at its infinite limits.

Table A.1 *Useful Fourier and Laplace Transform pairs.*

Fourier Transforms

Time domain	$f(t)$	Frequency domain	$\tilde{f}(\omega)$
Constant	a	Delta function	$2\pi a\delta(\omega)$
Harmonic	$ae^{i\omega_0 t}$	Delta function	$2\pi a\delta(\omega - \omega_0)$
Exponential	$ae^{-|t|/\tau}$	Lorentzian	$2a\dfrac{1/\tau}{\omega^2 + (1/\tau)^2}$
Gaussian	$ae^{-(t/\tau)^2}$	Gaussian	$\sqrt{\pi}\,a\tau e^{-(\omega\tau/2)^2}$

Laplace Transforms

Time domain	$f(t)$	Frequency domain	$\tilde{f}(s)$
Differentiation	$f'(t)$		$sF(s) - f(0)$
Second derivative	$f''(t)$		$s^2 F(s) - sf(0) - f'(0)$
	$tf(t)$	Differentiation	$F'(s)$
Linear (one sided)[1]	$t \cdot H(t)$		$\frac{1}{s^2}$
Exponential (one sided)	$e^{-at} \cdot H(t)$		$\frac{1}{s+a}$
Exponential	$e^{-a|t|}$		$\frac{a}{s^2 + a^2}$
Sine	$\sin \omega t$		$\frac{\omega}{s^2 + \omega^2}$
Cosine	$\cos \omega t$		$\frac{s}{s^2 + \omega^2}$
Power law[2]	t^p		$\frac{\Gamma(p+1)}{s^{p+1}}$

[1] $H(t)$ is the Heaviside step function.
[2] $\Gamma(x) = \int_0^\infty e^r r^{x-1}\,dr$ is the Gamma function. If x is a positive integer, then $\Gamma(x + 1) = x!$

A.1.1 Unilateral Fourier and Laplace Transform

The Laplace Transform is defined as

$$F(s) = \mathscr{L}\{f(t)\} = \int_0^\infty f(t)e^{-st}\,dt. \tag{A.7}$$

The transform converges in the upper limit by multiplying the function $f(t)$ a damping factor $\exp(-\sigma t)$ such that the Laplace Transform variable is a complex number $s = \sigma + i\omega$. The Laplace Transform is suited to *causal* functions for which the behavior of $f(t)$ for $t > 0$ is of interest.

The Unilateral Fourier Transform, also known as the Fourier–Laplace Transform or the one-sided Fourier Transform, is found by analytic continuation on the pure imaginary axis by the substitution

$$s = i\omega. \tag{A.8}$$

We denote the inverse Laplace Transform of the function as $f(t) = \mathcal{L}^{-1}\{\tilde{f}(s)\}$. Several useful Laplace Transforms are given in the Table A.1. The Laplace Transform also has the convolution theorem

$$f * g = \tilde{f}(s)\tilde{g}(s). \tag{A.9}$$

In Chapter 3 we use it to solve the Langevin equation.

A.1.2 Spatial Fourier Transform

The Fourier Transform may be generalized to functions defined in a three-dimensional space. The transform of a function is $f(\mathbf{r})$

$$\tilde{f}(\mathbf{q}) = \int f(\mathbf{r})e^{i\mathbf{q}\cdot\mathbf{r}}d\mathbf{r}, \tag{A.10}$$

and the inverse transform

$$f(\mathbf{r}) = \frac{1}{(2\pi)^3} \int \tilde{f}(\mathbf{q})e^{-i\mathbf{q}\cdot\mathbf{r}}d\mathbf{q}. \tag{A.11}$$

In Cartesian coordinates,

$$\tilde{f}(u, v, w) = \iiint_{-\infty}^{\infty} f(x, y, z)e^{-i(xu+yv+zw)}dx\,dy\,dz \tag{A.12}$$

and the inverse transform is

$$f(x, y, z) = \frac{1}{(2\pi)^3} \iiint_{-\infty}^{\infty} \tilde{f}(u, v, w)e^{i(ux+vy+wz)}du\,dv\,dw. \tag{A.13}$$

Again, using integration by parts, it is straightforward to show that the Fourier Transform of $\nabla f(\mathbf{r})$ is

$$\int \nabla f(\mathbf{r})e^{i\mathbf{q}\cdot\mathbf{r}}d\mathbf{r} = -i\mathbf{q}\tilde{f}(\mathbf{q}). \tag{A.14}$$

Likewise, the Fourier Tranform of $\nabla^2 f(\mathbf{r})$ is

$$\int \nabla^2 f(\mathbf{r})e^{i\mathbf{q}\cdot\mathbf{r}}d\mathbf{r} = -q^2\tilde{f}(\mathbf{q}). \tag{A.15}$$

These relationships are particularly useful for solving differential equations when the homogeneous solutions can be neglected (such as the long-time behavior of the Langevin equation).

Fourier Transforms are useful in the theory of spatial correlations of colloids (as well as molecular fluids and polymers,) especially in scattering experiments (x-ray, light and neutron).

A.1.3   Dirac delta function

In one dimension, the Dirac delta function is defined as the derivative of the Heaviside step function $H(x)$

$$\delta(x) = \frac{dH(x)}{dx} = \begin{cases} 0 & x \neq 0 \\ \infty & x = 0 \end{cases} \tag{A.16}$$

The "sifting" property of the Dirac delta function is expressed as

$$f(0) = \int_{-\infty}^{\infty} f(x)\delta(x)dx. \tag{A.17}$$

In Cartesian space, the Dirac delta function may defined such that $\delta(\mathbf{r}) = 0$ if $\mathbf{r} \neq 0$ and $\delta(\mathbf{r}) = \infty$ if $\mathbf{r} = 0$.

$$\int_{-\infty}^{\infty} \int_{-\infty}^{\infty} \int_{-\infty}^{\infty} \delta(\mathbf{r}) d^3 r = 1. \tag{A.18}$$

A useful relationship of the delta function is that its Fourier Transform is unity,

$$\int \delta(\mathbf{r}) e^{-i\mathbf{q}\cdot\mathbf{r}} d\mathbf{r} = 1. \tag{A.19}$$

A.2   Relating Fourier and Laplace Transforms

Consider a function $V(t)$ that is identically zero for $t < 0$, for which the Fourier Transform is

$$\tilde{V}(\omega) = \int_{-\infty}^{\infty} V(t) e^{-i\omega t} dt, \tag{A.20}$$

and inverse Fourier Transform

$$V(t) = \frac{1}{2\pi} \int_{-\infty}^{\infty} \tilde{V}(\omega) e^{i\omega t} d\omega. \tag{A.21}$$

Important properties of $\tilde{V}$ emerge when this is performed via contour integration. When $t < 0$ (for which $V(t < 0) = 0$), the exponential in the inverse transform becomes $e^{-i\omega|t|}$, meaning that any ω with positive real part grows exponentially for negative t as $|\omega| \to \infty$. Therefore, we must close the contour around the *negative* imaginary plane of ω for all $t < 0$, so that the countour at infinity vanishes.

Since that integral must be identically zero for $t < 0$, residue calculus requires $\tilde{V}(\omega)$ to be analytic on the lower-half plane.[3]

Taking the Laplace Transform of the inverse Fourier Transform will allow us to relate the two transforms.

$$\hat{V}(s) = \frac{1}{2\pi} \int_{-\infty}^{\infty} \int_{0}^{\infty} \tilde{V}(\omega) e^{i\omega t - st} \, dt \, d\omega \qquad (A.22)$$

which is given by

$$\hat{V}(s) = \frac{1}{2\pi} \int_{-\infty}^{\infty} \frac{\tilde{V}(\omega)}{i\omega - s} e^{-i\omega t - st} \Bigg|_{t=0}^{t=\infty} \, d\omega. \qquad (A.23)$$

$$\hat{V}(s) = -\frac{1}{2\pi i} \int_{-\infty}^{\infty} \frac{\tilde{V}(\omega)}{\omega + is} \, d\omega. \qquad (A.24)$$

Since $\tilde{V}(\omega)$ is analytic on the lower-half plane, the only singularity in the integrand is the pole at $\omega = -is$. Consequently, we can push the contour down, picking up only the residue from the pole at $\omega = -is$, to give

$$\hat{V}(s) = \tilde{V}(\omega \to -is). \qquad (A.25)$$

So, the Fourier and Laplace Transforms are related for causal functions (which are zero for $t < 0$).

Another way to show this is via analytic continuation. To see that, we start once again with the definition of the Fourier Transform

$$\tilde{V}(\omega) = \int_{-\infty}^{\infty} V(t) e^{-i\omega t} \, dt. \qquad (A.26)$$

Because $V(t < 0) = 0$, the bilateral Fourier Transform is identical to the unilaterial Fourier Transform,

$$\tilde{V}(\omega) = \int_{0}^{\infty} V(t) e^{-i\omega t} \, dt. \qquad (A.27)$$

We now allow ω to take a complex argument, with negative imaginary part (as required for eqn A.27 to converge)

$$\omega = a - is \qquad (A.28)$$

and in fact, take $a = 0$, then the Fourier Transform becomes

$$\tilde{V}(\omega \to -is) = \int_{0}^{\infty} V(t) e^{-st} \, dt = \hat{V}(s). \qquad (A.29)$$

Given a causal function (for which $V(t < 0) = 0$), the Laplace and Fourier Transforms are related. Namely, taking the Fourier Transform $\tilde{V}(\omega)$ and replacing $\omega = -is$ gives the Laplace Transform. This holds for all causal functions.

[3] If the Fourier Transform is defined with the opposite sign convention, then $\tilde{V}(\omega)$ must be analytic in the upper-half plane.

A.3 Kramers–Kronig relations

The Kramers–Kronig relations allow the real part of any Fourier-Transformed causal function to be determined from the imaginary part, and vice-versa. We will derive them for the complex modulus $G^*(\omega)$, which is the Fourier Transform of the memory function $m(t)$:

$$m(t) = \frac{1}{2\pi} \int_{-\infty}^{\infty} G^*(\omega) e^{i\omega t} \, d\omega \tag{A.30}$$

For all $t < 0$, the fact that $m(t)$ is causal requires

$$m(t < 0) = 0 = \frac{1}{2\pi} \int_{-\infty}^{\infty} G^*(\omega) e^{i\omega t} \, d\omega. \tag{A.31}$$

This, in turn, requires that $G^*(\omega)$ be analytic in the lower-half plane. We will now consider the integral

$$\int_C \frac{G^*(\omega)}{\omega - \omega_0} \, d\omega, \tag{A.32}$$

We will consider a closed contour that proceeds along the real axis, making an infinitesimally small semicircular path below the pole at ω_0, then returns to $-\infty$ via a semicircular arc around the lower-half plane at infinity. Because $G^*(\omega)$ is analytic in the lower-half plane, this contour contains no singularities, and the contour integral must be zero.

$$\int_{-\infty}^{\omega_0-\rho} \frac{G^*(\omega)}{\omega - \omega_0} \, d\omega + \int_{\omega_0+\rho}^{\infty} \frac{G^*(\omega)}{\omega - \omega_0} \, d\omega + \int_{\rho} \frac{G^*(\omega)}{\omega - \omega_0} \, d\omega = 0, \tag{A.33}$$

where the final integral is an infinitesimally small semicircle, wrapping around the pole at ω_0 in the positive direction, contributing half of that pole's residue. The first two integrals, in the limit $\rho \to 0$, represent the Cauchy Principle value of the integral, leaving

$$\mathscr{P} \int_{-\infty}^{\infty} \frac{G^*(\omega)}{\omega - \omega_0} \, d\omega + i\pi G^*(\omega_0) = 0. \tag{A.34}$$

Separating the real and imaginary parts of $G^*(\omega) = G'(\omega) + iG''(\omega)$ gives

$$\mathscr{P} \int_{-\infty}^{\infty} \frac{G'(\omega) + iG''(\omega)}{\omega - \omega_0} \, d\omega + i\pi G'(\omega_0) - \pi G''(\omega_0) = 0. \tag{A.35}$$

The real and imaginary parts of this equation must be satisfied independently, thus yielding the Kramers–Kronig relations

$$G'(\omega_0) = -\frac{1}{\pi} \mathcal{P} \int_{-\infty}^{\infty} \frac{G''(\omega)}{\omega - \omega_0} d\omega \qquad (A.36)$$

$$G''(\omega_0) = \frac{1}{\pi} \mathcal{P} \int_{-\infty}^{\infty} \frac{G'(\omega)}{\omega - \omega_0} d\omega. \qquad (A.37)$$

Note that choosing the opposite sign convention for Fourier Transforms, as Landau *et al.* (1986) do, renders $G^*(\omega)$ analytic in the upper-half plane, so that the contour must go above ω_0. This is in the negative direction, and would reverse the signs on the right-hand side of eqns A.36–A.37.

A.4 Vector harmonic solutions to Stokes equations

The use of harmonic functions is particularly elegant when deriving the solution to creeping-flow equations like Stokes flow around a sphere. Leal (2007) presents an excellent introduction to the topic, including solutions for rotating spheres and spheres in general linear flows. In this section, we derive the velocity and pressure fields around a sphere translating through a quiescent fluid.

A.4.1 Harmonic functions

Harmonic functions are solutions to the differential equation

$$\nabla^2 \psi = 0. \qquad (A.38)$$

The harmonic functions consist of *decaying* and *growing* harmonics. The decaying harmonics are conveniently represented by taking higher-order derivatives of $1/r$,

$$\frac{1}{r} \qquad (A.39)$$

$$\nabla\left(\frac{1}{r}\right) \rightarrow -\frac{x_i}{r^3} \qquad (A.40)$$

$$\nabla\left(\frac{\mathbf{x}}{r^3}\right) \rightarrow \frac{\delta_{ij}}{r^3} - 3\frac{x_i x_j}{r^5} \qquad (A.41)$$

$$\nabla\left(\frac{\delta}{r^3} - 3\frac{\mathbf{xx}}{r^5}\right) \rightarrow 15\frac{x_i x_j x_k}{r^7} - 3\frac{x_i \delta_{jk} + x_j \delta_{ik} + x_k \delta_{ij}}{r^5} \qquad (A.42)$$

Written in index notation, the functions are

$$\frac{1}{r} \tag{A.43}$$

$$\frac{x_i}{r^3} \tag{A.44}$$

$$\frac{x_i x_j}{r^5} - \frac{\delta_{ij}}{3r^3} \tag{A.45}$$

$$\frac{x_i x_j x_k}{r^7} - \frac{x_i \delta_{jk} + x_j \delta_{ik} + x_k \delta_{ij}}{5r^5} \tag{A.46}$$

or

$$\phi_{-(n+1)} = \frac{(-1)^n}{1 \cdot 3 \cdot 5 \cdots (2n-1)} \frac{\partial^n}{\partial x_i \partial x_j \partial x_k \cdots} \left(\frac{1}{r} \right), \quad n = 0, 1, 2, \ldots \tag{A.47}$$

The *growing harmonics* are

$$1 \tag{A.48}$$

$$x_i \tag{A.49}$$

$$x_i x_j - \frac{r^2}{3} \delta_{ij} \tag{A.50}$$

$$x_i x_j x_k - \frac{r^2}{5} x_i \delta_{jk} + x_j \delta_{ik} + x_k \delta_{ij}. \tag{A.51}$$

and may be expressed in terms of the decaying harmonics by

$$r^{2n+1} \phi_{-(n+1)}. \tag{A.52}$$

A.4.2 A sphere translating in a quiescent fluid

We seek solutions to the Stokes flow for an incompressible Newtonian fluid

$$\eta \nabla^2 \mathbf{v} - \nabla P = 0 \tag{A.53}$$

and

$$\nabla \cdot \mathbf{v} = 0. \tag{A.54}$$

First, we re-write the Stokes equations in a harmonic form. Taking the divergence of Stokes equation,

$$\nabla \cdot \left(\eta \nabla^2 \mathbf{v} - \nabla p \right) = 0 \tag{A.55}$$

where

$$\nabla^2 p = 0. \tag{A.56}$$

Next, we write the velocity field as

$$\mathbf{v} = \frac{\mathbf{x}}{2\eta} p + \mathbf{v}^H \tag{A.57}$$

which is a solution to eqn A.55 where $\mathbf{v}^H$ is a harmonic function,

$$\nabla^2 \mathbf{v}^H = 0. \tag{A.58}$$

Continuity requires that

$$\nabla \cdot \mathbf{v}^H = -\frac{1}{2\eta} (3p + \mathbf{x} \cdot \nabla p). \tag{A.59}$$

For a velocity of the sphere $\mathbf{V}$ we can construct a solution beginning with the pressure. The pressure is constructed from harmonic solutions that are linear in $\mathbf{V}$ and $\mathbf{x}$ only, therefore

$$p = C_1 \frac{\mathbf{V} \cdot \mathbf{x}}{r^3} \tag{A.60}$$

and now

$$\mathbf{v} = \frac{\mathbf{x}}{2\eta} \left(C_1 \frac{\mathbf{V} \cdot \mathbf{x}}{r^3} \right) + \mathbf{v}^H. \tag{A.61}$$

We are left to find solutions for $\mathbf{v}^H$. These must be decaying functions that are linear in $\mathbf{V}$ and are real vectors (same tensorial rank) and same tensorial parity. There are two terms constructed from $\mathbf{V}$ and the harmonic functions

$$\mathbf{v}^H = C_2 \frac{\mathbf{V}}{r} + C_3 \mathbf{V} \cdot \left(\frac{\mathbf{xx}}{r^5} - \frac{\delta}{3r^3} \right) \tag{A.62}$$

that satisfy the criteria.

Next, we find the constants C_1, C_2, and C_3 in the equation

$$\mathbf{v} = \frac{\mathbf{x}}{2\eta} \left(C_1 \frac{\mathbf{V} \cdot \mathbf{x}}{r^3} \right) + C_2 \frac{\mathbf{V}}{r} + C_3 \mathbf{V} \cdot \left(\frac{\mathbf{xx}}{r^5} - \frac{\delta}{3r^3} \right). \tag{A.63}$$

First, continuity requires that

$$C_2 = \frac{C_1}{2\eta}. \tag{A.64}$$

Now

$$v^H = \frac{C_1}{2\eta}\frac{\mathbf{V}}{r} + C_3\mathbf{V}\cdot\left(\frac{\mathbf{x}\mathbf{x}}{r^5} - \frac{\delta}{3r^3}\right) \tag{A.65}$$

and

$$\mathbf{v} = \frac{\mathbf{x}}{2\eta}\left(C_1\frac{\mathbf{V}\cdot\mathbf{x}}{r^3}\right) + \frac{C_1}{2\eta}\frac{\mathbf{V}}{r} + C_3\mathbf{V}\cdot\left(\frac{\mathbf{x}\mathbf{x}}{r^5} - \frac{\delta}{3r^3}\right) \tag{A.66}$$

which can be rearranged to

$$\mathbf{v} = \frac{\mathbf{x}(\mathbf{V}\cdot\mathbf{x})}{r^3}\left(\frac{C_1}{2\eta} + \frac{C_3}{r^2}\right) + \frac{\mathbf{V}}{r}\left(\frac{C_1}{2\eta} - \frac{C_3}{3r^2}\right). \tag{A.67}$$

Satisfying the boundary condition that $\mathbf{v} = \mathbf{V}$ at $\mathbf{x} = \hat{\mathbf{n}}a$, or equivalently, at $|\mathbf{x}| = r = a$, leads to

$$\mathbf{V} = a^2\frac{\hat{\mathbf{n}}(\mathbf{V}\cdot\hat{\mathbf{n}})}{a^3}\left(\frac{C_1}{2\eta} + \frac{C_3}{a^2}\right) + \frac{\mathbf{V}}{a}\left(\frac{C_1}{2\eta} - \frac{C_3}{3a^2}\right) \tag{A.68}$$

and the following two equations that determine the constants C_1 and C_3:

$$\frac{C_1}{2\eta} + \frac{C_3}{a^2} = 0 \tag{A.69}$$

$$\frac{C_1}{2\eta a} - \frac{C_3}{3a^3} = 1 \tag{A.70}$$

These give us

$$C_1 = \frac{3\eta a}{2} \tag{A.71}$$

$$C_3 = -\frac{3a^3}{4} \tag{A.72}$$

and

$$\mathbf{v} = \mathbf{x}(\mathbf{V}\cdot\mathbf{x})\left(\frac{3}{4}\frac{a}{r^3} - \frac{3}{4}\frac{a^3}{r^5}\right) + \mathbf{V}\left(\frac{3}{4}\frac{a}{r} + \frac{1}{4}\frac{a^3}{r^3}\right). \tag{A.73}$$

The corresponding pressure distribution is

$$p(\mathbf{x}) = \frac{3\eta a^2}{2}\frac{\mathbf{V}\cdot\mathbf{x}}{r^3}, \tag{A.74}$$

which appear as eqns 2.70 and 2.71 in Section 2.5.

A.5 Dynamics of an oscillating particle

The equation governing the motion of an optically trapped sphere (eqn 9.38) is

$$\zeta \dot{x} + \kappa_t x = \kappa_t x_t \tag{A.75}$$

where $x_t = A \cos \omega t$. Here, we show that the general solution is

$$x(t) = D(\omega) e^{i[\omega t - \delta(\omega)]}. \tag{A.76}$$

We rewrite the equation of motion,

$$\zeta \dot{x} + \kappa_t x = \kappa_t A e^{i \omega t} \tag{A.77}$$

recognizing that $x(t)$ is the real part of the solution.

We assume the solution $x = D' e^{i \omega t}$, which upon substituting into eqn A.77, gives

$$D'(\omega) = \frac{\kappa_t A}{\kappa_t + i \omega \zeta}. \tag{A.78}$$

In polar coordinates,

$$\kappa_t + i \omega \zeta = \sqrt{\kappa_t^2 + \omega^2 \zeta^2} e^{i \delta} \tag{A.79}$$

where

$$\tan \delta = \omega \zeta / \kappa_t \tag{A.80}$$

so

$$D'(\omega) = \frac{\kappa_t A}{\sqrt{\kappa_t^2 + \omega^2 \zeta^2}} e^{-i \delta}. \tag{A.81}$$

Thus, the solution is of the form

$$x(t) = D(\omega) e^{i[\omega t - \delta(\omega)]} \tag{A.82}$$

with

$$D(\omega) = \frac{\kappa_t A}{\sqrt{\kappa_t^2 + \omega^2 \zeta^2}}. \tag{A.83}$$

Taking the real part, we find

$$x(t) = D(\omega) \cos[\omega t - \delta(\omega)]. \tag{A.84}$$

Bibliography

Abdala, Ahmed A, Amin, Samiul, van Zanten, John H, and Khan, Saad A (2015). Tracer microrheology study of a hydrophobically modified comblike associative polymer. *Langmuir*, **31**(13), 3944–51.

Adamson, Arthur W and Gast, Alice P (1997). *Physical Chemistry of Surfaces* (6th edn). Wiley, New York.

Addas, Karrim M, Schmidt, Christoph F, and Tang, Jay X (2004). Microrheology of solutions of semiflexible biopolymer filaments using laser tweezers interferometry. *Phys. Rev. E*, **70**, 21503.

Adolf, Douglas and Martin, James E (1990). Time-cure superposition during cross-linking. *Macromolecules*, **23**, 3700–4.

Aichinger, Pierre Anton, Michel, Martin, Servais, Colin, Dillmann, Marie Lise, Rouvet, Martine, D'Amico, Nicola, Zink, Ralf, Klostermeyer, Henning, and Horne, David S (2003). Fermentation of a skim milk concentrate with Streptococcus thermophilus and chymosin: Structure, viscoelasticity and syneresis of gels. *Colloids Surfaces B Biointerfaces*, **31**(1–4), 243–55.

Alexander, Marcela and Dalgleish, Douglas G (2004). Application of transmission diffusing wave spectroscopy to the study of gelation of milk by acidification and rennet. *Colloids Surfaces B Biointerfaces*, **38**(1–2), 83–90.

Alexander, Marcela and Dalgleish, Douglas G (2007). Diffusing Wave Spectroscopy of aggregating and gelling systems. *Curr. Opin. Colloid Interface Sci.*, **12**(4–5), 179–86.

Amblard, François, Yurke, B, Pargellis, A, and Leibler, S (1996). A magnetic manipulator for studying local rheology and micromechanical properties of biological systems. *Rev. Sci. Instrum.*, **67**(3), 818–27.

Apgar, Joshua, Tseng, Yider, Fedorov, Elena, Herwig, Matthew B, Almo, Steve C, and Wirtz, Denis (2000). Multiple-particle tracking measurements of heterogeneities in solution of actin filaments and actin bundles. *Biophys. J.*, **79**, 1095–106.

Appleyard, D C, Vandermeulen, K Y, Lee, H, and Lang, M J (2007). Optical trapping for undergraduates. *Am. J. Phys.*, **75**(1), 5.

Arigo, Mark T and McKinley, G H (1998). An experimental investigation of negative wakes behind spheres settling in a shear-thinning viscoelastic fluid. *Rheol. Acta*, **37**(4), 307–27.

Asakura, S and Oosawa, F (1954). On interaction between two bodies immersed in a solution of macromolecules. *J. Chem. Phys.*, **22**, 1255–6.

Ashkin, Arthur (1970*a*). Acceleration and trapping of particles by radiation pressure. *Phys. Rev. Lett.*, **24**, 156–9.

Ashkin, Arthur (1970*b*). Optical levitation by radiation pressure. *Appl. Phys. Lett.*, **19**, 283–5.

Ashkin, Arthur (1992). Forces of a single-beam gradient laser trap on a dielectric sphere in the ray optics regime. *Biophys. J.*, **61**, 569–82.

Ashkin, Arthur, Dziedzic, J M, Bjorkholm, J E, and Chu, S (1986). Observation of a single-beam gradient force optical trap for dielectric particles. *Opt. Lett.*, **11**, 288–90.

Ashkin, Arthur, Dziedzic, J M, and Yamane, T (1987). Optical trapping and manipulation of single cells using infrared laser beams. *Nature*, **330**, 769–71.

Atakhorrami, M and Schmidt, C F (2006). High-bandwidth one- and two-particle microrheology in solutions of wormlike micelles. *Rheol. Acta*, **45**, 449–56.

Atakhorrami, M, Sulkowska, J I, Addas, Karrim M, Koenderink, G H, Tang, J X, Levine, A J, MacKintosh, F C, and Schmidt, C F (2006). Correlated fluctuations of microparticles in viscoelastic solutions: Quantitative measurement of material properties by microrheology in the presence of optical traps. *Phys. Rev. E*, **73**, 61501.

Aufderhorst-Roberts, Anders, Frith, William J, and Donald, Athene M (2012). Micro-scale kinetics and heterogeneity of a pH triggered hydrogel. *Soft Matter*, 8(21), 5940–6.

Banchio, Adolfo J, Nägele, Gerhard, and Bergenholtz, Johan (1999). Viscoelasticity and generalized Stokes–Einstein relations of colloidal dispersions. *J. Chem. Phys.*, **111**, 8721–40.

Barnes, Howard Anthony and Nguyen, Quoc Dzuy (2001). Rotating vane rheometry: A review. *J. Non-Newtonian Fluid Mech.*, **98**(1), 1–14.

Baselt, David R, Lee, Gil U, Natesan, Mohan, Metzger, Steven W, Sheehan, Paul E, and Colton, Richard J (1998). A biosensor based on magnetoresistance technology. *Biosens. Bioelectron.*, **13**(7–8), 731–9.

Basset, Alfred Barnard (1888). *A Treatise on Hydrodynamics*. Volume 2. Deighton, Bell and Co., London.

Batchelor, By G K (1976). Brownian diffusion of particles with hydrodynamic interaction. *J. Fluid Mech.*, **74**, 1–29.

Bausch, Andreas R, Ziemann, Florian, Boulbitch, Alexei A, Jacobson, Ken, and Sackmann, Erich (1998). Local measurements of viscoelastic parameters of adherent cell surfaces by magnetic bead microrheometry. *Biophys. J.*, **75**, 2038–49.

Berg-Sørensen, Kirstine, Oddershede, Lene, Florin, Ernst-Ludwig, and Flyvbjerg, Henrik (2003). Unintended filtering in a typical photodiode detection system for optical tweezers. *J. Appl. Phys.*, **93**, 3167–76.

Bergna, Horacia E. (1994). *Colloid Chemistry of Silica*. American Chemical Society, Washington, D.C.

Beris, A N, Tsamopoulos, J A, Armstrong, R C, and Brown, R A (1985). Creeping motion of a sphere through a Bingham plastic. *J. Fluid Mech.*, **158**, 219–44.

Berne, Bruce Jl and Pecora, Robert (2000). *Dynamic Light Scattering*. Dover Publications, New York.

Besseling, R, Isa, L, Weeks, E R, and Poon, W C K (2009). Quantitative imaging of colloidal flows. *Adv. Colloid Interface Sci.*, **146**(1–2), 1–17.

Bhardwaj, Avinash, Miller, Erik, and Rothstein, Jonathan P (2007). Filament stretching and capillary breakup extensional rheometry measurements of viscoelastic wormlike micelle solutions. *J. Rheol.*, **51**(4), 693–728.

Biancaniello, Paul L and Crocker, John C (2006). Line optical tweezers instrument for measuring nanoscale interactions and kinetics. *Rev. Sci. Instrum.*, **77**(11), 113702.

Bigg, Charlotte (2008). Evident atoms: visuality in Jean Perrin's Brownian motion research. *Stud. Hist. Philos. Sci. Part A*, **39**(3), 312–22.

Bigg, Charlotte (2011). A visual history of Jean Perrin's Brownian motion curves. In Histories of Scientific Observation (eds. Lorraine Daston and Elizabeth Lunbeck), Ch. 6, pp. 156–79. University Of Chicago Press, Chicago.

Blaakmeer, J, Cohen Stuart, M A, and Fleer, G J (1990). The adsorption of polyampholytes on negatively and positively charged polystyrene latex. *J. Colloid Interface Sci.*, **140**(2), 314–25.

Blake, J R (1971). A note on the image system for a Stokeslet in a no-slip boundary. *Math. Proc. Camb. Phil. Soc.*, **70**(2), 303–10.

Block, Ian D and Scheffold, Frank (2010). Modulated 3D cross-correlation light scattering: Improving turbid sample characterization. *Rev. Sci. Instrum.*, **81**(12), 123107.

Boddapati, Aparna, Rahane, Santosh B, Slopek, Ryan P, Breedveld, Victor, Henderson, Clifford L, and Grover, Martha A (2011). Gel time prediction of multifunctional acrylates using a kinetics model. *Polymer (Guildf).*, **52**(3), 866–73.

Bogush, G H, Tracy, M A, and Zukoski, C F (1988). Preparation of monodisperse silica particles: Control of size and mass fraction. *J. Non. Cryst. Solids*, **104**(1), 95–106.

Bogush, G H and Zukoski, C F (1991). Studies of the kinetics of the precipitation of uniform silica particles through the hydrolysis and condensation of silicon alkoxides. *J. Colloid Interface Sci.*, **142**(1), 1–18.

Booij, H C and Thoone, G P J M (1982, December). Generalization of Kramers-Kronig transforms and some approximations of relations between viscoelastic quantities. *Rheol Acta*, **21**, 15–24.

Born, Max and Wolf, Emil (1999). *Principles of Optics* (7th edn). Cambridge University Press, New York.

Bozorth, Richard M (1978). *Ferromagnetism*. IEEE Press, New York.

Bracewell, Ronald N (1986). *The Fourier Transform and Its Applications* (2nd edn). McGraw-Hill, San Diego.

Brady, John F (1993). The rheological behavior of concentrated colloidal dispersions. *J. Chem. Phys.*, **99**(1), 567–81.

Brady, John F and Morris, Jeffrey F (1997). Microstructure of strongly sheared suspensions and its impact on rheology and diffusion. *J. Fluid Mech.*, **348**, 103–39.

Brau, R R, Ferrer, J M, Lee, H, Castro, C E, Tam, B K, Tarsa, P B, Matsudaira, P, Boyce, M C, Kamm, R D, and Lang, M J (2007). Passive and active microrheology with optical tweezers. *J. Opt. A-Pure Appl. Opt.*, **9**, S103–12.

Bray, Dennis (2001). *Cell Movements*. Garland Science, New York.

Breedveld, V and Pine, D J (2003). Microrheology as a tool for high-throughput screening. *J. Mat. Sci.*, **38**, 4461–70.

Brown, Matthew A, Abbas, Zareen, Kleibert, Armin, Green, Richard G, Goel, Alok, May, Sylvio, and Squires, Todd M (2016). Determination of surface potential and electrical double-layer structure at the aqueous electrolyte-nanoparticle interface. *Phys. Rev. X.*, **6**(1), 011007–12.

Brown, Robert (1828). A brief account of microscopical observations made in the months of June, July and August, 1827, on the particles contained in the pollen of plants; and on the general existence of active molecules in organic and inorganic bodies. *Edinburgh New Philos. J.* (July–September), 358–71.

Brown, R G and Smart, A E (1997). Practical considerations in photon correlation experiments. *Appl. Opt.*, **36**(30), 7480–92.

Brown, Wyn (1993). *Dynamic Light Scattering*. Clarendon Press, New York.

Brown, William Fuller (1963). Thermal fluctuations of a single-domain particle. *Phys. Rev.*, **130**, 1677–86.

Buchanan, M, Atakhorrami, M, Palierne, J F, MacKintosh, F C, and Schmidt, C F (2005a). High-frequency microrheology of wormlike micelles. *Phys. Rev. E*, **72**, 11504.

Buchanan, M, Atakhorrami, M, Palierne, J F, and Schmidt, C F (2005b). Comparing macrorheology and one- and two-point microrheology in wormlike micelle solutions. *Macromolelcules*, **38**(21), 8840–44.

Burgess, Arthur E (1999). The Rose model, revisited. *J. Opt. Soc. Am. A*, **16**(3), 633–46.

Caggioni, M, Spicer, P T, Blair, D L, Lindberg, S E, and Weitz, D A (2007). Rheology and microrheology of a microstructured fluid: The gellan gum case. *J. Rheol.*, **51**(5), 851.

Candau, Françoise and Ottewill, Ronald H (1990). *Introduction to Polymer Colloids*. Kluwer, New York.

Cappallo, Nathan, Lapointe, Clayton, Reich, Daniel H, and Leheny, Robert L (2007). Nonlinear microrheology of wormlike micelle solutions using ferromagnetic nanowire probes. *Phys. Rev. E*, 76, 031505.

Cardinaux, F, Cipelletti, L, Scheffold, F, and Schurtenberger, P (2002). Microrheology of giant-micelle solutions. *Europhys. Lett.*, 57, 74483.

Carslaw, Horatio S and Jaeger, John C (1986). *Conduction of Heat in Solids* (2nd edn). Oxford University Press, New York.

Casanellas, Laura, Alves, Manuel A, Poole, Robert J, Lerouge, Sandra, and Lindner, Anke (2016). The stabilizing effect of shear thinning on the onset of purely elastic instabilities in serpentine microflows. *Soft Matter*, 12, 6167–75.

Chae, Byeong-Seok and Furst, Eric M (2005). Probe surface chemistry dependence and local polymer network structure in F-actin microrheology. *Langmuir*, 21, 3084–9.

Chaikin, Paul M and Lubensky, Thomas C (2000). *Principles of Condensed Matter Physics*. Cambridge University Press, New York.

Chandler, David (1987). *Introduction to Modern Statistical Mechanics*. Oxford University Press, New York.

Cheezum, Michael K, Walker, William F, and Guilford, William H (2001). Quantitative comparison of algorithms for tracking single fluorescent particles. *Biophys. J.*, 81(4), 2378–88.

Chen, D T, Weeks, E R, Crocker, J C, Islam, M F, Verma, R, Gruber, J, Levine, A J, Lubensky, T C, and Yodh, A G (2003). Rheological microscopy: Local mechanical properties from microrheology. *Phys. Rev. Lett.*, 90, 108301.

Chen, D T N, Chen, K, Hough, L A, Islam, M F, and Yodh, A G (2010). Rheology of carbon nanotube networks during gelation. *Macromolelcules*, 43(4), 2048–53.

Cheng, Z and Mason, T G (2003). Rotational diffusion microrheology. *Phys. Rev. Lett.*, 90, 18304.

Chevry, L, Sampathkumar, N K, Cebers, A, and Berret, J F (2013). Magnetic wire-based sensors for the microrheology of complex fluids. *Phys. Rev. E*, 88(6), 062306.

Choi, Gerald N and Krieger, Irvin M (1986a). Rheological studies on sterically stabilized dispersions of uniform colloidal spheres. I. Sample Preparation. *J. Colloid Interface Sci.*, 113(1), 94–100.

Choi, Gerald N and Krieger, Irvin M (1986b). Rheological studies on sterically stabilized dispersions of uniform colloidal spheres. II. Steady-Shear Viscosity. *J. Colloid Interface Sci.*, 113, 101–13.

Choi, S Y, Steltenkamp, S, Pascall, A J, Zasadzinski, J A, and Squires, T M (2011). Active microrheology of phospholipid monolayers: seeing stretching, flowing, yielding and healing. *Nat. Commun.*, 2, 312.

Chu, Benjamin (1991). *Laser Light Scattering: Basic Principles and Practice* (2nd edn). Academic Press, San Diego.

Churnside, J H (1982). Speckle from a rotating diffuse object. *J. Opt. Soc. Am.*, 72(11), 1464–9.

Cingil, Hande E, Rombouts, Wolf H, Van Der Gucht, Jasper, Cohen Stuart, Martien A, and Sprakel, Joris (2015). Equivalent pathways in melting and gelation of well-defined biopolymer networks. *Biomacromolecules*, 16(1), 304–10.

Cipelletti, Luca and Weitz, D A (1999). Ultralow-angle dynamic light scattering with a charge coupled device camera based multispeckle, multitau correlator. *Rev. Sci. Inst.*, 70, 3214–21.

Claesson, Stig, Malmrud, Sture, and Lundgren, Björn (1970). High pressure vessel for optical studies in the 1-8000 atm range. *Trans. Faraday Soc.*, 66, 3048–52.

Clarke, Andrew, Howe, Andrew M, Mitchell, Jonathan, Staniland, John, Hawkes, Laurence, and Leeper, Katherine (2015). Mechanism of anomalously increased oil displacement with aqueous viscoelastic polymer solutions. *Soft Matter*, 11, 3536–41.

Clasen, C, Eggers, J, Fontelos, M A, Li, J, and McKinley, G H (2006*a*). The beads-on-string structure of viscoelastic threads. *J. Fluid Mech.*, **556**, 283–308.

Clasen, Christian, Gearing, Brian P, and McKinley, Gareth H (2006*b*). The flexure-based microgap rheometer (FMR). *J. Rheol.*, **50**(6), 883.

Clasen, Christian and McKinley, Gareth H (2004). Gap-dependent microrheometry of complex liquids. *J. Non-Newtonian Fluid Mech.*, **124**, 1–10.

Cohen, Itai and Weihs, Daphne (2010). Rheology and microrheology of natural and reduced-calorie Israeli honeys as a model for high-viscosity Newtonian liquids. *J. Food Eng.*, **100**(2), 366–71.

Corrigan, A M and Donald, A M (2009*a*). Particle tracking microrheology of gel-forming amyloid fibril networks. *Eur. Phys. J. E*, **28**, 457–62.

Corrigan, Adam M and Donald, Athene M (2009*b*). Passive microrheology of solvent-induced fibrillar protein networks. *Langmuir*, **25**, 8599–605.

Corrigan, Adam M and Donald, Athene M (2010). Lengthscale dependence of critical exponents determined by vibration-corrected two-particle microrheology. *Soft Matter*, **6**(17), 4105–11.

Cousin, P and Smith, P (1994). Synthesis and characterization of styrene-based microbeads possessing amine functionality. *J. Appl. Polym. Sci.*, **54**, 1631–41.

Cordoba, Andres, Indei, Tsutomu, Schieber, Jay D, and Córdoba, Andrés (2012). Elimination of inertia from a Generalized Langevin Equation: Applications to microbead rheology modeling and data analysis. *J. Rheol.*, **56**(1), 185–212.

Cribb, Jeremy A, Meehan, Timothy D, Shah, Sheel M, Skinner, Kwan, and Superfine, Richard (2010). Cylinders vs. spheres: Biofluid shear thinning in driven nanoparticle transport. *Ann. Biomed. Eng.*, **38**(11), 3311–22.

Cribb, J A, Vasquez, P A, Moore, P, Norris, S, Shah, S, Forest, M G, and Superfine, R (2013). Nonlinear signatures in active microbead rheology of entangled polymer solutions. *J. Rheol.*, **57**(4), 1247–64.

Crick, F H C and Hughes, A F W (1950). The physical properties of cytopasm: A study by means of the magnetic particle method. *Exp. Cell Res.*, **1**, 37–50.

Crocker, John C and Grier, David G (1996). Methods of digital video microscopy for colloidal studies. *J. Colloid Interface Sci.*, **179**, 298–310.

Crocker, J C and Hoffman, B D (2007). Multiple particle tracking and two-point microrheology in cells. *Methods Cell Biol.*, **83**, 141–8.

Crocker, J C, Matteo, J A, Dinsmore, A D, and Yodh, A G (1999). Entropic attraction and repulsion in binary colloids probed with a line optical tweezer. *Phys. Rev. Lett.*, **82**(21), 4352–5.

Crocker, John C, Valentine, M T, Weeks, Eric R, Gisler, T, Kaplan, P D, Yodh, A G, and Weitz, D A (2000). Two-point microrheology of inhomogeneous soft materials. *Phys. Rev. Lett.*, **85**, 888–91.

Daoud, M (2000). Viscoelasticity near the sol-gel transition. *Macromolelcules*, **33**(8), 3019–22.

Dasgupta, Bivash R, Tee, Shang-You, Crocker, John C, Frisken, B J, and Weitz, D A (2002). Microrheology of polyethylene oxide using diffusing wave spectroscopy and single scattering. *Phys. Rev. E*, **65**, 51505.

Dasgupta, Bivash R and Weitz, D A (2005). Microrheology of cross-linked polyacrylamide networks. *Phys. Rev. E*, **71**, 21504.

de Gennes, Pierre-Gille (1982). Polymers at an interface: 2. Interaction between two plates carrying adsorbed polymer layers. *Macromolecules*, **500**(19), 492–500.

de Gennes, P. G. (1987). Polymers at an interface; a simplified view. *Adv. Colloid Interface Sci.*, **27**(3–4), 189–209.

Denk, W and Webb, W W (1990). Optical measurement of picometer displacements of transparent microscopic objects. *Appl. Opt.*, **29**(16), 2382–91.

Denn, Morton M. and Bonn, Daniel (2011). Issues in the flow of yield-stress liquids. *Rheol. Acta*, **50**(4), 307–15.

DePuit, Ryan J and Squires, Todd M (2012*a*). Micro-macro discrepancies in nonlinear microrheology: I. Quantifying mechanisms in a suspension of Brownian ellipsoids. *J. Phys. Condens. Matter*, **24**(46), 464106.

DePuit, Ryan J and Squires, Todd M (2012*b*). Micro-macro discrepancies in nonlinear microrheology: II. Effect of probe shape. *J. Phys. Condens. Matter*, **24**(46), 464107.

Derjaguin, B and Landau, L (1941). Theory of the stability of strongly charged lyophobic sols and of the adhesion of strongly charged particles in solutions of electrolytes. *Acta Phys. Chem. URSS*, **14**, 633–62.

D'Haene, P, Mewis, J, and Fuller, G C (1993). Scattering dichroism measurements of flow-induced structure of a shear thickening suspension. *J. Colloid Interface Sci.*, **156**, 350–8.

Dhar, P, Cao, Y, Fischer, T, and Zasadzinski, J A (2010). Active microrheology of aging protein films. *Phys. Rev. Lett.*, **104**, 016001.

Dichtl, M A and Sackmann, E (2002). Microrheometry of semiflexible actin networks through enforced single-filament reptation: Frictional coupling and heterogeneities in entangled networks. *Proc. Natl. Acad. Sci. USA*, **99**, 6533–8.

Dickinson, Eric and Eriksson, Leif (1991). Particle flocculation by adsorbing polymers. *Adv. Colloid Interface Sci.*, **34**(C), 1–29.

DiNoia, T P, Kirby, Christopher F, Van Zanten, John H, and McHugh, Mark a. (2000). SANS study of polymer-supercritical fluid solutions: Transitions from liquid to supercritical fluid solvent quality. *Macromolecules*, **33**, 6321–9.

Doi, Masao and Edwards, Samuel F (1986). *The Theory of Polymer Dynamics*. Clarendon Press, New York.

Domínguez-García, P, Cardinaux, Frédéric, Bertseva, Elena, Forró, László, Scheffold, Frank, and Jeney, Sylvia (2014). Accounting for inertia effects to access the high-frequency microrheology of viscoelastic fluids. *Phys. Rev. E*, **90**(6), 060301.

Drewel, M, Ahrens, J, and Podschus, U (1990). Decorrelation of multiple scattering for an arbitrary scattering angle. *J. Opt. Soc. Am. A*, **7**(2), 206.

Dufresne, E R and Grier, D G (1998). Optical tweezer arrays and optical substrates created with diffractive optics. *Rev. Sci. Instrum.*, **69**, 1974–7.

Dufresne, Eric R, Squires, Todd M, Brenner, Michael P, and Grier, David G (2000). Hydrodynamic coupling of two Brownian spheres to a planar surface. *Phys. Rev. Lett.*, **85**, 3317–20.

Durian, A D Gopal D J (1997). Fast thermal dynamics in aqueous foams. *J. Opt. Soc. Am. A*, **14**(1), 150–5.

Durian, D J, Weitz, D A, and Pine, D J (1991). Multiple light-scattering probes of foam structure and dynamics. *Science*, **252**(5006), 686–8.

Dzuy, Nguyen Quoc and Boger, D. V. (1983). Yield stress measurement for concentrated suspensions. *J. Rheol.*, **27**, 321–49.

Egres, Ronald G and Wagner, Norman J (2005). The rheology and microstructure of acicular precipitated calcium carbonate colloidal suspensions through the shear thickening transition. *J. Rheol.*, **49**, 719–746.

Einstein, Albert (1905). Über die von der molekularkinetischen theorie der warme geforderte bewegung von in ruhenden flussigkeiten suspenderten teilchen. *Ann. Phys.*, **17**, 549–560.

Einstein, Albert (1906). Eine neue Bestimmung der Moleküldimensionen. *Ann. Phys.*, **324**(2), 289–306.

Elster, C, Honerkamp, J, and Weese, J (1992). Using regularization methods for the determination of relaxation and retardation spectra of polymeric liquids. *Rheol. Acta*, **31**(2), 161–74.

Evans, Robert and Napper, D H (1973). Flocculation of latices by low molecular weight polymers. *Nature*, **246**(5427), 34–5.

Ewoldt, Randy H, Johnston, Michael T, and Caretta, Lucas M (2014). Experimental challenges of shear rheology: How to avoid bad data. In *Complex Fluids Biol. Syst.* (ed. S E Spagnolie), pp. 207–43. Springer-Verlag, New York.

Fabry, Ben, Maksym, Geoffrey N, Butler, James P, Glogauer, Michael, Navajas, Daniel, and Fredberg, Jeffrey J (2001). Scaling the microrheology of living cells. *Phys. Rev. Lett.*, **87**, 148102.

Farré, Arnau, Marsà, Ferran, and Montes-Usategui, Mario (2012). Optimized back-focal-plane interferometry directly measures forces of optically trapped particles. *Opt. Express*, **20**(11), 12270.

Faucheux, Luc P, Stolovitzky, Gustavo, Libchaber, Albert, Faucheaux, Luc P, Stolovitzky, Gustavo, and Libchaber, Albert (1995). Periodic forcing of a Brownian particle. *Phys. Rev. E*, **51**(6), 5239–50.

Faxén, Hilding (1922). Der Widerstand gegen die Bewegung einer starren Kugel in einer zähen Flüssigkeit, die zwischen zwei parallelen ebenen Wänden eingeschlossen ist. *Ann. Phys.*, **373**(10), 89–119.

Ferry, John D (1980). *Viscoelastic Properties of Polymers*. Wiley, New York.

Fert, A and Piraux, L (1999). Magnetic nanowires. *J. Magn. Magn. Mater.*, **200**(1–3), 338–58.

Feynman, Richard P, Leighton, Robert B, and Sands, Matthew (1964). *Lectures on Physics*. Addison Wesley, Reading, Massachusetts.

Finer, Jeffrey T, Simmons, Robert M, and Spudich, James A (1994). Single myosin molecule mechanics: Piconewton forces and nanometre steps. *Nature*, **368**, 113–9.

Fischer, Mario, Richardson, Andrew C, Reihani, S Nader S, Oddershede, Lene B, and Berg-Sørensen, Kirstine (2010). Active-passive calibration of optical tweezers in viscoelastic media. *Rev. Sci. Instrum.*, **81**(1), 15103.

Fischer, Thomas (2004). Comment on shear viscosity of Langmuir monolayers in the low-density limit. *Phys. Rev. Lett.*, **92**(13), 139603.

Fisher, J K, Cribb, J, Desai, K V, Vicci, L, Wilde, B, Keller, K, Taylor, R M, Haase, J, Bloom, K, O'Brien, E Timothy, and Superfine, R (2006). Thin-foil magnetic force system for high-numerical-aperture microscopy. *Rev. Sci. Instrum.*, **77**(2), 023702.

Flory, Paul J (1941). Molecular size distribution in three dimensional polymers. I. Gelation. *J. Am. Chem. Soc.*, **63**(11), 3083–90.

Flory, Paul J (1942). Constitution of three-dimensional polymers and the theory of gelation. *J. Phys. Chem.*, **46**(1), 132–40.

Flory, Paul J and Rehner, John (1943). Statistical mechanics of cross-linked polymer networks. *J. Adhes.*, **11**(11), 521–26.

Fonnum, G, Johansson, C, Molteberg, A, Morup, S, and Aksnes, E (2005). Characterisation of Dynabeads (R) by magnetization measurements and Mossbauer spectroscopy. *J. Magn. Magn. Mater.*, **293**(1), 41–7.

Fraden, Seth and Maret, Georg (1990). Multiple light scattering from concentrated, interacting suspensions. *Phys. Rev. Lett.*, **65**(4), 512–15.

Fresnais, J, Berret, J F, Frka-Petesic, B, Sandre, O, and Perzynski, R. (2008). Electrostatic co-assembly of iron oxide nanoparticles and polymers: Towards the generation of highly persistent superparamagnetic nanorods. *Adv. Mater.*, **20**(20), 3877–81.

Freundlich, H and Roder, H L (1938). Dilatancy and its relation to thixotropy. *Trans. Faraday Soc.*, **34**(1), 308–15.

Freundlich, Herbert and Seifriz, William (1923). Über die Elastizität von Solen und Gelen. *Z. Phys. Chem.*, **104**, 233–61.

Fuller, Gerald G and Vermant, Jan (2012). Complex fluid-fluid interfaces: Rheology and structure. *Annu. Rev. Chem. Biomol. Eng.*, **3**(1), 519–43.

Furst, Eric M (2005). Applications of laser tweezers in complex fluid rheology. *Curr. Opin. Colloid Interface Sci.*, **10**(1–2), 79–86.

Gal, Naama, Lechtman-Goldstein, Diana, and Weihs, Daphne (2013). Particle tracking in living cells: a review of the mean square displacement method and beyond. *Rheol. Acta*, **52**(5), 425–43.

Galvan-Miyoshi, J, Delgado, J, and Castillo, R (2008). Diffusing wave spectroscopy in Maxwellian fluids. *Eur. Phys. J. E*, **26**(4), 369–77.

Gardel, M L, Valentine, M T, Crocker, J C, Bausch, A R, and Weitz, D A (2003). Microrheology of entangled F-actin solutions. *Phys. Rev. Lett.*, **91**, 158302.

Gavara, Núria and Chadwick, Richard S (2010). Noncontact microrheology at acoustic frequencies using frequency-modulated atomic force microscopy. *Nat. Methods*, **7**(8), 650–4.

Genack, Ariel Z (1990). *Fluctuations, correlation and average transport of electromagnetic radiation in random media*. In Sheng (1990), pp. 207–311.

Gibson, Graham M, Leach, Jonathan, Keen, Stephen, Wright, Amanda J, and Padgett, Miles J (2008). Measuring the accuracy of particle position and force in optical tweezers using high-speed video microscopy. *Opt. Express*, **16**(19), 14561–70.

Gisler, T, Rüger, H, Egelhaaf, S U, Tschumi, J, Schurtenberger, P, and Rička, J (1995). Mode-selective dynamic light scattering: theory versus experimental realization. *Appl. Opt.*, **34**(18), 3546–53.

Gisler, T and Weitz, D A (1998). Tracer microrheology in complex fluids. *Curr. Opin. Colloid Interface Sci.*, **3**(6), 586–92.

Gisler, T and Weitz, D A (1999). Scaling of the microrheology of semidilute F-actin solutions. *Phys. Rev. Lett.*, **82**, 1606–9.

Gittes, F and MacKintosh, F C (1998). Dynamic shear modulus of a semiflexible polymer network. *Phys. Rev. E*, **58**, R1241–4.

Gittes, Frederick and Schmidt, Christoph F (1998). Interference model for back-focal-plane displacement detection in optical tweezers. *Opt. Lett.*, **23**, 7–9.

Gittes, F, Schnurr, B, Olmsted, P D, MacKintosh, F C, and Schmidt, C F (1997). Microscopic viscoelasticity: shear moduli of soft materials determined from thermal fluctuations. *Phys. Rev. Lett.*, **79**, 3286–9.

Gomez-Solano, J R and Bechinger, C (2014). Probing linear and nonlinear microrheology of viscoelastic fluids. *Europhys. Lett.*, **108**(5), 54008.

Goodwin, J W, Hearn, J, Ho, C, and Ottewill, R H (1973). The preparation and characterization of polymer latices formed in the absence of surface active agents. *Br. Polym. J.*, **5**, 347–62.

Gosse, Charlie and Croquette, Vincent (2002). Magnetic tweezers: Micromanipulation and force measurement at the molecular level. *Biophys. J.*, **82**, 3314–29.

Graham, Thomas (1861). Liquid diffusion applied to analysis. *Philos. Trans. R. Soc. London*, **151**, 183–224.

Graham, T (1864). On the properties of silicic acid and other analogous colloidal substances. *Proc. R. Soc. London*, **13**, 335–41.

Granick, Steve, Kumar, Sanat K, Amis, Eric J, Antonietti, Markus, Balazs, Anna C, Chakraborty, Arup K, Grest, Gary S, Hawker, Craig, Janmey, Paul, Kramer, Edward J, Nuzzo, Ralph, Russell, Thomas P, and Safinya, Cyrus R (2003). Macromolecules at surfaces: Research challenges and opportunities from tribology to biology. *J. Poly. Sci. B*, **41**, 2755–93.

Grier, David G (1997). Optical tweezers in colloid and interface science. *Curr. Opin. Colloid Interface Sci.*, **2**, 264–70.

Grier, David G (2003). A revolution in optical manipulation. *Nature*, **424**, 810–13.

Grimm, Matthias, Franosch, Thomas, and Jeney, Sylvia (2012). High-resolution detection of Brownian motion for quantitative optical tweezers experiments. *Phys. Rev. E*, **86**(2), 021912.

Grimm, Matthias, Jeney, Sylvia, and Franosch, Thomas (2011). Brownian motion in a Maxwell fluid. *Soft Matter*, **7**(5), 2076.

Guazzelli, Élisabeth and Morris, Jeffrey F (2012). *A Physical Introduction to Suspension Dynamics*. Cambridge University Press, New York.

Haber, Charbel and Wirtz, Denis (2000). Magnetic tweezers for DNA micromanipulation. *Rev. Sci. Instrum.*, **71**, 4561–70.

Happel, John and Brenner, Howard (1983). *Low Reynolds Number Hydrodynamics*. Springer, New York.

Harada, Yasuhiro and Asakura, Toshimitsu (1996). Radiation forces on a dielectric sphere in the Rayleigh scattering regime. *Opt. Commun.*, **124**(5–6), 529–41.

Harden, J and Viasnoff, V (2001). Recent advances in DWS-based micro-rheology. *Curr. Opin. Colloid Interface Sci.*, **6**, 438–45.

He, Feng, Becker, Gerald W, Litowski, Jennifer R, Narhi, Linda O, Brems, David N, and Razinkov, Vladimir I (2010). High-throughput dynamic light scattering method for measuring viscosity of concentrated protein solutions. *Anal. Biochem.*, **399**(1), 141–3.

He, Jun and Tang, Jay X (2011). Surface adsorption and hopping cause probe-size-dependent microrheology of actin networks. *Phys. Rev. E*, **83**(4), 41902.

Healy, Thomas W. (1994). *Stability of aqueous silica sols*, pp. 147–59. In Bergna (1994).

Healy, Thomas W and LaMer, Victor K (1962). The adsorption-flocculation reactions of a polymer with an aqueous colloidal dispersion. *J. Phys. Chem.*, **66**(10), 1835–8.

Hecht, Eugene (2001). *Opitcs* (4th edn). Addison-Wesley, New York.

Heertje, I (2014). Structure and function of food products: A review. *Food Struct.*, **1**(1), 3–23.

Heilbronn, A (1922). Eine Neue Methode zur Bestimmung der Viskosität lebender Protoplasten. *Jahrb. Wiss. Bot.*, **61**, 284–338.

Heilbrunn, L V (1924). The colloid chemistry of protoplasm III. The viscosity of protoplasm at various temperatures. *Am. J. Physiol.*, **68**(3), 645–8.

Heinemann, Cornelia, Cardinaux, Frédéric, Scheffold, Frank, Schurtenberger, Peter, Escher, Felix, and Conde-Petit, Béatrice (2004). Tracer microrheology of gamma-dodecalactone induced gelation of aqueous starch dispersions. *Carbohydr. Polym.*, **55**, 155–61.

Henderson, Stuart, Mitchell, Steven, and Bartlett, Paul (2001). Direct measurements of colloidal friction coefficients. *Phys. Rev. E*, **64**, 61403.

Henderson, Stuart, Mitchell, Steven, and Bartlett, Paul (2002). Propagation of hydrodynamic interactions in colloidal suspensions. *Phys. Rev. Lett.*, **88**, 88302.

Hiemenz, Paul C and Rajagopalan, Raj (1997). *Principles of Colloid and Surface Chemistry* (3rd edn). CRC Press, New York.

Hinch, E J (1975). Application of the Langevin equation to fluid suspensions. *J. Fluid Mech.*, **72**(03), 499–511.

Hiramoto, Y (1969). Mechanical properties of the protoplasm of the sea urchin egg: I. Unfertilized egg. *Exp. Cell Res.*, **56**, 201–08.

Hoffman, Brenton D, Massiera, Gladys, Van Citters, Kathleen M, and Crocker, John C (2006). The consensus mechanics of cultured mammalian cells. *Proc. Natl. Acad. Sci. USA*, **103**(27), 10259–64.

Höfling, Felix and Franosch, Thomas (2013). Anomalous transport in the crowded world of biological cells. *Rep. Prog. Phys.*, **76**, 046602.

Holliday, A K (1947). A method for the determination of viscosity at low rates of shear. *Trans. Faraday Soc.*, **43**, 630–5.

Holly, Erik E, Venkataraman, Sundar K, Chambon, Francois, and Henning Winter, H (1988). Fourier transform mechanical spectroscopy of viscoelastic materials with transient structure. *J. Non-Newtonian Fluid Mech.*, **27**, 17–26.

Honerkamp, J and Weese, J (1989). Determination of the relaxation spectrum by a regularization method. *Macromolecules*, **22**(11), 4372–7.

Honerkamp, J and Weese, J (1993). A nonlinear regularization method for the calculation of relaxation spectra. *Rheol. Acta*, **32**(1), 65–73.

Horn, F M, Richtering, W, Bergenholtz, J, Willenbacher, N, and Wagner, N J (2000). Hydrodynamic and colloidal interactions in concentrated charge-stabilized polymer dispersions. *J. Colloid Interface Sci.*, **225**(1), 166–78.

Hough, David B and White, Lee R (1980). The calculation of Hamaker constants from Liftshitz theory with applications to wetting phenomena. *Adv. Colloid Interface Sci.*, **14**(1), 3–41.

Hough, L A and Ou-Yang, H Daniel (2002). Correlated motions of two hydrodynamically coupled particles confined in separate quadratic potential wells. *Phys. Rev. E*, **65**, 21906.

Hough, L A and Ou-Yang, H Daniel (2006). Viscoelasticity of aqueous telechelic poly(ethylene oxide) solutions: Relaxation and structure. *Phys. Rev. E*, **73**, 31802.

Howard, Jonathan (2000). *Mechanics of Motor Proteins and the Cytoskeleton*. Sinauer, New York.

Huang, Hayden, Dong, Chen Y, Kwon, Hyuk-Sang, Sutin, Jason D, Kamm, Roger D, and So, Peter T C (2002). Three-dimensional cellular deformation analysis with a two-photon magnetic manipulator workstation. *Biophys. J.*, **82**(4), 2211–23.

Huh, Ji Yeon and Furst, Eric M (2006). Colloid dynamics in semiflexible polymer solutions. *Phys. Rev. E*, **74**(3), 31802.

Hunter, Robert J (2001). *Foundations of Colloid Science*. Oxford University Press, New York.

Ikada, Y (1994). Surface modification of polymers for medical applications. *Biomaterials*, **15**(10), 725–36.

Iler, Ralph K (1979). *The Chemistry of Silica*. Wiley, New York.

In, M and Prud'homme, R K (1993). Fourier transform mechanical spectroscopy of the sol-gel transition in zirconium alkoxide ceramic gels. *Rheol. Acta*, **32**(6), 556–65.

Indei, Tsutomu, Schieber, Jay D, and Córdoba, Andrés (2012a). Competing effects of particle and medium inertia on particle diffusion in viscoelastic materials, and their ramifications for passive microrheology. *Phys. Rev. E*, **85**, 041504.

Indei, Tsutomu, Schieber, Jay D, Cordoba, Andres, and Pilyugina, Ekaterina (2012b). Treating inertia in passive microbead rheology. *Phys. Rev. E*, **85**(2), 21504.

Ishimaru, Akira (1990). *Wave Propagation in Random Media*. Academic Press, San Diego.

Israelachvili, Jacob N (2011). *Intermolecular and Surface Forces* (3rd edn). Academic Press, New York.

Jackson, John D (1998). *Classical Electrodynamics*. Wiley, New York.

Jaishankar, A and McKinley, G H (2012). Power-law rheology in the bulk and at the interface: quasi-properties and fractional constitutive equations. *Proc. R. Soc. A Math. Phys. Eng. Sci.*, **469**(1989), 20120284.

Janesick, J R, Klaasen, K P, and Elliott, T (1987). Charge-coupled-device charge-collection efficiency and the photon-transfer technique. *Opt. Eng.*, **26**(10), 972–80.

Joanny, J F (1982). Flow birefringence at the sol-gel transition. *J. Phys.*, **43**, 467–73.

Johansson, C, Hanson, M, Pedersen, M S, and Mørup, S (1997). Magnetic properties of magnetic liquids with iron-oxide particles The influence of anisotropy and interactions. *J. Magn. Magn. Mater.*, **173**(1–2), 5–14.

Joosten, J G H, Gelade, E T F, and Pusey, P N (1990). Dynamic light-scattering by nonergodic media: Brownian particles trapped in polyacrylamide gels. *Phys. Rev. A*, **42**(4), 2161–73.

Josephson, Lilian Lam, Furst, Eric M, and Galush, William J. (2016a). Particle tracking microrheology of protein solutions. *J. Rheol.*, **60**(4), 531–40.

Josephson, Lilian Lam, Galush, William J, and Furst, Eric M (2016b). Parallel temperature-dependent microrheological measurements in a microfluidic chip. *Biomicrofluidics*, **10**, 043503.

Kaplan, P D, Dinsmore, A D, Yodh, A G, and Pine, D J (1994). Diffuse-transmission spectroscopy: A structural probe of opaque colloidal mixtures. *Phys. Rev. E*, **50**(6), 4827–35.

Keller, M, Schilling, J, and Sackmann, E (2001). Oscillatory magnetic bead rheometer for complex fluid microrheometry. *Rev. Sci. Instrum.*, **72**, 3626–33.

Kermis, Thomas W, Li, Dan, Guney-Altay, Ozge, Park, Il-hyun, van Zanten, John H, and Mchugh, Mark A (2004). High-pressure dynamic light scattering of poly(ethylene-co-1-butene) in ethane, propane, butane, and pentane at 130 c and kilobar pressures. *Macromolelcules*, **37**, 9123–31.

Khair, Aditya S and Brady, John F (2006). Single particle motion in colloidal dispersions: a simple model for active and nonlinear microrheology. *J. Fluid Mech.*, **557**, 73–117.

Kim, A J, Manoharan, V N, and Crocker, J C (2005). Swelling-based method for preparing stable, functionalized polymer colloids. *J. Am. Chem. Soc.*, **127**(6), 1592–3.

Kim, Sangtae and Karilla, Seppo J (1991). *Microhydrodynamics: Principles and Selected Applications*. Butterworth-Heinemann, Boston.

Kirby, C F and McHugh, M A (1997). Simple method for measuring refractive index of supercritical fluids. *Rev. Sci. Instrum.*, **68**(8), 3150–3.

Kloxin, Christopher J (2006). *Investigating Aqueous PEO-PPO-PEO Triblock Copolymer Dispersion Dynamics with Colloidal Sphere Thermal Motion*. Ph.D. thesis, North Carolina State University.

Kloxin, Christopher J and Van Zanten, John H (2009). Microviscoelasticity of adhesive hard sphere dispersions: Tracer particle microrheology of aqueous Pluronic L64 solutions. *J. Chem. Phys.*, **131**, 134904.

Kloxin, C J and van Zanten, J H (2010). High pressure phase diagram of an aqueous PEO-PPO-PEO triblock copolymer system via probe diffusion measurements. *Macromolelcules*, **43**(4), 2084–87.

Koenderink, G, Atakhorrami, M, MacKintosh, F, and Schmidt, C (2006). High-frequency stress relaxation in semiflexible polymer solutions and networks. *Phys. Rev. Lett.*, **96**(13), 138307.

Koenderink, Gijsberta H, Sacanna, Stefano, Pathmamanoharan, Chellapah, Rasa, Mircea, and Philipse, Albert P (2001). Preparation and properties of optically transparent aqueous dipsersions of monodisperse fluorinated colloids. *Langmuir*, **17**, 6086–93.

Krall, A H and Weitz, D A (1998). Internal dynamics and elasticity of fractal colloidal gels. *Phys. Rev. Lett.*, **80**(4), 778–81.

Kramers, H A (1940). Brownian motion in a field of force and the diffusion model of chemical reactions. *Physica*, 7, 284–304.

Krause, W E, Bellomo, E G, and Colby, R H (2001). Rheology of sodium hyaluronate under physiological conditions. *Biomacromolecules*, **2**(1), 65–9.

Kruithof, M, Chien, F, de Jager, M, and van Noort, J (2008). Subpiconewton dynamic force spectroscopy using magnetic tweezers. *Biophys. J.*, **94**(6), 2343–8.

Kubo, Rygo (1966). The fluctuation-dissipation theorem. *Rep. Prog. Phys.*, **29**, 255–84.

Kubo, R, Ichimura, H, Usui, T, and Hashitsume, N (1990). *Statistical Mechanics*. North Holland, Amsterdam, Netherlands.

Kubo, Ryogo, Toda, Morikazu, and Hashitsume, Natsuki (1991). *Statistical Physics* (2nd edn). Volume 2. Springer-Verlag, Amsterdam, Netherlands.

Kuipers, B W M, Bakelaar, I A, Klokkenburg, M, and Erń, B H (2008). Complex magnetic susceptibility setup for spectroscopy in the extremely low-frequency range. *Rev. Sci. Instrum.*, **79**(1), 0–7.

Landau, L D, Lifshitz, E M, Kosevich, A M, and Pitaevskii, L P (1986). *Theory of Elasticity*. Pergamon, Oxford.

Landau, L D, Lifshitz, E M, and Pitaevskii, L P (1984). *Electrodynamics of Continuous Media* (2nd edn). Pergamon, New York.

Langevin, P (1905). Sur la théorie du magnétisme. *J. Phys. Théorique Appliquée*, **4**(1), 678–93.

Larsen, Travis, Schultz, Kelly, and Furst, Eric M (2008). Hydrogel microrheology near the liquid-solid transition. *Korea-Australia Rheol. J.*, **20**(3), 165–173.

Larsen, Travis H, Branco, Monica C, Rajagopal, Karthikan, Schneider, Joel P, and Furst, Eric M (2009). Sequence-dependent gelation kinetics of β-hairpin peptide hydrogels. *Macromolelcules*, **42**(21), 8443–50.

Larsen, Travis H and Furst, Eric M (2008). Microrheology of the liquid-solid transition during gelation. *Phys. Rev. Lett.*, **100**(14), 146001.

Le Goff, Loïc, Amblard, François, and Furst, Eric M (2002). Motor-driven dynamics in actin-myosin networks. *Phys. Rev. Lett.*, **88**(1), 18101.

Leal, L Gary (2007, January). *Advanced Transport Phenomena*. Cambridge University Press, Cambridge.

Lee, Hyungsuk, Shin, Yongdae, Kim, Sun Taek, Reinherz, Ellis L, and Lang, Matthew J (2012). Stochastic optical active rheology. *Appl. Phys. Lett.*, **101**(3), 31902–4.

Lee, Myung Han, Reich, Danial H, Stebe, Kathleen J, and Leheny, Robert L (2010). Combined passive and active microrheology study of protein-layer formation at an air-water interface. *Langmuir*, **26**(4), 2650–8.

Lee, Woei Ming, Reece, Peter J, Marchington, Robert F, Metzger, Nikolaus K, and Dholakia, Kishan (2007). Construction and calibration of an optical trap on a fluorescence optical microscope. *Nat. Protoc.*, **2**(12), 3226–38.

Leheny, Robert L (2012). XPCS: Nanoscale motion and rheology. *Curr. Opin. Colloid Interface Sci.*, **17**(1), 3–12.

Lekkerkerker, Henk N W and Tuinier, Remco (2011). *Colloids and the Depletion Interaction*. Springer, New York.

Lele, Pushkar P, Swan, James W, Brady, John F, Wagner, Norman J, and Furst, Eric M (2011). Colloidal diffusion and hydrodynamic screening near boundaries. *Soft Matter*, 7(15), 6844.

Lentz, Harro (1969). A method of studying the behavior of fluid phases at high pressures and temperatures. *Rev. Sci. Instrum.*, **40**(2), 371–72.

Lesemann, M, Nathan, H, Dinoia, T P, Kirby, C F, Mchugh, M A, Van Zanten, J H, and Paulaitis, M E (2003). Self-assembly at high pressures: SANS study of the effect of pressure on microstructure of C8E5 micelles in water. *Ind. Eng. Chem. Res.*, **42**, 6425–30.

Levine, Alex J, Liverpool, T B, and MacKintosh, F C (2004). Dynamics of rigid and flexible extended bodies in viscous films and membranes. *Phys. Rev. Lett.*, **93**, 038102.

Levine, Alex J and Lubensky, T C (2000). One- and two-point microrheology. *Phys. Rev. Lett.*, **85**, 1774–7.

Levine, Alex J and Lubensky, T C (2001). Two-point microrheology and the electrostatic analogy. *Phys. Rev. E*, **65**, 11501.

Lin, Jun and Valentine, Megan T (2012*a*). High-force NdFeB-based magnetic tweezers device optimized for microrheology experiments. *Rev. Sci. Instrum.*, **83**(5), 053905.

Lin, Jun and Valentine, Megan T (2012*b*). Ring-shaped NdFeB-based magnetic tweezers enables oscillatory microrheology measurements. *Appl. Phys. Lett.*, **100**(20), 201902.

Lin, T H and Phillies, G D J (1984). Probe diffusion in poly(acrylic acid)-water. Effect of probe size. *Macromolelcules*, **17**, 1686–91.

Lipfert, Jan, Hao, Xiaomin, and Dekker, Nynke H (2009). Quantitative modeling and optimization of magnetic tweezers. *Biophys. J.*, **96**(12), 5040–9.

Liu, J, Gardel, M L, Kroy, K, Frey, E, Hoffman, B D, Crocker, J C, Bausch, A R, and Weitz, D A (2006). Microrheology probes length scale dependent rheology. *Phys. Rev. Lett.*, **96**, 118104.

Lu, Qiang and Solomon, Michael J (2002). Probe size effects on the microrheology of associating polymer solutions. *Phys. Rev. E*, **66**, 61504.

Lukić, B, Jeney, S, Tischer, C, Kulik, A, Forró, L, and Florin, E L (2005). Direct observation of nondiffusive motion of a Brownian particle. *Phys. Rev. Lett.*, **95**(16), 1–4.

MacKintosh, F C, Käs, J, and Janmey, P A (1995). Elasticity of semiflexible biopolymer networks. *Phys. Rev. Lett.*, **75**, 4425–8.

MacKintosh, F C, Zhu, J X, Pine, D J, and Weitz, D A (1989). Polarization memory of multiply scattered light. *Phys. Rev. E*, **40**(13), 9342–5.

Macosko, Christopher W. (1994). *Rheology: Principles, Measurements, and Applications*. Wiley, New York.

Macosko, Christopher W (2010). Joe Starita: Father of modern rheometry. *Rheol. Bull.*, **79**(2), 11.

Maestro, Armando, Bonales, Laura J, Ritacco, Hernan, Fischer, Thomas M, Rubio, Ramón G, and Ortega, Francisco (2011). Surface rheology: macro- and microrheology of poly(tert-butyl acrylate) monolayers. *Soft Matter*, **7**(17), 7761.

Maiocchi, R (1990). The case of Brownian motion. *Br. J. Hist. Sci.*, **23**, 257–83.

Maret, G and Wolf, P E (1987). Multiple light scattering from disordered media. The effect of Brownian motion of scatterers. *Zeitschrift für Phys. B Condens. Matter*, **65**(4), 409–13.

Martin, James E, Adolf, Douglas, and Wilcoxton, Jess P (1988). Viscoelasticity of near-critical gels. *Phys. Rev. Lett.*, **61**, 2620–3.

Martin, James E, Wilcoxon, Jess, and Adolf, Douglas (1987). Critical exponents for the sol-gel transistion. *Phys. Rev. A*, **36**, 1803–10.

Martin, P, Hudspeth, A J, and Ju, F (2001). Comparison of a hair bundle's spontaneous oscillations with its response to mechanical stimulation reveals the underlying. *Proc. Natl. Acad. Sci. USA*, **98**, 14380–5.

Mason, Thomas G (2000). Estimating the viscoelastic moduli of complex fluids using the generalized Stokes-Einstein equation. *Rheol. Acta*, **39**, 371–78.

Mason, T G, Ganesan, K, van Zanten, J H, Wirtz, D, and Kuo, S C (1997*a*). Particle tracking microrheology of complex fluids. *Phys. Rev. Lett.*, **79**, 3282–5.

Mason, T G, Gang, H, and Weitz, D A (1997*b*). Diffusing-wave-spectroscopy measurements of viscoelasticity of complex fluids. *J. Opt. Soc. Am.*, **14**, 139–49.

Mason, Thomas G, Gisler, T, Kroy, K, Frey, E, and Weitz, D A (2000). Rheology of F-actin solutions determined from thermally driven tracer motion. *J. Rheol.*, **44**, 917–28.

Mason, T G and Weitz, D A (1995). Optical measurements of frequency-dependent linear viscoelastic moduli of complex fluids. *Phys. Rev. Lett.*, **74**, 1250–3.

McGrath, James L, Hartwig, John H, and Kuo, Scot C (2000). The mechanics of F-actin microenvironments depend on the chemistry of probing surfaces. *Biophys. J.*, **79**, 3258–66.

McKinley, Gareth H and Sridhar, Tamarapu (2002). Filament-stretching rheometry of complex fluids. *Ann. Rev. Fluid Mech.*, **34**, 375–415.

McQuarrie, Donald Allan (2000). *Statistical Mechanics*. University Science Books, Sausalito, CA.

Meier, G, Vavrin, R, Kohlbrecher, Joachim, Buitenhuis, J, Lettinga, M P, and Ratajczyk, M (2008). SANS and dynamic light scattering to investigate the viscosity of toluene under high pressure up to 1800 bar. *Meas. Sci. Technol.*, **19**, 034017.

Mertz, Jerome (2010). *Introduction to Optical Microscopy*. Roberts, Greenwood Village, Colorado.

Meyer, Alexander, Marshall, Andrew, Bush, Brian G, and Furst, Eric M (2006). Laser tweezer microrheology of a colloidal suspension. *J. Rheol.*, **50**, 77–92.

Mezzenga, Raffaele, Schurtenberger, Peter, Burbidge, Adam, and Michel, Martin (2005). Understanding foods as soft materials. *Nat. Mater.*, **4**(10), 729–40.

Milner, Scott T (1993). Dynamical theory of concentration fluctuations in polymer solutions under shear. *Phys. Rev. E*, **48**, 3674–91.

Mio, C, Gong, T, Terray, A, and Marr, D W M (2000). Design of a scanning optical trap for multiparticle manipulation. *Rev. Sci. Inst.*, **71**, 2196–200.

Mizuno, Daisuke, Tardin, Catherine, Schmidt, C F, and MacKintosh, F C (2007). Nonequilibrium mechanics of active cytoskeletal networks. *Science*, **315**, 370–73.

Mohammadigoushki, Hadi and Muller, Susan J (2016). Sedimentation of a sphere in wormlike micellar fluids. *J. Rheol.*, **60**(4), 587–601.

Mohan, Lavanya, Cloitre, Michel, and Bonnecaze, Roger T (2014). Active microrheology of soft particle glasses. *J. Rheol.*, **58**(5), 1465–82.

Moller, P, Fall, A, Chikkadi, V, Derks, D, and Bonn, D (2009). An attempt to categorize yield stress fluid behaviour. *Phil. Trans. R. Soc. A*, **367**(1909), 5139–55.

Møller, Peder C F, Mewis, Jan, and Bonn, Daniel (2006). Yield stress and thixotropy: On the difficulty of measuring yield stresses in practice. *Soft Matter*, **2**, 274.

Morse, David C (1998*a*). Viscoelasticity of concentrated isotropic solutions of semiflexible polmers. 2. Linear response. *Macromolecules*, **31**, 7044–67.

Morse, David C (1998*b*). Viscoelasticity of tightly entangled solutions of semiflexible polymers. *Phys. Rev. E*, **58**(2), R1237–40.

Morse, David C (1998*c*). Viscoelasticity of tightly entangled solutions of semiflexible polymers. 1. Model and stress tensor. *Macromolelcules*, **31**, 7030–43.

Moschakis, Thomas (2013). Microrheology and particle tracking in food gels and emulsions. *Curr. Opin. Colloid Interface Sci.*, **18**(4), 311–23.

Mulyasasmita, W, Lee, J S, and Heilshorn, S C (2011). Molecular-level engineering of protein physical hydrogels for predictive solgel phase behavior. *Biomacromolecules*, **12**(10), 3406–1.

Muthukumar, M (1989). Screening effect on viscoelasticity near the gel point. *Macromolelcules*, **22**, 4656–8.

Muthukumar, M and Winter, H Henning (1986). Fractal dimensions of a cross-linking polymer at the gel point. *Macromolelcules*, **19**, 1284–5.

Napper, Donald H. (1983). *Polymeric Stabilization of Colloidal Dispersions*. Academic Press, New York.

Narita, Tetsuharu, Mayumi, Koichi, Ducouret, Guylaine, and Hébraud, Pascal (2013). Viscoelastic properties of poly(vinyl alcohol) hydrogels having permanent and transient cross-links studied by microrheology, classical rheometry, and dynamic light scattering. *Macromolelcules*, **46**(10), 4174–83.

Néel, M. Louis (1949). Théorie du traînage magnétique des ferromagnétiques en grains fins avec application aux terres cuites. *Ann. Geophys.*, **5**, 99–136.

Neuman, Keir C and Block, Steven M (2004). Optical trapping. *Rev. Sci. Instrum.*, **75**(9), 2787.

Neuman, K C and Nagy, A (2008). Single-molecule force spectroscopy: Optical tweezers, magnetic tweezers and atomic force microscopy. *Nat. Methods*, **5**(6), 491–505.

Nguyen, Q D and Boger, D V (1992). Measuring the flow properties of yield stress fluid. *Annu. Rev. Fluid Mech.*, **24**, 47–88.

Nisato, G, Hébraud, P, Munch, J P, and Candau, S J (2000). Diffusing-wave-spectroscopy investigation of latex particle motion in polymer gels. *Phys. Rev. E*, **61**(3), 2879–87.

Norris, Andrew N (2006). Impedance of a sphere oscillating in an elastic medium with and without slip. *J. Acoust. Soc. Am.*, **119**(4), 2062–5.

Nowak, A P, Breedveld, V, Pakstis, L, Ozbas, B, Pine, D J, Pochan, D J, and Deming, T J (2002). Rapidly recovering hydrogel scaffolds from self-assembling dibplock copolypeptide amphiphiles. *Nature*, **417**, 424–8.

Nyquist, H (1928). Thermal agitation of electric charge in conductors. *Phys. Rev.*, **32**(1), 110–13.

O'Brien, E Tim, Cribb, Jeremy, Marshburn, David, Taylor, Russell M, and Superfine, Richard (2008). Magnetic manipulation for force measurements in cell biology. In *Methods Cell Biol.*, Volume 89, pp. 433–450. Elsevier, New York.

Oelschlaeger, C, Coelho, M Cota Pinto, and Willenbacher, Norbert (2013). Chain flexibility and dynamics of polysaccharide hyaluronan in entangled solutions: a high frequency rheology and diffusing wave spectroscopy study. *Biomacromolecules*, **14**, 3689–96.

Oelschlaeger, C, Schopferer, M, Scheffold, F, and Willenbacher, N (2009). Linear-to-branched micelles transition: A rheometry and diffusing wave spectroscopy (DWS) study. *Langmuir*, **25**(2), 716–23.

Oelschlaeger, C, Suwita, P, and Willenbacher, N (2010). Effect of counterion binding efficiency on structure and dynamics of wormlike micelles. *Langmuir*, **26**(10), 7045–53.

Oelschlaeger, Claude and Willenbacher, Norbert (2012). Mixed wormlike micelles of cationic surfactants: Effect of the cosurfactant chain length on the bending elasticity and rheological properties. *Colloids Surfaces A*, **406**, 31–7.

Oestricher, Hans L (1951). Field and impedance of an oscillating sphere in a viscoelastic medium with an application to biophysics. *J. Acoust. Soc. Am.*, **23**, 707–14.

Ogunnaike, Babatunde A (2009). *Random Phenomena: Fundamentals of Probability and Statistics for Engineers*. CRC Press, New York.

Oppong, Felix K, Coussot, P, and De Bruyn, John R (2008). Gelation on the microscopic scale. *Phys. Rev. E*, 78(2), 1–10.

Oppong, Felix K and de Bruyn, John R (2007). Diffusion of microscopic tracer particles in a yield-stress fluid. *J. Non-Newtonian Fluid Mech.*, 142(1–3), 104–11.

Ortega, Francisco, Ritacco, Hernan, and Rubio, Ramón G (2010). Interfacial microrheology: Particle tracking and related techniques. *Curr. Opin. Colloid Interf. Sci*, 15(4), 237–45.

Ovarlez, G, Rodts, S, Chateau, X, and Coussot, P (2009). Phenomenology and physical origin of shear localization and shear banding in complex fluids. *Rheol Acta*, 48(8), 831–44.

Ozbas, Bulent, Rajagopal, Karthikan, Schneider, Joel P, and Pochan, Darrin J (2004). Semiflexible chain networks formed via self-assembly of beta-hairpin molecules. *Phys. Rev. Lett.*, 93, 268106.

Palmer, A, Xu, J, Kuo, S C, and Wirtz, D (1999). Diffusing wave spectroscopy microrheology of actin filament networks. *Biophys. J.*, 76(2), 1063–71.

Palmer, Andre, Xu, Jingyuan, and Wirtz, Denis (1998). High-frequency viscoelasticity of crosslinked actin filament networks measured by diffusing wave spectroscopy. *Rheol. Acta*, 37, 97–106.

Pashias, N, Boger, D V, Summers, J, and Glenister, D J (1996). A fifty cent rheometer for yield stress measurement. *J. Rheol.*, 40(6), 1179–12.

Pawley, James B (2006). *Biological Confocal Microscopy*. Springer, New York.

Perrin, Jean (1909). Mouvement Brownien et Réalité Moléculaire. *Ann. Chim. Phys.*, 18, 5–114.

Perrin, Jean (1913). *Les Atomes*. Librairie Félix Alcan, Paris.

Peterman, Erwin J G, van Dijk, Meindert A, Kapitein, Lukas C, and Schmidt, Christoph F (2003). Extending the bandwidth of optical-tweezers interferometry. *Rev. Sci. Instrum.*, 74, 3246–9.

Pham, K N, Egelhaaf, S U, Moussad, A, and Pusey, P N (2004). Ensemble-averaging in dynamic light scattering by an echo technique. *Rev. Sci. Instrum.*, 75(7), 2419–31.

Phillies, George D J (1981). Suppression of multiple scattering effects in quasielastic light scattering by homodyne cross-correlation techniques. *J. Chem. Phys.*, 74(1), 260.

Pichot, C (2004). Surface-functionalized latexes for biotechnological applications. *Curr. Opin. Colloid Interface Sci.*, 9(3–4), 213–21.

Piirma, Irja and Gardon, John L (ed.) (1976). *Emulsion Polymerization*. American Chemical Society, New York.

Pine, D, Weitz, D, Chaikin, P, and Herbolzheimer, E (1988). Diffusing wave spectroscopy. *Phys. Rev. Lett.*, 60(12), 1134–7.

Pine, D J, Weitz, D A, Zhu, J X, and Herbolzheimer, E (1990). Diffusing-wave spectroscopy: dynamic light-scattering in the multiple-scattering limit. *J. Phys.*, 51, 2101.

Pipe, Christopher J and McKinley, Gareth H. (2009). Microfluidic rheometry. *Mech. Res. Commun.*, 36(1), 110–20.

Pipkin, Allen Compere (1986). *Lectures on Viscoelasticity Theory* (2nd edn). Springer-Verlag, New York.

Poehlein, Gary W Ottweill, Ronald H. and Goodwin, James W. (1985). *Science and Technology of Polymer Colloids*. Springer, New York.

Pozrikidis, C (1992). *Boundary Integral and Singularity Methods for Linearized Viscous Flow*. Cambridge University Press, Cambridge.

Preece, Daryl, Warren, Rebecca, Evans, R M L, Gibson, Graham M, Padgett, Miles J, Cooper, Jonathan M, and Tassieri, Manlio (2011). Optical tweezers: Wideband microrheology. *J. Opt.*, 13(4), 44022.

Provencher, S W (1982*a*). A constrained regularization method for inverting data represented by linear algebraic or integral-equations. *Comput. Phys. Commun.*, **27**(3), 229–242.

Provencher, Stephen W (1982*b*). CONTIN: A general purpose constrained regularization program for inverting noisy linear algebraic and integral equations. *Comput. Phys. Commun.*, **27**, 229.

Puertas, A M and Voigtmann, Th (2014). Microrheology of colloidal systems. *J. Phys. Condens. Matter*, **26**(46), 243101.

Pusey, Peter N (2002). Dynamic Light Scattering. In *Neutrons, X-Rays Light* (ed. P. Lindner and T. Zemb), Chapter 9, pp. 203–20. Elsevier, New York.

Pusey, P N and Van Megen, W (1989). Dynamic light scattering by non-ergodic media. *Physica A*, **157**(2), 705–41.

Qiu, X, Wu, X L, Xue, J Z, Pine, D J, Weitz, D A, and Chaikin, P M (1990). Hydrodynamic interactions in concentrated suspensions. *Phys. Rev. Lett.*, **65**, 516–19.

Quake, Stephen R, Babcock, Hazen, and Chu, Steven (1997). The dynamics of partially extended single molecules of DNA. *Nature*, **388**, 151–54.

Rahman, A (1964). Correlations in the motion of atoms in liquid argon. *Phys. Rev.*, **136**(2A), 405–11.

Rallison, J M and Hinch, E J. (1976). The effect of particle interactions on dynamic light scattering from a dilute suspension. *J. Fluid Mech.*, **167**, 131–68.

Reddy, Sridhar, Moore, Lee R, Sun, Liping, Zborowski, Maciej, and Chalmers, Jeffrey J (1996). Determination of the magnetic susceptibility of labeled particles by video imaging. *Chem. Eng. Sci.*, **51**(6), 947–56.

Resnick, Andrew (2003). Use of optical tweezers for colloid science. *J Colloid Interface Sci.*, **262**, 55–9.

Rich, Jason P, Lammerding, Jan, McKinley, Gareth H, and Doyle, Patrick S. (2011*a*). Nonlinear microrheology of an aging, yield stress fluid using magnetic tweezers. *Soft Matter*, **7**(21), 9933.

Rich, Jason P, McKinley, Gareth H, and Doyle, Patrick S (2011*b*). Size dependence of microprobe dynamics during gelation of a discotic colloidal clay. *J. Rheol.*, **55**(2), 273.

Richards, C J and Fisch, M R (1994). High-pressure quasielastic and static light scattering apparatus. *Rev. Sci. Instrum.*, **65**(2), 335–8.

Richtering, H W, Gagnos, K D, Lenz, R W, Fuller, R C, and Winter, H H (1992). Physical gelation of a bacterial thermoplastic elastomer. *Macromolelcules*, **25**, 2429–33.

Ricka, J (1993). Dynamic light scattering with single-mode and multimode receivers. *Appl. Opt.*, **32**, 2860–75.

Rojas-Ochoa, L, Romer, S, Scheffold, F, and Schurtenberger, P (2002). Diffusing wave spectroscopy and small-angle neutron scattering from concentrated colloidal suspensions. *Phys. Rev. E*, **65**, 051403.

Rojas-Ochoa, Luis Fernando, Lacoste, David, Lenke, Ralf, Schurtenberger, Peter, and Scheffold, Frank (2004). Depolarization of backscattered linearly polarized light. *J. Opt. Soc. Am. A*, **21**(9), 1799–1804.

Romer, Sara, Scheffold, Frank, and Schurtenberger, Peter (2000). Sol-gel transition of concentrated colloidal suspensions. *Phys. Rev. Lett.*, **85**(23), 4980–3.

Rose, Albert (1948). The sensitivity performance of the human eye on an absolute scale. *J. Opt. Soc. Am.*, **38**(2), 196–208.

Rubinstein, Michael and Colby, Ralph H (2003). *Polymer Physics*. Oxford University Press, New York.

Russ, John C (2011). *Image Processing Handbook* (6th edn). CRC Press, Boca Raton, FL.

Russel, W B, Saville, D. A., and Schowalter, W R. (1989). *Colloidal Dispersions*. Cambridge University Press, New York.

Sainis, Sunil K, Germain, Vincent, and Dufresne, Eric R (2007). Statistics of particle trajectories at short time intervals reveal fN-scale colloidal forces. *Phys. Rev. Lett.*, **99**, 18303.

Samaniuk, Joseph R and Vermant, Jan (2014). Micro and macrorheology at fluid-fluid interfaces. *Soft Matter*, **10**, 7023–33.

Sarmiento-Gomez, Erick, Lopez-Diaz, David, and Castillo, Rolando (2010). Microrheology and characteristic lengths in wormlike micelles made of a zwitterionic surfactant and SDS in brine. *J. Phys. Chem. B*, **114**(38), 12193–202.

Sarmiento-Gomez, E, Montalvan-Sorrosa, D, Garza, C, Mas-Oliva, J, and Castillo, R (2012). Rheology and DWS microrheology of concentrated suspensions of the semiflexible filamentous fd virus. *Eur. Phys. J. E*, **35**(5), 35.

Sarmiento-Gomez, Erick, Morales-Cruzado, Beatriz, and Castillo, Rolando (2014). Absorption effects in diffusing wave spectroscopy. *Appl. Opt.*, **53**(21), 4675–82.

Sato, J and Breedveld, V (2006). Transient rheology of solvent-responsive complex fluids by integrating microrheology and microfluidics. *J. Rheol.*, **50**(1), 1–19.

Sato, M, Wong, T Z, and Allen, R D (1983). Rheological properties of living cytoplasm: Endoplasm of Physarum plasmodium. *J. Cell Biol.*, **97**(4), 1089–97.

Savin, Thierry and Doyle, Patrick S (2005). Role of a finite exposure time on measuring an elastic modulus using microrheology. *Phys. Rev. E*, **71**, 41106.

Savin, T and Doyle, P S (2007a). Electrostatically tuned rate of peptide self-assembly resolved by multiple particle tracking. *Soft Matter*, **3**(9), 1194–202.

Savin, Thierry and Doyle, Patrick S (2007b). Statistical and sampling issues when using multiple particle tracking. *Phys. Rev. E*, **76**, 21501.

Scanlan, James C and Winter, Henning W (1991). Composition dependence of the viscoelasticity of end-linked poly(dimethylsiloxane) at the gel point. *Macromolelcules*, **24**, 47–54.

Schätzel, Klaus (1991). Suppression of multiple scattering by photon cross-correlation techniques. *J. Mod. Opt.*, **38**(9), 1849–65.

Schätzel, Klaus (1993). *Single-photon correlation techniques*, In Dynamic Light Scattering (ed. Wyn Brown) , Ch. 2, pp. 76–148. Oxford University Press, New York.

Scheffold, F and Maret, G (1998). Universal conductance fluctuations of light. *Phys. Rev. Lett.*, **81**(26), 5800–03.

Scheffold, Frank and Schurtenberger, Peter (2003). Light scattering probes of viscoelastic fluids and solids. *Soft Mater.*, **1**(2), 139–65.

Scheffold, F, Skipetrov, S E, Romer, S, and Schurtenberger, P (2001). Diffusing-wave spectroscopy of nonergodic media. *Phys. Rev. E*, **63**, 061404.

Schieber, Jay D, Córdoba, Andrés, and Indei, Tsutomu (2013). The analytic solution of Stokes for time-dependent creeping flow around a sphere: Application to linear viscoelasticity as an ingredient for the generalized Stokes-Einstein relation and microrheology analysis. *J. Non-Newtonian Fluid Mech.*, **200**, 3–8.

Schneider, Joel P, Pochan, Darrin J, Ozbas, Bulent, Rajagopal, Karthikan, Pakstis, Lisa, and Kretsinger, Juliana (2002). Responsive hydrogels from the intramolecular folding and self-assembly of a designed peptide. *J. Am. Chem. Soc.*, **124**, 15030–7.

Schnurr, B, Gittes, F, MacKintosh, F C, and Schmidt, C F (1997). Determining microscopic viscoelasticity in flexible and semiflexible polymer networks from thermal fluctuations. *Macromolelcules*, **30**, 7781–92.

Schultz, Kelly M and Anseth, Kristi S (2013). Monitoring degradation of matrix metalloproteinases-cleavable PEG hydrogels via multiple particle tracking microrheology. *Soft Matter*, 9(5), 1570.

Schultz, Kelly M, Baldwin, Aaron D, Kiick, Kristi L, and Furst, Eric M (2009a). Gelation of covalently cross-linked PEG-heparin hydrogels. *Macromolelcules*, 42, 5310–16.

Schultz, Kelly M, Baldwin, Aaron D, Kiick, Kristi L, and Furst, Eric M (2009b). Rapid rheological screening to identify conditions of biomaterial hydrogelation. *Soft Matter*, 5, 740–42.

Schultz, Kelly M, Baldwin, Aaron D, Kiick, Kristi L, and Furst, Eric M (2012a). Measuring the modulus and reverse percolation transition of a degrading hydrogel. *ACS Macro Lett.*, 1(6), 706–8.

Schultz, Kelly M, Campo-Deano, Laura, Baldwin, Aaron D, Kiick, Kristi L, Clasen, Christian, and Furst, Eric M (2012b). Electrospinning covalently cross-linking biocompatible hydrogelators. *Polymer (Guildf).*, 54, 363–71.

Schultz, Kelly M and Furst, Eric M (2011). High-throughput rheology in a microfluidic device. *Lab Chip*, 11(22), 3802–9.

Schultz, Kelly M, Kyburz, Kyle A, and Anseth, Kristi S (2015). Measuring dynamic cell-material interactions and remodeling during 3D human mesenchymal stem cell migration in hydrogels. *Proc. Natl. Acad. Sci. USA*, 112(29), E3757–64.

Segrè, P N, Megen, W, Pusey, Peter N, Schaëtzl, K, and Peters, W (2005). Two-colour dynamic light scattering. *J. Mod. Opt.*, 42, 1929–52.

Seifriz, William (1924). An elastic value of protoplasm, with further observations on the viscosity of protoplasm. *Brit. J. Exp. Biol.*, 2, 1–11.

Seifriz, William (1928). How Do the Life Processes Work? *Sci. Am.* (January), 18–21.

Shankar, V, Pasquali, Matteo, and Morse, David C (2002). Theory of linear viscoelasticty of semiflexible rods in dilute solution. *J. Rheol.*, 46, 1111–54.

Shaqfeh, Eric S G (1996, February). Purely elastic instabilities in viscometric flows. *Ann. Rev. Fluid Mech.*, 28, 129–85.

Sharma, V, Jaishankar, A, Wang, Y C, and McKinley, G H (2011). Rheology of globular proteins: Apparent yield stress, high shear rate viscosity and interfacial viscoelasticity of bovine serum albumin solutions. *Soft Matter*, 7(11), 5150–60.

Sheng, Ping (ed.) (1990). *Scattering and Localization of Classical Waves in Random Media*. World Scientific, Singapore.

Shevkoplyas, Sergey S, Siegel, Adam C, Westervelt, Robert M, Prentiss, Mara G, and Whitesides, George M (2007). The force acting on a superparamagnetic bead due to an applied magnetic field. *Lab Chip*, 7(10), 1294–302.

Shikata, Toshiyuki and Pearson, Dale S (1993). Viscoelastic behavior of concentrated spherical suspensions. *J. Rheol.*, 38(3), 601–16.

Shindel, Matthew M and Furst, Eric M (2015). Frequency modulated microrheology. *Lab Chip*, 15, 2460–6.

Shindel, Matthew M, Swan, James W, and Furst, Eric M (2013). Calibration of an optical tweezer microrheometer by sequential impulse response. *Rheol. Acta*, 52(5), 455–65.

Simmons, Robert M, Finer, Jeffrey T, Chu, Steven, and Spudich, James A (1996). Quantitative measurements of force and displacement using an optical trap. *Biophys. J.*, 70(4), 1813–22.

Slomkowski, Stanislaw, Alemán, José V, Gilbert, Robert G, Hess, Michael, Horie, Kazuyuki, Jones, Richard G, Kubisa, Przemyslaw, Meisel, Ingrid, Mormann, Werner, Penczek, Stanisław, and Stepto, Robert F T (2011). Terminology of polymers and polymerization processes in dispersed systems (IUPAC Recommendations 2011). *Pure Appl. Chem.*, 83(12), 2229–59.

Smith, Douglas E and Chu, Steven (1998). Response of flexible polymers to a sudden elongational flow. *Science*, **281**, 1335–40.

Solomon, Michael J and Lu, Qiang (2001). Rheology and dynamics of particles in viscoelastic media. *Curr. Opin. Colloid Interface Sci.*, **6**, 430–37.

Spaldin, Nicola A (2011). *Magnetic Materials* (2nd edn). Cambridge University Press, New York.

Spero, Richard Chasen, Vicci, Leandra, Cribb, Jeremy, Bober, David, Swaminathan, Vinay, O'Brien, E Timothy, Rogers, Stephen L, and Superfine, R (2008). High throughput system for magnetic manipulation of cells, polymers, and biomaterials. *Rev. Sci. Instrum.*, **79**, 083707.

Squires, Todd M (2008). Nonlinear microrheology: Bulk stresses versus direct interactions. *Langmuir*, **24**, 1147–59.

Squires, T M and Quake, S R (2005). Microfluidics: Fluid physics at the nanoliter scale. *Rev. Mod. Phys.*, **77**(3), 977–1026.

Sriram, Indira, DePuit, Ryan, Squires, Todd M, and Furst, Eric M (2009). Small amplitude active oscillatory microrheology of a colloidal suspension. *J. Rheol.*, **53**, 357–81.

Sriram, Indira, Meyer, Alexander, and Furst, Eric M. (2010). Active microrheology of a colloidal suspension in the direct collision limit. *Phys. Fluids*, **22**(6), 62003.

Starrs, Laura and Bartlett, Paul (2003a). Colloidal dynamics in polymer solutions: Optical two-point microrheology measurements. *Faraday Discuss. Chem. Soc.*, **123**, 323–34.

Starrs, L and Bartlett, P (2003b). One- and two-point micro-rheology of viscoelastic media. *J. Phys. Condens. Matter*, **15**, S251–6.

Stauffer, D, Coniglio, A, and Adam, M (1982). Gelation and critical phenomena. *Adv. Polym. Sci*, **44**, 103–58.

Stöber, Werner, Fink, Arthur, and Bohn, Ernst (1968). Controlled growth of monodisperse silica spheres in the micron size range. *J. Colloid Interface Sci.*, **26**(1), 62–9.

Stockmayer, Walter H (1943). Theory of molecular size distribution and gel formation in branched-chain polymers. *J. Chem. Phys.*, **11**(2), 45.

Stokes, George Gabriel (1850, August). On the effect of the internal friction of fluids on the motion of pendulums. *Transactions of the Cambridge Philosophical Society*, **9**, 8.

Stradner, Anna, Romer, Sara, Urban, Claus, and Schurtenberger, Peter (2001). Aggregation and gel formation in biopolymer solutions. *Chimia (Aarau).*, **55**(3), 155–9.

Sutherland, William (1905). A dynamical theory of diffusion for non-electrolytes and the molecular mass of albumin. *Philos. Mag.*, **9**(49–54), 781–5.

Svoboda, Karel and Block, Steven M (1994a). Biological applications of optical forces. *Annu. Rev. Biophys. Biomol. Struct.*, **23**, 247–285.

Svoboda, Karel and Block, Steven M (1994b). Optical trapping of metallic Rayleigh particles. *Opt. Lett.*, **19**, 930–32.

Swan, J, Shindel, M, and Furst, Eric M. (2012). Measuring thermal rupture force distributions from an ensemble of trajectories. *Phys. Rev. Lett.*, **109**(19), 198302.

Swan, James W and Brady, John F (2010). Particle motion between parallel walls: Hydrodynamics and simulation. *Phys. Fluids*, **22**(10), 103301.

Swindells, J F, Snyder, C F, Hardy, R C, and Golden, P. E. (1958). *Viscosities of Sucrose Solutions at Various Temperatures: Tables of Recalculated Values*. National Bureau of Standards Circular 440, Washington, D.C.

Tadros, Th F. (1993). Industrial applications of dispersions. *Adv. Colloid Interface Sci.*, **46**(C), 1–47.

Tanase, Monica, Bauer, Laura Ann, Hultgren, Anne, Silevitch, Daniel M., Sun, Li, Reich, Daniel H., Searson, Peter C., and Meyer, Gerald J. (2001). Magnetic alignment of fluorescent nanowires. *Nano Lett.*, **1**(3), 155–8.

Tanner, Shaun A, Amin, Samiul, Kloxin, Christopher J., and Van Zanten, John H (2011). Microviscoelasticity of soft repulsive sphere dispersions: Tracer particle microrheology of triblock copolymer micellar liquids and soft crystals. *J. Chem. Phys.*, **134**, 174903.

Thompson, Russell E, Larson, Daniel R, and Webb, Watt W (2002). Precise nanometer localization analysis for individual fluorescent probes. *Biophys. J.*, **82**(5), 2775–83.

Thomson, W (1848, May). Note on the integration of the equations of equilibrium of an elastic solid. *Cambr. Dubl. Math. J.*, **3**, 87–9.

Tseng, Yider, An, Kwang M, and Wirtz, Denis (2002). Microheterogeneity controls the rate of gelation of actin filament networks. *J. Biol. Chem.*, **277**, 18143–50.

Tu, R S and Breedveld, V (2005). Microrheological detection of protein unfolding. *Phys. Rev. E*, **72**(4), 41914.

Uhde, Jorg, Feneberg, Wolfgang, Ter-Oganessian, N, Sackmann, Erich, and Boulbitch, Alexei (2005). Osmotic force-controlled microrheometry of entangled actin networks. *Phys. Rev. Lett.*, **94**, 198102.

Vale, Ronald D and Milligan, Ronald A (2000). The way things move: Looking under the hood of molecular motor proteins. *Science*, **288**, 88–95.

Valentine, M T, Kaplan, P D, Thota, D, Crocker, J C, Gisler, T, Prud'homme, R K, Beck, M, and Weitz, D A (2001). Investigating the microenvironments of inhomogeneous soft materials with multiple particle tracking. *Phys. Rev. E*, **64**, 61506.

Valentine, M T, Perlman, Z E, Gardel, M L, Shin, J H, Matsudaira, P, Mitchison, T J, and Weitz, D A (2004). Colloid surface chemistry critically affects multiple particle measurements of biomaterials. *Biophys. J.*, **86**, 4004–14.

van Blaaderen, Alfons and Vrij, A. (1992). Synthesis and characterization of colloidal dispersions of fluorescent, monodisperse silica spheres. *Langmuir*, 8, 2921–31.

Van Citters, Kathleen M, Hoffman, Brenton D, Massiera, Gladys, and Crocker, John C (2006). The role of F-actin and myosin in epithelial cell rheology. *Biophys. J.*, **91**(10), 3946–56.

Van Helden, A K, Jansen, J Wo, and Vrij, A (1981). Preparation and characterization of spherical monodisperse silica dispersions in nonaqueous solvents. *J. Colloid Interface Sci.*, **81**(2), 354–68.

Van Hove, Léon (1954). Correlations in space and time and Born approximation scattering in systems of interacting particles. *Phys. Rev.*, **95**(1), 249.

Van Wazer, J R, Lyons, J W, Kim, K I, and Colwell, R E (1963). *Viscosity and Flow Measurement: A Laboratory Handbook of Rheology*. Interscience, New York.

van Zanten, J H, Amin, S, and Abdala, A A (2004). Brownian motion of colloidal spheres in aqueous PEO solutions. *Macromolelcules*, **37**(10), 3874–80.

Vasbinder, Astrid J. and De Kruif, Cornelis G. (2003). Casein-whey protein interactions in heated milk: The influence of pH. *Int. Dairy J.*, **13**(8), 669–77.

Vasbinder, Astrid J., Van Mil, Peter J J M, Bot, Arjen, and De Kruif, Kees G. (2001). Acid-induced gelation of heat-treated milk studied by diffusing wave spectroscopy. *Colloids Surfaces B Biointerfaces*, **21**(1–3), 245–50.

Veerman, Cecile, Rajagopal, Karthikan, Palla, Chandra Sekhar, Pochan, Darrin J, Schneider, Joel P, and Furst, Eric M (2006). Gelation kinetics of beta-hairpin peptide hydrogel networks. *Macromolelcules*, **39**, 6608–14.

Verwey, E J W and Overbeek, J Th G (1948). *Theory of the Stability of Lyophobic Colloids*. Elsevier, Amsterdam.

Viasnoff, Virgile, Jurine, Stéphane, and Lequeux, François (2003). How are colloidal suspensions that age rejuvenated by strain application? *Faraday Discuss. Chem. Soc.*, **123**(0), 253–66.

Viasnoff, Virgile, Lequeux, Francois, and Pine, D J (2002). Multispeckle diffusing-wave spectroscopy: A tool to study slow relaxation and time-dependent dynamics. *Rev. Sci. Inst.*, **73**, 2336–44.

Viravathana, P and Marr, D W M (2000). Optical trapping of titania/silica core-shell particles. *J Colloid Interface Sci.*, **221**, 301–7.

Visscher, K, Schnitzer, M J, and Block, S M (1999). Single kinesin molecules studied with a molecular force clamp. *Nature*, **400**(6740), 184–9.

Voorn, Dirk J, Ming, W W, and Van Herk, Alex M (2005). Control of charge densities for cationic latex particles. *Macromolecules*, **38**(9), 3653–62.

Wagner, Norman J and Brady, John F (2009). Shear thickening in colloidal dispersions. *Phys. Today*, **62**(10), 27–32.

Weeks, Eric R, Crocker, J C, Levitt, Andrew C, Schofield, Andrew, and Weitz, D A (2000). Three-dimensional direct imaging of structural relaxation near the colloidal glass transition. *Science*, **287**, 627–31.

Weese, Jürgen (1992). A reliable and fast method for the solution of Fredhol integral equations of the first kind based on Tikhonov regularization. *Comput. Phys. Commun.*, **69**(1), 99–111.

Weese, Jürgen (1993). A regularization method for nonlinear ill-posed problems. *Comput. Phys. Commun.*, **77**(3), 429–40.

Weitz, David A and Pine, David J (1993). *Diffusing wave spectroscopy*. In Dynamic Light Scattering (ed. Wyn Brown), Ch. 16, pp. 652–720. Oxford University Press, New York.

Weitz, D A, Pine, D J, Pusey, P N, and Tough, R J A (1989). Nondiffusive Brownian motion studied by diffusing wave spectroscopy. *Phys. Rev. Lett.*, **63**(16), 1747–50.

Weitz, D A, Zhu, J X, Durian, D J, and Pine, D. J (1992). *Principles and applications of diffusing-wave spectroscopy*. In Structure and Dynamics of Strongly Interacting Colloids and Supramolecular Aggregates in Solution (eds. Sow-Hsin Chen, John S. Huang, Piero Tartaglia) NATO ASI Series, Volume 369, pp. 688–91. Springer, New York.

White, G W (1980). Micro-rheology. *Microscope*, **28**, 3–8.

Wilking, James N and Mason, Thomas G (2008). Optically driven nonlinear microrheology of gelatin. *Phys. Rev. E*, **77**, 055101(R).

Willenbacher, N and Oelschlaeger, C (2007). Dynamics and structure of complex fluids from high frequency mechanical and optical rheometry. *Curr. Opin. Colloid Interface Sci.*, **12**, 43–49.

Willenbacher, N, Oelschlaeger, C, Schopferer, M, Fischer, P, Cardinaux, F, and Scheffold, F (2007). Broad bandwidth optical and mechanical rheometry of wormlike micelles. *Phys. Rev. Lett.*, **99**, 68302.

Wilson, L G, Harrison, A W, Schofield, A B, Arlt, J, and Poon, W C K (2009). Passive and active microrheology of hard-sphere colloids. *J. Phys. Chem. B*, **113**(12), 3806–12.

Wilson, Laurence G and Poon, Wilson C K (2011). Small-world rheology: An introduction to probe-based active microrheology. *Phys. Chem. Chem. Phys.*, **13**(22), 10617–14.

Winter, H Henning and Chambon, François (1986). Analysis of linear viscoelasticity of a crosslinking polymer at the gel point. *J. Rheol.*, **30**, 367–82.

Winter, H Henning and Chambon, François (1987). Linear viscoelasticity at the gel point of a crosslinking PDMS with imbalanced stoichiometry. *J. Rheol.*, **31**, 683–97.

Winter, Horst Henning and Mours, Marian (1997). Rheology of polymers near liquid-solid transitions. *Adv. Polym. Sci*, **134**, 165–234.

Wong, A P Y and Wiltzius, P (1993). Dynamic light scattering with a CCD camera. *Rev. Sci. Instrum.*, **64**, 2547–49.

Wong, I Y, Gardel, M L, Reichman, D R, Weeks, E R, Valentine, M T, Bausch, A R, and Weitz, David A (2004). Anomalous diffusion probes microstructure dynamics of entangled F-actin networks. *Phys. Rev. Lett.*, **92**, 178101.

Wyss, Hans M, Romer, Sara, Scheffold, Frank, Schurtenberger, Peter, and Gauckler, Ludwig J (2001). Diffusing-wave spectroscopy of concentrated alumina suspensions during gelation. *J Colloid Interface Sci.*, **240**, 89–97.

Xu, Chunyu, Breedveld, Victor, and Kopecek, Jindich (2005). Reversible hydrogels from self-assembling genetically engineered protein block copolymers. *Biomacromolecules*, **6**(3), 1739–49.

Xu, Jingyuan, Viasnoff, Virgile, and Wirtz, Denis (1998*a*). Compliance of actin filament networks measured by particle-tracking microrheology and diffusing wave spectroscopy. *Rheol. Acta*, **37**(4), 387–98.

Xu, Jingyuan, Wirtz, Denis, and Pollard, Thomas D (1998*b*). Dynamic cross-linking by alpha-actinin determines the mechanical properties of actin filament networks. *J. Biol. Chem.*, **273**, 9570–6.

Xue, J Z, Pine, D J, Milner, S T, Wu, X L, and Chaikin, P M (1992*a*). Nonergodicity and light-scattering from polymer gels. *Phys. Rev. A*, **46**(10), 6550–63.

Xue, J-Z, Wu, X-L, Pine, D J, and Chaikin, P M (1992*b*). Hydrodynamic interactions in hard sphere suspensions. *Phys. Rev. A*, **45**, 989–93.

Yagi, K (1961). The mechanical and colloidal properties of Amoeba protoplasm and their relations to the mechanism of amoeboid movement. *Comp. Biochem. Physiol.*, **3**, 73–91.

Yamada, Soichiro, Wirtz, Denis, and Kuo, Scot C (2000). Mechanics of living cells measured by laser tracking microrhology. *Biophys. J.*, **78**, 1736–47.

Yamaguchi, Nori, Zhang, Le, Chae, Byeong Seok, Palla, Chandra S, Furst, Eric M, and Kiick, Kristi L (2007). Growth factor mediated assembly of cell-receptor-responsive hydrogels. *J. Am. Chem. Soc.*, **129**, 3040–41.

Yan, M, Fresnais, J, Sekar, S, Chapel, J P, and Berret, J F (2011). Magnetic nanowires generated via the waterborne desalting transition pathway. *ACS Appl. Mater. Interfaces*, **3**(4), 1049–54.

Yang, Yali, Lin, Jun, Meschewski, Ryan, Watson, Erin, and Valentine, Megan T. (2011). Portable magnetic tweezers device enables visualization of the three-dimensional microscale deformation of soft biological materials. *Biotechniques*, **51**(1), 29–34.

Yao, Alison, Tassieri, Manlio, Padgett, Miles, and Cooper, Jonathan (2009). Microrheology with optical tweezers. *Lab Chip*, **9**(17), 2568–75.

Zacchia, Nicholas A and Valentine, Megan T. (2015). Design and optimization of arrays of neodymium iron boron-based magnets for magnetic tweezers applications. *Rev. Sci. Instrum.*, **86**(5), 053704.

Zakharov, P, Cardinaux, F, and Scheffold, F (2006). Multispeckle diffusing wave spectroscopy with a single-mode detection scheme. *Phys. Rev. E*, **73**, 11413.

Zaner, K S and Valberg, P A (1989). Viscoelasticity of F-actin measured with magnetic microparticles. *J. Cell Biol.*, **109**, 2233–2243.

Zhang, Jinlong, Boyd, C., and Luo, Weili (1996). Two mechanisms and a scaling relation for dynamics in ferrofluids. *Phys. Rev. Lett.*, **77**, 390–93.

Ziemann, F, Radler, J, and Sackmann, E (1994). Local measurements of viscoelastic moduli of entangled actin networks using an oscillating magnetic bead micro-rheometer. *Biophys. J.*, **66**, 2210–16.

Zimenkov, Y, Dublin, S N, Ni, R, Tu, R S, Breedveld, V, Apkarian, Robert P, and Conticello, V P (2006). Rational design of a reversible pH-responsive switch for peptide self-assembly. *J. Am. Chem. Soc.*, **128**(21), 6770–71.

Zwanzig, Robert and Bixon, Mordechai (1970). Hydrodynamic theory of the velocity correlation function. *Phys. Rev. A*, **2**, 2005–12.

Index